CRC Handbook
of
Basic Tables
for
Chemical Analysis

Authors

Thomas J. Bruno, Ph.D.
Research Chemist
Thermophysics Division
National Institute of Standards and Technology
Boulder, Colorado

Paris D. N. Svoronos, Ph.D.
Associate Professor
Department of Chemistry
Queensborough Community College
City University of New York
Bayside, New York

CRC Press, Inc.
Boca Raton, Florida

Library of Congress Cataloging-in-Publication Data

Bruno, Thomas J.
    Handbook of basic tables for chemical analysis / authors, Thomas
J. Bruno, Paris D. N. Svoronos.
      p. cm.
    Bibliography: p.
    Includes index.
    ISBN 0-8493-3935-9
    1. Chemistry, Analytic--Tables. I. Svoronos, Paris D. N.
II. Title.
QD78.B78   1989
543'.00212--dc19

           88-7595
           CIP

This book represents information obtained from authentic and highly regarded sources. Reprinted material is quoted with permission, and sources are indicated. A wide variety of references are listed. Every reasonable effort has been made to give reliable data and information, but the author and the publisher cannot assume responsibility for the validity of all materials or for the consequences of their use.

Certain commercial equipment, instruments, or materials are identified in this handbook in order to provide an adequate description. Such identification does not imply recommendation or endorsement by the National Institute of Standards and Technology, the City University of New York, Georgetown University, or Los Alamos National Laboratory, nor does it imply that the materials or equipment identified are necessarily the best available for the purpose. The authors, publisher, and their respective institutions are not responsible for the use of which this handbook is made. Occasional use is made of non-SI units, in order to conform to the standard and accepted practice in modern analytical chemistry.

Those portions of the volume that are the work of Thomas J. Bruno were prepared in his capacity as an employee of the U.S. government and are not subject to copyright. (The sections on gas chromatography, high performance liquid chromatography, supercritical fluid chromatography, ultraviolet spectrophotometry, qualitative tests, and miscellaneous tables make up this material.)

Direct all inquiries to CRC Press, Inc., 2000 Corporate Blvd., N.W., Boca Raton, Florida, 33431.

© 1989 by CRC Press, Inc.

International Standard Book Number 0-8493-3935-9

Library of Congress Card Number 88-7595

Printed in the United States of America   3 4 5 6 7 8 9 0

PREFACE

This work began as a slim booklet prepared by one of the authors (TJB) to accompany a course on chemical instrumentation presented at the then National Bureau of Standards, Boulder Laboratories. The booklet contained tables on chromatography, spectroscopy, and chemical (wet) methods, and was intended to provide students with enough basic data to design their own analytical methods and procedures. Shortly thereafter, with the co-authorship of Professor Paris D. N. Svoronos, it was expanded into a more extensive compilation entitled *Basic Tables for Chemical Analysis,* published as a National Bureau of Standards Technical Note (Number 1096). That work has now been expanded and updated into the present body of tables.

Although there have been considerable changes since the first version of these tables, the aim has remained essentially the same. We have tried to provide a single source of information for those practicing scientists and research students who must use various aspects of chemical analysis in their work. In this respect, it is geared less toward the researcher in analytical chemistry than to those practitioners in other chemical disciplines who must make routine use of chemical analysis. We have given special emphasis to those "instrumental techniques" which are most useful in solving common analytical problems. In many cases, the tables contain information gleaned from the most current research papers and provide data not easily obtainable elsewhere. In some cases, data are presented which are not available at all in other sources. An example is the section covering supercritical fluid chromatography, in which a tabular P-ρ-T surface for carbon dioxide has been calculated (specifically for this work), using an accurate equation of state.

While the authors have endeavored to include data which they perceive to be most useful, there will undoubtedly be areas which have been slighted. We therefore ask you, the user, to assist us in this regard by informing the corresponding author (TJB) of any topics or tables which should be included in future editions.

The authors would like to acknowledge some individuals who have been of great help during the preparation of this work. Stephanie Outcalt and Juli Schroeder, chemical engineers at the National Institute of Standards and Technology, provided invaluable assistance in searching the literature and compiling a good deal of the data included in this book. Teresa Yenser, manager of the NIST word processing facility, provided excellent copy, despite occasional disorganization on the part of the authors. We owe a great debt to our board of reviewers, who provided insightful comments on the manuscript: Professors D. W. Armstrong, S. Chandrasegaran, G. D. Christian, D. Crist, C. F. Hammer, K. Nakanishi, C. F. Poole, E. Sarlo, Drs. R. Barkley, W. Egan, D. G. Friend, S. Ghayourmanesh, J. W. King, R. F. X. Klein, M. L. Loftus, J. E. Mayrath, G. W. A. Milne, R. Reinhardt, R. Tatken, and D. Wingeleth. The authors would like to acknowledge the financial support of the Gas Research Institute and the U.S. Department of Energy, Office of Basic Energy Sciences (TJB), and the National Science Foundation and the City University of New York (PDNS). Finally, we must thank our wives, Clare and Soraya, for their patience throughout the period of hard work and late nights.

THE AUTHORS

Thomas J. Bruno, Ph.D., is a research chemist in the Thermophysics Division at the National Institute of Standards and Technology (formerly the National Bureau of Standards), Boulder, Colorado. He is also on the adjunct faculty in the Department of Chemical Engineering at the Colarado School of Mines. Dr. Bruno received his B.S. in chemistry from the Polytechnic Institute of Brooklyn and his M.S. and Ph.D. in physical chemistry from Georgetown University. He served as a National Academy of Sciences-National Research Council postdoctoral associate at NIST, and was later appointed to the staff.

He currently conducts research on thermophysical properties of fluids, and manages the division analytical chemistry laboratory. He was awarded the Department of Commerce Bronze Medal in 1986 for his work on the thermophysics of reacting fluids.

Paris D. N. Svoronos, Ph.D., is Professor of Chemistry at QCC of the City University of New York. In addition, he holds a continuing appointment as visiting professor in the Department of Chemistry at Georgetown University. Dr. Svoronos obtained a B.S. in chemistry and a B.S. in physics at the American University of Cairo, and his M.S. and Ph.D. in organic chemistry at Georgetown University.

Among his research interests are synthetic sulfur chemistry, organic electrochemistry, and organic structure determination and trace analysis. He also maintains a keen interest in chemical education, and has authored several widely used undergraduate laboratory manuals.

CONTRIBUTOR
(Atomic Absorption and Emission Spectrometry)

Michael L. Parsons, Ph.D.
Pacific Scientific
HTL / KinTech Division
Duarte, California

We dedicate this work to our wives,
Clare and Soraya.

TABLE OF CONTENTS

VII. NUCLEAR MAGNETIC RESONANCE SPECTROSCOPY

VIII. MASS SPECTROSCOPY

IX. ATOMIC ABSORPTION AND EMISSION SPECTROMETRY

I. Gas Chromatography

CARRIER GAS PROPERTIES

The following table gives the properties of common gas chromatographic carrier gases. These properties are those used most often in designing separation and optimizing detector performance. The density values are determined at 0 °C and 0.101 MPa (760 torr).[1] The thermal conductivity values are determined at 48.9 °C (120°F).[1] The viscosity values are determined at the temperatures listed, and at 0.101 MPa (760 torr).[1] The heat capacity (constant pressure) values are determined at 15 °C and 0.101 MPa (760 torr).[2]

REFERENCES

1. **Weast, R. C., Ed.,** *Handbook of Chemistry and Physics*, 63rd ed., CRC Press, Boca Raton, FL., 1983.
2. **Dal Nogare, S. and Juvet, R. S.,** *Gas-Liquid Chromatography: Theory and Practice*, Interscience, New York, 1962.

CARRIER GAS PROPERTIES

Carrier gas	Density (kg/m³)	Thermal conductivity ×10⁻² (W/[m·K])	Thermal conductivity differences δλ (He)	δλ (N₂)	δλ (Ar)	Viscosity ×10⁻⁵ (Pa·s)	Heat capacity (J/[kg·K])	Relative molecular mass
Hydrogen	0.08988	19.71	3.97	16.96	17.81	0.876 (20.7 °C) 1.086 (129.4 °C) 1.381 (299.0 °C)	14,112.7	2.016
Helium	0.17847	15.74	—	12.99	13.84	1.941 (20.0 °C) 2.281 (100.0 °C) 2.672 (200.0 °C)	5,330.6	4.003
Methane	0.71680	3.74	−12.00	0.99	1.84	1.087 (20.0 °C) 1.331 (100.0 °C) 1.605 (200.5 °C)	2,217.2	16.04
Oxygen	1.42904	2.85	−12.89	0.10	0.95	2.018 (19.1 °C) 2.568 (127.7 °C) 3.017 (227.0 °C)	915.3	32.00
Nitrogen	1.25055	2.75	−12.99	—	0.85	1.781 (27.4 °C) 2.191 (127.2 °C) 2.559 (226.7 °C)	1,030.5	28.016
Carbon monoxide	1.25040	2.67	−13.07	−0.08	0.77	1.753 (21.7 °C) 2.183 (126.7 °C) 2.548 (227.0 °C)	1,030.7	28.01
Ethane	1.35660	2.44	−13.30	−0.31	0.54	0.901 (17.2 °C) 1.143 (100.4 °C) 1.409 (200.3 °C)	1,614.0	30.07
Ethene	1.26040	2.30	−13.44	−0.45	0.40	1.008 (20.0 °C) 1.257 (100.0 °C) 1.541 (200.0 °C)	—	28.05
Propane	2.00960	2.03	−13.71	−0.72	0.13	0.795 (17.9 °C) 1.009 (100.4 °C)	—	44.09
Argon	1.78370	1.90	−13.84	−0.85	—	1.253 (199.3 °C) 2.217 (20.0 °C) 2.695 (100.0 °C) 3.223 (200.0 °C)	523.7	39.94
Carbon dioxide	1.97690	1.83	−13.91	−0.92	−0.07	1.480 (20.0 °C) 1.861 (99.1 °C) 2.221 (182.4 °C)	836.6	44.01
n-Butane	2.51900	1.82	−13.92	−0.93	−0.08	0.840 (14.7 °C)	—	58.12
Sulfur hexafluoride	6.50 (20°C)	1.63	−14.11	−1.12	−0.27	1.450 (21.1 °C)	674.0	146.05

MESH SIZES AND PARTICLE DIAMETERS

The following tables give the relationship between particle size diameter (in μm) and several standard sieve sizes. The standards are as follows: (1) U.S. Standard Sieve Series, ASTM E-11-61; (2) Canadian Standard Sieve Series, 8-GP-16; (3) British Standards Institution, London, BS-410-62; (4) Japanese Standard Specification, JI S-Z-8801; (5) French Standard, AFNOR X-11-501; and (6) German Standard, DIN-4188.

MESH SIZES AND PARTICLE DIAMETERS

Particle size (μm)	U.S. sieve size	Tyler mesh size	British sieve size	Japanese sieve size	Canadian sieve size
4000	5	—	—	—	—
2000	10	9	8	9.2	8
1680	12	10	—	—	—
1420	14	12	—	—	—
1190	16	14	—	—	—
1000	18	16	—	—	—
841	20	20	18	20	18
707	25	24	—	—	—
595	30	28	25	28	25
500	35	32	—	—	—
420	40	35	36	36	36
354	45	42	—	—	—
297	50	48	52	52	52
250	60	60	60	55	60
210	70	65	72	65	72
177	80	80	85	80	85
149	100	100	100	100	100
125	120	115	120	120	120
105	140	150	150	145	150
88	170	170	170	170	170
74	200	200	200	200	200
63	230	250	240	250	240
53	270	—	300	280	300
44	325	—	350	325	350
37	400	—	—	—	—

MESH-SIZE RELATIONSHIPS

Mesh range	Top screen opening (μm)	Bottom screen opening (μm)	Micron screen (μm)	Range ratio
10/20	2000	841	1159	2.38
10/30	2000	595	1405	3.36
20/30	841	595	246	1.41
30/40	595	420	175	1.41
35/80	500	177	323	2.82
45/60	354	250	104	1.41
60/70	250	210	40	1.19
60/80	250	177	73	1.41
60/100	250	149	101	1.68
70/80	210	177	33	1.19
80/100	177	149	28	1.19
100/120	149	125	24	1.19
100/140	149	105	44	1.42
120/140	125	105	20	1.19
140/170	105	88	17	1.19
170/200	88	74	14	1.19
200/230	74	63	11	1.17
230/270	63	53	10	1.19
270/325	53	44	9	1.20
325/400	44	37	7	1.19

FRENCH AND GERMAN SIEVE SIZES

Particle size (μm)	Sieve size
2000	34
800	30
500	28
400	27
315	26
250	25
200	24
160	23
125	22
100	21
80	20
63	19
50	18
40	17

GAS CHROMATOGRAPHIC SUPPORT MATERIALS FOR PACKED COLUMNS

The following table lists the more common solid supports used in packed column gas chromatography, along with relevant properties.[1-4] The performance of several of these materials can be improved significantly by acid washing and treatment with DMCS (dimethyldichlorosilane) to further deactivate the surface. The non-acid-washed materials can be treated with hexamethyl disilazane to deactivate the surface; however, the deactivation is not as great as that obtained by an acid wash followed by DMCS treatment. Most of the materials are available in several particle size ranges. The use of standard sieves will help ensure reproducible sized packings from one column to the next. Data are provided for the Chromosorb family of supports since they are among the most well characterized. It should be noted that other supports are available to the chromatographer, with a similar range of properties provided by the Chromosorb series.

REFERENCES

1. **Poole, C. F. and Schuette, S. A.,** *Contemporary Practice of Chromatography,* Elsevier, Amsterdam, 1984.
2. **Gordon, A. J. and Ford, R. A.,** *The Chemist's Companion,* John Wiley & Sons, New York, 1972.
3. **Heftmann, E., Ed.,** *Chromatography: A Laboratory Handbook of Chromatographic and Electrophoretic Methods,* 3rd ed., Van Nostrand Reinhold, New York, 1975.
4. **Grant, D. W.,** *Gas-Liquid Chromatography,* Van Nostrand Reinhold, London, 1971.

GAS CHROMATOGRAPHIC SUPPORT MATERIALS FOR PACKED COLUMNS

Support name	Support type	Density (free fall; g/mL)	Density (packed; g/mL)	pH	Surface area (m²/g)	Max liquid loading (%)	Color	Notes
Chromosorb A	Diatomite	0.40	0.48	7.1	2.7	25	Pink	Most useful for preparative gas chromatography; high strength; high liquid phase capacity; low surface activity
Chromosorb G	Diatomite	0.47	0.58	8.5	0.5	5	Oyster white	High mechanical strength; low surface activity; high density
Chromosorb P	Diatomite firebrick	0.38	0.47	6.5	4.0	30	Pink	High mechanical strength; high liquid capacity; moderate surface activity; for separations of moderately polar compounds
Chromosorb W	Diatomite	0.18	0.24	8.5	1.0	15	White	Lower mechanical strength than pink supports; very low surface activity; for polar compound separation
Chromosorb 750	Diatomite	0.33	0.49		0.75	7	White	Highly inert surface; useful for biomedical and pesticide analyses; mechanical strength similar to chromosorb G
Chromosorb R-6470-1	Diatomite				5—6	Low	White	Ultra-fine particle size used to coat inside walls of capillary columns; typical particle size is 1—4 µm

Chromosorb T[a]	Polytetrafluoroethylene	0.42	0.49	7.5	5	White	Maximum temperature of 240 °C; handling is difficult due to static charge; tends to deform when compressed; useful for analysis of high-polarity compounds
Kel-F[a]	Chlorofluorocarbon			2.2	20	White	Hard, granular chlorofluorocarbon; mechanically similar to chromosorbs; generally gives poor efficiency; use below 160 °C; very rarely used
Fluoropak-80[a]	Fluorocarbon resin			1.3	5	White	Granular fluorocarbon with sponge-like structure; low liquid-phase capacity; use below 275 °C
Teflon-6[a]	Polytetrafluoroethylene			10.5	20	White	Usually 40—60 (U.S.) mesh size; for relatively nonpolar liquid phases; low mechanical strength; highly inert surface; difficult to handle due to static charge; difficult to obtain good coating of polar phases due to highly inert surface
T-Port-F[a]	Polytetrafluoroethylene	0.5		2—500, type dependent	40	White	Use below 150 °C
Porasil (Types A through F)	Silica					White	Rigid, porous silica bead; controlled pore size varies from 10—150 mm; highly inert; also used as a solid adsorbent

[a] The fluorocarbon supports can be difficult to handle since they develop electrostatic charge easily. It is generally advisable to work with them below 19 °C (solid transition point), using polyethylene laboratory ware.

PACKED COLUMN SUPPORT MODIFIERS

During the analysis of strongly acidic or basic compounds, peak tailing is almost always a problem, especially when using packed columns. Pretreatment of support materials, such as acid-washing and treatment with DMCS, will usually result in only modest improvement in performance. A number of modifiers can be added to the stationary phase (in small amounts, 1 to 3%) in certain situations to achieve a reduction in peak tailing. The following table provides several such reagents.[1] It must be remembered that the principal liquid phase must be compatible with any modifier being considered. Thus, the use of potassium hydroxide with polyester or polysiloxane phases would be inadvisable since this reagent can catalyze the depolymerization of the stationary phase. It should also be noted that the use of a tail-reducing modifier may lower the maximum working temperature of a particular stationary phase.

REFERENCES

1. **Poole, C. F. and Schuette, S. A.,** *Contemporary Practice of Chromatography,* Elsevier, Amsterdam, 1984.

PACKED COLUMN SUPPORT MODIFIERS

Compound class	Modifier reagent	Notes
Acids	Phosphoric acid, FFAP (Carbowax-20M-terephthalic acid ester), trimer acid	These modifiers will act as subtractive agents for basic components in the sample; FFAP will selectively abstract aldehydes; phosphoric acid may convert amides to the nitrile (of the same carbon number) or desulfonate sulfur compounds and may esterify or dehydrate alcohols
Bases	Potassium hydroxide, polyethyleneimine, polypropyleneimine, N,N'-bis-1-methylheptyl-p-phenylenediamine, sodium metanilate, THEED (tetrahydroxyethylenediamine)	These modifiers will act as subtractive agents for acidic components in the sample; polypropyleneimine will selectively abstract aldehydes, polyethyleneimine will abstract ketones

PROPERTIES OF CHROMATOGRAPHIC COLUMN MATERIALS

The following table provides physical, mechanical, electrical, and (where appropriate) optical properties of materials commonly used as chromatographic column tubing.[1-5] The data will aid the user in choosing the appropriate tubing material for a given application. The mechanical properties are measured at ambient temperature unless otherwise specified. The chemical incompatibilities cited are usually important only when dealing with high concentrations which are normally not encountered in gas chromatography. Caution is urged nevertheless.

REFERENCES

1. Materials Engineering — Materials Selector, Penton/IPC, Cleveland, 1986.
2. **Khol, R., Ed.,** *Mach. Des.,* (Materials Reference Issue), 58(8), 1986.
3. **Polar, J. P.,** *A Guide to Corrosion Resistance,* Climax Molybdenum Co., Greenwich, 1981.
4. **Fontana, M. G. and Green, N. D.,** *Corrosion Engineering,* McGraw-Hill, New York, 1967.
5. **Shand, E. B.,** *Glass Engineering Handbook,* McGraw-Hill, New York, 1958.
6. **Fuller, A.,** Science Products Division, Corning Glass Works, Corning, N.Y., private communication, 1987.

PROPERTIES OF CHROMATOGRAPHIC COLUMN MATERIALS

Aluminum (alloy 3003)

Density	2.74 g/mL
Hardness (Brinell)	28—55
Melting range	643.3—654.4 °C
Coefficient of expansion (20 to 100°C)	2.32×10^{-5} °C^{-1}
Thermal conductivity (20°C, annealed)	193.14 W/(m·K)
Specific heat (100 °C)	921.1 J/(kg·K)
Tensile strength (hard)	152 MPa
Tensile strength (annealed)	110 MPa

Note: Soft and easily formed into coils; high thermal conduction; incompatible with strong bases, nitrates, nitrites, carbon disulfide, and diborane.

Actual alloy composition (%): Mn, 1.5; Cu, 0.05—0.20; balance is Al.

Copper (alloy C12200)[a]

Density	8.94 g/mL
Hardness (Rockwell-F)	40—45
Melting point	1082.8 °C
Coefficient of expansion (20 to 300 °C)	1.76×10^{-5} °C^{-1}
Thermal conductivity (20 °C)	339.22 W/(m·K)
Specific heat (20 °C)	385.11 J/(kg·K)
Tensile strength (hard)	379 MPa
Tensile strength (annealed)	228 MPa
Elongation (in 0.0508 m annealed)	45%

Note: Copper columns often cause adsorption problems; incompatible with amines, anilines, acetylenes, terpenes, steroids, and strong bases.

PROPERTIES OF CHROMATOGRAPHIC COLUMN
MATERIALS (continued)

Borosilicate glass

Density	2.24 g/mL
Hardness (Knh 100)[b]	418
Young's modulus (25 °C)	62 GPA
Poisson's ratio (25 °C)	0.20
Softening point	806.9 °C
Annealing point	565 °C
Melting point	1600 °C
Strain point	520 °C
Coefficient of expansion (av)	$3 \times 10^{-6} \ °C^{-1}$
Thermal conductivity	1.26 W/(m·K)
Specific heat	710 J/(kg·K)
Refractive index[c]	1.473
Normal service temperature (annealed)	215 °C
Extreme service temperature (annealed)	476 °C
Critical surface tension	750 mN/m

Note: Has been used for both packed columns and capillary columns; incompatible with fluorine, oxygene difluoride, and chlorine trifluoride.

Fused silica (SiO_2)

Density	2.15 g/mL
Hardness (Moh)	6
Young's modulus (25 °C)	72 GPa
Poisson's ratio (25 °C)	0.14
Softening point	1590 °C
Annealing point	1105 °C
Melting point	1704 °C
Strain point	1000 °C
Coefficient of expansion (av)	$5 \times 10^{-7} \ °C^{-1}$
Thermal conductivity	1.5 W/(m·K)
Specific heat	1000 J/(kg·K)
Refractive index (588 nm)	1.458
Normal service temperature (annealed)	886 °C
Extreme service temperature (annealed)	1086 °C
Critical surface tension	760 mN/m

Note: Used for capillary columns; typical inside diameters range from 5 to 530 μm; coated on outside surface by polyimide or aluminum to prevent surface damage; incompatible with fluorine, oxygen difluoride, chlorine trifluoride, and hydrogen fluoride.

Nickel (Monel R-405)

Density	8.83 g/mL
Hardness (Brinell, 21 °C)	110—245
Melting range	1298.89—1348.89 °C
Coefficient of expansion (21 to 537 °C)	$1.64 \times 10^{-5} \ °C^{-1}$
Thermal conductivity (21 °C)	21.81 W/(m·K)
Specific heat (21 °C)	427.05 J/(kg·K)
Tensile strength (hard)	483 MPa
Tensile strength (annealed)	793 MPa
Elongation (in 2 in., 21.1 °C)	15—50%

Note: Provides excellent corrosion resistance, no major chemical incompatibilities.

Actual alloy composition (%): Ni, 66; Cu, 31.5; Fe, 1.35; C, 0.12; Mn, 0.9; S, 0.005; Si, 0.15.

PROPERTIES OF CHROMATOGRAPHIC COLUMN
MATERIALS (continued)

Polytetrafluorethylene (Teflon)

Specific gravity	2.13—2.24
Hardness (Rofkwell-D)	52—65
Melting range	1298.89—1348.89 °C
Coefficient of expansion	1.43×10^{-4} °C^{-1}
Thermal conductivity (21 °C)	2.91 W/m·K
Specific heat (21 °C)	1046.7 J/kg·K
Tensile strength	17—45 MPa
Refractive index[d]	1.35

Note: Flexible and easy to use; cannot be used above 230 °C; thermal decomposition products are toxic; tends to adsorb many compounds which may increase tailing. No major chemical incompatibilities.

Stainless steel (304)

Density	7.71 g/mL
Hardness (Rockwell-B)	149
Melting range	1398.9—1421.1 °C
Coefficient of expansion (0—100) °C	1.73×10^{-5} °C^{-1}
Thermal conductivity (0 °C)	16.27 W/(m·K)
Specific heat (0—100 °C)	502.42 J/(kg·K)
Tensile strength (hard)	758 MPa
Tensile strength (annealed)	586 MPa
Elongation (in 2 in.)	60%

Note: Good corrosion resistance; easily brazed using silver-bearing alloys; high nickel content may catalyze some reactions at elevated temperatures. No major chemical incompatibilities.

Actual alloy composition (%): C, 0.08; Mn, 2 (max); Si, 1 (max); P, 0.045 (max); S, 0.030 (max); Cr, 18—20; Ni, 8—12; balance is Fe. The low-carbon alloy, 304L, is similar except for 0 = 0.3% max, and is more suitable for applications involving welding operations and where high concentrations of hydrogen are used.

Stainless steel (316)

Density	7.71 g/mL
Hardness (Rockwell-B)	149
Melting range	1371.1—1398.9 °C
Coefficient of expansion (0—100 °C)	7.17×10^{-5} °C^{-1}
Thermal conductivity (0 °C)	16.27 W/(m·k)
Specific heat (0—100 °C)	502.42 J/(kg·k)
Tensile strength (annealed)	552 MPa
Elongation (in 2 in.)	60%

Note: Best corrosion resistance of any standard stainless steel, including the 304 varieties, especially in reducing and high-temperature environments.

Actual alloy composition (%): C, 0.08 max; Mn, 2 (max); Si, 1 (max); P, 0.045 (max); S, 0.030 (max); Cr, 16—18; Ni, 10—14; Mo, 2—3; balance is Fe. The low carbon alloy, 316L, is similar except for C = 0.03% max, and is more suitable for applications involving welding operations and where high concentrations of hydrogen are used.

[a] High-purity phosphorus deoxidized copper.
[b] The hardness value listed is on the Knoop scale, using a 100-g load.
[c] Clear grade, at 588 nm.
[d] Using sodium D line, as per ASTM Standard test D542-50.

PROPERTIES OF SOME LIQUID PHASES FOR PACKED COLUMNS

The following table lists some of the more common gas-chromatographic liquid phases, along with some relevant data and notes.[1-3] Many of these phases have been superceded by silicone phases used in capillary columns, but these liquid phases still find application in many instances. This is especially true with work involving established protocols, such as ASTM or AOAC methods. The minimum temperatures, where reported, indicate the point at which some of the phases approach solidification, or when the viscosity increases to the extent that performance is adversely affected. The maximum working temperatures are determined by vapor pressure (liquid-phase bleeding) and chemical stability considerations. The liquid phases are listed by their most commonly used names. Where appropriate, chemical names or common generic names are provided in the notes.

The McReynolds constants (a modification of the Rohrschneider constant) tabulated here are based on the retention characteristics of the following test probe samples:

Constant	Test Probe
X	Benzene
Y	1-Butanol
Z	3-Pentanone
U	1-Nitropropane
S	Pyridine

Compounds which are chemically similar to these probe solutes will show similar retention characteristics. Thus, benzene can be thought of as representing lower aromatic or olefinic compounds. Higher values of the McReynolds constant usually indicate a longer retention time (higher retention volume) for a compound represented by that constant, for a given liquid (stationary) phase.

REFERENCES

1. **McReynolds, W. O.**, Characterization of some liquid-phases, *J. Chromatogr. Sci.*, 8, 685, 1970.
2. **McNair, H. M. and Bonelli, E. J.**, *Basic Gas Chromatography*, Varian Aerograph, Palo Alto, CA, 1968.
3. **Heftmann, E.**, *Chromatography: A Laboratory Handbook of Chromatographic and Electrophoretic Methods*, 3rd ed., Van Nostrand Reinhold, New York, 1975.

PROPERTIES OF SOME LIQUID PHASES FOR PACKED COLUMNS

Liquid phase	T_{min}(°C)	T_{max}(°C)	Polarity[a]	Solvents[b]	McReynolds constant					Notes
					X	Y	Z	U	S	
acetonyl acetone (2,5-hexanedione)	−4	25	I	Ace						
acetyl tributyl citrate	25	180	I	Ace	135	268	202	314	233	
adiponitrile	5	50	I	Chlor, MeCl						1,4-Dicyanobutane
Alka terge-T, amine surfactant	59	75	I	Chlor, MeCl, MeOH						60% Oxazoline, weakly cationic
Amine 220	0	180	P	Cholor, MeCl	117	380	181	293	133	2-(8-Heptadecenyl)-2-imidazoline-ethanol
Ansul ether		80	P	MeOH						Tetraethylene glycol dimethyl ether, used for hydrocarbons
Apiezon H	50	275	N	Chlor	59	56	81	151	129	Low vapor pressure hydrocarbon oil
Apiezon J	50	300	N	Chlor, MeCl	38	36	27	49	57	Low vapor pressure hydrocarbon oil
Apiezon L	50	300	N	Chlor, MeCl	32	22	15	32	42	Low vapor pressure hydrocarbon oil
Apiezon M	50	275	N	Chlor, MeCl	31	22	15	30	40	Low vapor pressure hydrocarbon oil
Apiezon N	50	300	N	Chlor, MeCl	38	40	28	52	58	Low vapor pressure hydrocarbon oil
Apiezon K	50	300+	N	Chlor						Low vapor pressure hydrocarbon oil
Apiezon W	50	275	N	Chlor	82	135	99	155	154	Low vapor pressure hydrocarbon oil
Apolane-87	30	280	N	Tol	21	10	3	12	35	24,24-Diethyl-19,29-dioctadecyl heptatetracontane, C-87 hydrocarbon
Armeen SD		100	P, S, HB	Chlor, MeCl						Primary aliphatic amine
Armeen 12D		100	P, HB	Chlor, MeCl						
Armeen		125	P, HB	Tol						Secondary aliphatic amine
Armeen 2HT		100	P, HB	Chlor						
Armeel DD		100	P	MeOH						Aliphatic nitrile

PROPERTIES OF SOME LIQUID PHASES FOR PACKED COLUMNS (continued)

Liquid phase	T_{min}(°C)	T_{max}(°C)	Polarity[a]	Solvents[b]	McReynolds constant					Notes
					X	Y	Z	U	S	
Arochlor 1242		125		Ace						Chlorinated polyphenyl, used for gases; may be carcinogenic
Asphalt		300	N	Chlor, MeCl						Complex mixture of aliphatic, aromatic, and heterocyclic compounds
Atpet 80			I	Chlor						Sorbitan partial fatty acid esters
p,p-azoxydiphenetol	130	140	I	Chlor						
Baymal		300		Tol						Colloidal alumina
Beeswax		200		Chlor	43	110	61	88	122	For essential oils
Bentone-34	20	200	S	Tol						Dimethyl dioctadecylammonium bentonite
7,8-benzoquinoline		150	I	Chlor						For hydrocarbons, aromatics, heterocycles, and sulfur compounds
benzylamine adipate		125	I	Chlor						
benzyl cellosolve		50	I	Ace						2-(Benzyloxy ethanol), for hydrocarbons
benzyl cyanide		50	I	MeOh						Phenyl acetonitrile
benzyl cyanide-AgNO$_3$		25	S	MeCl						
benzyl diphenyl		100	I	Ace						
benzyl ether		50	I	Chlor, MeCl						Dibenzyl ether
bis(2-butoxyethyl) phthalate		175	I	MeOh	151	282	227	338	267	
bis(2-ethoxyethyl) phthalate					214	375	305	446	364	
bis(2-ethoxyethyl) sebacate				Ace	151	306	211	320	274	
N,N-bis(2-cyanoethyl formamide)	0	125	I	MeOH	690	991	853	110	000	
bis(2-ethoxyethyl) adipate	0	150	I	Ace						
bis(2-methoxyethyl) adipate	20	150	I	Ace, Chlor						
bis(2-ethylhexyl) tetrachlorophthalate	0	150	I	Chlor, MeCl	112	150	123	108	181	

butanediol adipate	60	225	I,P	Chlor, MeCl						
butanediol 1,4-succinate		225	I,P	Chlor	370	571	488	651	611	(BDS) Craig polyester, for alcohols, aromatics, heterocycles, fatty acids and esters, and hydrocarbons
bis[2-(2-methoxyethoxyethyl]ether		50	I	Chlor						Tetraethylene glycol dimethylether
Carbitol		100	P	Ace						Glycol ether (mol mass 134) for aldehydes, ketones
Carbowax 300	10	100	P	MeCl	333	653	405			Polyethylene glycol, av mol mass <380
Carbowax 400	10	125	P	MeCl						Polyethylene glycol, av mol mass 380—420
Carbowax 400 monooleate	10	125	P	MeCl						
Carbowax 550	20	125	P	MeCl						
Carbowax 600	30	125	P	MeCl	323	583	382			Polyethylene glycol, av mol mass 570—630
Carbowax 600 monostearate		125	P	MeCl						
Carbowax 750	25	150	P	MeCl	347	607	418	626	589	Methoxy polyethylene glycol, av mol mass 715—785
Carbowax 1000	40	175	P	MeCL						Polyethylene glycol, av mol mass 950—1,050
Carbowax 1500 (or Carbowax 540)	40	200	P	MeCl	371	639	453	666	641	Polyethylene glycol, av mol mass 500—600
Carbowax 1540	40	200	P	MeCl						Polyethylene glycol, av mol mass 1,300—1,600
Carbowax 4000 (or 3350)	60	200	P	MeCl	317	545	378	578	521	Polyethylene glycol, av mol mass 3,000—3,700
Carbowax 4000 TPA		175	P	MeCl, MeOH	282	496	331	517	467	Terminated with terephthalic acid
Carbowax 4000 monostearate	60	220	P	MeCl						
Carbowax 6000	60	200	P	MeCl	322	540	369	577	512	Polyethylene glycol, av mol mass 6,000—7,500
Carbowax 8000	60	120	P	Chlor	322	540	369	577	512	Polyethylene glycol, av mol mass 7,000—8,500

PROPERTIES OF SOME LIQUID PHASES FOR PACKED COLUMNS (continued)

Liquid phase	T_{min}(°C)	T_{max}(°C)	Polarity[a]	Solvents[b]	McReynolds constant					Notes
					X	Y	Z	U	S	
Carbowax 20M	60	250	P	MeCl	322	536	368	572	510	Polyethylene glycol, av mol mass 15,000—20,000
Carbowax 20M-TPA	60	250	P	MeCl	321	537	367	573	520	Terminated with terephthalic acid
Castorwax	90	200	P	MeCl	108	265	175	229	246	Triglyceride of 12-hydroxy-steric acid (hydrogenated castor oil)
Citroflex A-4		150	I	MeOH	135	286	213	324	262	Tributyl citrate
Chlorowax 70		130	P	MeCl						Chlorinated parafin, 70% (wt/wt) Cl; for hydrocarbons
1-chloronaphthalene		75	I	Tol						
cyanoethyl sucrose	20	175	P	Ace	647	919	043	976		Vitrifies at −10 °C
cyclodextrin acetate		250	I	Ace						For fatty acids and esters
cyclohexane di-methanol succinate	100	210	I	Chlor	269	446	328	498	481	
n-decane		30	N	MeCl						For inorganic and organometallic compounds
di(ethoxyethoxyethyl) phthalate					233	408	317	470	389	
di(butoxyethyl) adipate	−10	150	P	Ace	137	278	198	300	235	
di(butoxyethyl) phthalate	−30	300	P	Tol	157	292	233	348	272	
di-n-butyl cyanamide		50	P,I	MeOH						For gases
di-n-butyl maleate	0	50	I	Tol						For halogenated compounds
di-n-butyl phthalate	−20	100	I	Tol						For aldehydes, ketones, halogenated compounds, hydrocarbons and phosphorus compounds
dibutyl tetrachlorophthalate	0	150	I	Tol						
didecyl phthalate	20	150	I	Tol	136	255	213	320	235	
dicyclohexyl phthalate			I		146	257	206	316	245	
diethylene glycol adipate	0	200	I	MeCl	378	603	460	665	658	DEGA; for aldehydes, ketones, esters, fatty acids, and pesticides

Name				Solvent						Notes
diethylene glycol glutarate	80	225	I	MeCl						
diethylene glycol sebacate		190	I	MeCl						
diethylene glycol succinate	20	190	P	MeCl	496	746	590	837	835	DEGSB DEGS; for alcohols, aldehydes, ketones, amino acids, essential oils, steroids, esters, and phosphorus and sulfur compounds
diethylene glycol stearate					64	193	106	143	191	
dit(2-ethylhexyl) phthalate	20	150	P	Tol	135	254	213	320	235	
di(2-ethylhexyl) adipate	−30	250	P	Ace	76	181	121	197	134	Dioctyl adipate
di(2-ethylhexyl) sebacate	−20	125	I	Tol	72	168	108	180	125	For alcohols, drugs, alkaloids, esters, fatty acids, halogenated compounds, and blood gases
diethyl-D-tartarate		125	P,S	MeCl						For alcohols
diglycerol	20	120	HB	MeCl, MeOH	371	826	560	676	854	For alcohols, aldehydes, ketones, aromatics, heterocycles, and hydrocarbons
dilauryl phthalate		150	I	Tol	79	158	120	192	158	
diisodecyl adipate	−10	175	P	Ace	71	171	113	185	128	
diisooctyl adipate	90	150	P	Ace	78	187	126	204	140	
diisodecyl phthalate	0	150	I	Tol, Ace	84	173	137	218	155	For alcohols, aromatics, heterocycles, essential oils, esters, halogen and sulfur compounds, and hydrocarbons
diisooctyl sebacate		175	I	Ace						For aldehydes, ketones, and hydrocarbons
2,4-dimethyl sulfolane	0	50	P	Chlor						For hydrocarbons and inorganic and organometallic compounds
dimer acid		100	I	MeCl						C$_{36}$ dicarboxylic acid
diisooctyl phthalate	0	175	I	Tol	94	193	154	243	202	
dimethyl formamide	−20	20	P	Ace						DMF
dimethyl sulfoxide	20	30	P	MeCl						DMSO; for gases
dinonyl phthalate	20	150	I	Tol	83	183	147	231	159	For aromatics, heterocycles, and halogen compounds
dioctyl phthalate		150	I	Tol	92	186	150	230	167	
dioctyl sebacate	−20	100	I	MeCl						For aromatics, heterocycles, and halogen compounds
diphenyl formamide	75	100	I	Tol	72	168	108	180	123	

PROPERTIES OF SOME LIQUID PHASES FOR PACKED COLUMNS (continued)

Liquid phase	T_{min}(°C)	T_{max}(°C)	Polarity[a]	Solvents[b]	McReynolds constant					Notes
					X	Y	Z	U	S	
di-*n*-propyl tetrachlorophthalate	10	75	I	Tol						
ditridecyl phthalate	−10	225	P	Tol	75	156	122	195	140	
Emulphor ON-870	0	200	I	Chlor	202	395	251	395	344	Aryloxy polyethylene oxyethanol; for aromatics, heterocycles, essential oils, and halogen compounds
EPON 1001	60	225	P	MeCl (hot)	284	489	406	539	601	Epichlorohydrin-bisphenol A resin, av mol mass 900; for steroids and pesticides
Ethofat 60/25	50	125	I	MeCl (hot)	191	382	244	380	333	Polyethylene oxyglycol stearate; for aldehydes and ketones
Ethomeen S/25		75	P	MeCl	186	395	242	370	339	Polyethoxylated aliphatic amine
ethyl benzoate		150	I	MeOH						For hydrocarbons
ethylene glycol adipate	100	225	I,P	MeCl	372	576	453	655	617	For alcohols, aromatics, heterocycles, bile/urinary compounds, drugs, alkaloids, essential oils, and nitrogen and sulfur compounds
ethylene glycol phthalate	100	200	I,P	Tol	453	697	602	816	872	For nitrogen compounds and steroids
ethylene glycol succinate	100	200	I,P	Ace	537	787	643	903	889	
ethylene glycol glutarate		225	I,P	MeCl						
ethylene glycol sebacate		200	I,P	MeCl (hot)						
ethylene glycol tetrachlorophthalate	120	200	P	Tol	307	345	318	428	466	
ethylene glycol		30	HB	MeOH						
ethylene glycol silver nitrate		30	S	Ace						
eutectic (LiNO₃—NaNO₃/KNO₃/ 27.3—18.2—54.5)		400	—	H₂O						For aromatic hydrocarbons and heterocycles

Stationary phase										
eutectic (KCl—CdCl₂/ 33—67)		400	—	H₂O						For aromatic hydrocarbons and heterocycles
eutectic (NaCl—AgCl/ 41—59)		400	—	Water						For aromatic hydrocarbons and heterocycles
eutectic (BiCl₃—PbCl₃/ 89—11)		400	—	Water						For aromatic hydrocarbons and heterocycles
FFAP	50	250	P,S	Chlor	340	580	397	602	627	Carbowax 20M nitro terephthalic acid ester; for aldehydes and ketones
Flexol 8N8		180	P	Ace	96	254	164	260	179	2,2'-(2-Ethyl hexynamido)-di-ethyl-di-2-ethylhexanoate, for alcohols and nitrogen compounds
Fluorolube HG-1200		100	I	Ace	51	68	114	144	116	Polymers of trifluoro vinyl chloride; for halogenated compounds
formamide	20	50	I	MeOH						For alcohols
glycerol	20	100	HB	MeOH						
Fluorad FC-431	40	200		EAC	281	423	297	509	360	Fluorocarbon surfactant
Hallcomid M-18	40	150	I	MeCl	79	580	397	602	627	Dimethylsteramide; for alcohols, ketones, aldehydes, and esters
Hallcomid M-18-OL	8	150	I	MeCl	89	280	143	239	165	Dimethyloleamide; for alcohols, ketones, aldehydes, and fatty acids
Halocarbon 20-25	20	100	I	Chlor	47	70	108	113	111	
Halocarbon K 352	0	250			47	70	73	238	146	
Halocarbon W9X(600)	50	150		Ace	55	71	116	143	123	
Halocarbon-1321	0	100		Ace						
Halocarbon-11-14	0	100		Ace						
HMPA	20	35	P	Chlor	378	603	460	665	658	Hexamethylphosphoramide
Hi-Eff-1 AP	20	210	I,P	Chlor	372	576	453	655	617	Diethylene glycol adipate
Hi-Eff-2 AP	100	210	I,P	Chlor	271	444	333	498	463	
Hi-Eff-8 BB	100	250	I,P	Chlor	499	751	593	840	889	Cyclohexane dimethanol succinate
Hi-Eff-1 BP	20	200	I,P	Chlor	537	787	643	903		Diethylene glycol succinate
Hi-Eff-2 BP	100	200	I,P	Chlor						Ethylene glycol succinate
Hi-Eff-3 AP	50	230	I,P	Chlor						Neopentyl glycol adipate
Hi-Eff-8 AP	100	250	I,P	Chlor						Cyclohexane dimethanol adipate
Hi-Eff-9 AP	100	250	I,P	Chlor						Tetramethyl cyclobutanediol adipate
Hi-Eff-3 BP		230	I,P	Chlor						Neopentyl glycol succinate
Hi-Eff-4 PB	50	230	I,P	Chlor						Butane-1,4-diol succinate
Hi-Eff-10 BP	20	230	I,P	Chlor						Phenyl diethanolamine succinate

PROPERTIES OF SOME LIQUID PHASES FOR PACKED COLUMNS (continued)

Liquid phase	T_{min}(°C)	T_{max}(°C)	Polarity[a]	Solvents[b]	McReynolds constant					Notes
					X	Y	Z	U	S	
Hi-Eff-2 CP	100	200	I,P	Chlor						Ethylene glycol sebacate
Hi-Eff-3 CP	50	230	I,P	Chlor						Neopentyl glycol sebacate
Hi-Eff-2 EP	100	210	I,P	Chlor						Ethylene glycol isophthalate
Hi-Eff-26 P	100	210	I,P	Chlor						Ethylene glycol phthalate
Hyprose-SP-80		225	P	MeOH	336	742	492	639	727	Octakis (2-hydroxy propyl) sucrose
1,2,3,4,5,6 hexakis-(2-cyanoethoxycyclo hexane)	125	150	I,P	Tol	567	825	713	978	901	
Hercoflex 600		150	P	MeCl	112	234	168	261	194	High-boiling ester of pentaerythritol and a saturated aliphatic acid
n-hexadecane	20	50	N	Pent						Isomeric mixture
hexadecene	20	50	N	Pent						
1-hexadecanol		35	I	MeOH						Cetyl alcohol; for halogenated compounds and hydrocarbons
hexatricontane	80	150	N	MeCl	12	2	−3	1	11	$C_{36}H_{74}$
IGEPAL CO-880	100	200	I	MeCl (hot)	259	461	311	482	426	Nonyl phenoxypolyethyleneoxy-ethanol, n = 30; for alcohols
IGEPAL CO-990	100	200	I	MeCl (hot)	298	508	345	540	475	Nonylphenoxypoly (ethyl-eneoxyethanol), n = 100; for alcohols
IGEPAL CA-630	100	200	I	MeCl (hot)	192	381	253	382	344	Nonylphenoxy poly (ethyl-eneoxyethanol), n = 9; for alcohols
IGEPAL CO-730			I		224	418	279	428	379	
IGEPAL CO-710	100	200	I		205	397	266	401	361	
β,β-iminodiprop-ionitrile		110	I	MeOH						For halogenated compounds
isoquinoline		50	I,P	MeCl	19	58	14	21	47	For hydrocarbons
Lexan	220	270	P	DMP (hot)						Polycarbonate resin
mannitol	170	200	HB	H_2O						For sugars
Montan wax		175	I	Chlor						For halogenated compounds
naphthylamine		150	I	Chlor						For aromatics and heterocycles
neopentylglycol adipate	50	240	I	MeCl	234	425	312	402	438	NPGA; for amino acids, drugs, alkaloids, pesticides, and steroids

neopentylglycol isophthalate	50	240	I	MeCl	172	327	225	344	326	NPGSB; for amino acids and steroids
neopentylglycol sebacate	50	225	I	MeCl	272	469	366	539	474	NPGS; for amino acids, bile and urinary compounds, esters and inorganics
neopentylglycol succinate	50	225	I	MeCl						
nitrobenzene		150	I	MeOH						For hydrocarbons and inorganic and organometallic compounds
Nujol	30	100	N	Pent	9	5	2	6	11	Paraffin oil, mineral oil; for hydrocarbons
n-octadecane		55	N	Pent						For inorganic and organo-metallic compounds
octyl decyl adipate		175		Ace	79	119	193	134		
Oronite NIW		170	P		180	370	242	370	327	Complex mixture of petroleum liquids
β,β'-oxydipropionitrile		100	P	Ace						For halogenated compounds
phenyl diethanolamine succinate		225	P	Ace	386	555	472	674	654	For drugs, alkaloids, and hydrocarbons
polyethylene imine	0	250	P	MeOH	322	800	—	573	524	
poly-*m*-phenylxylene	125	375	I	Tol	257	355	348	433	—	PPE-20
poly-*m*-phenyl ether	0	250	I	Tol	176	227	224	306	283	Five rings; for aromatics and heterocycles
poly-*m*-phenyl ether	0	300	I	Ace, Tol	182	233	228	313	293	Six rings; for alcohols, essential oils, and esters
poly-*m*-phenyl ether	50	400	I	Tol						High polymer
poly-*m*-phenyl ether with squalane	50	100	I	MeCl						Six rings
polypropylene glycol	0	150	HB	MeOH	128	294	173	264	226	av mol. mass 2,000; for drugs, alcohols, and alkaloids
polypropylene glycol sebacate	20	225	I	Chlor	196	345	251	381	328	
polypropylene glycol silver nitrate	20	75	S	MeCl						PEG/AgNO$_3$—3/1; for unsaturated hydrocarbons
polypropylene imine	0	200	I, P	Chlor	122	425	168	263	224	
propylene carbonate	0	60	P	MeCl						1,2 Propanediol cyclic carbonate; for gases and hydrocarbons
polysulfone	0	315	I	Ace, MeOH						
polyvinyl pyrrolidone	80	225	HB							
Quadrol	0	150	HB	Chlor	214	571	357	472	489	N,N,N',N'-tetrakis(2-hydroxypropyl); ethylenediamine; for alcohols, aldehydes, ketones, amino acids, and essential oils

PROPERTIES OF SOME LIQUID PHASES FOR PACKED COLUMNS (continued)

Liquid phase	T_{min}(°C)	T_{max}(°C)	Polarity[a]	Solvents[b]	McReynolds constant					Notes
					X	Y	Z	U	S	
Reoplex 400	0	200	I	MeCl	364	619	449	647	671	Poly(propylene glycol adipate); for aromatics, heterocycles, vitamins, and sulfur and phosphorus compounds
Reoplex 100	0	200	I	MeCl						Poly(propylene glycol sebacate)
Renex-678				MeOH	223	417	278	427	381	Ethylene oxide-nonylphenol surfactant; for alcohols
sebaconitrile		150	P							
squalane	20	100	N	Pent	0	0	0	0	0	For hydrocarbons, organic vapors, nitrogen, and sulfur and phosphorus compounds
squalene	0	100	N,I	Pent	152	341	238	329	344	For hydrocarbons, gases, nitrogen, and sulfur and phosphorus compounds
sorbitol	15	150	P	Chlor	232	582	313			Hexahydric alcohol, $C_6H_6(OH)_6$
STAP	100	255	P	Chlor	345	586	400	610	627	Steroid analysis phase
Siponate-DS-10	20	210	I,P	MeOH						Sodium dodecylbenzene sulfonate
sorbitan monooleate	20	150	P	Chlor	97	266	170	216	268	SPAN-80
sorbitol hexaacetate					335	553	449	652	543	
sucrose acetate isobutyrate	0	200	I,P	MeCl	172	330	251	378	295	SAIB; for alcohols and essential oils
sucrose octaacetate	90	250	I,P	Ace	344	570	461	671	569	
Tergitol Nonionic NP-35	10	175	P	Chlor	197	380	258	389	351	Surfactant mixture
TCEPE	30	175	P,S	MeCl	526	782	677	920	837	Tetracyanoethylated pentaerythritol; for fatty acids and esters
terephthalic acid	100	250	P,I	Tol						
tetraethylene glycol		70	P	MeCl						For hydrocarbons
tetraethylenepentamine		150	HB	MeOH						For nitrogen compounds
1,2,3,4-tetrakis(2-cyanoethyl) butane	110	200	I,P	Chlor	617	860	773	048	941	
THEED	0	125	HB	Chlor	463	942	626	801	893	Tetrahydroxyethylenediamine; for alcohols, hydrocarbons, and nitrogen compounds
β,β'-thiodipropionitrile		100	P	MeOH						For hydrocarbons

	Min	Max	Type	Solvent						
triacetin	20	60	P	MeOH	176	321	250	374	299	For gases
tributyl phosphate	20	125	I	Ace						For gases
tricresyl phosphate		125	I	MeOH						Tritolyl phosphate
triethanolamine		100	HB	MeOH						For alcohols and gases
trimer acid	20	200	HB	MeOH	94	271	163	182	378	C_{54} tricarboxylic acid; for alcohols
1,2,3-tris(2-cyanoethoxy)-propane	30	150	P	MeOH	594	857	759	1031	917	For alcohols, aldehydes, ketones, halogen compounds, and inorganic and organometallic compounds
tris (tetrahydrofurfuryl) phosphate	20	125	I	Ace						
tris(2-cyanoethyl)-nitromethane	20	140	I,P	Chlor						
Triton X-100	20	190	P	MeCl	203	399	268	402	362	Octylphenoxypolyethyl ethanol for aromatics and heterocycles
Triton X-305	20	250	P	Ace	262	467	314	488	430	Octylphenoxypolyethyl ethanol for alcohols
trixylol phosphate	20	250	I,P	Ace						
Tween 20	20	150	P	MeOH						Polyethoxysorbitan monolaurate; for essential oils
Tween 80	20	160	P	MeOH	227	430	283	438	396	Polyethoxysorbitan monooleate; for fatty acids, esters, and pesticides
UCON LB-550-X	0	200	P	Chlor	118	271	158	243	206	10% Polyethylene glycol, 90% propylene glycol
UCON 50-HB-280-X	0	200	P	Chlor	177	362	227	351	302	30% Polyethylene glycol, 70% propylene glycol; for alcohols, fatty acids, and esters
UCON 50-HB-2000	0	200	P	Chlor	202	394	253	392	341	40% Polyethylene glycol, 60% propylene glycol; for alcohols, aldehydes, and ketones
UCON 50-HB-5100	20	200	P	MeCl	214	418	278	421	375	50% Polyethylene glycol, 50% propylene glycol
UCON LB-1715	20	200	I	MeCl	132	297	180	275	235	For alcohols, ketones, and nitrogen compounds
UCON 75-H-90,000	20	200	P	MeCl	255	452	299	470	406	80% Polyethylene glycol, 10% propylene glycol
Versamide 900	190	250	P	MeCl	109	314	145	212	209	Polyamide resin; for alcohols
Versamide 940	115	200	P	MeCl						Polyamide resin; for alcohols

PROPERTIES OF SOME LIQUID PHASES FOR PACKED COLUMNS (continued)

Liquid phase	$T_{min}(°C)$	$T_{max}(°C)$	Polarity[a]	Solvents[b]	McReynolds constant					Notes
					X	Y	Z	U	S	
Versamide 930	115	150	P	MeCl	109	313	144	211	209	Polyamide resin
Versamide 940		200	P	See notes	109	314	145	212	209	Soluble in hot chloroform butanol, 50/50 v/v, for aromatics, heterocycles, pesticides, and nitrogen compounds
xylenyl phosphate		175	I	MeCl						
Zonyl E7		200	I	MeCl	223	359	468	549	465	Fluoroalkyl ester
Zonyl E91		200	I	MeCl	130	250	320	377	293	Fluoroalcohol camphorate
zinc stearate	135	175	I	Ace (warm)	61	231	59	98	544	

[a] N, nonpolar; P, polar; I, intermediate polarity; HB hydrogen bonding; S, specific interaction.
[b] Ace, acetone; Chlor, chloroform; DMP, dimethylpentane; EAC, ethyl acetate; H$_2$O, water; MeCl, methylene chloride; MeOH, methanol; Pent, *n*-pentane; Tol, toluene.

ADSORBENTS FOR GAS-SOLID CHROMATOGRAPHY

The following table lists the more common adsorbents used in gas-solid chromatography, along with relevant information on separation and technique.[1-3] The adsorbents are used chiefly for the analysis of gaseous mixtures. The maximum temperatures listed represent the point of severe resolution loss. The materials are often chemically stable to much higher temperatures. The 60 to 100 mesh sizes (U.S.) are most useful for chromatographic applications. All of these materials must be activated before being used, and the degree of activation will influence the retention behavior. The user should also be aware that the adsorption of water during use will often change retention characteristics dramatically, sometimes resulting in a reversal of positions of adjacent peaks. Due to surface adsorption of solutes, some experimentation with temperature may be necessary to prevent tailing or to avoid statistical correlation (or a propagating error) among the replicate analyses.[4]

REFERENCES

1. **Jeffery, P. G. and Kipping, P. J.,** *Gas Analysis by Gas Chromatography,* Pergamon Press, Oxford, 1972.
2. **Cowper, C. J. and DeRose, A. J.,** *The Analysis of Gases by Chromatography,* Pergamon Press, Oxford, 1983.
3. **Breck, D. W.,** *Zeolite Molecular Sieves,* John Wiley & Sons, New York, 1973.
4. **Bruno, T. J.,** An apparatus for direct fugacity measurements on mixtures containing hydrogen, *J. Res. Natl. Bur. Stand.,* 90(2), 1127, 1985.

ADSORBENTS FOR GAS-SOLID CHROMATOGRAPHY

Packing name	Max temp (°C)	Separation effected	Notes
Silica gel	300	H_2, air, CO, C_1 to C_4 normal hydrocarbons, alkenes and alkynes	Used often as a second column (with a molecular sieve); very hydrophilic; requires activation; can be unpredictable; largely replaced by porous polymers
Porous silica	300	Same as silica gel	Higher surface area than silica gel: often used with a humidified carrier gas; can be coated with a conventional liquid phase; Spherosil and Porasil are examples.
Alumina	300	Light hydrocarbons at ambient temperature (C_1 to C_5), H_2 and light hydrocarbons at subambient temperature	Often useful with controlled water preadsorption after activation; can be coated with a conventional liquid phase
Activated carbon	300	H_2, CO, CO_2, C_1 to C_3 alkanes, alkenes and alkynes	Requires oxygen-free carrier gas; largely replaced by porous polymers
Graphite	300	Light hydrocarbons, H_2S, SO_2, CH_3SH, sour gas	Often modified with small quantities (1.5 to 5%) of conventional liquid phases; requires oxygen-free carrier
Carbon molecular sieve	300	H_2 (O_2, N_2 co-elute), CO, CH_4, H_2O, CO_2, C_1 to C_3 alkanes, alkenes, alkynes	High affinity for hydrocarbons; requires oxyen-free carrier
Molecular sieve, 5A	225	Air and light-gas analysis; H_2, O_2, N_2, (CH_4, CO, NO, SF_6 co-elute)	Synthetic calcium aluminosilicate (zeolite) having an effective pore diameter of 0.5 nm CO_2 is adsorbed strongly; 5A usually gives the best results of all synthetic zeolites; should be activated before use, and used above critical adsorption temperature; 21.6% (wt/wt) water capacity
Molecular sieve, 13X	200	Same as 5A, but with C_1 to C_4, alkanes, alkenes, and alkynes being separated as well	Sodium aluminosilicate (zeolite), having a larger pore size than 0.5 nm, thus producing lower retention times and less resolution; 28.6% (wt/wt) water capacity
Molecular sieve, 3A	200	Light permanent gases	Potassium aluminosilicate (zeolite); 20% (wt/wt) water capacity; smaller pore size than 0.5 nm, thus different retention characteristics
Molecular sieve, 4A	200	Light permanent gases	Sodium aluminosilicate (zeolite); 22% (wt/wt) water capacity; retention characteristics differ from 5A due to smaller pore size

POROUS POLYMER PHASES

Porous polymer phases, first reported by Hollis,[1] are of great value for a wide variety of separations. They are usually white in color, but may darken during use especially at higher temperatures. This darkening does not affect their performance. High-temperature conditioning is required to drive off solvent and residual monomer. The polymers may either swell or shrink with heating; thus, flow-rate changes must be anticipated. The retention indices reported here are from the work of Dave.[2] The use of these indices is the same as for the packed column liquid phases, provided in an earlier table.

Index	Test probe
W	Benzene
X	t-Butanol
Y	2-Butanone
Z	Acetonitrile

The physical property data were taken from the work of Poole and Schuette.[3]

REFERENCES

1. **Hollis, O. L.**, Separation of gaseous mixtures using porous polyaromatic polymer beads, *Anal. Chem.*, 38, 309, 1966.
2. **Dave, S.**, A comparison of the chromatographic properties of porous polymers, *J. Chromatogr. Sci.*, 7, 389, 1969.
3. **Poole, C. F. and Schuette, S. A.**, *Contemporary Practice of Chromatography*, Elsevier, Amsterdam, 1984.

POROUS POLYMER PHASES

Packing name	Max temp (°C)	Material type	Free-fall density (g/cm³)	Surface area (m²/g)	Pore diam av (μm)	Retention indices				Separation effected	Notes
						W	X	Y	Z		
Chromosorb 101	275	Styrene-divinylbenzene copolymer	0.30	<50	0.3—0.4	745	565	645	580	Free fatty acids, glycols, alcohols, alkanes, esters, aldehydes, ketones, ethers	Hydrophobic; condition at 250°C; not recommended for amines or anilines, lower retention times than obtained with Chromosorb 102
Chromosorb 102	250	Styrene-divinylbenzene copolymer	0.29	300—500	0.0085	650	525	570	460	Subambient temperature: H_2, O_2, N_2, Ar, NO, CO; ambient temperature: H_2, (air + NO + CO), CH_4, CO_2, H_2O, N_2O, C_2H_6; above ambient temperature: C_1—C_4 hydrocarbons, H_2S, COS, SO_2, esters, ethers, alcohols, ketones, aldehydes, glycols	May entrain some species; hydrophobic; condition at 225°C; not recommended for amines or nitriles; little tailing of water or oxygenated hydrocarbons
Chromosorb 103	275	Polystyrene cross-linked	0.32	15—25	0.3—0.4	720	575	640	565	Ammonia, light amines, light amides, alcohols, aldehydes, hydrazines	Hydrophobic; high affinity for basic species, not recommended for acidic species, glycols, nitriles, nitroalkanes
Chromosorb 104	250	Acrylontrile divinylbenzene copolymer	0.32	100—200	0.06—0.08	845	735	860	885	Sulfur gases, ammonia, nitrogen oxides, nitriles nitroalkanes, xylenols, water in benzene	Hydrophobic; condition at 225°C; not recommended for glycols and amines; moderately polar

	Max temp (°C)	Composition		Surface area						Applications	Characteristics
Chromosorb 105	250	Acrylic ester (polyaromatic)	0.34	600—700	0.04—0.06	635	545	580	480	Permanent and light hydrocarbon gases; aqueous solutions of light organics such as formalin	Hydrophobic; less polar than Chromosorb 104; condition at 225 °C; Not recommended for acidic species, glycols, amines and amides
Chromosorb 106	250	Polystyrene cross-linked	0.28	700—800	0.05	605	505	540	405	Fatty acids from fatty alcohols, up to C_5; benzene from nonpolar organic compounds	Hydrophobic; not recommended for glycols and amines
Chromosorb 107	250	Acrylic ester cross-linked	0.30	400—500	0.8	660	620	650	550	Aqueous solutions of formaldehyde; alkynes from alkanes	Hydrophobic; moderately polar; not recommended for glycols and amines
Chromosorb 108	250	Acrylic ester cross-linked	0.30	100—200	0.25	710	645	675	605	Polar materials such as water, alcohols, aldehydes, glycols	Hydrophobic; condition at 250 °C
Haysep A	165	Divinylbenzene/ethylene glycol dimethacrylate (high purity)	0.356	526	—	—[a]	—[a]	—[a]	—[a]	Separates permanent gases at ambient temperatures, and is useful for hydrocarbons to C_2, H_2S, H_2O at elevated temperatures	Relatively high polarity
Haysep B	190	Divinylbenzene/polyethyleneimine	0.330	608	—	—[a]	—[a]	—[a]	—[a]	Separates C_1 and C_2 amines, and trace levels of NH_3 and H_2O	High polarity
Haysep C	250	Divinylbenzene/acrylonitrile	0.322	442	—	—[a]	—[a]	—[a]	—[a]	Separates polar hydrocarbons; also HCN, NH_3, H_2S, H_2	Moderate polarity, with separation characteristics similar to Chromosorb 104
Haysep D	290	Divinylbenzene (high purity)	0.3311 (av)	795 (av)	0.0308—0351	—[a]	—[a]	—[a]	—[a]	Separates light gases; CO, CO_2, C_2H_2, C_2 hydrocarbons, H_2S, H_2O	Low-polarity polymer available in four formulations of different surface area (771—803 m^2/g), density (0.3283—0.3834 g/ml), and porosity (64.2—70.4%)

POROUS POLYMER PHASES (continued)

Packing name	Material type	Max temp (°C)	Free-fall density (g/cm³)	Surface area (m²/g)	Pore diam av (μm)	Retention indices				Separation effected	Notes
						W	X	Y	Z		
Haysep N	Divinylbenzene/ethylene glycol dimethacrylate (high purity)	165	0.355	405	—	—[a]	—[a]	—[a]	—[a]	Separation similar to Porapak materials; moderately high H_2O retention; see retention table	Low polarity polymer
Haysep P	Divinylbenzene/styrene	250	0.420	165	—	—[a]	—[a]	—[a]	—[a]	Separation of low molecular mass materials containing halogens, sulfur, water, aldehydes, ketones alcohols, esters, and fatty acids	Moderate to low polarity
Haysep Q	Divinylbenzene	275	0.351	582	—	—[a]	—[a]	—[a]	—[a]	Separation similar to Haysep P; see retention table	Low polarity
Haysep R	Divinylbenzene/N-vinyl-2-pyrollidinone	250	0.324	344	—	—[a]	—[a]	—[a]	—[a]	Separation similar to Haysep P; see retention table	Moderate polarity
Haysep S	Divinylbenzene/4-vinyl-pyridine	250	0.334	583	—	—[a]	—[a]	—[a]	—[a]	Separation similar to Haysep P; see retention table	Moderate polarity
Haysep T	Ethylene glycol dimethacrylate (high purity)	165	0.381	250	—	—[a]	—[a]	—[a]	—[a]	See retention table	High polarity
Porapak-Q	Ethylvinylbenzene-divinylbenzene copolymer	250	0.35	500—700	0.0075	630	538	580	450	Similar to Chromosorb 102	Similar to Chromosorb 102 condition at 250°C; most popular of all porous polymer phases

Porapak-P	250	Styrene-divinylbenzene copolymer	0.28	100—200	—	765	560	650	590	Similar to Porapak-Q	Hydrophobic; low polarity; larger size than Porapak-Q, thus lower retention times are observed; not recommended for amines or anilines; condition at 250 °C
Porapak-N	200	Vinylpyrolidone	0.39	225—350	—	735	605	705	595	Similar to Chromosorb 105; high water retention; CO_2, NH_3, H_2O, C_2H_2, from light hydrocarbons	Condition at 175°C; not recommended for glycols, amines, or acidic species
Porapak-R	250	Vinylpyrolidone	0.33	300—450	0.0076	645	545	580	455	Ethers, esters, H_2O from chlorine gases (HCl, Cl_2), nitriles, nitroalkanes	Moderately polar; condition at 250°C; not recommended for glycols and amines
Porapak-S	250	Vinylpyridine	0.35	300—450	0.0076	645	550	575	465	Normal and branched alcohols, aldehydes, ketones, halocarbons	High polarity; not recommended for acidic species and amines; condition at 250°C
Porapak-T	200	Ethylene glycol dimethacrylate	0.44	250—300	0.009	—	675	700	635	Water in formalin (and other aqueous organic mixtures), retention characteristics similar to Chromosorb 107	Condition at 180°C; highest polarity of Porapak series; not recommended for glycols and amines
Porapak-QS	250	Ethylvinylbenzene-divinylbenzene copolymer	—	—	—	625	525	565	445	Similar to Porapak-Q at lower operating temperatures, but useful for higher-molecular-weight solutes	Silanized Porapak-Q, reduces tailing of high-polarity compounds; condition at 250°C
Porapak-PS	250	Styrene-divinylbenzene copolymer	—	—	—	—	—	—	—	Similar to Porapak-P	Silanized Porapak-P, condition at 250°C
Tenax-GC	375	p-2,6 Diphenylphenylene oxide polymer	0.37	18.6	—	—	—	—	—	Similar to Porapak-Q	Highest thermal stability of all porous polymers

a Retention indices are not available for these porous polymers, but a table of relative retentions on some representative solutes is included at the end of this section.

RELATIVE RETENTION ON SOME HAYSEP POROUS POLYMERS

The following table provides relative retention values for Haysep polymers N, Q, R, S, T. These data were obtained using a 2-m long, 0.32-cm O.D. stainless steel column, using helium as the carrier gas.

HAYSEP POLYMER

Compound	N	Q	R	S	T
Hydrogen	0.19	0.143	0.17	0.19	0.21
Air	0.23	0.186	0.2	0.21	0.25
Nitric oxide	0.25	0.217	0.21	0.23	0.33
Methane	0.30	0.256	0.28	0.3	0.35
Carbon dioxide	0.71	1.15	0.50	0.52	0.85
Nitrous oxide	0.80	1.43	0.59	0.59	—
Ethylene	0.83	0.74	0.78	0.78	0.9
Acetylene	1.41	0.74	1.0	0.87	2.11
Ethane	1.0	1.0	1.0	1.0	1.0
Water	10.1	1.45	0.68	4.12	19.1
Hydrogen sulfide	2.1	1.40	1.73	1.87	2.88
Hydrogen cyanide	1.93	2.31	15.6	8.26	28.8
Carbonyl sulfide	2.82	2.33	2.46	2.63	3.4
Sulfur dioxide	12.0	3.05	9.78	17.8	19.0
Proplyene	4.66	3.20	3.45	3.65	4.91
Propane	4.66	3.67	3.88	4.1	4.63
Propadiene	6.50	4.12	4.39	4.7	7.55
Methylacetylene	9.5	4.12	4.84	5.14	11.3
Methyl chloride	7.43	3.93	4.67	4.92	9.2
Vinyl chloride	14.9	6.04	9.04	9.7	17.3
Ethylene oxide	17.7	6.06	8.78	9.7	23.3
Ethyl chloride	35.0	12.25	19.3	20.7	43.2
Carbon disulfide	—	32.4	—	—	40.7

SILICONE LIQUID PHASES

The following table lists the chromatographic properties of some of the more popular polysiloxane-based liquid phases.[1-8] The polysiloxanes are the most widely used stationary phases in gas chromatography, and are especially applicable to capillary columns. The listing provided here is far from exhaustive. Since it is impractical to present the structures of all polysiloxane-based phases, the OV phases have been chosen as representative since their properties are among the most well characterized. The phases which are listed in the notes as "similar phases" have thermal and chromatographic properties which are similar to the phase described.

The McReynolds constants are indices with respect to the following test probe compounds:

McReynolds constant	Test probe
1	Benzene
2	1-Butanol
3	2-Pentanone
4	1-Nitropropane
5	Pyridine
6	2-Methyl-2-pentanol
7	1-Iodobutane
8	2-Octyne
9	1,4-Dioxane
10	*cis*-Hydrindane

The use of these constants is described in the table entitled "Properties of Some Liquid Phases for Packed Columns". The viscosity data, where available, are presented in cSt, which is 10^{-6} m^2/s. Cross-linked silicone phases based on the silicones are especially valuable for capillary gas chromatography. They are not specifically treated in this table since the differences in many properties are quite often subtle. The cross-linked phases have much longer lifetimes due to the effective immobilization.

REFERENCES

1. **Yancey, J. A.**, Liquid phases used in packed gas chromatographic columns, *J. Chromatogr. Sci.*, 23, 161, 1985.
2. **McReynolds, W. O.**, Chrararacterization of some liquid-phases, *J. Chromatogr. Sci.*, 8, 685, 1970.
3. **Mann, J. R. and Preston, S. T.**, Selection of preferred liquid-phases, *J. Chromatogr. Sci.*, 11, 216, 1973.
4. **Trash, C. R.**, Methyl silicones—their chemistry and use as gas-chromatographic liquid-phases, *J. Chromatogr. Sci.*, 11, 196, 1973.
5. **McNair, H. M. and Bonelli, E. J.**, *Basic Gas Chromatography*, Varian Aerograph, Palo Alto, CA, 1969.
6. **Heftmann, E.**, *Chromatography: A Laboratory Handbook of Chromatographic and Electrophoretic Methods*, 3rd ed., Van Nostrand Reinhold, New York, 1975.
7. **Grant, D. W.**, *Gas Liquid Chromatography*, Van Nostrand Reinhold, London, 1971.
8. **Coleman, A. E.**, Chemistry of Liquid-phases—Other silicones, *J. Chromatogr. Sci.*, 11, 198, 1973.

SILICONE LIQUID PHASES

Liquid phase[a]	Solvent[b]	Av mol mass	Viscosity	T_{min} (°C)	T_{max} (°C)	Polarity[c]	McReynolds constants										Notes
							1	2	3	4	5	6	7	8	9	10	
OV-1, dimethyl silicone (gum)	Tol	$>10^6$	Gum	100	350	N	16	55	44	65	42	32	4	23	45	−1	100% Methyl, low selectivity, boiling-point separations; similar phases: UCC-L45, UCC-W-98, SE-30
OV-101, dimethyl silicone fluid	Tol	3×10^4	1,500	20	350	N	17	57	45	67	43	33	4	23	46	−2	100% Methyl, low selectivity, boiling-point separations; similar phases: DC-11, DC-200, DC-550, SF-96, SP 2100, STAP
OV-3, phenylmethyl dimethyl silicone	Ace	2×10^4	500	20	350	I	44	86	81	124	88	55	39	46	84	17	10% Phenylmethyl; similar to SE-52

Name		Solvent						Fragments	Description
OV-7, phenylmethyl dimethyl silicone	$\left[-\overset{CH_3}{\underset{CH_3}{Si}}-O- \right]_n \left[-\overset{CH_3}{\underset{\phi}{Si}}-O- \right]_m$	Ace	1×10^4	500	20	350	I	69 113 111 171 128 77 68 66 120 35	20% Phenylmethyl
OV-11, phenylmethyl dimethyl silicone	$\left[-\overset{CH_3}{\underset{CH_3}{Si}}-O- \right]_n \left[-\overset{CH_3}{\underset{\phi}{Si}}-O- \right]_m$	Ace	7×10^3	500	0	350	I	102 142 145 219 178 100 103 92 164 59	35% Phenylmethyl; similar phases: DC-710
OV-17, phenylmethyl silicone	$\left[-\overset{CH_3}{\underset{\phi}{Si}}-O- \right]_n$	Ace	4×10^3	1,300	20	350	I	119 158 162 243 202 112 119 105 184 69	50% Methyl; similar phases: SP-2250
OV-22, phenylmethyl dimethyl silicone	$\left[-\overset{CH_3}{\underset{\phi}{Si}}-O- \right]_n \left[-\overset{\phi}{\underset{\phi}{Si}}-O- \right]_m$	Ace	8×10^3	>50,000	20	350	I	160 188 191 283 253 133 152 132 228 99	65% Phenyl

SILICONE LIQUID PHASES (continued)

Liquid phase[a]	Solvent[b]	Av mol mass	Viscosity	T_{min} (°C)	T_{max} (°C)	Polarity[c]	McReynolds constants										Notes
							1	2	3	4	5	6	7	8	9	10	
OV-25, phenylmethyl dimethyl silicone	Ace	1×10^4	100,000	20	350	I	178	204	208	305	280	144	169	147	215	113	75% Phenyl
OV-61, diphenyldimethyl silicone	Tol	4×10^4	>50,000	20	350	I	101	143	142	213	174	99	—	86	—	—	33% Phenyl
OV-73, diphenyldimethyl silicone gum	Tol	8×10^5	Gum	20	350	I	40	86	76	114	85	57	—	39	—	—	5.5% Phenyl; similar phases: SE-52, SE-54

OV-105, cyanopropylmethyl dimethyl silicone

$$\left[\begin{array}{c} CH_3 \\ -Si-O- \\ C_2H_4 \\ C\equiv N \end{array}\right]_n \left[\begin{array}{c} CH_3 \\ -Si-O- \\ CH_3 \end{array}\right]_m$$

Ace   1,500   20   250   N,I   36   108   93   139   86   74   —   29   —   —

OV-202, trifluoropropyl methyl silicone

$$\left[\begin{array}{c} CH_3 \\ -Si-O- \\ C_2H_4 \\ CF_3 \end{array}\right]_n$$

Chlor 1×10^4 500   0   275   I,P   146   238   358   468   310   202   139   56   283   60

50% Trifluoropropyl fluid; similar phases: SP-2401

OV-210, trifluoropropyl methyl silicone

$$\left[\begin{array}{c} CH_3 \\ -Si-O- \\ C_2H_4 \\ CF_3 \end{array}\right]_n$$

Chlor 2×10^5 10,000   20   275   I,P   146   238   358   468   310   206   139   56   283   60

50% Trifluoropropyl; similar phases; QF-1, FS-1265, SD-2401

SILICONE LIQUID PHASES (continued)

Liquid phase[a]	Solvent[b]	Av mol mass	Viscosity	T_{min} (°C)	T_{max} (°C)	Polarity[c]	McReynolds constants 1	2	3	4	5	6	7	8	9	10	Notes
OV-215, trifluoropropyl methyl silicone gum			Gum			I,P	149	240	363	478	315	208	—	56	—	—	50% Trifluoropropyl
OV-225, cyanopropyl-methylphenyl methylsilicone	Ace	8×10^3	9,000	20	275	I,P	228	369	338	492	386	282	226	150	342	117	25% Phenyl, 25% cyanopropylmethyl; similar phases: EX-60, AN-600
OV-275, dicyanoallyl silicone	Ace	5×10^3	20,000	20	275	P	781	1,006	885	1,177	1,089	—	—	—	—	—	

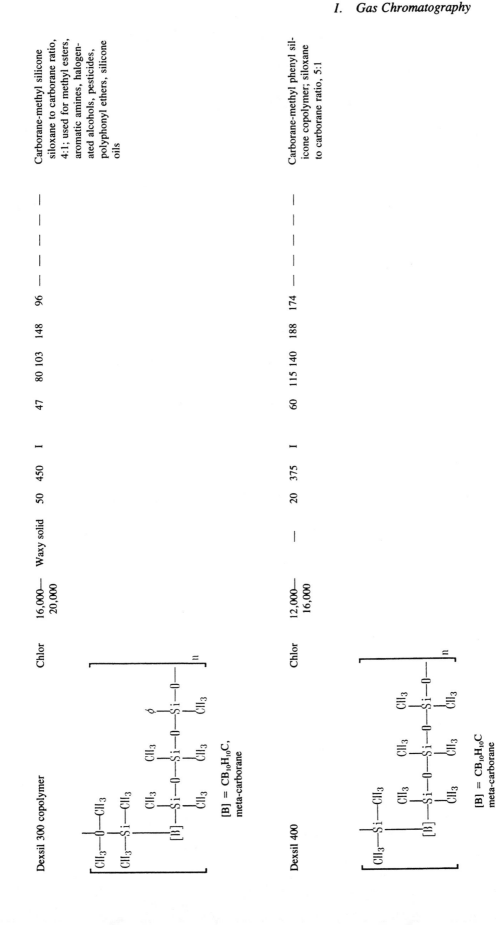

| Dexsil 300 copolymer | Chlor | 16,000—20,000 | Waxy solid | 50 | 450 | I | 47 | 80 | 103 | 148 | 96 | — | — | — | Carborane-methyl silicone siloxane to carborane ratio, 4:1; used for methyl esters, aromatic amines, halogenated alcohols, pesticides, polyphonyl ethers, silicone oils |
| Dexsil 400 | Chlor | 12,000—16,000 | — | 20 | 375 | I | 60 | 115 | 140 | 188 | 174 | — | — | — | Carborane-methyl phenyl silicone copolymer; siloxane to carborane ratio, 5:1 |

[B] = $CB_{10}H_{10}C$, meta-carborane

[B] = $CB_{10}H_{10}C$ meta-carborane

SILICONE LIQUID PHASES (continued)

Liquid phase[a]	Solvent[b]	Av mol mass	Viscosity	T_{min} (°C)	T_{max} (°C)	Polarity[c]	McReynolds constants										Notes
							1	2	3	4	5	6	7	8	9	10	
Dexsil 410	Chlor	9,000—12,000		20	375	I	85	165	170	240	180	—	—	—	—	—	Carborane-methyl-β-silicone cyanoethyl copolymer; siloxane to carborane ratio, 5:1

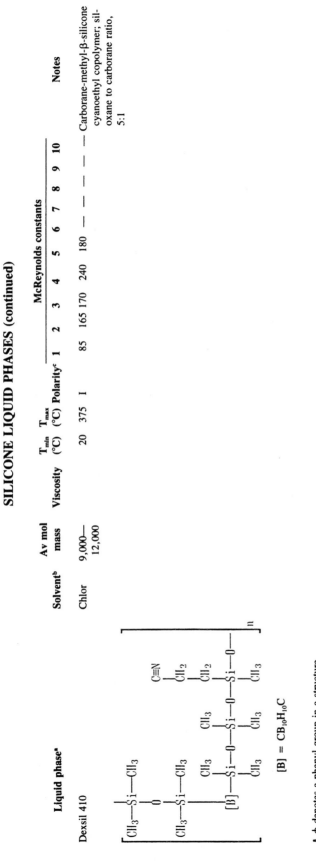

[B] = $CB_{10}H_{10}C$

[a] φ denotes a phenyl group in a structure.

[b] Ace, acetone; Tol, toluene; Chlor, chloroform. When a Silicone fluid is cross-linked, it will be insoluble.

[c] N, nonpolar; I, intermediate polarity; P, polar.

MESOGENIC STATIONARY PHASES

The following table lists the liquid crystalline materials which have found usefulness as gas chromatographic stationary phases. In each case, the name, structure, and transition temperatures are provided (where available), along with a description of the separations which have been done using these materials. The table has been divided into two sections. The first section contains information on phases which have either smectic or nematic phases or both, while the second section contains mesogens which have a cholesteric phase. It should be noted that each material may be used for separations other than those listed, but the listing contains the applications reported in the literature.

It should be noted that many of the mesogens listed in this table are not commercially available, and must be prepared synthetically for laboratory use. The reader is referred to the appropriate citation for details.

REFERENCES

1. **Panse, D. G., Naikwadi, K. P., Bapat, B. V., and Ghatge, B. B.,** Applications of laterally mono and disubstituted liquid crystals as stationary phases in gas liquid chromatography, *Indian J. Technol.,* 19, 518, 1981.
2. **Grushka, E. and Solsky, J. F.,** *p*-Azoxyanisole liquid crystal as a stationary phase for capillary column gas chromatography, *Anal. Chem.,* 45(11), 1836, 1973.
3. **Witkiewicz, Z. and Stanislaw, P.,** Separation of close-boiling compounds on liquid-crystalline stationary phases, *J. Chromatogr.,* 154, 60, 1978.
4. **Naikwadi, K. P., Panse, D. G., Bapat, B. V., and Ghatge, B. B.,** Synthesis and application of stationary phases in gas-liquid chromatography, I, *J. Chromatogr.,* 195, 309, 1980.
5. **Dewar, M. and Schroeder, J. P.,** Liquid crystals as solvents. I. The use of nematic and smectic phases in gas-liquid chromatography, *J. Am. Chem. Soc.,* 86, 5235, 1964.
6. **Dewar, M., Schroeder, J. P., and Schroeder, D.,** Molecular order in the nematic mesophase of 4,4'-di-*n*-hexyloxyazoxybenzene and its mixture with a 4,4'-dimethoxyazoxybenzene, *J. Org. Chem.,* 32, 1692, 1967.
7. **Naikwadi, K. P., Rokushika, S., and Hatano, H.,** New liquid crystalline stationary phases for gas chromatography of positional and geometrical isomers having similar volatilities, *J. Chromatogr.,* 331, 69, 1985.
8. **Richmond, A. B.,** Use of liquid crystals for the separation of position isomers of disubstituted benzenes, *J. Chromatogr. Sci.,* 9, 571, 1971.
9. **Witkiewicz, Z. Pietrzyk, M., and Dabrowski, R.,** Structure of liquid crystal molecules and properties of liquid-crystalline stationary phases in gas chromatography, *J. Chromatogr.,* 177, 189, 1979.
10. **Ciosek, M., Witkiewicz, Z., and Dabrowski, R.,** Direct gas-chromatrographic determination of 2-paphthylamine n 1-napthylamine on liquid-crystalline stationary phases, *Chem. Anal. (Warsaw),* 25, 567, 1980.
11. **Jones, B. A., Bradshaw, J. S., Nishioka, M., and Lee, M. L.,** Synsthesis of smectic liquid-crystalline polysiloxanes from biphenylcarboxylate esters and their use as stationary phases for high-resolution gas chromatography, *J. Org. Chem.,* 49, 4947, 1984.
12. **Porcaro, P. J. and Shubiak, P.,** Liquid crystals as substrates in the GLC of aroma chemicals, *J. Chromatogr. Sci.,* 9, 689, 1971.
13. **Witkiewicz, Z., Suprynowicz, Z., Wojcik, J., and Dabrowski, R.,** Separation of the isomers of some disubstituted benzenes on liquid crystalline stationary phases in small-bore packed micro-columns, *J. Chromatogr.,* 152, 323, 1978.
14. **Dewar, M. and Schroeder, J. P.,** Liquid-crystals as solvents. II. Further studies of liquid crystals as stationary phases in gas-liquid chromatography, *J. Org. Chem.,* 30, 3485, 1965.
15. **Witkiewicz, Z., Szule, J., Dabrowski, R., and Sadowski, J.,** Properties of liquid crystalline cyanoazoxybenzene alkyl carbonates as stationary phases in gas chromatography, *J. Chromatogr.,* 200, 65, 1980.
16. **Witkiewicz, Z., Suprynowicz, Z., and Dabrowski, R.** Liquid crystalline cyanoazoxybenzene alkyl carbonates as stationary phases in small-bore packed micro-columns, *J. Chromatogr.,* 175, 37, 1979.
17. **Lochmüller, C. H. and Souter, R. W.,** Direct gas chromatographic resolution of enantiomers on optically active mesophases, *J. Chromatogr.,* 88, 41, 1974.
18. **Markides, K. E., Nishioka, M., Tarbet, B. J., Bradshaw, J. S., and Lee, M. L.,** Smectic biphenylcarboxylate ester liquid crystalline polysiloxane stationary phase for capillary gas chromatography, *Anal. Chem.,* 57, 1296, 1985.

19. **Vetrova, Z. P., Karabanov, N. T., Shuvalova, T. N., Ivanova, L. A., and Yashin, Ya. I.,** The use of p-n-butyl oxybenzoic acid as liquid crystalline sorbent in gas chromatography, *Chromatographia,* 20, 41, 1985.

20. **Cook, L. E. and Spangelo, R. C.,** Separation of monosubstituted phenol isomers using liquid crystals, *Anal. Chem.,* 46(1), 122, 1974.

21. **Kong, R. C. and Milton, L. L.,** Mesogenic polysiloxane stationary phase for high resolution gas chromatography of isomeric polycyclic aromatic compounds, *Anal. Chem.,* 54, 1802, 1982.

22. **Bartle, K. D., El-Nasri, A. I., and Frere, B.,** *Identification and Analysis of Organic Pollutants in Air,* Ann Arbor Science, Ann Arbor, MI, 1984, 183.

23. **Apfel, M. A., Finkelmann, H., Janini, G. M., Laub, R. J., Lühmann, B.-H., Price, A., Roberts, W. L., Shaw, T. J., and Smith, C. A.,** Synthesis and properties of high-temperature mesomorphic polysiloxane solvents: biphenyl- and terphenyl-based nematic systems, *Anal. Chem.,* 57, 651, 1985.

24. **Janini, G. M.,** Recent usage of liquid crystal stationary phases in gas chromatography, *Adv. Chromatogr.,* 17, 231, 1979.

25. **Witkiewicz, Z. and Waclawczyk, A.,** Some properties of high-temperature liquid crystalline stationary phases, *J. Chromatogr.,* 173, 43, 1979.

26. **Zielinski, W. L., Johnston, R., and Muschik, G. M.,** Nematic liquid crystal for gas-liquid chromatographic separation of steroid epimers, *Anal. Chem.,* 48, 907, 1976.

27. **Smith, Wozny, M. E.,** Gas chromatographic separation of underivatized steroids using BPhBT liquid crystal stationary phase, *J. High Resol. Chromatogr. Chromatogr. Commun.,* 3, 333, 1980.

28. **Barrall, E. M., Porter, R. S., and Johnson, J. F.,** Gas chromatograhy using cholesteryl ester liquid phase, *J. Chromatogr.,* 21, 392, 1966.

29. **Heath, R. R. and Dolittle, R. E.,** Derivatives of cholesterol cinnamate. A comparison of the separations of geometerical isomers when used as gas chromatographic stationary phases, *J. High Resol. Chromatogr. Chromatogr. Commun.,* 6, 16, 1983.

30. **Sonnet, P. E. and Heath, R. R.,** Aryl substituted diastereomeric alkenes: gas chromatographic behaviour on a non-polar versus a liquid crystal phase, *J. Chromatogr.,* 321, 127, 1985.

MESOGENIC STATIONARY PHASES

Name: 2-Chloro-4'-*n*-butyl-4-(4-*n*-butoxybenzoyloxy)azobenzene
Structure:

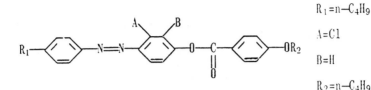

$R_1 = n-C_4H_9$

$A = Cl$

$B = H$

$R_2 = n-C_4H_9$

Thermophysical properties:
  Solid   → nematic 87.2 °C
  Nematic → isotropic 168 °C
Analytical properties: Separation of close-boiling disubstituted benzenes
Reference: 1

Name: *p*-Azoxyanisole (4,4'-dimethoxyazoxybenzene)
Structure:

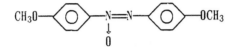

Thermophysical properties:
  Solid   → nematic 118 °C
  Nematic → isotropic 135 °C

Note: Supercooling has been noted at 110 °C by observing nematic-like properties. Liquid crystalline behavior can sometimes persist to 102 °C.

Analytical properties: Separation of xylenes, separation of lower-molecular-weight aromatic hydrocarbon isomers, especially at the lower area of the nematic region.
Reference: 2

Name: 2-Chloro-4'-*n*-butyl-4-(4-methoxybenzoyloxy)azobenzene
Structure:

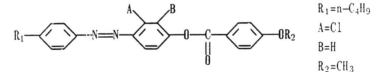

$R_1 = n-C_4H_9$

$A = Cl$

$B = H$

$R_2 = CH_3$

Thermophysical properties:
  Solid   → nematic 92.5 °C
  Nematic → isotropic 176 °C
Analytical properties: Separation of close-boiling disubstituted benzenes
Reference: 1

Name: 2-Chloro-4'-ethyl-4-(4-*n*-butoxybenzoyloxy)azobenzene
Structure:

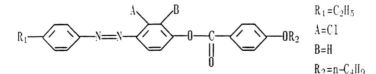

$R_1 = C_2H_5$

$A = Cl$

$B = H$

$R_2 = n-C_4H_9$

Thermophysical properties:
  Solid   → nematic 117 °C
  Nematic → isotropic 172 °C

MESOGENIC STATIONARY PHASES (continued)

Analytical properties: Separation of close-boiling disubstituted benzenes
Reference: 1

Name: 2-Chloro-4'-n-butyl-4-(4-ethoxybenzoyloxy)azobenzene
Structure:

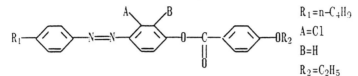

$R_1 = n-C_4H_9$
$A = Cl$
$B = H$
$R_2 = C_2H_5$

Thermophysical properties:
  Solid   → nematic 89.7 °C
  Nematic → isotropic 170 °C

Analytical properties: Separation of close-boiling disubstituted benzenes
Reference: 1

Name: 2-Chloro-4'-methyl-4(4-butoxybenzoyloxy)azobenzene
Structure:

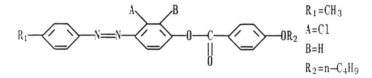

$R_1 = CH_3$
$A = Cl$
$B = H$
$R_2 = n-C_4H_9$

Thermophysical properties:
  Solid   → nematic 112 °C
  Nematic → isotropic 165 °C

Analytical properties: Separation of close-boiling disubstituted benzenes
Reference: 1

Name: 2-Chloro-4'-n-methyl-4-(4-ethoxybenzoyloxy)azobenzene
Structure:

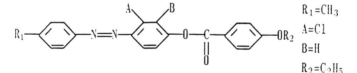

$R_1 = CH_3$
$A = Cl$
$B = H$
$R_2 = C_2H_5$

Thermophysical properties:
  Solid   → nematic 128.3 °C
  Nematic → isotropic 185 °C

Analytical properties: Separation of close-boiling disubstituted benzenes
Reference: 1

Name: *p*-Cyano-*p'*-pentoxyazoxybenzene
Structure:

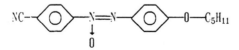

Thermophysical properties:
  Solid   → nematic 124 °C
  Nematic → isoptropic 153 °C

Analytical properties: Complete separation of ethyltoluenes, chlorotoluenes, bromotoluenes, and dichlorobe
also, ethylbenzene from xylenes and propylbenzene from ethyltoluenes
Reference: 3

MESOGENIC STATIONARY PHASES (continued)

Name: *p*-Cyano-*p'*-pentoxyazoxybenzene
Structure:

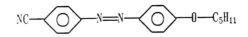

Thermophysical properties:
  Solid → nematic 106 °C
  Nematic → isotropic 116.5 °C
Analytical properties: Separation of ethyltoluenes, chlorotoluenes, bromotoluenes, and dichlorobenzenes, also; ethylbenzenes from xylenes and propylbenzenes from ethylbenzenes
Reference: 3

Name: *p*-Cyano-*p'*-pentoxyazoxybenzene (mixed isomers)
Structure:

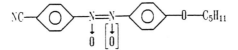

Thermophysical properties:
  Solid → nematic 93.5 °C
  Nematic → isotropic 146.5 °C
Analytical properties: Complete separation of ethyltoluenes, chlorotoluenes, bromotoluenes, and dichlorobenzenes; also, ethylbenzene from xylenes and propylbenzene from ethyltoluenes
Reference: 3

Name: *p*-Cyano-*p'*-octoxyazoxybenzene (mixed isomers)
Structure:

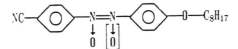

Thermophysical properties:
  Solid → smectic 71 °C
  Smectic → nematic 117 °C
  Nematic → isotropic 135 °C
Analytical properties: Separation of ethyltoluenes, chlorotoluenes, bromotoluenes, and dichlorobenzenes; also, ethylbenzene from xylenes and propylbenzene from ethylbenzenes
Reference: 3

Name: *p*-Cyano-*p'*-octoxyazoxybenzene
Structure:

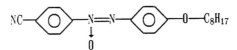

Thermophysical properties:
  Solid → smectic 100.5 °C
  Smectic → nematic 138.5 °C
  Nematic → isotropic 148.5 °C
Analytical properties: Separation of ethyltoluenes, chlorotoluenes, bromotoluenes, and dichlorobenzenes; also, ethylbenzene from xylenes and propylbenzene from ethylbenzenes
Reference: 3

MESOGENIC STATIONARY PHASES (continued)

Name: 4'-*n*-Butyl-4(4-*n*-butoxybenzoyloxy)azobenzene
Structure:

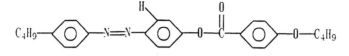

Thermophysical properties:
 Solid   → nematic 94 °C
 Nematic → isotropic 234 °C
Analytical properties: Separation of chlorinated biphenyls
Reference: 4

Name: 4,4'-Di-*n*-heptyloxyazoxybenzene
Structure:

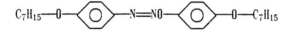

Thermophysical properties:
 Solid   → nematic 95 °C
 Nematic → isotropic 127 °C
Analytical properties: Separation of *meta*- and *para*-xylene in nematic region
Reference: 5

Name: 4,4'-Di-*n*-hexyloxyazoxybenzene
Structure:

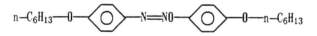

Thermophysical properties:
 Solid   → nematic 81 °C
 Nematic → isotropic 129 °C
Analytical properties: Separation of *meta*- and *para*-xylene using gas chromatography
Reference: 5,6

Name: 4'-Methoxy-4-(-4-*n*-butoxybenzoyloxy)azobenzene
Structure:

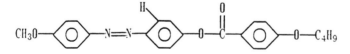

Thermophysical properties:
 Solid   → nematic 116 °C
 Nematic → isotropic 280 °C
Analytical properties: Separation of chlorinated biphenyls
Reference: 4

MESOGENIC STATIONARY PHASES (continued)

Name: 2-Methyl-4'-*n*-butyl-4-(4-*n*-butoxybenzoyloxy)azobenzene
Structure:

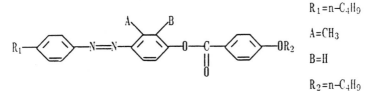

$$R_1 = n\text{-}C_4H_9$$
$$A = CH_3$$
$$B = H$$
$$R_2 = n\text{-}C_4H_9$$

Thermophysical properties:
  Solid → nematic 90 °C
  Nematic → isotropic 175 °C
Analytical properties: Separation of close-boiling disubstituted benzenes
Reference: 1

Name: 2-Methyl-4'-*n*-butyl-4-(*p*-methoxycinnamoyloxy)azobenzene
Structure:

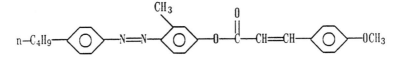

Thermophysical properties:
  Solid → nematic 109 °C
  Nematic → isotropic 253 °C
Analytical properties: Separation of positional isomers of aromatic hydrocarbons
Reference: 7

Name: 2-Methyl-4'-ethoxy-4-(*p*-methoxycinnamoyloxy)azobenzene
Structure:

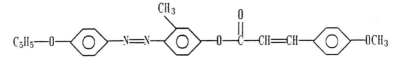

Thermophysical properties:
  Solid → nematic 126 °C
  Nematic → isotropic 262 °C
Analytical properties: Separation of insect sex pheromones
Reference: 7

Name:
2-methyl-4'-ethyl-4-(*p*-methoxycinnamoyloxy)azobenzene
Structure:

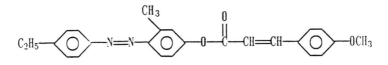

Thermohysical properties:
  Solid → nematic 126 °C
  Nematic → isotropic 262 °C
Analytical properties: Separation of insect sex pheromones
Reference: 7

MESOGENIC STATIONARY PHASES (continued)

Name: 2-Methyl-4′-methoxy-4-(4-ethoxybenzoyloxy)azobenzene
Structure:

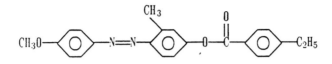

Thermophysical properties:
   Solid   → nematic 125 °C
   Nematic   → isotropic 244 °C
Analytical properties: Separation of chlorinated biphenyls
Reference: 4

Name: 2-Methyl-4′-methoxy-4-(4-methoxybenzoyloxy)azobenzene
Structure:

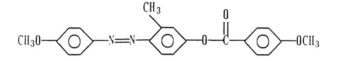

Thermophysical properties:
   Solid   → nematic 160 °C
   Nematic   → isotropic 253 °C
Analytical properties: Separation of chlorinated biphenyls
Reference: 4

Name: 2-Methyl-4′-ethyl-4-(4″-methoxycinnamyloxy)azobenzene
Structure:

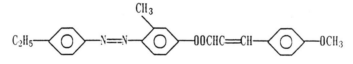

Thermophysical properties:
   Solid   → nematic 126 °C
   Nematic   → isotropic 262 °C
Analytical properties: Separation of polyaromatic hydrocarbons and insect sex pheromones
Reference: 5

Name: 2-Methyl-4′-methoxy-4-(*p*-methoxycinnamoyloxy)azobenzene
Structure:

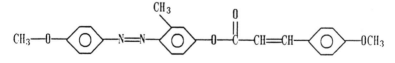

Thermophysical properties:
   Solid   → nematic 149 °C
   Nematic   → isotropic 298 °C
Analytical properties: Separation of positional isomers of aromatic compounds, and geometrical isomers of sex pheromones
Reference: 7

MESOGENIC STATIONARY PHASES (continued)

Name: 2-Methyl-4'-methyl-4-(4-ethoxybenzoyloxy)azobenzene
Structure:

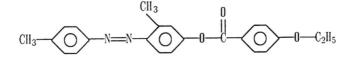

Thermophysical properties:
   Solid   →   nematic 125 °C
   Nematic   →   isotropic 220 °C
Analytical properties: Separation of chlorinated biphenyls
Reference: 4

Name: 4,4'-Azoxydianisole
Structure: See entry for *p*-azoxyanisole
Reference: 8

Name: 4,4'-Azoxyphenetole
Structure:

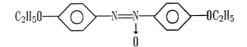

Thermophysical properties:
   Solid   →   nematic 138 °C
   Nematic   →   isotropic 168 °C
Analytical properties: Separation of *meta* and *para* isomers of disubstituted benzenes
Reference: 8

Name: 4,4-Biphenylylene-bis-[*p*-(heptyloxy)benzoate]
Structure:

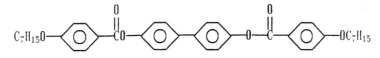

Thermophysical properties:
   Solid   →   smectic 150 °C
   Smectic   →   nematic 211 °C
   Nematic   →   isotropic 316 °C
Analytical properties: Separation of *meta* and *para* isomers of disubstituted benzenes
Reference: 8

Name: *p*'-Ethylazoxybenzene *p*-cyanobenzoate (mixed isomers)
Structure:

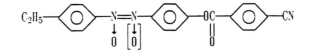

Thermophysical properties:
   Melting range   →   114—136 °C
   Nematic   →   isotropic >306 °C
Analytical properties: Separation of substituted xylenes
Reference: 9

MESOGENIC STATIONARY PHASES (continued)

Name: *p*'-Ethylazoxybenzene *p*-cyanobenzoate (pure isomer)
Structure:

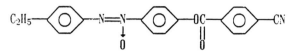

Thermophysical properties:
  Solid   →   nematic 115 °C
  Nematic   →   isotropic 294 °C
Analytical properties: Separation of nitronaphthalenes
Reference: 10

Name: *p*'-Ethylazobenzene *p*-cyanobenzoate
Structure:

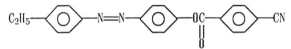

Thermophysical properties:
  Solid   →   138—140 °C
  Nematic   →   isotropic 292 °C
Analytical properties: Separation of substituted xylenes
Reference: 9

Name: *p*'-Ethylazobenzene *p*-methylbenzoate
Structure:

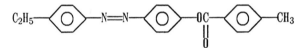

Thermophysical properties:
  Solid   →   nematic 108 °C
  Nematic   →   isotropic 230 °C
Analytical properties: Separation of nitronaphthalenes and substituted xylenes
Reference: 10

Name: *p*-Ethylazoxybenzene *p*'-methylbenzoate (mixed isomers)
Structure:

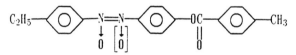

Thermophysical properties:
  Directly after crystallization: After melting and cooling:
    Crystal   →   nematic 97.5 °C   Crystal   →   nematic 87.5—97.5 °C
    Nematic   →   isotropic 250.5 °C   Nematic   →   isotropic 250.5 °C
Analytical properties: Separation of substituted xylenes
Reference: 9

Name: 4'-Methoxybiphenyl-4-y/4-[-(allyloxy)phenyl]benzoate
Structure:

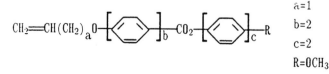

a=1
b=2
c=2
R=OCH$_3$

MESOGENIC STATIONARY PHASES (continued)

Thermophysical properties:
Solid   → nematic 214 °C
Nematic   → isotropic 290 °C
Analytical properties: Suggested for separation of polycyclic aromatic compounds
Reference: 11

Name: (S)-4-[(2-methyl-1-butoxy)carbonyl]phenyl 4-[4-(4-pentenyloxy)phenyl] benzoate
Structure:

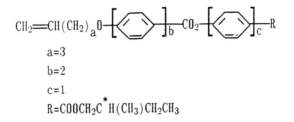

a=3

b=2

c=1

$R=COOCH_2\overset{*}{C}H(CH_3)CH_2CH_3$

Thermophysical properties:
Solid   → smectic 105 °C
Smectic   → isostropic 198 °C
Analytical properties: Suggested for separation of polycyclic aromatic compounds
Reference: 11

Name: 4-Methoxyphenyl(4-[4-(allyloxy)phenyl]benzoate
Structure:

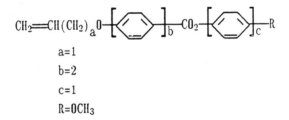

a=1

b=2

c=1

$R=OCH_3$

Thermophysical properties:
Solid   → nematic 137 °C
Nematic   → isotropic 243 °C
Analytical properties: Suggested for the separation of polycyclic aromatic compounds
Reference: 11

Name: 4-Methoxyphenyl-4-[4-(4-pentenyloxy)phenyl]benzoate
Structure:

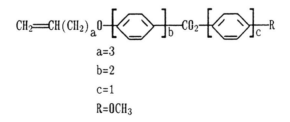

a=3

b=2

c=1

$R=OCH_3$

Thermophysical properties:
Solid   → smectic 133 °C
Smectic   → nematic 172 °C
Nematic   → isotropic 253 °C
Analytical properties: Suggested for separation of polycyclic aromatic compounds
Reference: 11

MESOGENIC STATIONARY PHASES (continued)

Name: *p*-Phenylene-bis-4-*n*-heptyloxybenzoate
Structure:

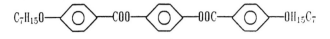

Thermophysical properties:
  Solid   → smectic 83 °C
  Smectic   → nematic 125 °C
  Nematic   → isotropic 204 °C

Analytical properties: Separation of 1- and 2-ethylnapthalene; baseline separation of pyrazines
Reference: 12

Name: 4[(4-Dodecyloxyphenyl)azoxy]-benzonitrile
Structure:

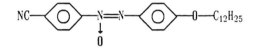

Thermophysical properties:
  Solid   → smectic 106 °C
  Smectic   → isotropic 147 °C

Analytical properties: Marginal effectiveness in separating disubstituted benzene isomers
Reference: 13

Name: 4-[(4-Pentyloxyphenyl)azoxy]-benzonitrile (mixed isomers)
Structure:

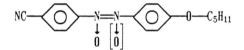

Thermophysical properties:
  Solid   → nematic 94 °C
  Nematic   → isotropic 141.5 °C

Analytical properties: Does not separate diethylbenzene (DEB) isomers; good separation of disubstituted benzene isomers
Reference: 13

Name: 4-[(4-octyloxyphenyl)azoxy]-benzonitrile (pure isomer)
Structure:

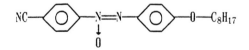

Thermophysical properties:
  Solid   → smectic 101.5 °C
  Smectic   → nematic 137 °C
  Nematic   → isotropic 151.5 °C

Analytical properties: Separates diethylbenzene isomers
Reference: 13

MESOGENIC STATIONARY PHASES (continued)

Name: 4-[(4-Pentyloxyphenyl) azoxy]-benzonitrile (pure isomer)
Structure:

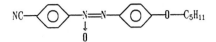

Thermophysical properties:
   Solid   →   nematic 124 °C
   Nematic   →   isotropic 153 °C
Analytical properties: Complete separation of dichlorobenzene or bromotoluene isomers at 126 °C; complete separation of chlorotoluene isomers at 87 °C; partial separation of *m* and *p* xylenes at 87 °C
Reference: 13

Name: 4,4'-Bis(*p*-methoxybenzylidene amino)-3,3'-dichlorobiphenyl
Structure:

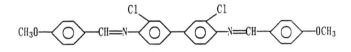

Thermophysical properties:
   Solid   →   nematic 154 °C
   Nematic   →   isostropic 344 °C
Analytical properties: Separationof dimethylbenzene isomers, dihalobenzene isomers (Cl, Br), halo-ketone benzene isomers, and dimethoxybenzene isomers
Reference: 14

Name: Azoxybenzene *p*-cyano-*p'*-heptyl carbonate
Structure:

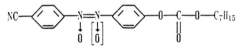

Thermophysical properties:
   Solid   →   nematic 66 °C
Analytical properties: Separation of disubstituted benzene isomers
Reference: 15

Name: Azoxybenzene *p*-cyano-*p'*-octyl carbonate (mixed isomers)
Structure:

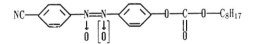

Thermophysical properties:
   Solid   →   smectic 60.5 °C
   Smectic   →   nematic 119.5 °C
Analytical properties: Separation of ethyltoluenes, chlorotoluemnes, bromotoluenes, and dichlorobenzenes; also, ethylbenzenes from xylenes and propylbenzene from ethylbenzenes
Reference: 15

Name: Azoxybenzene *p*-cyano-*p'*-pentyl carbonate (pure isomer)
Structure:

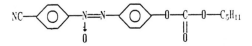

MESOGENIC STATIONARY PHASES (continued)

Thermophysical properties:
Solid   →   nematic 60.5 °C
Nematic   →   isostropic 132 °C.
Analytical properties: Separation of ethyltoluenes, chlorotoluenes, bromotoluenes, and dichlorobenzenes; also, ethylbenzenes from xylenes and propylbenzene from ethylbenzenes
Reference: 3

Name: Azoxybenzene *p*-cyano-*p'*-pentyl carbonate (mixed isomers)
Structure:

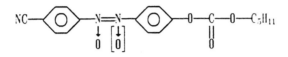

Thermophysical properties:
Solid   →   nematic 96—100 °C
Analytical properties: Separation of disubstituted benzene isomers
Reference: 15

Name: Cyanoazoxybenzene decyl carbonate (mixed isomers)
Structure:

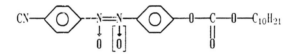

Thermophysical properties:
Solid   →   smectic 74 °C
Smectic   →   isotropic 125.5 °C
Analytical properties: Separation of polycyclic hydrocarbons
Reference: 16

Name: Cyanoazoxybenzene hexyl carbonate (mixed isomers)
Structure:

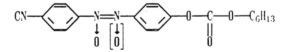

Thermophysical properties:
Solid   →   nematic 73—76 °C
Nematic   →   isotropic 137°C
Analytical properties: Separation of xylene and ethyltoluene isomers
Reference: 16

Name: Cyanoazoxybenzene nonyl carbonate (mixed isomers)
Structure:

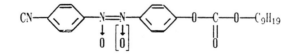

Thermophysical properties:
Solid   →   smectic 61 °C
Smectic   →   nematic 124°C
Nematic   →   isotropic 127 °C
Analytical properties: Separation of polycyclic hydrocarbons
Reference: 16

MESOGENIC STATIONARY PHASES (continued)

Name: *p*-(*p*-Ethoxphenylazo)phenyl crotonate
Structure:

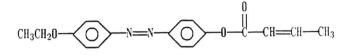

Thermophysical properties:
   Solid   →   nematic 110 °C
   Nematic   →   isotropic 197 °C
Analytical properties: Separation of aromatic isomers
Reference: 12

Name: Carbonyl-bis-(D-leucine isopropyl ester)
Structure:

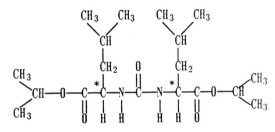

Thermophysical properties:
   Solid   →   smectic 55 °C
   Smectic   →   isotropic 110 °C
Analytical properties: Baseline and near-baseline separations of racemic mixtures of *N*-perfluoroacyl-2-aminoethyl benzenes, trifluoroacetyl (TFA), pentafluoropropionyl (PFP), and heptafluorobutyl, (HFB)
Reference: 17

Name: Carbonyl-bis-(L-valine isopropyl ester)
Structure:

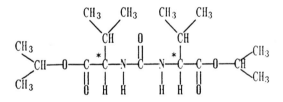

Thermophysical properties:
   Solid   →   smectic[1] 91 °C
   Smectic[1]   →   smectic;[2] 99 °C
   Smectic[2]   →   isotropic 109 °C
(This compound exhibits two stable smectic states prior to melting)
Analytical properties: Separation of enantiomers
Reference: 17

MESOGENIC STATIONARY PHASES (continued)

Name: Carbonyl-bis-(L-valine *t*-butyl ester)
Structure:

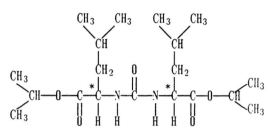

Thermophysical properties:

Solid   → smectic 98 °C
Smectic   → isotropic 402 °C
Analytical properties: Separation of enantiomers
Reference: 17

Name: Carbonyl-bis-(L-valine ethyl ester)
Structure:

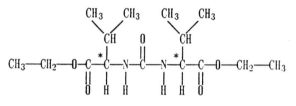

Thermophysical properties:

Solid   → smectic 88 °C
Smectic   → isotropic 115 °C
Analytical properties: Separation of enantiomers
Reference: 17

Name: Carbonyl-bis-(L-valine methyl ester)
Structure:

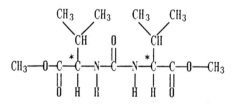

Thermophysical properties:

Solid   → smectic 382 °C
Smectic   → isotropic 415 °C
Analytical properties: Separation of enantiomers
Reference: 17

MESOGENIC STATIONARY PHASES (continued)

Name: Phenylcarboxylate ester (systematic name not available)
Structure:

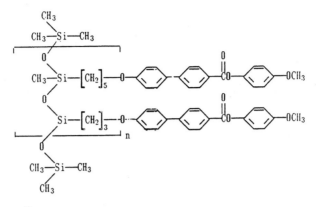

Thermophysical properties:

 Solid   → smectic 118 °C
 Smectic   → isotropic 300 °C

Analytical properties: Separation of three- and four-member methalated polycyclic aromatic hydrocarbons (PAH), on basis of length-to-breadth ratio (l/b); as l/b increases, retention time decreases; cross-linking increases retention times and separation of methylcrypene isomers

Reference: 18

Name: *p*-Cyano-*p'*-octoxyazobenzene
Structure:

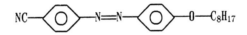

Thermophysical properties:

 Solid   → nematic 101 °C
 Nematic   → isotropic 111 °C

Analytical properties: Separation of ethyltoluenes, chlorotoluenes, bromotoluenes, and dichlorobenzenes; also, ethylbenzenes from xylenes and propylbenzene from ethylbenzenes

Reference: 3

Name: *p-n*-Butoxy benzoic acid
Structure:

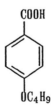

Thermophysical properties:

 Solid, 100 °C
 Mesomorphous, 150 °C (not well characterized)
 Isotropic, 160 °C

Analytical properties: Separation of methyl and monoalkyl substituted benzenes as well as organoelemental compounds (for example, dimethyl mercury)

Reference: 19

MESOGENIC STATIONARY PHASES (continued)

Name: *p*-[*p*-Methoxybenzylidene)-amino]phenylacetate
Structure:

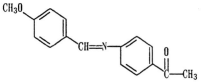

Thermophysical properties:
  Solid   → nematic 80 °C
  Nematic   → isotropic 108 °C
Analytical properties: Separation of substituted phenols; selectivity is best at the lower end of the nematic range.
Reference: 20

Name: Poly (mesogen/methyl) siloxane (PMMS) — compound has not been named
Structure:

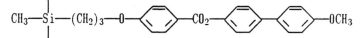

Thermophysical properties:
  Solid   → nematic 70 °C
  Nematic   → isotropic 300 °C
  High thermal stability
Analytical properties: Separation of methylchrysene isomers
Reference: 21

Name: *N,N'*-Bis-(*p*-butoxybenzylidene)-bis-*p*-toluidine (BBBT)
Structure:

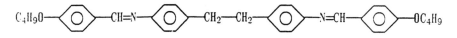

Thermophysical properties:
  Solid   → smectic 159 °C
  Smectic   → nematic 188 °C
  Nematic   → isotropic 303 °C
Analytical properties: Separation of polycyclic aromatic hydrocarbons on the basis of length-to-breadth ratio
Reference: 23

Name: *N,N'*-Bis(ethoxy-benzylidene)-α,α'-bi-*p*-toluidine (BEBT)
Structure:

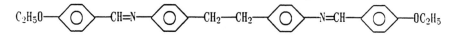

Thermophysical properties:
  Solid   → nematic 173 °C
  Nematic   → isotropic 341 °C
Analytical properties: Separation of polynuclear aromatic hydrocarbons
Reference: 24

Name: *N,N'*-Bis(*n*-heptoxy-benzylidene)-α,α'-bi-*p*-toluidine (BHpBT)
Structure:

$$C_7H_{15}O—\phi—CH{=}N—\phi—CH_2—CH_2—\phi—N{=}CH—o—OC_7H_{15}$$

$$\phi = \langle\bigcirc\rangle$$

MESOGENIC STATIONARY PHASES (continued)

Thermophysical properties:

Solid   → smectic 119 °C
Smectic   → nematic 238 °C
Nematic   → isotropic 262 °C

Analytical properties: Separation of polynuclear aromatic hydrocarbons
Reference: 24

Name: N,N'-Bis(n-hexoxy-benzylidene)-α,α'-bi-p-toluidine (BHxBT)
Structure:

$$H_{13}C_6O-\phi-CH{=}N-\phi-CH_2-CH_2-\phi-N{=}CH-\phi-OC_6H_{13}$$

$$\phi = \langle\bigcirc\rangle$$

Thermophysical properties:

Solid   → smectic 127 °C
Smectic   → nematic 229 °C
Nematic   → isotropic 276 °C

Analytical properties: Separation of methyl- and nitro-derivatives of naphthalene; separation of higher hydrocarbons
Reference: 25

Name: N,N'-Bis (p-methoxybenzylidene)-α,α'-bi-p-toluidine (BMBT)
Structure:

$$CH_3O-\phi-CH{=}N-\phi-CH_2-CH_2-\phi-N{=}CH-\phi-OCH_3$$

$$\phi = \langle\bigcirc\rangle$$

Thermophysical properties:

Solid   → nematic 181 °C
Nematic   → isotropic 320 °C

Analytical properties: Separation of androstane and cholestane alcohols and ketones; good separation of azaheterocyclic compounds; column bleed of BMBT can occur during prolonged periods of operation at elevated temperatures
Reference: 26

Name: N,N-Bis(n-octoxybenzylidene)-α,α'-bi-p-toluidine (BoBT)
Structure:

$$C_8H_{17}O-\phi-CH{=}N-\phi-CH_2-CH_2-\phi-N{=}CH-\phi-OC_8H_{17}$$

$$\phi = \langle\bigcirc\rangle$$

Thermophysical properties:

Solid   → smectic 118 °C
Smectic   → nematic 244 °C
Nematic   → isotropic 255 °C

Analytical properties: Separation of polynuclear aromatic hydrocarbons
Reference: 24

MESOGENIC STATIONARY PHASES (continued)

Name: *N*,*N*-Bis(*n*-pentoxybenzylidene)-α,α'-bi-*p*-toluidine (BPeBT)
Structure:

$$C_5H_{11}O—\phi—CH\!=\!\!N—\phi—CH_2—CH_2—\phi—N\!=\!\!CH—\phi—OC_5H_{11}$$

$$\phi = \langle\!\bigcirc\!\rangle$$

Thermophysical properties:

Solid	→	smectic 139 °C
Smectic	→	nematic 208 °C
Nematic	→	isotropic 283 °C

Analytical properties: Separation of polynuclear aromatic hydrocarbons
Reference: 24

Name: *N*,*N'*-Bis(*p*-phenylbenzylidene)-α,α'-bi-*p*-toluidine (BphBT)
Structure:

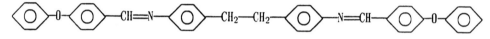

Thermophysical properties:

| Solid | → | nematic 257 °C |
| Nematic | → | isotropic 403 °C |

Analytical properties: Separation of unadulterated steroids; used chromatographically in the temperature range of 260 to 270 °C.
Reference: 27

Name: *N*,*N'*-Bis(*n*-propoxybenzylidene)-α,α'-bi-*p*-toluidine (BPrBT)
Structure:

$$C_3H_7O—\phi—CH\!=\!\!N—\phi—CH_2—CH_2—\phi—N\!=\!\!CH—\phi—OC_3H_7$$

$$\phi = \langle\!\bigcirc\!\rangle$$

Thermophysical properties:

Solid	→	smectic 169 °C
Smectic	→	nematic 176 °C
Nematic	→	isotropic 311 °C

Analytical properties: Separation of polynuclear aromatic hydrocarbons
Reference: 24

CHOLESTERIC PHASES

Name: Cholesteryl acetate
Structure:

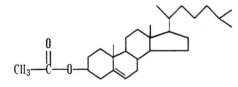

Thermophysical properties:

| Solid | → | cholesteric 94.5 °C |
| Cholesteric | → | isostropic 116.5 °C |

Analytical properties: Separation of aromatics and paraffins
Reference: 28

CHOLESTERIC PHASES (continued)

Name: (S)-4'-[(2-Methyl-1-butoxy)carbonyl]biphenyl-4-yl-4(allyloxy)benzoate
Structure:

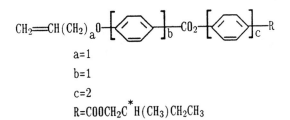

a=1
b=1
c=2
R=COOCH$_2$C*H(CH$_3$)CH$_2$CH$_3$

Thermophysical properties:
Solid   →   smectic 100 °C
Smectic   →   cholesteric 150 °C
Cholesteric   →   isotropic 188 °C
Analytical properties: Suggested for separation of polycyclic aromatic compounds
Reference: 11

Name: (S)-4'-[(2-Methyl-1-butoxy)carbonyl]biphenyl-4-yl-4-[4-(allyloxy)phenyl]benzoate
Structure:

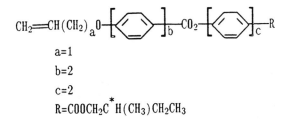

a=1
b=2
c=2
R=COOCH$_2$C*H(CH$_3$)CH$_2$CH$_3$

Thermophysical properties:
Solid   →   smectic 152 °C
Smectic   →   cholesteric 240 °C
Cholesteric   →   isotropic 278 °C
Analytical properties: Suggested for separation of polycyclic aromatic compounds
Reference: 11

Name: (S)-4'-[(2-Methyl-1-butoxy)biphenyl-4-yl 4-[4-4-pentenyloxy)phenyl]benzoate
Structure:

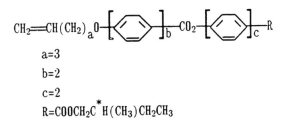

a=3
b=2
c=2
R=COOCH$_2$C*H(CH$_3$)CH$_2$CH$_3$

Thermophysical properties:
Solid   →   smectic 135 °C
Smectic   →   cholesteric 295 °C
Cholesteric   →   isotropic 315 °C
Analytical properties: Suggested for separation of polycyclic aromatic compounds
Reference: 11

CHOLESTERIC PHASES (continued)

Name: (*S*)-4-[(2-Methyl-1-butoxy)carbonyl]phenyl 4-[4-(allyloxy)phenyl]benzoate
Structure:

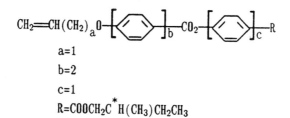

a=1

b=2

c=1

$R=COOCH_2C^*H(CH_3)CH_2CH_3$

Thermophysical properties:

Solid	→	smectic 118 °C
Smectic	→	cholesteric 198 °C
Cholesteric	→	isotropic 213 °C

Analytical properties: Suggested for separation of polycyclic aromatic compounds
Reference: 11

Name: Cholesterol cinnamate
Structure:

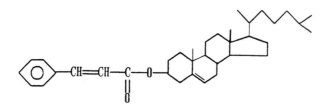

Thermophysical properties:

Solid	→	cholesteric 160 °C
Cholesteric	→	isotropic 210 °C

Analytical properties: Separation of olefinic positional isomers
Reference: 12, 29

Name: Cholesterol-*p*-chlorocinnamate (CpCC)
Structure:

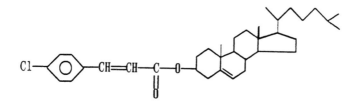

Thermophysical properties:

Solid	→	cholesteric 144 °C
Cholesteric	→	isotropic 268 °C

Analytical properties: Separation of diastereomeric amides and carbamates; the separation of olefinic geometrical isomers is dependent upon the position of the double bond
Reference: 29, 30

CHOLESTERIC PHASES (continued)

Name: Cholesterol-*p*-methylcinnamate
Structure:

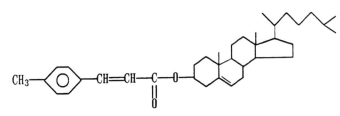

Thermophysical properties:
Solid   → cholesteric 157 °C
Cholesteric   → isotropic 254 °C
Analytical properties: Separation of olefinic positional isomers
Reference: 29

Name: Cholesterol-*p*-methoxycinnamate
Structure:

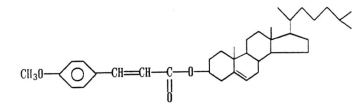

Thermophysical properties:
Solid   → cholesteric 165 °C
Cholesteric   → isotropic 255 °C
Analytical properties: Separation of olefinic positional isomers
Reference: 29

Name: Cholesterol *p*-nitro cinnamate
Structure:

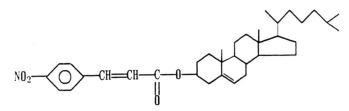

Thermophysical properties:
Solid   → cholesteric 167 °C
Cholesteric   → isotropic 265 °C
Analytical properties: Separation of geometrical isomers (two- and three-octadecene) using p-substituted choles-terols; (best separation) *p*-NO₂ > *p*-MeO > cholesterol cinnamate > *p*-Me > *p*-Cl (worst separation) for unsaturation occurring within four carbon atoms from the terminal methyl; the above order holds for separations of tetradecen-1-ol acetates; for unsaturation on carbons 5 to 12 from the terminal methyl of the tetradecen-1-ol of acetates, the best separation is the reverse of the above
Reference: 29

CHOLESTERIC PHASES (continued)

Name: Cholesteryl nonanoate
Structure:

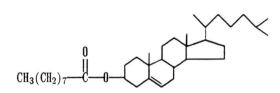

Thermophysical properties:

Solid	→	smectic 77.5 °C
Smectic	→	cholesteric 80.5 °C
Cholesteric	→	isotropic 92 °C

Analytical properties: Separation of aromatics and paraffins
Reference: 28

Name: Cholesteryl valerate
Structure:

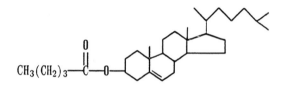

Thermophysical properties:

Solid	→	cholesteric 93 °C
Cholesteric	→	isotropic 101.5 °C

Analytical properties: Separation of aromatics and paraffins
Reference: 28

STATIONARY PHASES FOR PACKED COLUMN GAS CHROMATOGRAPHY

The following stationary phases have been shown to be of value in the separation of these major classes of compounds, using packed columns of typical dimensions (4 to 10 m in length, 0.32 cm in diameter).[1-11] The resolution will undoubtedly be lower than that obtainable with capillary columns, which have superceded packed columns in many applications. The two main exceptions are in the analysis of permanent gases and preparative-scale gas chromatography. Data on the packed column stationary phases are included since they still find use in many laboratories. This table is meant to provide only a rough guide. The additional data which can be found in the preceding stationary phase data table will aid in determining the final choice.

REFERENCES

1. **Heftmann, E., Ed.,** *Chromatography — A Laboratory Handbook of Chromatographic and Electrophoretic Methods,* 3rd. ed., Van Nostrand Reinhold, New York, 1975.
2. **Grant, D. W.,** *Gas-Liquid Chromatography,* Van Nostrand Reinhold, London, 1971.
3. **McNair, H. M. and Bonelli, E. J.,** *Basic Gas Chromatography,* Varian Aerography, Palo Alto, Ca, 1969.
4. **Grob, R. L., Ed.,** *Modern Practice of Gas Chromatography,* 2nd ed., Interscience, New York, 1985.
5. **Poole, C. F. and Schuette, S. A.,** *Contemporary Practice of Chromatography,* Elsevier, Amsterdam, 1984.
6. **Mann, J. R. and Preston, S. T.,** Selection of preferred liquid phases, *J. Chromatogr. Sci.,* 11, 216, 1973.
7. **Coleman, A. E.,** Chemistry of liquid phases, other silicones, *J. Chromatogr. Sci.,* 11, 198, 1973.
8. **Yancey, J. A.,** Liquid phases used in packed gas chromatographic columns. I. Polysiloxane phases, *J. Chromatogr. Sci.,* 23, 161, 1985.
9. **Yancey, J. A.,** Liquid phases used in packed gas chromatographic columns. II. Use of liquid phases which are not polysiloxanes, *J. Chromatogr. Sci.,* 23, 370, 1985.
10. Analabs Guide to Chromatographic Phases, Foxboro Co., North Haven, CT, 1981.
11. **McReynolds, W. O.,** Characterization of some liquid phases, *J. Chromatogr. Sci.,* 8, 685, 1970.

STATIONARY PHASES FOR PACKED COLUMN GAS CHROMATOGRAPHY

Compound	Suggested stationary phases
Alcohols	
C_1—C_5	Apiezon L; Apiezond M; benzyldiphenyl; butane diol succinate (Craig polyester), Carbowax 400, 600, 750, 1000, 1000 (monostearate), diethylene glycolsuccinate; di-(2-ethylexyl) sebacate; diethyl-*d*-tartrate; di-*n*-decylphthalate; diglycerol; diisodecyl phthalate; dinonyl phthalate; ethylene glycol succinate; Flexol 8N8; Hallcomid M-18-OL; quadrol; Renex 678; sorbitol; tricresyl phosphate; triethanolamine
C_5—C_{18}	Butane diol succinate (Craig polyester); Carbowax 1500, 1540, 4000, 4000 (dioleate), 4000 (monostearate), 6000, 20M, 20M-TPA; ethylene glycol adipate; IGEPAL series; Ucon series, Versamid series
Di-poly	FFAP; QF-1, Porapak-Q, Porapak-QS
Aldehydes (and ketones)	Apiezon L, M; Carbowax 400, 750, 1000, 1500, 1540; di-*n*-butyl phthalate; diethylene glycol succinate; ethylene glycol succinate; Hallcomid M18; squalene; tricresyl phosphate; 1,2,3-*tris* (2-cyanoethoxy)propane; Ucon series
Alkaloids (includes drugs and vitamins)	Apiezon L; Carbowax 20M; di (2-ethylhexyl) sebacate; ethylene glycol adipate; ethylene glycol succinate, neopentyl glycol adipate; phenyldiethanolamine succinate; SE-30 (methyl silicone phases)
Amides	Carbowax 600 (on Chromosorb T); diethylene glycol succinate; ethylene glycol succinate; neopentyl glycol sebacate; Versamid 900; SE-30 (methyl silicone phases)
Amino acids (and derivatives)	Carbowax 600; diethylene glycol succinate (stabilized); Ethofat (on Chromososrb T); ethylene glycol succinate; neopentyl glycol adipate; SE-30; XE-60 (methyl silicone phases)
Amines	Penwalt 213; Chromosorb 103 (see support modifiers)
Boranes	Apiezon L; beeswax; Carbowax 400, 1540, 4000, 20M; castorwax; diethylene glycol succinate; di-*n*-decyl phthalate; diisodecyl phthalate; Emulphor-ON-870; ethylene glycol adipate; FFAP, polyphenyl ether (five- or six-ring); quadrol; reoplex 400; SE-30; XE-60; sucrose acetate isobutyrate; tricresyl phosphate; Ucon series
Esters	Apiezon L; benzyldiphenyl; Carbowax 20M; cyclodextrin acetate; diethylene glycol adipate; di(2-ethylhexyl) sebacate; diisodecyl phthalate; dimer acid/OV-1 (50/50, v/v); Hallcomid M18, neopentyl glycol succinate; propylene glycol; SE-30; SE-52; XE-60; Friton X-100; Tween-80
Ethers	Apiezon L; Carbowax 1500, 1540, 4000, 20M; diethylene glycol sebacate, ethylene glycol adipate
Glycols	Porapak-Q, Porapak-QS; QF-1
Halogenated compounds	Bentone 34; benzyldiphenyl; butanediol succinate (Craig polyester); Carbowax 400, 1000, 4000, 20M; dibutyl phthalate; diethylene glycol succinate; di(2-ethylhexyl) sebacate; di-*n*-decyl phthalate; dinonyl phthalate; dioctyl phthalate; β,β^1-iminodipropionitrile; β,β^1-oxydipropionitrile; SE-30; squalane; Tween-80
Inorganic compounds (includes organometallic compounds)	*n*-Decane; di-*n*-decyl phthalate; dimethyl sulfolane; neopentyl glycol succinate; 1,2,3-Trix (2-cyanoethoxy) propane; SE-30 (methyl silicone phases)
Hydrocarbons	
C_1—C_5 (aliphatic)	Carbowax 400 to 1500; most branched and substituted phthalate, sebacate, succinate, and adipate phases; octadecane; squalane (boiling-point separations); methyl silicones
Above C_5 (aliphatic)	Apiezon phases; Carbowax 1500, 1540, 4000, 6000, 20M; most of the high-temperature substituted adipates, phthalates, succinates, and sebacates (boiling-point separations); methyl silicones

STATIONARY PHASES FOR PACKED COLUMN GAS CHROMATOGRAPHY
(continued)

Compound	Suggested stationary phases
Alcohols	
Above C_5 (aromatic)	Apiezon phases; bentone-34; Carbowax phases; substituted adipates, phthalates, succinates, and sebacates; tetra cyanoethylated pentaerythritol; liquid crystalline phases; phenyl silicone phases
Nitrogen compounds	Apiezon L; Armeen SD; butanediol succinate (Craig polyester); Carbowax 400, 1500, 20M; ethylene glycol adipate; propylene glycol; tetraethylene glycol dimethyl ether; THEED, Ucon phases
Pesticides	Carbowax-20M; diethylene glycol adipate; Epon 1001; neopentyl glycol adipate; methyl silicone phases, including gum viscosities
Phosphorous compounds	Apiezon L; Carbowax 20M; di-*n*-butyl phthalate; diethylene glycol succinate; Reoplex-400; methyl silicone phases, including gum viscosities; squalane, STAP
Silanes	Methyl silicone phases; STAP
Sugars	Apiezon L; butanediol succinate; Carbowax 4000, Hyprose SP80, mannitol; methyl silicone phases
Sulfur compounds	Apiezon L; 7,8-benzoquinoline; Carbowax 1500, 20M; diethylene glycol succinate; diisodecyl phthalate; methyl silicone phases; Reoplex-400; tricresyl phosphate
Urinary and bile compounds	Ethylene glycol adipate; methyl silicone-nitrile phases

TRAPPING SORBENTS

The following table provides a listing of the major types of sorbents used in sampling, concentrating, odor profiling, and air and water pollution research.[1-6] These materials are useful in a wide variety of research and control applications. Many can be obtained commercially in different sizes, depending upon the application involved. The purpose of this table is to aid in the choice of a sorbent for a given analysis.

REFERENCES

1. **Borgstedt, H. U., Emmel, H. W., Koglin, E., Melcher, R. G., and Peters, Sequaris, J. M. L.,** *Analytical Problems,* Springer-Verlag, Berlin, 1986.
2. **Averill, W. and Purcell, J. E.,** Concentration and gc determination of organic compounds from air and water, *Chromatogr. Newslett.,* 6(2), 30, 1978.
3. **Gallant, R. F., King, J. W., Levins, P. L., and Piecewicz, J. F.,** Characterization of Sorbent Resins for Use in Environmental Sampling, Report EPA-600/7-78-054, Office of Research and Development, Environmental Protection Agency, Washington, D.C., March 1978.
4. **Chladek, E. and Marano, R. S.,** Use of bonded phase silica sorbents for the sampling of priority pollutants in waste waters, *J. Chromatogr. Sci.,* 22, 313, 1984.
5. **Good, T. J.,** Applications of bonded-phase materials, *Am. Lab.,* July 1981, p. 36.
6. **Beyermann, K.,** *Organic Trace Analysis,* Halsted Press, New York, 1984.

TRAPPING SORBENTS

Sorbent	Desorption solvents	Applications
Activated carbon	Carbon disulfide, methylene chloride, diethyl ether, diethyl either with 1% methanol, diethyl ether with 5% 2-propanol (caution: CS₂ and CH₃OH can react in the presence of charcoal)	Used for common volatile organics; examples include methylene choloride, vinyl chloride, chlorinated aliphatics, aromatics, and acetates; more data are provided in the table entitled "Adsorbents for Gas Chromatography".

Note: Metallic or salt impurities in the sorbent can sometimes cause the irreversible adsorption of electron-rich oxygen functionalities; examples include 1-butanol, 2-butanone, and 2-ethoxyacetate; recovery rate is often poor for polar compounds.

Graphitized carbon-black	Carbon disulfide, methylene chloride, diethyl ether (or thermal desorption can be used)	Used for common volatile aliphatic and aromatic compounds, organic acids and alcohols, and chlorinated aliphatics; more data are provided in the table entitled "Adsorbents for Gas Chromatography".

Note: These sorbents are hydrophobic and are not very sensitive to moisture; the possibility of thermal desorption make them of value for "trace-level" analyses.

Silica gel	Methanol, ethanol, water, diethyl ether	Used for polar compound collection and concentration; examples include alcohols, phenols, chlorophenols, chlorinated aromatics, aliphatic and aromatic amines, and nitrogen dioxide; more data are provided in the table entitled "Adsorbents for Gas Chromatography".

Note: Useful for compounds which cannot be recovered from the charcoal sorbents; the most serious problem with silica is the effect of water, which can cause desorption of the analytes of interest; the heating effect involved can sometimes initiate reactions such as polymerization of the anlayte.

TRAPPING SORBENTS (continued)

Sorbent	Desorption solvents	Applications
Activated alumina	Water, diethyl ether, methanol	Used for polar compounds such as alcohols, glycols, ketones, and aldehydes; has also been used for polychlorinated biphenyls and phthalates; more data are provided in the table entitled "Adsorbents for Gas Chromatography".

Note: Similar in application to silica gel.

Porous polymers	Hexane, diethyl ether, alcohols (thermal desorption also possible in some cases)	Used for a wide range of compounds which include phenols, acidic and basic organics, pesticides, and priority pollutants; more data are provided in the table entitled "Porous Polymer Phases".

Note: The most commonly used porous polymer sorbent is Tenax-GC, although the Porpak and Chromosorb Century series have also been used; Tenax-GC has been used with thermal desorption methods, but can release toluene, benzene, and trichloroethylene residues at the higher temperatures; in addition to Tenax-GC, XAD 2-8, Porapak-N, and Chromosorbs 101, 102, 103, and 106 have found applications, sometimes in "stacked" sampling devices (for example, a sorbent column of Tenax-GC/Chromosorb 106 in tandem); Chromosorb 106, a very low-polarity polymer, has the lowest retention of water with respect to organic materials, and is well suited for use as a back-up sorbent.

Bonded phases	Methanol, hexane, diethyl ether	Used for specialized applications in pesticides, herbicides, and polynuclear aromatic hydrocarbons

Note: Most expensive of the common sorbents; useful for the collection of organic samples from water.

Molecular sieves	Carbon disulfide, hexane diethyl ether	Have been used for the collection of aldehydes and alcohols, and for acrolein

Note: Molecular sieve 13-X is the main molecular sieve to be used as a trapping adsorbent; these sorbents will also retain water.

DERIVATIZING REAGENTS FOR GAS CHROMATOGRAPHY

The following table lists some of the more common derivatizing reagents used in gas chromatography for the purposes of (1) increasing sample volatility, (2) increasing sample thermal stability, (3) reducing sample-support interactions, and (4) increasing sensitivity toward a particular detector. The table is divided into reagents for acylation, alkylation, esterification, pentafluoro-phenylation, and silylation. The conditions and concentrations used in derivatization must be carefully considered, since one can often cause more problems than one cures using these methods. Such problems include poor peak resolution, incomplete reactions and side products, and less than stoichometric yields of products. The reader is referred to the citation list for more detail on the reagents, conditions, and difficulties.

REFERENCES

General References
1. **Blau, K. and King, G. S., Eds.**, *Handbook of Derivatives for Chromatography*, Hayden, London, 1978.
2. **Knapp, D. R.**, *Handbook of Analytical Derivatization Reactions*, John Wiley & Sons, New York, 1979.
3. **Drozd, J.**, *Chemical Derivatization in Gas Chromatography*, Elsevier, Amsterdam, 1981.
4. **Poole, C. F. and Schutte, S. A.**, *Comtemporary Practice of Chromatography*, Elsevier, Amsterdam, 1984.
5. **Grob, R. L.**, *Modern Practice of Gas Chromatography*, John Wiley & Sons, New York, 1985.
6. **Braithwaite, A. and Smith, F. J.**, *Chromatographic Methods*, Chapman and Hall, London, 1985.
7. **Merritt, C.**, in *Ancillary Techniques of Gas Chromatography*, Ettre, L. S. and McFadden, W. H., Eds., Interscience, New York, 1969.
8. **Hammarstrand, K. and Bonelli, E. J.**, *Derivative Formation in Gas Chromatography*, Varian Aerograph, Walnut Creek, CA, 1968.
9. **Vanden Heuvel, W. J. A.**, *Gas Chromatography of Steroids in Biological Fluids*, Plenum Press, New York, 1965.

Acylating Reagents
1. **Brooks, C. J. W. and Horning, E. C.**, Gas chromatographic studies of catecholamines, tryptamines, and other biological amines, *Anal. Chem.*, 36(8), 1540, 1964.
2. **Imai, K., Sugiura, M., and Tamura, Z.**, Catecholamines in rat tissues and serum determined by gas chromatographic method, *Chem. Pharm. Bull.*, 19, 409, 1971.
3. **Scoggins, M. W., Skurcenski, L., and Weinberg, D. S.**, Gas chromatographic analysis of geometric diamine isomers as tetramethyl derivatives, *J. Chromatogr. Sci.*, 10, 678, 1972.

Esterification Reagents
1. **Shulgin, A. T.**, Separation and analysis of methylated phenols as their trifluoroacetate ester derivatives, *Anal. Chem.*, 36(4), 920, 1964.
2. **Argauer, R. J.**, Rapid procedure for the chloroacetylation of microgram quantities of phenols and detection by electron-capture gas chromatography, *Anal. Chem.*, 40(1), 122, 1968.
3. **Vanden Heuvel, W. J. A., Gardiner, W. L., and Horning, E. C.**, Characterization and separation of amines by gas chromatography, *Anal. Chem.*, 36(8), 1550, 1964.
4. **Änggård, E. and Göran, S.**, Gas chromatography of catecholamine metabolites using electron capture detection and mass spectrometry, *Anal. Chem.*, 41(10), 1250, 1969.
5. **Alley, C. C., Brooks, J. B., and Choudhary, G.**, Electron capture gas-liquid chromatography of short chain acids as their 2,2,2-trichloroethyl esters, *Anal. Chem.*, 48(2), 387, 1976.
6. **Godse, D. D., Warsh, J. J., and Stancer, H. C.**, Analysis of acidic monoamine metabolites by gas chromatography-mass spectrometry, *Anal. Chem.*, 49(7), 915, 1977.
7. **Matin, S. B. and Rowland, M.**, Electron-capture sensitivity comparison of various derivatives of primary and secondary amines, *J. Pharm. Sci.*, 61(8), 1235, 1972.
8. **Bertani, L. M., Dziedzic, S. W., Clarke, D. D., and Gitlow, S. E.**, A gas-liquid chromatographic method for the separation and quantitation of nomethanephrine and methanephrine in human urine, *Clin. Chem. Acta*, 30, 227, 1970.
9. **Kawai, Tamura, Z.**, Gas chromatography of catecholamines as their trifluoroacetates, *Chem. Pharm. Bull.*, 16(4), 699, 1968.
10. **Moffat, A. C. and Horning, E. C.**, A new derivative for the gas-liquid chromatography of picogram quantities of primary amines of the catecholamine series, *Biochem. Biophys. Acta*, 222, 248, 1970.

11. **Lamparski, L. I. and Nestrick, T. J.**, Determination of trace phenols in water by gas chromatographic analysis of heptafluorobutyl derivatives, *J. Chromatogr.*, 156, 143, 1978.
12. **Mierzwa, S. and Witek, S.**, Gas-liquid chromatographic method with electron-capture detection for the determination of residues of some phenoxyacetic acid herbicides in water as their 2,2,2-trichloroethyl esters, *J. Chromatogr.*, 136, 105, 1977.
13. **Hoshika, Y.**, Gas chromatographic separation of lower aliphatic primary amines as their sulphur-containing schiff bases using a glass capillary column, *J. Chromatogr.*, 136, 253, 1977.
14. **Brooks, J. B., Alley, C. C., and Liddle, J. A.**, Simultaneous esterification of carboxyl and hydroxyl groups with alcohols and heptafluorobutyric anhydride for analysis by gas chromatography, *Anal. Chem.*, 46(13), 1930, 1974.
15. **Deyrup, C. L., Chang, S. M., Weintraub, R. A., and Moye, H. A.**, Simultaneous esterification and acylation of pesticides for analysis by gas chromatography. I. Derivatization of glyphosate and (aminomethyl) phosphonic acid with fluoronated alcohols-perfluoronated anhydrides, *J. Agric. Food Chem.*, 33(5), 944, 1985.
16. **Samar, A. M., Andrieu, J. L., Bacconin, A., Fugier, J. C., Herilier, H., and Faucon, G.**, Assay of lipids in dog myocardium using capillary gas chromatography and derivatization with boron trifluoride and methanols, *J. Chromatogr.*, 339(1), 25, 1985.

Pentafluoro Benzoyl Reagents

1. **Mosier, A. R., Andre, C. E., and Viets, F. G., Jr.**, Identification of aliphatic amines volatilized from cattle feedyard, *Environ. Sci. Technol.*, 7(7), 642, 1973.
2. **DeBeer, J., Van Petegham, C., and Heyndridex, Al.**, Electron caapture-gas-liquid chromatography (EC-GLC) determination of the herbicidal monohalogenated phenoxyalkyl acid mecoprop in tissues, urine and plasma after derivatization with pentafluorobenzylbromide, *Vet. Hum. Toxicol.*, 21, 172, 1979.
3. **Davis, B.**, Crown ether catalyzed derivatization of carboxylic acids and phenols with pentafluorobenzyl bromide for electron capture gas chromatography, *Anal. Chem.*, 49(6), 832, 1977.
4. **Avery, M. J. and Junk, G. A.**, Gas chromatography/mass spectrometry determination of water-soluble primary amines as their pentafluorobenzaldehyde imines, *Anal. Chem.*, 57(4), 790, 1985.

Silylating Reagents

1. **Metcalfe, L. D. and Martin, R. J.**, Gas chromatography of positional isomers of long chain amines and related compounds, *Anal. Chem.*, 44(2), 403, 1972.
2. **Sen, H. P. and McGeer, P. L.**, Gas chromatography of phenolic and catecholic amines as the trimethylsilyl ethers, *Biochem. Biophys. Res. Commun.*, 13(5), 390, 1963.
3. **Fogelgvist, E., Josefsson, B., and Ross, C.**, Determination of carboxylic acids and phenols in water by extractive alkylation using pentafluorobenzylation, glass capillary g.c. and electron capture detection, *J. High Resol. Chromatogr. Chromatogr. Commun.*, 3, 568, 1980.
4. **Poole, C. F., Sye, W. F., Singhawangcha, S., Hsu, F., Zlatkis, A., Arfwidsson, A., and Vessman, J.**, New electron-capturing pentafluorophenyldialkylchlorosilanes as versatile derivatizing reagents for gas chromatography, *J. Chromatogr.*, 199, 123, 1980.
5. **Quilliam, M. A., Ogilvie, K. K., Sadana, K. L., and Westmore, J. B.**, Study of rearrangement reactions occurring during gas chromatography of tert-butyl-dimethylsilyl ether derivatives of uridine, *J. Chromatogr.*, 194, 379, 1980.
6. **Poole, C. F. and Zlatkis, A.**, Trialkylsilyl ether derivatives (other than TMS) for gas chromatography and mass spectrometry, *J. Chromatogr. Sci.*, 17(3), 115, 1979.
7. **Francis, A. J., Morgan, E. D., and Poole, C. F.**, Flophemesyl derivatives of alcohols, phenols, amines and carboxylic acids and their use in gas chromatography with electron-capture detection, *J. Chromatogr.*, 161, 111, 1978.
8. **Harvey, D. J.**, Comparison of fourteen substituted silyl derivatives for the characterization of alcohols, steroids and cannabinoids by combined gas-liquid chromatography and mass spectrometry, *J. Chromatogr.*, 147, 291, 1978.
9. **Quilliam, M. A. and Yaraskavitch, J. M.**, Tertbutyldiphenylsilyl derivatization for liquid chromatography and mass spectrometry, *J. Liq. Chromatogr.*, 8(3), 449, 1985.

DERIVATIZING REAGENTS FOR GAS CHROMATOGRAPHY

Derivatizing reagent	Structure/formula	Notes

Acylating Reagents

Derivatizing reagent	Structure/formula	Notes
Acetic anhydride	$(CH_3CO)_2O$	Used for amino acids, steroids, urinary sugars, pesticides and herbicides, and narcotics
Chloracetic anhydride	$(CH_2ClCO)_2O$	Useful for electron-capture detection of lower aliphatic primary amines
2,4′-Dibromoacetophenone		Used for short- and medium-chain aliphatic carboxylic acids
Heptafluorobutyric anhydride	$(CF_3CF_2CF_2CO)_2O$	Used in basic solution for alcohols, amines, nitrosamines, amino acids, and steroids; heptafluorobutyrlimidazole is used in a similar fashion in the analysis of phenols
Pentafluorobenzaldehyde		Useful for electron-capture detection of several primary amines
Pentafluorobenzoyl chloride		Useful for electron-capture detection of several primary amines
Pentafluoropropionic anhydride	$(CF_3CF_2CO)_2O$	Used for aromatic monoamines and their metabolites
Propionic anhydride	$(CH_3CH_2CO)_2O$	Used for amines, amino acids, and narcotics
Pivalic anhydride	$[(CH_3)_3CCO]_2O$	Used for hormone analysis
2-Thiophene aldehyde		Used for electron-capture detection of lower aliphatic primary amines
Trifluoroacetic anhydride	$(CF_3CO)_2O$	Used for phenols, amines, amino acids, amino phosphonic acids, saccharides, and vitamins
N-Trifluoroacetylimidazole		Useful for the relatively straightforward acylation of hydroxyl groups, and secondary or tertiary amines.

DERIVATIZING REAGENTS FOR GAS CHROMATOGRAPHY (continued)

Derivatizing reagent	Structure/formula	Notes

Acylating Reagents

Derivatizing reagent	Structure/formula	Notes
Diazomethane	$CH_2=N=N$ + −	Used as a common alkylating agent; acts on acidic and enolic groups rapidly, and more slowly on other groups with replaceable hydrogens (the use of a Lewis acid catalyst such as BF_3 is sometimes helpful); all diazoalkanes are toxic and sometimes explosive, and are used in microscale operations only
Trimethylanilinium hydroxide (TMAH; in methanol)	$\overset{+}{N}{-}(CH_3)_3$ (attached to benzene ring)	Useful for methylation of amines
Pentafluorobenzyl bromide	$H_2C{-}Br$ (pentafluorophenyl ring, F substituents)	Useful for the derivation of acids, amids, and phenols, providing great increase in sensitivity toward electron-capture detection

Esterification Reagents

Derivatizing reagent	Structure/formula	Notes
Boron trifluoride + methanol	$BF_3 + CH_3OH$	Useful for carboxylic acids (aromatic and aliphatic), fatty acids, fatty acid esters, and Krebs cycle acids
Boron trifluoride + *n*-propanol	$BF_3 + CH_3(CH_2)_2OH$	Useful for fatty acids, lactic acid, and succinic acid
N,N-Dimethylformamide dimethyl acetal	OCH_3 CH_3 $C{-}CH{-}N$ OCH_3 CH_3	Useful in the formation of fatty acid esters, and for N-protected amino acids, sulfonamides, barbiturates
2-Bromopropane	$(CH_3)_2CHBr$	Used for amino acids and amides
Butanol	$CH_3(CH_2)_3OH$	Used for carboxylic acids and amino acids
Hydrogen chloride + methanol	$HCl + CH_3OH$	Useful for carboxylic acids, branched-chain fatty acids, oxalic acid, amino acids, and lipids; HCl serves as a catalytic agent
Sodium methoxide	CH_3ONa in CH_3OH	Used for the transesterification of lipids
Sulfuric acid + methanol	$H_2SO_4 + CH_3OH$	Useful for carboxylic and fatty acids
Tetramethyl ammonium hydroxide	$(CH_3)_4NOH$ in CH_3OH	Useful for carboxylic acids, fatty acids, and alkyd and polyester resins
Thionyl chloride	$SOCl_2$	Useful in the formation of esters of carboxylic acids and other acidic functional groups
2,2,2-Trichloroethanol	$CCl_3{-}\overset{\displaystyle H}{\underset{\displaystyle H}{C}}{-}OH$	Useful in the esterification of short-chain acids following electron-capture detection; sometimes used with trifluoracetic anhydride in the presence of H_2SO_4

DERIVATIZING REAGENTS FOR GAS CHROMATOGRAPHY (continued)

Derivatizing reagent	Structure/formula	Notes

Esterification Reagents

Triethyl orthoformate — $HC(OC_2H_5)_3$ — Used for aminophosphonic acids

Trimethylphenyl-ammonium hydroxide

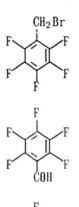

$(CH_3)_3N^+$ —⟨O⟩ in CH_3OH

OH^-

Used for fatty acids, aromatic acids, herbicides, and pesticides

Pentafluorophenyl Reagents

α-Bromopentafluoro-toluene — Used to etherify sterols and phenols, in diethyl ether with the presence of potassium *t*-butanolate

Pentafluorobenzaldehyde — Used in derivatizing primary amines; greatly enhances electron-capture detector response (to the picogram level)

Pentafluorobenzyl alcohol — Used in derivatizing carboxylic acids

Pentafluorobenzyl bromide — Used in the derivatization of carboxylic acids, phenols, mercaptans, and sulfamides; lachrymator; potentially unstable; high sensitivity for electron-capture detection; not usuable for formic acid

Pentafluorobenzyl chloride — Used in the derivatization of amines, phenols, and alcohols; used in a solution of NaOH

Pentafluorobenzyl chloroformate — Used in derivatization of tertiary amines

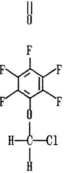

DERIVATIZING REAGENTS FOR GAS CHROMATOGRAPHY (continued)

Derivatizing reagent	Structure/formula	Notes

Pentafluorophenyl Reagents

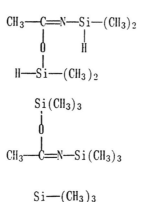

Pentafluorobenzyl hydroxylamine — Used in derivatization of ketones; can form both syn- and anti-isomers (two peaks)

Pentafluorophenacetyl chloride — Used in derivatization of alcohols, phenols, and amines

Pentafluorophenyl-hydrazine — Used in derivatization of ketones; can form both the syn- and anti-isomers, resulting in two peaks

Pentafluorophenoxyacetyl chloride — Used in derivatization of alcohols, phenols, and amines

Bis(dimethylsilyl)-acetamide (BSDA) — Similar in use and application to DMCS (see below)

$$CH_3-C=N-Si-(CH_3)_2$$

with O and H below, and $H-Si-(CH_3)_2$

N,N-bis (trimethylsylyl)-acetamide (BSA) — More reactive than HMDS (see below) or TMCS, but forming essentially similar derivatives; useful for alcohols, amines, amino acids, carboxylic acids, penicillic acid, and purine and pyrimidene bases

$$Si(CH_3)_3$$
$$|$$
$$O$$
$$|$$
$$CH_3-C=N-Si(CH_3)_3$$

Bis(trimethylsilyl) trifluoroacetamide (BSTFA) — Similar in use and application to BSA, but the derivatives are more volatile; byproducts often elute with the solvent front; reacts more strongly than HMDS or TMCS; may promote enol-TMS formation unless ketone groups are protected

$$Si-(CH_3)_3$$
$$|$$
$$O$$
$$|$$
$$CF_3-C=N-Si(CH_3)_3$$

DERIVATIZING REAGENTS FOR GAS CHROMATOGRAPHY (continued)

Derivatizing reagent	Structure/formula	Notes

Pentafluorophenyl Reagents

Derivatizing reagent	Structure/formula	Notes
Dimethylchlorosilane (DMCS)	$(CH_3)_2$—Si(H)—Cl	Similar in use and application to TMCS and HMDS, but usually forming more volatile and less thermally stable derivatives; also finds use in surface deactivation of chromatographic columns and injectors
1,1,1,3,3,3,-Heramethyl disilizane (HMDS)	$(CH_3)_3$-Si-NH-Si$(CH_3)_3$	Useful for such compounds as sugars, phenols, alcohols, amines, thiols, steroids; especially recommended for citric-acid cycle compounds and amino acids; reaction is often carried out in pyridine or dimethyl formamide (the latter being preferred for 17-keto steroids); care must be taken to eliminate moisture; lowest silyl-donating strength of all common silylating reagents
1,1,1,3,3,3,-hexamethyl disiloxane (HMDSO)	$(CH_3)_3$Si-O-Si-$(CH_3)_3$	Similar in use and application to HMDS (see above)
N-Methyl-N-(trimethylsilyl)-acetamide (MSTA)	CH_3—C(=O)—N(CH_3)—Si$(CH_3)_3$	Similar in use and application to HMDS, but somewhat higher "silyl-donating" strength
N-Methyl-N-(tri)methylsilyl) trifluoroacetamide (MSTFA)	CF_3—C(=O)—N(CH_3)—Si$(CH_3)_3$	Similar to MSTA, but produces the most volatile derivatives of all common silylating agents; particularly useful with low-molecular-mass derivatives
Tetramethyldisilazane (TMDS)	$(CH_3)_2$—Si(H)—NH—Si(H)—$(CH_3)_2$	Similar in use and application to DMCS
N-Trimethylsilyl diethylamine (TMSDEA)	$(CH_3)_3$-Si-N-$(C_2H_5)_2$.	Similar in use and application to DMCS
N-Trimethylsilyl imidazole (TMSIM)	$(CH_3)_3$—Si—N(imidazole)	Generally useful reagent with a high silyl-donor ability; will not react with amino groups; will not cause formation of enol-ether on unprotected ketone groups; especially useful for ecdysones, norepinephrine, dopamime, steroids, sugars, sugar phosphates, and ketose isomers

DERIVATIZING REAGENTS FOR GAS CHROMATOGRAPHY (continued)

Derivatizing reagent	Structure/formula	Notes

Pentafluorophenyl Reagents

Derivatizing reagent	Structure/formula	Notes
Trimethylchlorosilane (TMCS)	$(CH_3)_3SiCl$	Similar properties and applications as for HMDS; useful for amino-acid analyses; provides good response for electron-capture detection; has relatively low silyl-donating ability, and is usually used in the presence of a base such as pyridine; may cause enol-ether formation with unprotected ketone groups; often used as a catalyst with other silylating reagents
Halomethyl-flophemesyl reagents	$C_6F_5-\overset{\overset{\displaystyle CH_3}{\vert}}{\underset{\underset{\displaystyle R}{\vert}}{Si}}-Y$ $R = CH_2Cl$ $Y = Cl$	Similar in use and applications to the flophemesyl and alkylflophemesyl reagents
Halomethyldimethyl silyl reagents	$CH_2X-\overset{\overset{\displaystyle CH_3}{\vert}}{\underset{\underset{\displaystyle CH_3}{\vert}}{Si}}-Y$ $X = Cl, Br, I$ $Y = Cl, N(C_2H_5)_2, NHSi(CH_3)_2CH_2X$	Family of derivatizing agents which improve sensitivity of analyte to the electron-capture detector; the response enhancement is in the order expected; $I > Br > Cl \gg F$, the reverse order of the volatility of these compounds; the iodo-methyl dimethylsilyl reagents are unstable, and these derivatives are usually prepared *in situ*
Flophemesyl reagents	$C_6F_5-\overset{\overset{\displaystyle CH_3}{\vert}}{\underset{\underset{\displaystyle R}{\vert}}{Si}}-Y$ $R = CH_3$ $Y = Cl, NH_2, N(C_2H_5)_2$	Family of reagents forming derivatives which have stabilities similar to those produced by TMSIM, BSA, MSTFA, and BSTFA, with additional electron-capture detection sensitivity enhancement; usually used in pyridine as a solvent; reactions subject to steric considerations
Alkylflophemesyl reagents	$C_6F_5-\overset{\overset{\displaystyle CH_3}{\vert}}{\underset{\underset{\displaystyle R}{\vert}}{Si}}-Y$ $R = CH(CH_3)_2, C(CH_3)_3$ $Y = Cl$	Family of reagents forming derivatives of somewhat higher stability than the flophemesyl reagents; reactions subject to steric considerations

Miscellaneous Reagents

Derivatizing reagent	Structure/formula	Notes
Boronation reagents	$(OH)_2B-R$ $R=CH_3, -C(CH_3)_3$	Used to block two vicinal hydroxy groups; derivatives have very distinctive mass spectra which are usually identified
Carbon disulfide	CS_2	Used to derivatize primary amines to yield isothiocyanates
Dansyl chloride		Used for derivatization of tripeptides; provides high sensitivity toward spectrofluorimetric detection
Dimethyldiacetoxysilane	$(Cl)_2Si(CH_3)_2$	Used in similar applications to the boronation reagents in pyridene or trimethylamine solvent

DERIVATIZING REAGENTS FOR GAS CHROMATOGRAPHY (continued)

Derivatizing reagent	Structure/formula	Notes

Miscellaneous Reagents

2,4-Dinitrophenyl-hydrazone

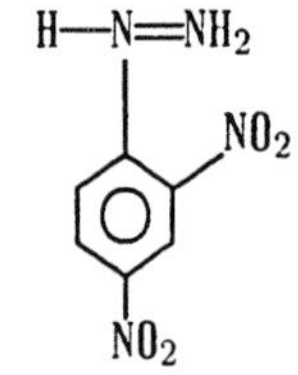

$H-N=NH_2$

NO_2

NO_2

Useful in derivatizing carbonyl compounds; also provides a "spot test" for these compounds

---

1-Fluoro-2,4-dinistrofluoro-benzene

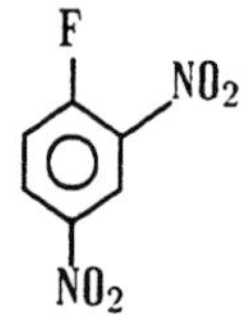

F

NO_2

NO_2

Useful for derivatizing C_1—C_4 primary and secondary amines, providing high electron-capture detector response; this reagent is also useful for primary alicyclic amines

---

Girard reagent T

$$(CH_3)_3 N - Cl - CH_2 C\overset{O}{\overset{\|}{C}}NHNH_2$$

Useful for derivatization of saturated aldehydes

Hydrazine

NH_2NH_2

Used for the analysis of C-terminal peptide residue species

Methyl iodide + silver oxide

$CH_3I + Ag_2O$ (in dimethylformamide)

Used to convert polyhydroxy compounds to the methyl ethers

Methyloxamine hydrochloride

$CH_3-O-NH_2 \cdot HCl$

Used in derivatization of steroids and carbohydrates

2-Methylthioanaline

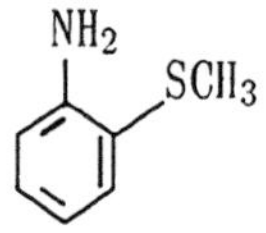

NH_2

SCH_3

Used to form sulfur-bearing derivatives of benzaldehydes

Phenyl isocyanate

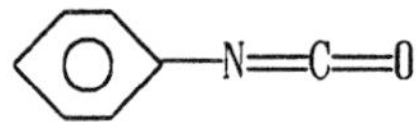

$N=C=O$

Used for derivatization of N-terminal peptide residue

2,4,6-Trichlorophenylhydrazine

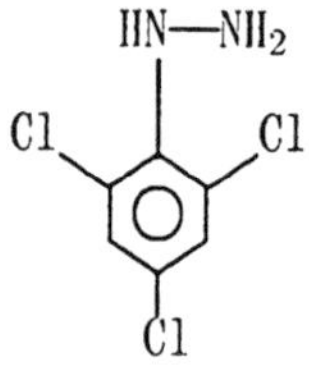

$HN-NH_2$

Cl   Cl

Cl

Used for derivatization of carbonyl compounds

CRYOGENS FOR SUBAMBIENT TEMPERATURE GAS CHROMATOGRAPHY

The following table lists properties of common cryogenic fluids used to produce subambient temperatures for gas chromatographic columns.[1-5] These properties are of value in designing low temperature chromatographic experiments efficiently and safely. Due to the potential dangers in handling extremely low temperatures and high pressures, appropriate precautions must be observed. These precautions must include protective clothing and shielding to prevent frostbite. Most cryogenic fluids can create a health hazard if they are vaporized in an inhabited area. Even small quantities can contaminate and displace air in a relatively short period of time. It may be advisable to locate a self-contained breathing apparatus immediately outside the laboratory in which the cryogens are being used. The effect of low temperatures on construction materials (of GC ovens and columns, for example) should also be considered. In this respect, differential expansion and tensile strength changes are pertinent issues. A dew point vs. moisture content table is also provided to allow the user to estimate the effects of ambient and impurity water. The viscosity data are provided in cP, which is equivalent to mPa·s, the appropriate SI unit. The freezing points are reported at 0.101315 MPa (1 atm), and the expansion ratios are reported at STP.

If temperatures no lower than approximately $-40°C$ are required, the use of the Ranque-Hilsch vortex tube should be considered.[6-8] This device requires a source of clean, dry compressed air at a pressure of approximately 0.70 MPa (100 psi) for proper operation. The flow-rate of air that is required depends on the volume of space to be cooled.

REFERENCES

1. **Zabetakis, M. G.,** *Safety with Cryogenic Fluids,* Plenum Press, New York, 1967.
2. **Cook, G. A., Ed.,** *Argon, Helium and the Rare Gases,* John Wiley & Sons (Interscience), New York, 1961.
3. **Brettell, T. A. and Grob, R. L.,** *Am. Laboratory,* 17(10), 19, 1985.
4. **Cowper, C. J. and DeRose, A. J.,** *The Analysis of Gases by Chromatography,* Pergamon Press, Oxford, 1983.
5. *Matheson Gas Data Book,* 4th ed., The Matheson Company, East Rutherford, 1966.
6. **Bruno, T. J.,** Vortex cooling for subambient temperature gas chromatography, *Anal. Chem.,* 58(7), 1596, 1986.
7. **Bruno, T. J.,** LC, Vortex refrigeration of HPLC components, *Liq. Chromatogr. HPLC Mag.,* 4(2), 134, 1986.
8. **Bruno, T. J.,** Laboratory applications of the vortex tube, *J. Chem. Educ.,* 64(11), 987, 1987.

CRYOGENS FOR SUBAMBIENT TEMPERATURE GAS CHROMATOGRAPHY

Cryogen name	Rel molecular mass	Freezing point [°C$_t$(K)]	Heat of fusion (J/g)	Normal bp [°C$_t$(K)]	Heat of vaporization (J/g)	Crit temp [°C$_t$(K)]	Crit pressure (MPa)	Crit density (g/L)	Vapor pressure (MPa)	Gas density (g/L)	Liquid/gas expansion ratio	Heat capacity C_p [J/(kg·k)]	Heat capacity C_v [J/(kg·k)]	Thermal conductivity $[\times 10^{-2} w/(m \cdot k)]$	Viscosity $Pa \cdot S \times 10^5$ (cP)	Solubility in water (v/v)
Argon Ar	39.948	−189.4 (83.8)	27.6	−185.9 (87.3)	163.2	−122.3 (150.9)	4.89	530.5	−[a]	1.63	860	523.8 (21°C)	313.8 (15.6°C)	1.44 (233 K)	2.21 (21°C)	0°C, 0.56
Carbon dioxide CO_2	44.01	−78.5[1] (194.7)	198.7	−56.6 (216.6)	151.5	31.1 (304.2)	7.38	468	5.72 (21°C)	1.98	790	831.8 (15.6°C)	638.8 (15.6°C)	1.17 (233 K)	1.48 (21°C)	20°C, 0.90
Helium He	4.003	−272[b] (1)	−[b]	−269.0 (4.2)	23.0 (15°C)	−268.0 (5.2)	0.23	69.3	−[a]	0.16	780	5221.6 (21°C)	3146.4 (15.6°C)	12.76 (233 K)	1.96 (21°C)	21°C, 0.0086
Methane CH_4	16.04	−182.6 (90.6)	58.6	−161.5 (87.3)	510.0	−82.1 (190.1)	4.64	162.5	−[a]	0.7174	650	2205.4 (15.6°C)	1687.0 (15.6°C)	2.57 (233 K)	1.20 (21°C)	—
Nitrogen N_2	28.013	−210.1 (63.1)	25.5	−195.81 (77.3)	199.6	−146.9 (150.9)	3.4	311	−[a]	1.14	710	1030.6 (21°C)	738.6 (21°C)	2.11 (233 K)	1.744 (15°C)	0°C, 0.023
Oxygen O_2	31.999	−218.8 (54.4)	13.8	−183.0 (90.2)	213.0	−118.4 (154.8)	5.04	410	−[a]	1.3	875	910.9 (15°C)	650.2 (15°C)	2.11 (233 K)	2.06 (20°C)	0°C, 0.0489

[a] Fluid is supercritical at ambient temperature.
[b] Helium will not solidify at 1 atmosphere pressure (0.101325 mPa). The approximate pressure at which solidification can occur is calculated to be 2535 kPa.

DEW POINT - MOISTURE CONTENT

Dew Point		Moisture, ppm (v/v)
°F	°C	
−130	−90.0	0.1
−120	−84.4	0.25
−110	−78.9	0.63
−105	−76.1	1.00
−104	−75.6	1.08
−103	−75.0	1.18
−102	−74.4	1.29
−101	−73.9	1.40
−100	−73.3	1.53
−99	−72.8	1.66
−98	−72.2	1.81
−97	−71.7	1.96
−96	−71.7	2.15
−95	−70.6	2.35
−94	−70.0	2.54
−93	−69.4	2.76
−92	−68.9	3.00
−91	−68.3	3.28
−90	−67.8	3.53
−89	−67.2	3.84
−88	−66.7	4.15
−87	−66.1	4.50
−86	−65.6	4.78
−85	−65.0	5.3
−84	−64.4	5.7
−83	−63.9	6.2
−82	−63.3	6.6
−81	−62.8	7.2
−80	−62.2	7.8
−79	−61.7	8.4
−78	−61.1	9.1
−77	−60.6	9.8
−76	−60	10.5
−75	−59.4	11.4
−74	−58.9	12.3
−73	−58.3	13.3
−72	−57.8	14.3
−71	−57.2	15.4
−70	−56.7	16.6
−69	−56.1	17.9
−68	−55.6	19.2
−67	−55.0	20.6
−66	−54.4	22.1
−65	−53.9	23.6
−64	−53.3	25.6
−63	−52.8	27.5
−62	−52.2	29.4
−61	−51.7	31.7
−60	−51.1	34.0

II. High Performance Liquid Chromatography

MODES OF LIQUID CHROMATOGRAPHY

The following flow chart provides a rough guide among the various liquid chromatographic techniques, based on sample properties.

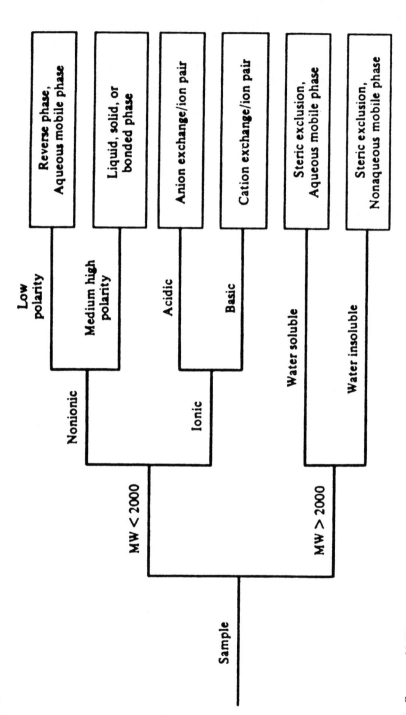

Courtesy of Millipore Corporation, Waters Chromatography Division.

SOLVENTS FOR LIQUID CHROMATOGRAPHY

The following table provides the important physical properties for the selection of solvent systems for high-performance liquid chromatography (HPLC).[1-7] These properties are required for proper detector selection, and the prediction of expected column pressure gradients. The values of the dielectric constant aid in estimating the relative solubilities of solutes and other solvents. Data on adsorption energies of useful HPLC solvents on silica and alumina (the elutropic series) can be found in the chapter on thin layer chromatography. The data presented were measured at 20 °C, unless otherwise indicated (in parentheses).

REFERENCES

1. **Willard, H. H., Merritt, L. L., Dean, J. A., and Settle, F. A.,** *Instrumental Methods of Analysis,* 6th ed., Wadsworth Publishing, Belmont, CA, 1981.
2. **Snyder, L. R. and Kirkland, J. J.,** *Introduction to Modern Liquid Chromatography,* 2nd ed., John Wiley & Sons, New York, 1979.
3. **Dreisbach, R. R.,** *Physical Properties of Chemical Compounds,* No. 22, Advances in Chemistry Series, American Chemical Society, Washington D.C., 1959.
4. **Krstulovic, A. M. and Brown, P. R.,** *Reverse Phase High Performance Liquid Chromatography,* John Wiley & Sons, New York, 1982.
5. **Weast, R. C., Ed.,** *Handbook of Chemistry and Physics,* 69th ed., CRC Press, Boca Raton, FL, 1987.
6. **Poole, C. F. and Shuttle, S. A.,** *Contempory Practice of Chromatography,* Elsevier, Amsterdam, 1984.
7. **Braithwaite, A. and Smith, F. J.,** *Chromatographic Methods,* 4th ed., Chapman and Hall, London, 1985.

SOLVENTS FOR LIQUID CHROMATOGRAPHY

Solvent	Viscosity (mPa·s; 20 °C)	UV cutoff (nm)	Refractive index (20 °C)	Normal boiling point (°C)	Dielectric constant (20 °C)
Acetic acid	1.31(15)		1.372	117.9	6.15
Acetone	0.30(25)	330	1.359	56.3	20.7(25)
Acetonitrile	0.34(25)	190	1.344	81.6	37.5
Benzene	0.65	278	1.501	80.1	2.284
1-Butanol	2.95	215	1.399	117.7	17.8
2-Butanol	4.21	260	1.397	99.6	15.8(25)
n-Butyl acetate	0.73	254	1.394	126.1	
n-Butyl chloride	0.47(15)	220	1.402	78.4	
Carbon tetrachloride	0.97	263	1.460	76.8	2.238
Chlorobenzene	0.80	287	1.525	131.7	2.708
Chloroform	0.58	245	1.446	61.2	4.806
Cyclohexane	0.98	200	1.426	80.7	2.023
Cyclopentane	0.44	200	1.406	49.3	1.965
o-Dichlorobenzene	1.32(25)	295	1.551	180.5	9.93(25)
N,N-Dimethylacetamide	2.14	268	1.438	166.1	37.8
Dimethylformamide	0.92	268	1.430	153.0	36.7
Dimethyl sulfoxide	2.20	286	1.478	189.0	4.7
Dioxane	1.44(15)	215	1.422	101.3	2.209(25)
2-Ethoxyethanol	2.05	210	1.408	135.6	
Ethyl acetate	0.46	256	1.372	77.1	6.02(25)
Ethyl ether	0.24	218	1.352	34.6	4.335
Glyme (ethylene glycol dimethyl ether)	0.46(25)	220	1.380	93:0	
Heptane	0.42	200	1.388	98.4	1.92
Hexadecane	3.34	200	1.434	287.0	
Hexane	0.31	200	1.375	68.7	1.890
Isobutyl alcohol	4.70(15)	200	1.396	107.7	15.8(25)
Methanol	0.55	205	1.328	64.7	32.63(25)
2-Methoxyethanol	1.72	210	1.402	124.6	16.9
2-Methoxyethyl acetate		254	1.402	144.5	
Methylene chloride	0.45(15)	233	1.424	39.8	9.08
Methylethylketone	0.42(15)	329	1.379	79.6	18.5
Methylisoamylketone		330	1.406	−144.0	
Methylisobutylketone	0.54(25)	334	1.396	116.5	
N-Methyl-2-pyrrolidone	1.67(25)	285	1.488	202.0	32.0
Nonane	0.72	200	1.405	150.8	1.972
Pentane	0.24	200	1.357	36.1	1.84
Petroleum ether	0.30	226		30—60	
β-Phenethylamine		285	1.529(25)	197—198	
1-Propanol	2.26	210	1.386	97.2	20.1(25)
2-Propanol	2.86(15)	205	1.377	82.3	18.3(25)
Propylene carbonate			1.419	240.0	
Pyridine	0.95	330	1.510	115.3	12.3(25)
Tetrachloroethylene	0.93(15)	295	1.506	121.2	
Tetrahydrofuran	0.55	212	1.407	66.0	7.6
Tetramethyl urea		265	1.449(25)	175.2	23.0
Toluene	0.59	284	1.497	110.6	2.379(25)
Trichloroethylene	0.57	273	1.477	87.2	3.4(16)
1,2,2-Trichloro-1,2,2-trifluoroethane	0.71	231	1.356(25)	47.6	
2,2,4-Trimethylpentane	0.50	215	1.391	99.2	1.94
Water	1.00	<190	1.333	100.0	78.54
o-Xylene	0.81	288	1.505	144.4	2.568
p-Xylene		290	1.496	138.5	2.270

REVERSE PHASE MATERIALS AND SOLVENTS

The following chart provides a rough guide to basic solvent systems for bonded reverse phase packings. These packings are the most widely used in high performance liquid chromatography.

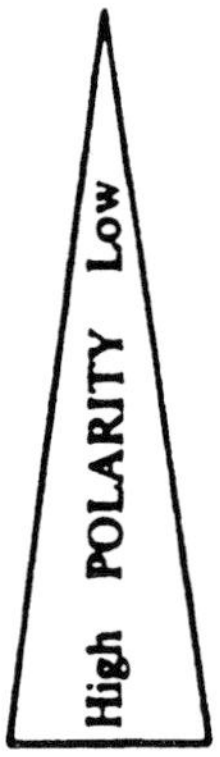

Sample	Column packing	Mobile phase
Low/moderate polarity (soluble in aliphatic hydrocarbons)	Bonded C-18	Methanol/water
Moderate polarity (soluble in methyl ethyl ketone)	Bonded C-8	Acetonitrile/water
High polarity (soluble in lower alcohols)	Bonded C-2	1,4-Dioxane/water

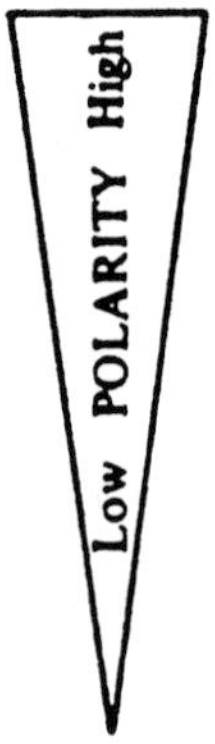

(Left arrow: High POLARITY Low; Right arrow: Low POLARITY High)

From INSTRUMENTAL METHODS OF ANALYSIS, Sixth Edition, by Hobart H. Willard, Lynne L. Merritt, Jr., John A. Dean, and Frank A. Settle, Jr.,© 1981 by Litton Educational Publishing, Inc. Reprinted by permission of Wadsworth, Inc.

MORE COMMON HIGH-PERFORMANCE LIQUID CHROMATOGRAPHY COLUMN PACKINGS

The following table provides a summary of the general characteristics of the most popular stationary phases used in modern high-performance liquid chromatography (HPLC).[1-3] The most commonly used phases are the bonded reverse phase materials, in which separation control is a function of the mobile (liquid) phase. The selection of a particular phase and solvent system is an empirical procedure involving survey analyses. The references provided below will assist the reader in this procedure.

REFERENCES

1. **Snyder, L. R. and Kirkland, J. J.**, *Introduction to Modern Liquid Chromatography*, 2nd ed., John Wiley & Sons, New York, 1979.
2. **Poole, C. F. and Schuette, S. A.**, *Contemporary Practice of Chromatography*, Elsevier, Amsterdam, 1984.
3. **Krstulovic, A. M. and Brown, P. R.**, *Reverse-Phase High Performance Liquid Chromatography*, Interscience, New York, 1982.
4. **Berridge, J. C.**, *Techniques for the Automated Optimization of HPLC Separations*, John Wiley & Sons, Chichester, U.K., 1985.
5. **Braithwaite, A. and Smith, F. J.**, *Chromatographic Methods*, 4th ed., Chapman and Hall, London, 1985.

MORE COMMON HPLC COLUMN PACKINGS

Phase type	Bond type	Functional group	Separation mode	Notes and application
Silica (pure)	SiO_2	—	Adsorption	Usually used with nonpolar mobile phase; selectivity is based on differences in number and location of polar groups; results can be unpredictable due to changes in the surface due to adsorption; water or acetic acid is often added (in low concentrations) to the mobile phase to better control surface characteristics
Alumina (pure)	Al_2O_3	—	Adsorption—normal phase	Similar in characteristics and application to silica
Bonded amine	Si–C	$-NH_2$	Polar-bonded phase	Selectivity is modified with respect to silica, although highly polar; useful for sugar and carbohydrate separations; not recommended for samples which contain aldehydes and ketones
Bonded nitrile	Si–C	$-C{\equiv}N$	Polar-bonded phase	Highly polar phase, but with selectivity modified with respect to silica; less sensitive to mobile-phase impurities than silica
Bonded diol	Si–C	OH OH | | –C –C	Polar-bonded phase	Useful in size-exclusion chromatography and in the analysis of glycols and glycerol and related compounds

MORE COMMON HPLC COLUMN PACKINGS (continued)

Phase type	Bond type	Functional group	Separation mode	Notes and application
Bonded nitro	Si–C	$-NO_2$	Polar-bonded phase	
ODS	Si–C	Octadecyl, n–C_{18} hydrocarbon chain	Bonded, reverse phase	Octadecylsilane; most common material used in HPLC; high resolution possible; pH must be maintained between 2 and 7
OS	Si–C	Octyl, n–C_8 hydrocarbon chain	Bonded, reverse phase	Octylsilane; lower resolution than the octadecyl bonded phase; useful when separations involve species of greatly different polarity
TMS	Si–C	Methyl, CH_3	Bonded, reverse phase	Tetramethylsilane; lowest resolution of reverse-phase packings, useful for "survey" separations and for large molecules
Phenyl	Si–C	$-\phi$[a]	Normal or reverse phase	Lower efficiency than other bonded phases; more polar than ODS, OS, and TMS phases; used with normal and reverse-phase solvent systems
Ion exchange	Varies, usually Si–O–Si	Sulfonic acid	Cation exhange	Separates cations, with divalent ions more strongly retained than monovalent ions, phosphate buffer systems are often used, sometimes with low concentrations of polar nonaqueous modifiers added
Ion exchange	Varies	Quaternary ammonium	Anion exchange	Separates anions; separation and retention of organic ions is due to both ionic- and reverse-phase effects
Ion exchange	Varies	$-N(CH_3)_2$	Anion exchange	Similar to the quaternary ammonium phase, with different retention and selectivity

[a] ϕ denotes a phenyl group.

ELUOTROPIC VALUES OF SOLVENTS ON OCTADECYLSILANE

The following table provides, for comparative purposes, eluotropic values on ODS for common solvents.[1]

Solvent	Eluotropic Value
Acetone	8.8
Acetonitrile	3.1
Dioxane	11.7
Methanol	1.0
Propanol-2	8.3
Tetrahydrofuran	3.7

REFERENCE

1. **Krieger, P. A.**, *High Purity Solvent Guide*, Burdick and Jackson Laboratories, McGaw Park, Il, 1984.

MESH-SIZE RELATIONSHIPS

The following table provides the relationship between particle sizes and standard sieve mesh sizes. It should be noted, however, that the trend in HPLC has been toward shorter columns containing much finer particles than the standard sieves will separate.

Mesh range	Top screen opening (μm)	Bottom screen opening (μm)	Micron screen (μm)	Range ratio
80/100	177	149	28	1.19
100/120	149	125	24	1.19
100/140	149	105	44	1.42
120/140	125	105	20	1.19
140/170	105	88	17	1.19
170/200	88	74	14	1.19
200/230	74	63	11	1.19
230/270	63	53	10	1.19
270/325	53	44	9	1.20
325/400	44	37	7	1.19

STATIONARY PHASES FOR LIQUID CHROMATOGRAPHY

The following table provides information on the properties and application of some of the more specialized bonded, adsorbed, and polymeric phases used in modern HPLC. In many cases the phases are not commercially available, and the reader is referred to the appropriate literature citation for details on the synthesis.

REFERENCES

1. **Pietrzyk, D. J. and Cahill, W. J.,** Amberlite XAD-4 as a stationary phase for preparative liquid chromatography in a radially compressed column, *J. Liq. Chromatogr.*, 5(4), 781, 1982.
2. **Nikolov, Z., Meagher, M., and Reilly, P.,** High-performance liquid chromatography of trisaccharides on amine-bonded silica columns, *J. Chromatogr.*, 321, 393, 1985.
3. **Ascalone, V. and Dal Bo, L.,** Determination of Ceftriaxone, a novel cephalosporin, in plasma, urine and saliva by high-performance liquid chromatography on an NH_2 bonded-phase column, *J. Chromatogr.*, 273, 357, 1983.
4. **Pharr, D. Y., Uden, P. C., and Siggia, S.,** A 3-(*p*-acetylphenoxy)propylsilane bonded phase for liquid chromatography of basic amines and other nitrogen compounds, *J. Chromatogr. Sci.*, 23, 391, 1985.
5. **Felix, G. and Bertrand, C.,** Separation of polyaromatic hydrocarbons on caffeine-bonded silica gel, *J. Chromatogr.*, 319, 432, 1985.
6. **Felix, G., Bertrand, C., and Van Gastel, F.,** A new caffeine bonded phase for separation of polyaromatic hydrocarbons and petroleum asphaltenes by high-performance liquid chromatography, *Chromatographia*, 20(3), 155, 1985.
7. **Bruner, F., Bertoni, G., and Ciccioli, P.,** Comparison of physical and gas chromatographic properties of Sterling FT and Carbopack-C graphitized carbon blacks, *J. Chromatogr.*, 120, 307, 1976.
8. **Bruner, F., Ciccioli, P., Crescentini, G., and Pistolesi, M. T.,** Role of the liquid phase in gas-liquid-solid chromatography and its influence on column performance — an experimental approach, *Anal. Chem.*, 45(11), 1851, 1973.
9. **Ciccioli, P. and Liberti, A.,** Microbore columns packed with graphitized carbon black for high-performance liquid chromatography, *J. Chromatogr.*, 290, 173, 1984.
10. **DiCorcia, A., Liberti, A., and Samperi, R.,** Gas-liquid-solid chromatography — theoretical aspects and analysis of polar compounds, *Anal. Chem.*, 45(7), 1228, 1973.
11. **Hatada, K., Kitayama, T., Shimizu, S-I., Yuki, H., Harris, W., and Vogl, O.,** High-performance liquid chromatography of aromatic compounds on polychloral, *J. Chromatogr.*, 248, 63, 1982.
12. **Abe, A., Tasaki, K., Inomata, K., and Vogl, O.,** Confirmational ridigity of polychloral: effect of bulky substituents on the polymerization mechanism, *Macromolecules*, 19, 2707, 1986.
13. **Kubisa, P., Corley, L. S., Kondo, T., Jacovic, M., and Vogl, O.,** Haloaldehyde polymers. XXIII. Thermal and mechanical properties of chloral polymers, *Pol. Eng. Sci.*, 21(13), 829, 1981.
14. **Veuthey, J.-L., Bagnoud, M.-A., and Haerdi, W.,** Enrichment of amino and carboxylic acids using copper-loaded silica pre-columns coupled on-line with HPLC, *Int. J. Environ. Anal. Chem.*, 26, 157, 1986.
15. **Guyon, F., Foucault, A., Caude, M., and Rosset, R.,** Separation of sugars by h.p.l.c. on copper silicate gel, *Carbohydr. Res.*, 140, 135, 1985.
16. **Leonard, J. L., Guyon, F., and Fabiani, P.,** High-performance liquid chromatography of sugars on copper (11) modified silica gel, *Chromatographia*, 18, 600, 1984.
17. **Miller, N. T. and Shieh, C. H.,** Preparative hydrophobic interaction chromatography of proteins using ether based chemically bonded phases, *J. Liq. Chromatogr.*, 9(15), 3269, 1986.
18. **Williams, R. C., Vasta-Russell, J. F., Glajch, J. L., and Golebiowski, K.,** Separation of proteins on a polymeric fluorocarbon high-performance liquid chromatography column packing, *J. Chromatogr.*, 371, 63, 1986.
19. **Hirayama, C., Ihara, H., Yoshinga, T., Hirayama, H., and Motozato, Y.,** Novel packing for high pressure liquid chromatography. Partially alkylated and cross-linked PMLG spherical particles, *J. Liq. Chromatogr.*, 9(5), 945, 1986.
20. **Kawasaki, T., Kobayashi, W., Ikeda, K., Takahashi, S., and Monma, H.,** High-performance liquid chromatography using spherical aggregates of hydroxyapatite micro-crystals as adsorbent, *Eur. J. Biochem.*, 157, 291, 1986.
21. **Kawasaki, T. and Kobayashi, W.,** High-performance liquid chromatography using novel square tile-shaped hydroxyapatite crystals as adsorbent, *Biochem. Int.*, 14(1), 55, 1987.
22. **Funae, Y., Wada, S., Imaoka, S., Hirotsune, S., Tominaga, M., Tanaka, S., Kishimoto, T., and Maekawa, M.,** Chromatographic separation of α-acid glycoprotein from α-antitrypsin by high-performance liquid using a hydroxyapatite column, *J. Chromatogr.*, 381, 149, 1986.

23. **Kadoya, T., Isobe, T., Ebihara, M., Ogawa, T., Sumita, M., Kuwahara, H., Kobayashi, A., Ishikawa, T., and Okuyama, T.,** A new spherical hydroxyapatite for high performance liquid chromatography of proteins, *J. Liq. Chromatogr.,* 9(16), 3543, 1986.

24. **Kawasaki, T., Nikura, M., Takahashi, S., and Kobayashi, W.,** High-performance liquid chromatography using improved spherical hydroxyapatite particles as adsorbent: efficiency and durability of the column, *Biochem. Int.,* 13(6), 969, 1986.

25. **Bernardi, G.,** Chromatography of nucleic acids on hydroxyapatite columns, *Methods Enzymol.,* 21D, 95, 1971.

26. **Bernardi, G.,** Chromatography of proteins on hydroxyapatite, *Methods Enzymol.,* 27, 471, 1973.

27. **Bernardi, G.,** Chromatography of proteins on hydroxyapatite, *Methods Enzymol.,* 22, 325, 1971.

28. **Figueroa, A., Corradini, C., Feibush, B., and Karger, B.,** High-performance immobilized-metal affinity chromatography of proteins on iminodiacetic acid silica-based bonded phases, *J. Chromatogr.,* 371, 335, 1986.

29. **Danielson, N. D., Ahmed, S., Huth, J. A., and Targrove, M. A.,** Characterization of organomagnesium modified Kel-f polymers as column packing, *J. Liq. Chromatogr.,* 9(4), 727, 1986.

30. **Taylor, P. J. and Sherman, P. L.,** Liquid crystals as stationary phases for high performance liquid chromatography, *J. Liq. Chromatogr.,* 3(1), 21, 1980.

31. **Taylor, P. J. and Sherman, P. L.,** Liquid crystals as stationary phases for high performance liquid chromatography, *J. Liq. Chromatogr.,* 2(9), 1271, 1979.

32. **Felix, G. and Bertrand, C.,** HPLC on pentafluorobenzamidopropyl silica gel, *J. High Resol. Chromatogr. Chromatogr. Commun.,* 8, 362, 1985.

33. **Kurosu, Y., Kawasaki, H., Chen, X.-C., Amano, Y., Fang, Y.-I., Isobe, T., and Okuyama, T.,** Comparison of retention times of polypeptides in reversed phase high performance liquid chromatography on polystyrene resin and on alkyl bonded silica, *Bunseki Kagaku,* 33, E301, 1984.

34. **Yang, Y.-B. and Verzele, M.,** New water-compatible modified polystyrene as a stationary phase for high-performance liquid chromatography, *J. Chromatogr.,* 387, 197, 1987.

35. **Nieminen, N. and Heikkila, P.,** Simultaneous determination of phenol, cresols and xylenols in workplace air, using a polystyrene-devinylbenzene column and electrochemical detection, *J. Chromatogr.,* 360, 271, 1986.

36. **Yang, Y. B., Nevejans, F., and Verzele, M.,** Reversed-phase and cation-exchange chromatography on a new poly(styrenedivinylbenzene) high capacity, weak cation-exchanger, *Chromatographia,* 20(12), 735, 1985.

37. **Tweeten, K. A. and Tweeten, T. N.,** Reversed-phase chromatography of proteins on resin-based wide-pore packings, *J. Chromatogr.,* 359, 111, 1986.

38. **Lee, D. P. and Lord, A. D.,** A high performance phase for the organic acids, *LC-GC,* 5(3), 261, 1987.

39. **Miyake, K., Kitaura, F., Mizuno, N., and Terada, H.,** Determination of partition coefficient and acid dissociation constant by high-performance liquid chromatography on porous polymer gel as a stationary phase, *Chem. Pharm. Bull.,* 35(1), 377, 1987.

40. **Joseph, J. M.,** Selectivity of poly(styrene-divinylbenzene) columns, *ACS Symp. Ser.,* 297, 83, 1986.

41. **Cope, M. J. and Davidson, I. E.,** Use of macroporous polymeric high-performance liquid chromatographic columns in pharmaceutical analysis, *Analyst,* 112, 417, 1987.

42. **Werkhoven-Goewie, C. E., Boon, W. M., Praat, A. J. J., Frei, R. W., Brinkman, U. A. Th., and Little, C. J.,** Preconcentration and LC analysis of chlorophenols, using a styrene-divinyl-benzene copolymeric sorbent and photochemical reaction detection, *Chromatographia,* 16, 53, 1982.

43. **Smith, R. M.,** Selectivity comparisons of polystyrenedivinylbenzene columns, *J. Chromatogr.,* 291, 372, 1984.

44. **Köhler, J.,** Poly(vinylpyrrolidone)-coated silicia: a versatile, polar stationary phase for h.p.l.c., *Chromatographia,* 21, 573, 1986.

45. **Murphy, L. J., Siggia, S., and Uden, P. C.,** High-performance liquid chromatography of nitroaromatic compounds on N-propylaniline bonded stationary phase, *J. Chromatogr.,* 366, 161, 1986.

46. **Felix, G. and Bertrand, C.,** HPLC on n-propyl picryl ether silica gel, *J. High Resolut. Chromatogr. Chromatogr. Commun.,* 7, 714, 1984.

47. **Risner, C. H. and Jezorek, J. R.,** The chromatographic interaction and separation of metal ions with 8-quinolinol stationary phases in several aqueous eluents, *Anal. Chem. Acta,* 186, 233, 1986.

48. **Shahwan, G. J. and Jezorek, J. R.,** Liquid chromatography of phenols on an 8-quinolinol silica gel-iron (III) stationary phase, *J. Chromatogr.,* 256, 39, 1983.

49. **Krauss, G.-J.,** Ligand-exchange h.p.l.c. of uracil derivatives on 8-hydroxyquinoline-silica-polyol, *J. High Resolut. Chromatogr. Chromatogr. Commun.,* 9, 419, 1986.

50. **Hansen, S. H., Helboe, P., and Thomsen, M.,** High-performance liquid chromatography on dynamically modified silica, *J. Chromatogr.,* 360, 53, 1986.

51. **Helboe, P.,** Separation of corticosteroids by high-performance liquid chromatography on dynamically modified silica, *J. Chromatogr.,* 366, 191, 1986.

52. **Hansen, D. H., Helboe, P., and Thomsen, M.,** Dynamically modified silica-the use of bare silica in reversed-phase high-performance liquid chromatography, *Trends Anal. Chem.,* 4(9), 233, 1985.

53. **Flanagan, R. J.,** High-performance liquid chromatographic analysis of basic drugs on silica columns using non-aqueous ionic eluents, *J. Chromatogr.,* 323, 173, 1985.

54. **Vespalec, R., Ciganková, M., and Viska, J.,** Effect of hydrothermal treatment in the presence of salts on the chromatographic properties of silica gel, *J. Chromatogr.,* 354, 129, 1986.

55. **Unger, K. K., Jilge, G., Kinkel, J. N., and Hearn, M. T. W.,** Evaluation of advanced silica packings for the separation on biopolymers by high-performance liquid chromatography, *J. Chromatogr.,* 359, 61, 1986.

56. **Lullmann, C., Genieser, H.-G., and Jastorff, B.,** Structural investigations on reversed-phase silicas, *J. Chromatogr.,* 354, 434, 1986.

57. **Schou, O. and Larsen, P.,** Preparation of 6,9,12-trioxatridecylmethylsilyl substituted silica, a new stationary phase for liquid chromatography, *Acta Chem. Scand.,* B35, 337, 1981.

58. **Desideri, P. G., Lepri, L., Merlini, L., and Cecchini, L.,** High-performance liquid chromatography of amino acids and peptides on silica coated with ammonium tungstophosphate, *J. Chromatogr.,* 370, 75, 1986.

STATIONARY PHASES FOR LIQUID CHROMATOGRAPHY

Name: Amberlite XAD-4
Structure: Macroporous polystyrene-divinylbenzene nonpolar adsorbent, 62 to 177 μm particle size
Analytical properties: Used mainly in preparative-scale HPLC; stable over entire pH range; (1 to 13) sometimes difficult to achieve column-to-column reproducibility due to packing the irregular particles; relatively lower efficiency than alkyl bonded phases; particles tend to swell as the organic content of the mobile phase increases
Reference: 1

Name: Amine-bonded phase
Structure: NH_2 functionality with a Si–O–Si–C or Si–C linkage
Analytical properties: Polar phase useful for sugar and carbohydrate separation; not recommended for samples which contain aldehydes and ketones
Reference: 2, 3

Name: 3-(*p*-Acetophenoxy)propyl-bonded phase
Structure:

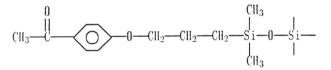

Analytical properties: Selective for aromatic amines, with the selectivity being determined by the interactions with the carbonyl group
Reference: 4

Name: Caffeine-bonded phase
Structure:

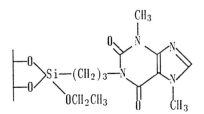

STATIONARY PHASES FOR LIQUID CHROMATOGRAPHY (continued)

Analytical properties: Separation of polynuclear aromatic hydrocarbons (of the type often encountered in petroleum residue work) by donor-acceptor complex formation
Reference: 5, 6

Name: Graphitized carbon black
Structure: Carbon subjected to +1300 °C in helium atmosphere, resulting in a graphite-like structure in the form of polyhedra, with virtually no unsaturated bonds, ions, lone electron pairs, or free radicals
Analytical properties: Especially for use in microbore columns; suggested for lower aromatics but with some potential for higher-molecular-mass compound separations
Reference: 7—10

Name: Polychloral (polytrichloroacetaldehyde)
Structure:

Analytical properties: Separation of lower aromatic hydrocarbons and smaller fused-ring systems using toluene and hexane/methanol as the stationary phases; the relatively low pressure rating on the polymeric phase limits solvent flow rate
Reference: 11—13

Name: Cyclam-copper-silica
Structure:

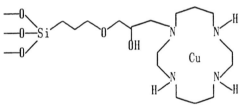

Analytical properties: This phase has found use in preconcentrating carboxylic acids on precolumns
Reference: 14

Name: Bis-dithiocarbamate-copper-silica
Structure:

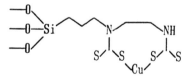

Analytical properties: This phase has found use in preconcentrating amino acids on precolumns.
Reference: 14

Name: Copper(II)-coated silica gel
Structure:

$$(\equiv Si\text{-}O)_2Cu(NH_3)_x(H_2O)_y$$
$$x = 1 \text{ or } 2$$

Analytical properties: Separation of sugars and amino sugars by ligand exchange or partitioning interactions using water + acetonitrile + ammonia liquid phases; the phase is usually prepared by treating silica gel with ammoniacal copper sulfate solution prior to packing
Reference: 15, 16

STATIONARY PHASES FOR LIQUID CHROMATOGRAPHY (continued)

Name: Ether-bonded phase
Structure: On 15- to 20-μm-wide-pore silica:

$$CH_3(0—CH_2—CH_2)_2—0—(CH_2)_3—Si—$$

Analytical properties: Separation by hydrophobic interaction chromatography, using aqueous salt solutions near pH = 7; used primarily in protein work
Reference: 17

Name: Fluorocarbon polymer phase
Structure: Proprietary information of E.I. du Pont de Nemours Corporation
Analytical properties: Similar separations to those obtained using C_3-bonded silica, with a much larger pH stability range than silica-based phases; useful for protein and peptide separations using TFA (trifluoroacetic acid) as a mobile phase modifier; less mechanical stability with silica-based phases
Reference: 18

Name: Poly(γ-methyl-L-glutamate)(PMLG)
Structure: Partially cross-linked, with long-chain alkyl branches:

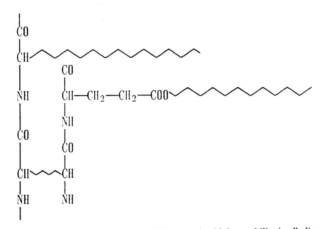

Analytical properties: Separation similar to ODS, but with somewhat higher stability in alkaline solutions; particles are spherical and macroporous
Reference: 19

Name: Hydroxyapatite adsorbent
Structure: Crystalline, nonstoichiometric mineral rich in surface ions (primarily carbonate):

$$Ca(PO_4)_6(OH)_2$$

Analytical properties: Separation of proteins; overcomes some difficulties associated with ion exchange; selectivity and efficiency depend to some extent on particle geometry (i.e., sphere, plate, etc.)
Reference: 21—28

Name: Iminodiacetic acid-bonded phase
Structure:

$$\begin{array}{c} O \\ \| \\ CH_2COH \\ | \\ N—(CH_2)_2O(CH_2)_2O(CH_2)_2—Si— \\ | \\ CH_2COH \\ \| \\ O \end{array}$$

STATIONARY PHASES FOR LIQUID CHROMATOGRAPHY (continued)

Analytical properties: Separation of proteins by immobilized-metal affinity chromatography (HPIMAC) with Cu(II) or Zn(II) present in the mobile phase
Reference: 28

Name: Kel-F (polychlorotrifluoroethylene)
Structure: Exact structure is proprietary (3-M Co.).
Analytical properties: Highly inert, even more nonpolar than hydrocarbon phases, with sufficient mechanical integrity to withstand high pressures; can be functionalized with $-CH_3$, $CH_3(CH_2)_5^-$, and phenyl (using Grignard reactions) to increase selectivity
Reference: 29

Name: 2-Ethylhexyl carbonate coated or bonded on cholesteryl silica (room-temperature liquid crystal)
Structure:

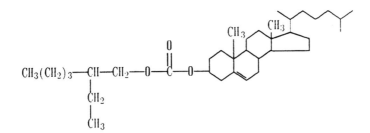

Analytical properties: Has been used for the separation of estrogens and corticoid steroids; liquid crystal phase retains some order when coated on an active substrate
Reference: 30, 31

Name: Cholesteryl-2-ethylhexanoate (room-temperature liquid crystal) on silica
Structure:

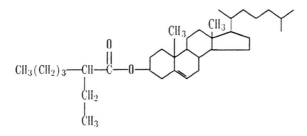

Analytical properties: Has been used for the separation of androstenediones and testosterone
Reference: 30, 31

Name: Pentafluorobenzamidopropyl silica gel
Structure:

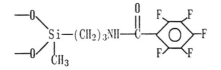

Analytical properties: Separations via interactions with π-electrons of solutes; can be used in both normal and reverse phase for such π-donor systems as polynuclear aromatic hydrocarbons
Reference: 32

STATIONARY PHASES FOR LIQUID CHROMATOGRAPHY (continued)

Name: Hydroxymethyl polystyrene
Structure:

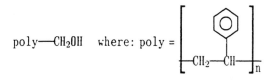

Analytical properties: Separation of polypeptides; usually gives shorter retention times than ODS; hydrophobic interactions not as strong as with ODS
Reference: 33

Name: Polystyrene
Structure:

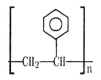

Analytical properties: Separation of polypeptides with results similar to those obtainable with ODS; higher stability at high pH levels (to allow the phase to be washed); stronger hydrophobic interactions than ODS in reverse-phase mode
Reference: 34

Name: Polystyrene-divinylbenzene (PS-DVB)
Structure:

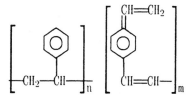

(Exact structure is proprietary.)
Analytical properties: Useful for the separation of relatively polar compounds such as phenols, carboxylic acids, organic anions, nucleosides, alkylarylketones, chlorophenols, barbiturates, and thimine derivatives, good stability under high and low pH; reasonable mechanical integrity at high carrier pressure; compatible with buffered liquid phases
Reference: 35—43

Name: Poly(vinylpyrrolidone) or (PVP) on silica
Structure:

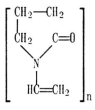

Analytical properties: Separates aromatic and polynuclear aromatic hydrocarbons; can be used in normal-phase mode (commonly using *n*-heptane or *n*-heptane + dichloromethane liquid phases) or reverse-phase mode (commonly using methanol + water acetonitrile + water or phosphate-buffered liquid phases)
Reference: 44

STATIONARY PHASES FOR LIQUID CHROMATOGRAPHY (continued)

Name: Bonded *n*-propylaniline
Structure:

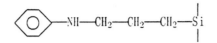

Analytical properties: Selectivity is based on charge transfer interactions; nitroaromatic compounds are separated essentially according to the number of nitro groups, with the higher number compounds being most strongly retained when using methanol/water mobile phases
Reference: 45

Name: *n*-Propylpicryl ether-bonded phase
Structure:

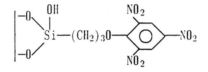

Analytical properties: Separation of aromatic species, including polynuclear aromatic species, by charge transfer interactions
Reference: 46

Name: 8-Quinolinol-bonded phases
Structure:

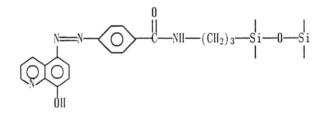

Analytical properties: Separates phenols and EPA priority polutants; often used with metal ions (such as iron [III] as chelate ligands; 8-quinolinol has a high affinity for oxygen moieties, and will form complexes with upwards of 60 metal ions; often with an acidic aqueous mobile phase
Reference: 47—49

Name: Cetyltrimethyl ammonium bromide adsorbed on silica
Structure:

$$(CH_3)_3\overset{+}{N}(n-C_{16}H_{33})Br^-$$

Analytical properties: Has been used to separate aromatic hydrocarbons, heterocyclic compounds, phenols and aryl amines using methanol/water/phosphate buffer; extent of adsorption affects retention times; also used as a mobile-phase modifier to provide a dynamically modified silica
Reference: 50—56

STATIONARY PHASES FOR LIQUID CHROMATOGRAPHY (continued)

Name: 6,9,12-Trioxatridecylmethyl-bonded phase
Structure:

$$CH_3—(OCH_2CH_2)_3(CH_2)_3Si(CH_3)(OH)OSi$$

Analytical properties: Phase is very well wetted by water, allowing mobile phases with high water concentrations to be used, somewhat higher efficiency and selectivity than ODS, but with similar separation properties
Reference: 57

Name: Ammonium tungstophosphate on silica
Structure: Tungstophosphoric acid with ammonium nitrate
Analytical properties: Separation of compounds containing the NH_4^+ group, such as amino acids and peptides; the coated silica also behaves as a reversed phase for the separation of aliphatic and aromatic acids; high selectivity for glycine and tyrosine oligomers
Reference: 58

CHIRAL STATIONARY PHASES FOR LIQUID CHROMATOGRAPHY

The following table provides information on the properties and application of some of the more specialized stationary phases used to carry out the separation of enantiomeric mixtures. In many cases the phases are not commercially available, and the reader is referred to the appropriate literature citation for details on the synthesis.

REFERENCES

General References

1. **Armstrong, D. W.**, Chiral stationary phases for high performance liquid chromatographic separation of enantiomers: a mini-review, *J. Liq. Chromatogr.*, 7(S-2), 353, 1984.
2. **Dappen, R., Arm, H., and Meyer, V. R.**, Applications and limitations of commercially available chiral stationary phases for high-performance liquid chromatography, *Chromatogr. Rev. CHREV*, p. 1, 1986.

Cited References

1. **Erlandsson, P., Hansson, L., and Isaksson, R.**, Direct analytical and preparative resolution of enantiomers using albumin adsorbed to silica as a stationary phase, *J. Chromatogr.*, 370, 475, 1986.
2. **Kuesters, E. and Giron, D.**, Enantiomeric separation of the beta-blocking drugs pindolol and bipindolol using a chiral immobilized protein stationary phase *J. High Resolut. Chromatogr. Chromatogr. Commun.*, 9(9), 531, 1986.
3. **Hermannson, J. and Eriksson, M.**, Direct liquid chromatographic resolution of acidic drugs using a chiral 1-acid glycoprotein column (Enantiopac), *J. Liquid Chromatogr.*, 9(2 and 3), 621, 1986.
4. **Naobumi, O. and Hajimu, K.**, HPLC separation of amino acid enantiomers on urea derivatives of L-valine bonded to silica gel, *J. Chromatogr.*, 285, 198, 1984.
5. **Okamoto, Y., Sakamoto, H., Hatada, K., and Irie, M.**, Resolution of enantiomers by HPLC on cellulose trans- and cis-tris (4-phenylazophenylcarbamate), *Chem. Lett.*, p. 983, 1986.
6. **Ichid, A., Shibata, T., Okamoto, I., Yuki, Y., Namikoshi, H., and Toga, Y.**, Resolution of enantiomers by HPLC on cellulose derivatives, *Chromatographia*, 19, 280, 1984.
7. **Tagahara, K., Koyama, J., Okatani, T., and Suzuta, Y.**, Chromatographic resolution of racemic tetrahydroberbeine alkaloids by using cellulose tris (phenylcarbamate) stationary phase, *Chem. Pharm. Bull.*, 34, 5166, 1986.
8. **Klemisch, W. and von Hodenberg, A.**, Separation on crosslinked acetylcellulose, *J. High Resolut. Chromatogr. Chromatogr. Commun.*, 9(12), 765, 1986.
9. **Rimboock, K., Kastner, F., and Mannschreck, A.**, Microcrystallinetribenzoyl cellulose: a high-performance‑liquid chromatographic sorbent for the separation of enantiomers, *J. Chromatogr.*, 351, 346, 1986.
10. **Lindner, K. and Mannschreck, A.**, Separation of enantiomers by high-performance liquid chromatography on triacetylcellulose, *J. Chromatogr.*, 193, 308, 1980.
11. **Gubitz, G., Jellenz, W., and Schonleber, D.**, High performance liquid chromatographic resolution of the optical isomers of D,L-tryptophane, D,L-5-hydroxytryptophan and D,L-dopa on cellulose columns, *J. High Resolut. Chromatogr. Chromatogr. Commun.*, 3, 31, 1980.
12. **Takayanagi, H., Hatano, O., Fujimura, K., and Ando, T.**, Ligand-exchange high-performance liquid chromatography of dialkyl sulfides, *Anal. Chem.*, 57, 1840, 1985.
13. **Armstrong, D. W.**, U. S. Patent 4,539,399, 1985; assigned to Advanced Separation Technologies, Inc., Whippany, N.J.
14. **Armstrong, D. W. and DeMond, W.**, Cyclodextrin bonded phases for the liquid chromatographic separation of optical, geometrical, and structural isomers, *J. Chromatogr. Sci.*, 22, 411, 1984.
15. **Armstrong, D. W., Ward, T. J., Armstrong, R. D., and Beesley, T. J.**, Separation of drug stereoisomers·by the formation of β-cyclodextrin inclusion complexes, *Science*, 232, 1132, 1986.
16. **Armstrong, D. W., DeMond, W., and Czech, B. P.**, Separation of metallocene enantiomers by liquid chromatography: chiral recognition via cyclodextrin bonded phases, *Anal. Chem.*, 57, 481, 1985.
17. **Armstrong, D. W., DeMond, W., Alak, A., Hinze, W. L., Riehl, T. E., and Bui, K. H.**, Liquid chromatographic separation of diastereomers and structural isomers on cyclodextrin-bonded phases, *Anal. Chem.*, 57, 234, 1985.
18. **Weaver, D. E. and van Lier, R.**, Coupled β-cyclodextrin and reverse-phase high-performance liquid chromatography for assessing biphenyl hydroxylase activity in hepatic 9000 g supernatant, *Anal. Biochem.*, 154, 590, 1986.
19. **Armstrong, D. W.**, Optical isomer separation by liquid chromatography, *Anal. Chem.*, 59, 84A, 1987.

20. **Chang, C. A., Wu, Q., and Tan, L.,** Normal-phase high-performance liquid chromatographic separations of positional isomers of substituted benzoic acids with amine and b-cyclodextrin bonded-phase columns, *J. Chromatogr.,* 361, 199, 1986.

21. **Cline-Love, L. and Arunyanart, M.,** Cyclodextrin mobile-phase and stationary-phase liquid chromatography, *ACS Symp. Ser.,* 297, 226, 1986.

22. **Feitsma, K., Bosman, J., Drenth, B., and DeZeeuw, R.,** A study of the separation of enantiomers of some aromatic carboxylic acids by high-performance liquid chromatography on a p-cyclodextrin-bonded stationary phase, *J. Chromatogr.,* 333, 59, 1985.

23. **Fujimura, K., Ueda, T., and Ando, T.,** Retention behavior of some aromatic compounds on chemically bonded cyclodextrin silica stationary phase in liquid chromatography, *Anal. Chem.,* 55, 446, 1983.

24. **Hattori, K., Takahashi, K., Mikami, M., and Watanabe, H.,** Novel high-performance liquid chromatographic adsorbents prepared by immobilization of modified cyclodextrins, *J. Chromatogr.,* 355, 383, 1986.

25. **Ridlon, C. D. and Issaq, H. J.,** Effect of column type and experimental parameters on the HPLC separation of dipeptides, *J. Liq. Chromatogr.,* 9(15), 3377, 1986.

26. **Sybilska, D., Debowski, J., Jurczak, J., and Zukowski,** The α and β-cyclodextrin complexation as a tool for the separation of *o-, m-* and *p*-nitro-*cis*- and *trans*-cinnamic acids by reversed-phase high-performance liquid chromatography, *J. Chromatgr.,* 286, 163, 1984.

27. **Chang, C. A., Wu, Q., and Eastman, M. P.,** Mobile phase effects on the separation of substituted anilines with a-cyclodextrin-bonded column, *J. Chromatogr.,* 371, 269, 1986.

28. **Maguire, J. H.,** Some structural requirements for resolution of hydantoin enantiomers with α-cyclodextrin liquid chromatography column, *J. Chromatogr.,* 387, 453, 1987.

29. **Sinibaldi, M., Carunchio, V., Coradini, C., and Girelli, A. M.,** High-performance liquid chromatographic resolution of enantiomers on chiral amine bonded silica gel, *Chromatographia,* 18(81), 459, 1984.

30. **Pettersson, C. and Stuurman, H. W.,** Direct separation of enantiomer of ephedrine and some analogues by reversed-phase liquid chromatography using (+)-di-*n*-butytartrate as the liquid stationary phase, *J. Chromatogr. Sci.,* 22, 441, 1984.

31. **Weems, H. B., Mushtaq, M., and Yank, S. K.,** Resolution of epoxide enantiomers of polycyclic aromatic hydrocarbons by chiral stationary-phase high-performance liquid chromatography, *Anal. Biochem.,* 148, 328, 1985.

32. **Weems, H., Mushtaq, M., Fu, P., and Yank, S.,** Direct separation of non-k-region monool and diol enantiomers of phenanthrene, benz[*a*]anthracene, and chrysene by high-performance liquid chromatography with chiral stationary phases, *J. Chromatogr.,* 371, 211, 1986.

33. **Yang, S., Mushtaq, M., and Fu, P.,** Elution order-absolute configuration of k-region dihydrodiol enantiomers of benz[*a*]anthracene derivatives in chiral stationary phase high performance liquid chromatography, *J. Chromatogr.,* 371, 195, 1986.

34. **Wainer, I.,** Applicability of HPLC chiral stationary phases to pharmacokinetic and disposition studies on enantiomeric drugs, *Methodol. Serv. Biochem. Anal. Subser. A,* 16, 243, 1986.

35. **Vaughan, G. T. and Millborrow, B. V.,** The resolution by HPLC of RS-[2-^{14}C] Me 1',4'-cis-diol of abscisic acid and the metabolism of (−)-R− and −S− abscisic acid, *J. Exp. Bot.,* 35 (150), 110, 1984.

36. **Tambute, A., Gareil, P., Caude, M., and Rosset, R.,** Preparative separation of racemic tertiary phosphine oxides by chiral high-performance liquid chromatography, *J. Chromatog.,* 363, 81, 1986.

37. **Wainer, I. and Doyle, T.,** The direct enantiomeric determination of (−) and (+)-propranolol in human serum by high-performance liquid chromatography on a chiral stationary phase, *J. Chromatogr.,* 306, 405, 1984.

38. **Okamoto, Y., Mohri, H., Ishikura, M., Hatada, K., and Yuki, H.,** Optically active poly(diphenyl-2-pyridylmethyl methacrylate): asymmetric synthesis, stability of helix, and chiral recognition ability, *J. Polym. Sci. Polym. Symp.,* 74, 125, 1986.

39. **Yamazaki, S., Omori, H., and Eon Oh, C.,** High-performance liquid chromatography of alkaline-earth metal ions using reversed-phase column coated with N-*n*-dodecyliminodiacetic acid, *J. High Resolut. Chromatogr. Chromatogr. Commun.,* 9, 765, 1986.

40. **Gubitz, G. and Milhellyes, S.,** Direct separation of 2-hydroxy acid enantiomers by high-performance liquid chromatography on chemically bonded chiral phases, *Chromatographia,* 19, 257, 1984.

41. **Schulze, J. and Konig, W.,** Enantiomer separation by high-performance liquid chromatography on silica gel with covalently bound mono-saccahrides, *J. Chromatogr.,* 355, 165, 1986.

42. **Kip, J., Van Haperen, P., and Kraak, J. C.,** *R-N*-(pentafluorobenzoyl) phenylglycine as a chiral stationary phase for the separation of enantiomers by high-performance liquid chromatography, *J. Chromatogr.,* 356, 423, 1986.

43. **Gelber, L. R., Karger, B. L., Neumeyer, J. L., and Feibush, B.,** Ligant exchange chromatography of amino alcohols. Use of schiff bases in enantiomer resolution, *J. Am. Chem. Soc.,* 106, 7729, 1984.

44. **Dabashi, Y. and Hara, S.,** Direct resolution of enantiomers by liquid chromatography with the novel chiral stationary phase derived from (*R,R*)-tartamide, *Tetrahedron. Lett.,* 26(35), 4217, 1985.

45. **Facklam, C., Pracejus, H., Oehme, G., and Much, H.,** Resolution of enantiomers of amino acid derivatives by high-performance liquid chromatography in a silica gel bonded chiral amide phase, *J. Chromatogr.*, 257, 118, 1983.

46. **Okamoto, Y., Honda, S., Hatada, K., and Yuki, H.,** High-performance liquid chromatographic resolution of enantiomers on optically active poly(tri-phenylmethyl methacrylate). IX, *J. Chromatogr.*, 350, 127, 1985.

47. **Okamoto, Y. and Hatada, K.,** Resolution of enantiomers by HPLC on optically active poly(triphenylmethyl methacrylate), *J. Liq. Chromatogr.*, 9(2 and 3), 369, 1986.

CHIRAL STATIONARY PHASES FOR HPLC

Name: Bovine serum albumin (covalently fixed to silica gel)
Structure: Prolate ellipsoid 14 × 4 nm, with a molecular mass of 66,500; amount absorbed is dependent on buffer pH, with the maximum at pH = 4.9
Analytical properties: Separation of bopindolol and also separation of pindolol after derivatization with isopropyl isocyanate, separation of DL mixtures of enantiomers; can be used on both the analytical and preparative scales, changes in pH will cause this phase to leach from the column; storage at 4 °C is recommended
Reference: 1, 2

Name: α-Acid glycoprotein
Structure: Structure is proprietary (Enantiopac, LKB Co.)
Analytical properties:
(1) Separation of the drugs ibuprofen, ketoprofen, naproxen, 2-phenoxypropionic acid, bendroflumethiazide, ethotoin, hexobarbital, disopyramide, and RAC 109; (2) retention and selectivity of the solutes can be regulated by addition of the tertiary amine N,N-dimethyloctylamine (DMOA) to the mobile phase; (3) DMOA decreases retention time and the enantioselectivity of the weaker acids, but has opposite effects on the stronger acids
Reference: 3

Name: N-(t-Butylamino carbonyl-L-valine)-bonded silica
Structure:

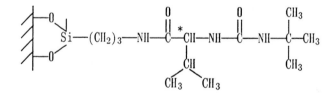

Analytical properties: Separation of amino acid enantiomers; most effective of the L-valine urea derivatives; depends on hydrogen bond interactions usually prepared on LiChrosorb (10 μm); hexane plus isopropanol modifier has been used as the liquid phase
Reference: 4

Name: Cellulose *cis* and *trans* tris(4-phenylazophenyl carbamate) (CPAPC)
Structure:

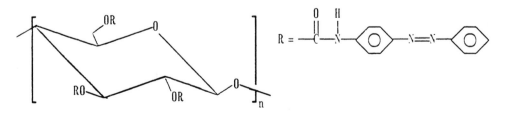

Analytical properties: *Trans* isomer provides excellent resolution of racemic mixtures such as atropine, pindolol, and flavanone; resolution decreases quickly with increasing *cis* isomer concentration; the *cis*/*trans* equilibrium is controlled by UV-radiation, and the phase is adsorbed to silica gel; liquid phase of hexane with 10% 2-propanol has been found useful
Reference: 5

CHIRAL STATIONARY PHASES FOR HPLC (continued)

Name: Cellulose triacetate
Structure:

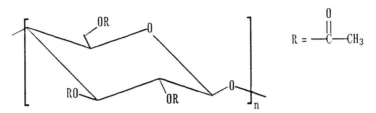

Analytical properties: Shows chiral recognition for many racemates and is especially effective for substrates with a phosphorous atom at an assymetric center; however, the degree of chiral recognition is not so high in general
Reference: 6

Name: Cellulose tribenzoate
Structure:

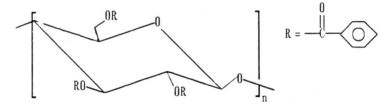

Analytical properties: Demonstrates good chiral recognition for the racemates with carbonyl group(s) in the neighborhood of an asymmetric center
Reference: 6

Name: Cellulose tribenzyl ether
Structure:

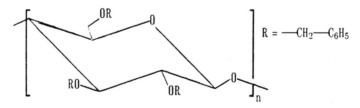

Analytical properties: Effective when protic solvents are used as mobile phases
Reference: 6

Name: Cellulose tricinnamate
Structure:

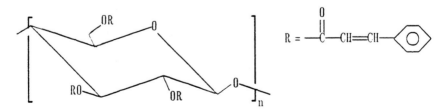

Analytical properties: Shows high chromatographic retention times and a good chiral recognition for many aromatic racemates and barbiturates
Reference: 6

CHIRAL STATIONARY PHASES FOR HPLC (continued)

Name: Cellulose tris(phenylcarbamate)
Structure:

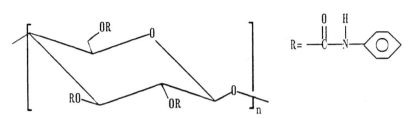

Analytical properties: (1) Separation of racemic mixtures of alkaloids, using ethanol as the eluent; *l*-isomers tend to be more strongly retained than *d*-isomers; (2) good chiral recognition of sulfoxides and high affinity for racemates having an −OH or −NH group, through hydrogen bonding
Reference: 6, 7

Name: Cross-linked acetylcellulose
Structure: Cellulose with one of the OH groups acylated
Analytical properties: Separation of enantiomers (such as etozolin, piprozolin, ozolinon, and bunolol) using an ethanol/water, 95/5 (v/v%) liquid phase.
Reference: 8

Name: Microcrystalline tribenzoylcellulose
Structure: Same structure as cellulose tribenzoate
(Coated on microporous silica gel)
Analytical properties: Resolution of *trans*-1,2-diphenyloxirane, 2-methyl-3-(2′-methylphenyl)-4(3*H*)-quinazolinone and some aromatic hydrocarbons
Reference: 9

Name: Triacetylcellulose
Structure: Same structure as cellulose triacetate
Analytical properties: Microcrystalline triacetylcellulose swells in organic solvents; separation of racemic thioamides, sulfoxides, organophosphorus compounds, drugs, and amino-acid derivatives; separations of these racemates were achieved at pressures at or above 4.9 MPa.
Reference: 10

Name: Untreated cellulose (average particle size of 7μm)
Structure:

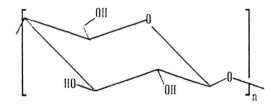

Analytical properties: Complete resolution of D,L-tryptophane, and D,L-5-hydroxytryptophane
Reference: 11

Name: Copper(II) 2-amino-1-cyclopentene-1-dithiocarboxylate
Structure:

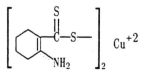

CHIRAL STATIONARY PHASES FOR HPLC (continued)

(Bonded to silica)
Analytical properties: Separation of dialkyl sulfides when hexane containing methanol or acetonitrile was used as the mobile phase
Reference: 12

Name: Cyclodextrin-bonded phases
Structure:

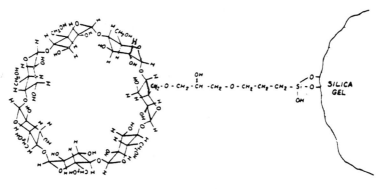

(This is the structure of β-cyclodextrin)
Analytical properties: (1) α-cyclodextrin (cyclohexamolyose), reversed-phase separation of barbiturates and other drugs, and aromatic amino acids; (2) β-cyclodextrin (cycloheptamylose), normal-phase separation of positional isomers of substituted benzoic acids; reversed-phase separation of dansyl and napthyl amino acids, several aromatic drugs, steroids, alkaloids, metallocenes, binaphthyl crown ethers, aromatic acids, aromatic amines, and aromatic sulfoxides; (3) γ-cyclodextrin (cyclooctamylose), reversed-phase separation of sterioisomers of polycyclic aromatic hydrocarbons, (4) acetylated β-cyclodextrin, reversed-phase separation of steroids and polycyclic compounds
Reference: 13—28

Name: (−)*trans*-1,2-Cyclohexanediamine
Structure:

(Bonded to silica gel)
Analytical properties: Resolution of such enantiomeric compounds as 2,2′-dihydroxy-1,1′binaphthyl and *trans*-1,2-cyclo hexandiol
Reference: 29

Name: (+)-Di-*n*-butyltartate
Structure:

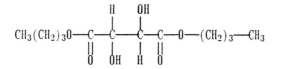

(Adsorbed on phenyl bonded silica)
Analytical properties: Resolution of ephedrine and nonephedrine
Reference: 30

CHIRAL STATIONARY PHASES FOR HPLC (continued)

Name: (*S*)-*N*-(3,5-Dinitrobenzoyl)leucine or (*S*)-DNBL
Structure:

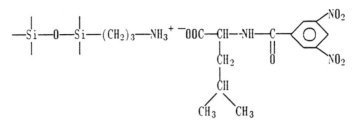

Ionically bonded X≡isopropyl; Y≡O⁻ + NH₃   Bonded to α-amino (propylsilanized silica)
Covalently bonded X≡isopropyl; Y = NH
Analytical properties: (1) Resolution of several enantiomers of polycyclic aromatic hydrocarbons, e.g., chrysene 5,6-expoxide, dibenz[*a*,*h*]anthracene 5,6 epoxide, 7-methyl benz[a]anthracene 5,6 epoxide; (2) resolution of barbiturates, mephenytoin, benzodiazepinones, and succinimides; (3) direct separation of some mono-ol and diol enantiomers of phenanthrene, benz[*a*]anthrene and chrysene; (4) ionically bonded to silica gel, this phase provides resolution of enantiomers of *cis*-dihydrodiols of unsubstituted and methyl- and bromo-substituted benz[*a*]-anthracene derivatives having hydroxyl groups that adopt quasiequatorial-quasiaxial and/or quasiaxial-quasiequatorial conformations
Reference: 31—35

Name: (*R*)-*N*-(3,5-Dinitrobenzyl) phenylglycine
Structure:

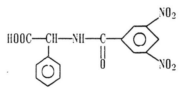

Analytical properties: (1) Ionically bonded to silica, this phase provides good resolution of enantiomeric quasiequatorial *trans*-dehydriols of unsubstituted and methyl and bromo-substituted benz[*a*]anthracene derivatives; (2) covalently bonded to silica, this phase provides good resolution of enantiomeric pairs of quasidiaxial *trans*-dihydrodiols of unsubstituted and methyl- and bromo-substituted benz[*a*]anthracene derivatives; (3) by addition of a third solvent (chloroform) to the classical binary mixture (hexane-alcohol) of the mobile phase, resolution of enantiomers of tertiary phosphine oxides is possible
Reference: 31—33, 36, 37

Name: Poly (Diphenyl-2-pyridylmethyl methacrylate) or PD2PyMa
Structure:

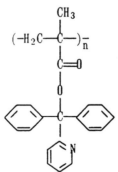

(Coated on macroporous silica gel)
Analytical properties: Resolution of such compounds as racemic 1,2-diphenol-ethanol, 2,2'-dihydroxyl-1,1'-binaphthyl, 2,3-diphenyloxirane, and phenyl-2-pyrid-o-toly-1-methanol
Reference: 38

CHIRAL STATIONARY PHASES FOR HPLC (continued)

Name: *N-n*-Dodecyliminodiacetic acid (coated on silica gel)
Structure:

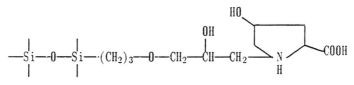

Analytical properties: Separation of alkaline-earth metal ions
Reference: 39

Name: L-Hydroxyproline
Structure:

$$-Si-O-Si-(CH_2)_3-O-CH_2-\overset{\overset{\displaystyle OH}{|}}{CH}-CH_2-N-\text{(ring)}-COOH$$

(As a fixed ligand on a silica gel; Cu(II) used as complexing agent
Analytical properties: Chiral phases containing L-hydroxy-proline as a fixed ligand show high enantioselectivity for 2-hydroxy acids; these phases have resolved some aromatic as well as aliphatic 2-hydroxy acids
Reference: 40

Name: 1-Isothiocyanato-D-glucopyranosides
Structure:

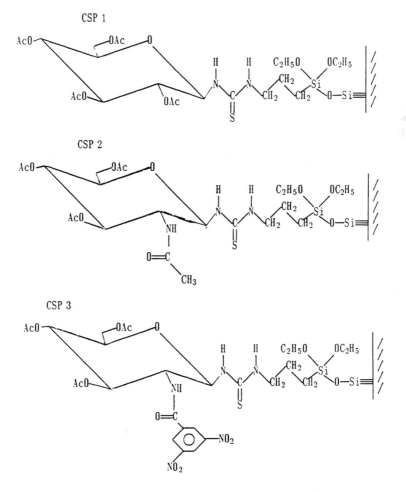

CHIRAL STATIONARY PHASES FOR HPLC (continued)

Analytical properties: (1) CSP 1 (chiral stationary phase); separates some chiral binapthyl derivatives when mixtures of hexane diethyl ether, dichloromethane, or dioxane are used as the mobile phase; (2) CSP 2; separates compounds with carbamate or amide functions (mixtures of *n*-hexane and 2-propanol can be used as mobile phase); (3) CSP 3; separation of compounds separated by CSP 2; in addition, separates compounds with carbanoyl or amide functions and some amino alcohols which have pharmaceutical relevance (β-blockers)
Reference: 41

Name: *N*-Isopropylaminocarbonyl-L-valine-bonded silica
Structure:

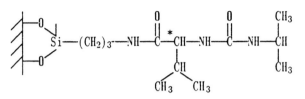

Analytical properties: Separation of amino acid enantiomers; good chiral recognition of *N*-acetyl amino acid methyl esters; depends on hydrogen bond interactions; hexane with isopropanol modifier has been used as the liquid phase; usually prepared on LiChrosorb (10 μm)
Reference: 4

Name: (*R*)-or (*S*)-*N*-(2-Naphthyl)alanine
Structure:

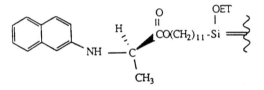

(*R*)-*N*-(2-naphthyl-alanine)

Analytical properties: High selectivities for a variety of dinitrobenzoyl derivatized compounds
Reference: 19

Name: (*S*)-1-(α-Naphthyl) ethylamine
Structure:

Analytical properties: Separation of 3,5-dinitrobenzoyl derivatives of amino acids; 3,5-dinitroanilide derivatives of carboxylic acids
Reference: 19

Name: *R*-*N*-(Pentafluorobenzoyl)phenylglycine-bonded silica
Structure:

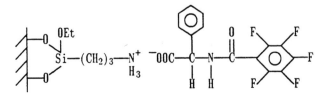

CHIRAL STATIONARY PHASES FOR HPLC (continued)

Analytical properties: Higher selectivity for nitrogen-containing racemates than *R-N*-(3,5-dinitrobenzoyl)phenylglycine; examples of nitrogen-containing racemates include succinimides, hydantoins, and mandelates
Reference: 42

Name: *N*-phenylaminocarbonyl-L-valine-bonded silica
Structure:

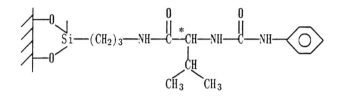

Analytical properties: Separation of amino-acid enantiomers by hydrogen-bond interactions; usually prepared on LiChrosorb (10 μm); hexane plus isopropanol modifier commonly used as liquid phase
Reference: 4

Name: (L-Proline) copper(II)
Structure:

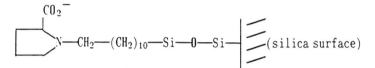

Analytical properties: Separation of primary α-amino alcohols, e.g., β-hydroxyphenethylamines and catecholamines
Reference: 43

Name: Derivative of (*R,R*)-Tartramide
Structure:

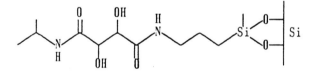

Analytical properties: Resolution of a series of β-hydroxycarboxylic acids as *tert*-butylamide derivatives
Reference: 44

Name: *tert*-Butylvalinamide
Structure:

$$(CH_3)_2CHCH\text{---}CONHC(CH_3)_3$$
$$|$$
$$NH_2$$

Analytical properties: Resolution of heavier amino acid derivatives
Reference: 45

CHIRAL STATIONARY PHASES FOR HPLC (continued)

Name: (+)-Poly(triphenylmethyl methacrylate) or (+)-PTrMA
Structure:

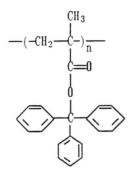

Analytical properties: Resolution of enantiomers such as *trans*-1,3-cyclohexene dibenzoate, 3,5-pentylene di-benzoate, 3,5-dichlorobenzoate, trisacetylacetonates, and racemic compounds having phosphorous as a chiral center; this phase will also resolve achiral compounds
Reference: 46, 47

ULTRAVIOLET DETECTION OF CHROMOPHORIC GROUPS

The following table is provided to aid the user in the application of the ultraviolet spectrophotometer as a detector.

REFERENCES

1. **Willard, H. H., Merritt, L. L., Dean, J. A., and Settle, F. A.,** *Instrumental Methods of Analysis,* 6th ed., Wadsworth Publishing, Belmont, CA, 1981.
2. **Silverstein, R. M., Bassler, G. C., and Morrill, T. C.,** *Spectrometric Identification of Organic Compounds,* 4th ed., John Wiley & Sons, New York, 1981.
3. **Lambert, J. B., Shuruell, H. F., Verbit, L., Cooks, R. G,, and Stout, G. H.,** *Organic Structural Analysis,* MacMillan, New York, 1976.

ULTRAVIOLET DETECTION OF CHROMOPHORIC GROUPS

Chromophore	System	λ_{max} (nm)	ϵ_{max}	λ_{max} (nm)	ϵ_{max}	λ_{max}	ϵ_{max}
Ether	–O–	185	1,000				
Thioether	–S–	194	4,600	215	1,600		
Amine	–NH$_2$–	195	2,800				
Thiol	–SH	195	1,400				
Disulfide	–S–S–	194	5,500	255	400		
Bromide	–Br	208	300				
Iodide	–I	260	400				
Nitrile	–C≡N	160					
Acetylide	–C≡C–	175—180	6,000				
Sulfone	–SO$_2$–	180					
Oxime	–NOH	190	5,000				
Azido	>C=N–	190	5,000				
Ethylene	–C=C–	190	8,000				
Ketone	>C=O	195	1,000	270—285	18—30		
Thioketone	>C=S	205	Strong				
Esters	–COOR	205	50				
Aldehyde	–CHO	210	Strong	280—300	11—18		
Carboxyl	–COOH	200—210	50—70				
Sulfoxide	>S→O	210	1,500				
Nitro	–NO$_2$	210	Strong				
Nitrite	–ONO	220—230	1,000—2,000	300—4,000	10		
Azo	–N=N–	285—400	3—25				
Nitroso	–N=O	302	100				
Nitrate	–ONO$_2$	270 (Shoulder)	12				
↑	–(C=C)$_2$– (Acyclic)	210—230	21,000				
	–(C=C)$_3$–	260	35,000				
	–(C=C)$_4$–	300	52,000				
Conjugated systems	–(C=C)$_5$–	330	118,000				
	–(C=C)$_2$– (Alicyclic)	230—260	3,000—8,000				
	C=C–C≡C	219	6,500				
	C=C–C=N	220	23,000				
	C=C–C=O	210—250	10,000—20,000			300—3500	Weak
↓	C=C–NO$_2$	229	9,500				
Benzene	φ[a]	184	46,700	202	6,900	255	170

ULTRAVIOLET DETECTION OF CHROMOPHORIC GROUPS (continued)

Chromophore	System	λ_{max} (nm)	ϵ_{max}	λ_{max} (nm)	ϵ_{max}	λ_{max}	ϵ_{max}
Diphenyl				246	20,000		
Naphthalene		220	112,000	275	5,600	312	175
Anthracene		252	199,000	375	7,900		
Pyridine		174	80,000	195	6,000	251	1,700
Quinoline		227	37,000	270	3,600	314	2,750
Isoquinoline		218	80,000	266	4,000	317	3,500

[a] ϕ denotes a phenyl group.

SOME USEFUL ION-PAIRING AGENTS

The following table provides a short list of ion-pair chromatographic modifiers, for use in the separation of ionic or ionizable species.[1-3] The use of these modifiers can often greatly improve the chromatographic performance of both normal- and reverse-phase systems.

REFERENCES

1. **Poole, C. F. and Schuette, S. A.**, *Contemporary Practice of Chromatography*, Elsevier, Amsterdam, 1984.
2. **Snyder, L. R. and Kirkland, J. J.**, *Introduction to Modern Liquid Chromatography*, John Wiley & Sons, New York, 1979.
3. **Krstulović, A. M. and Brown, P. R.**, *Reversed-Phase High Performance Liquid Chromatography*, John Wiley & Sons, New York, 1982.

SOME USEFUL ION-PAIRING AGENTS

Ion type/examples	Applications/notes
Perchloric acid	Used for a wide range of basic analytes; typically used at 0.1 M concentration in reverse-phase solvent system and at approximately the same concentration in a water-buffer system on the stationary phase in normal mode
Bis-(2-ethylhexyl) phosphate	Used for cationic species of intermediate polarity, such as phenols; typically used in reverse phase, on bis-(2-ethylhexyl) phosphoric acid/chloroform stationary phase at a pH $\approx$ 3.8
N,N-Dimethyl protriptyline	Used for carboxylic acids; typically used in normal phase, with a basic (pH $\approx$ 9) buffered stationary phase and on organic mobile phase
Quaternary amines: tetramethyl, tetrabutyl, and palmityltrimethylammonium salts, usually in chloride or phosphate forms	Used for strong and weak acids, sulfonated dyes, carboxylic acids; in normal-phase applications, typical buffer pH values are between 6 and 8.5, with an organic mobile phase; in reverse phase, the mobile phase is typically aqueous plus a polar organic modifier at nearly neutral pH values
Tertiary amines: tri-n-octyl amine	Used for carboxylic acids and sulfonates; used in reverse-phase mode with a water + buffer + $\approx$ 0.05 M perchloric acid mobile phase
Sulfonates, alkyl and aromatic: methane or heptanesulfonate, camphorsulfonic acid	Used for strong and weak bases, benzalkonium salts, and catecholamines
Alkyl sulfates: octyl, decyl, dodecyl, and lauryl sulfates	Used in similar applications as the sulfonates, but provide a different selectivity; typically used in reverse-phase mode, often using a water + methanol + sulfuric acid mobile phase

DERIVATIZING REAGENTS FOR HPLC

The following table provides a listing of the common reagents used in HPLC. Most of these reagents are used to impart a chromophoric or fluorescent group in the sample to enhance the detectability. Occasionally, a derivitization procedure is done in order to enhance selectivity, but this is the exception.

REFERENCES

1. **Umagat, H., Kucera, P., and Wen, L.-F.,** Total amino acid analysis using pre-column fluorescence derivatization, *J. Chromatogr.,* 239, 463, 1982.
2. **Nimura, N. and Kinoshita, T.,** Fluorescent labeling of fatty acids with 9-anthryldiazomethane (ADAM) for high performance liquid chromatography, *Anal. Lett.,* 13(A3), 191, 1980.
3. **Lehrfeld, J.,** Separation of some perbenzoylated carbohydrates by high performance liquid chromatography, *J. Chromatogr.,* 120, 141, 1976.
4. **Durst, H. D., Milano, M., Kikta, E. J., Connelly, S. A., and Grushka, E.,** Phenacyl esters of fatty acids via crown ether catalysts for enhanced ultraviolet detection in liquid chromatography, *Anal. Chem.,* 47(11), 1797, 1975.
5. **Farinotti, R., Siard, Ph., Bourson, J., Kirkiacharian, S., Valeur, B., and Mohuzier, G.,** 4-Bromo-methyl-6,7-dimethoxycoumarin as a fluorescent label for carboxylic-acids in chromatographic detection, *J. Chromatogr.,* 269(2), 81, 1983.
6. **Lam, S. and Grushka, E.,** Labeling of fatty acids with 4-bromomethyl-7-methoxycoumarin via crown ether catalyst for fluorimetric detection in high-performance liquid chromatography, *J. Chromatogr.,* 158, 207, 1978.
7. **Korte, W. D.,** 9-(Chloromethyl) anthracene: a useful derivatizing reagent for enhanced ultraviolet and fluorescence detection of carboxylic acids with liquid chromatography, *J. Chromatogr.,* 243, 153, 1982.
8. **Ahnoff, M., Grundevik, I., Arfwidsson, A., Fonselius, J., and Persson, B.-A.,** Derivatization with 4-chloro-7-nitrobenzofurazan for liquid chromatographic determination of hydroxyproline in collagen hydro-lysate, *Anal. Chem.,* 53, 485, 1981.
9. **Linder, W.,** *N*-Chloromethyl-4-nitro-phthalimide as a derivatizing reagent for HPLC, *J. Chromatogr.,* 198, 367, 1980.
10. **Lawrence, J. F.,** *Organic Trace Analysis by Liquid Chromatography,* Academic Press, New York, 1981.
11. **Avigad, G.,** Dansyl hydrazine as a fluorimetric reagent for thin-layer chromatographic analysis of reducing sugars, *J. Chromatogr.,* 139, 343, 1977.
12. **Lloyd, J. B. F.,** Phenanthramidazoles as fluorescent derivatives in the analysis of fatty acids by high-performance liquid chromatography, *J. Chromatogr.,* 189, 359, 1980.
13. **Frei, R. W. and Lawrence, J. F., Eds.,** *Chemical Derivatization in Analytical Chemistry,* Vol. 1, Plenum Press, New York, 1981.
14. **Goto, J., Komatsu, S., Goto, N., and Nambara, T.,** A new sensitive derivatization reagent for liquid chromatographic separation of hydroxyl compounds, *Chem. Pharm. Bull.,* 29(3), 899, 1981.
15. **Musson, D. G. and Sternson, L. A.,** Conversion of arylhydroxylamines to electrochemically active derivatives suitable for high-performance liquid chromatographic analysis with amperometric detection, *J. Chromatogr.,* 188, 159, 1980.
16. **Lankmayr, E. P., Budna, K. W., Müller, K., Nachtmann, F., and Rainer, F.,** Determination of d-penicillamine in serum by fluorescence derivatization and liquid column chromatography, *J. Chromatogr.,* 222, 249, 1981.
17. **Mopper, K., Stahovec, W. L., and Johnson, L.,** Trace analysis of aldehydes by reversed-phase high-performance liquid chromatography and precolumn fluorogenic labeling with 5,5-dimethyl-1,3-cyclohex-anedione, *J. Chromatogr.,* 256, 243, 1983.
18. **Lawrence, J. F. and Frei, R. W.,** *Chemical Derivatization in Liquid Chromatography,* Elsevier, Amsterdam, 1976.
19. **Carey, M. A. and Persinger, H. E.,** Liquid chromatographic determination of traces of aliphatic carbonyl compounds and glycols as derivatives that contain the dinitrophenyl group, *J. Chromatogr. Sci.,* 10, 537, 1972.
20. **Fitzpatrick, F. A., Siggia, S., and Dingman, J., Sr.,** High speed liquid chromatography of derivatized urinary 17-keto steroids, *Anal. Chem.,* 44(13), 2211, 1972.
21. **Pietrzyk, D. J. and Chan, E. P.,** Determination of carbonyl compounds by 2-diphenylacetyl-1,3-indan-dione-1-hydrazone, *Anal. Chem.,* 42(1), 37, 1970.
22. **Braun, R. A. and Mosher, W. A.,** 2-Diphenylacetyl-1,3-indandione 1-hydrazone—a new reagent for carbonyl compounds, *J. Am. Chem. Soc.,* 80, 3048, 1958.

23. **Schäfer, M. and Mutschuler, E.**, Fluorimetric determination of oxprenolol in plasma by direct evaluation of thin-layer chromatograms, *J. Chromatogr.*, 164, 247, 1979.

24. **Moye, H. A. and Boning, A. J., Jr.**, A versatile fluorogenic labelling reagent for primary and secondary amines: 9-fluorenylmethyl chloroformate, *Anal. Lett.*, 12(B1), 25, 1979.

25. **Lehninger, A. L.**, *Biochemistry*, 2nd ed., Worth Publishers, New York, 1978.

26. **Roos, R. W.**, Determination of conjugated and esterified estrogens in pharmaceutical tablet dosage forms by high-pressure, normal-phase partition chromatography, *J. Chromatogr. Sci.*, 14, 505, 1976.

27. **DeLeenheer, A., Sinsheimer, J. E., and Burckhalter, J. H.**, Fluorometric determination of primary and secondary aliphatic amines by reaction with 9-isothiocyanatoacridine, *J. Pharm. Sci.*, 62(8), 1370, 1973.

28. **Clark, C. R. and Wells, M. M.**, Precolumn derivatization of amines for enhanced detectability in liquid chromatography, *J. Chromatogr. Sci.*, 16, 332, 1978.

29. **Hulshoff, A., Roseboom, H., and Renema, J.**, Improved detectability of barbiturates in high-performance liquid chromatography by pre-column labelling and ultraviolet detection, *J. Chromatogr.*, 186, 535, 1979.

30. **Matthees, D. P. and Purdy, W. C.**, Napthyldiazomethane as a derivatizing agent for the high-performance liquid chromatography detection of bile acids, *Anal. Chim. Acta.*, 109, 161, 1979.

31. **Kuwata, K., Uebori, M., and Yamazaki, Y.**, Determination of phenol in polluted air as *p*-nitrobenzeneazophenol derivative by reversed phase high performance liquid chromatography, *Anal. Chem.*, 52, 857, 1980.

32. **Nachtmann, F., Spitzy, H., and Frei, R. W.**, Rapid and sensitive high-resolution procedure for digitalis glycoside analysis by derivatization liquid chromatography, *J. Chromatogr.*, 122, 293, 1976.

33. **Knapp, D. R. and Krueger, S.**, Use of *o-p*-nitrobenzyl-*N,N*-diisopropylisourea as a chromogenic reagent for liquid chromatographic anlaysis of carboxylic acids, *Anal. Lett.*, 8(9), 603, 1975.

34. **Jupille, T.**, UV-Visible absorption derivatization in liquid chromatography, *J. Chromatogr. Sci.*, 17, 160, 1979.

35. **Dunlap, K. L., Sandridge, R. L., Keller, J.**, Determination of isocyanates in working atmospheres by high speed liquid chromatography, *Anal. Chem.*, 48(3), 497, 1976.

36. **Politzer, I. R., Griffin, G. W., Dowty, B. J., and Laseter, J. L.**, Enhancement of ultraviolet detectability of fatty acids for purposes of liquid chromatographic-mass spectrometric analyses, *Anal. Lett.*, 6(6), 539, 1973.

37. **Cox, G. B.**, Estimation of volatile N-nitrosamines by high-performance liquid chromatography, *J. Chromatogr.*, 83, 471, 1973.

38. **Borch, R. F.**, Separation of long chain fatty acids as phenacyl esters by high pressure liquid chromatography, *Anal. Chem.*, 47(14), 2437, 1975.

39. **Poole, C. F., Singhawangcha, S., Zlatkis, A., and Morgan, E. D.**, Polynuclear aromatic boronic acids as selective fluorescent reagents for HPTLC and HPLC, *J. High Resolut. Chromatogr. Chromatogr. Commun.*, 1, 96, 1978.

40. **Björkqvist, B. and Toivonen, H.**, Separation and determination of aliphatic alcohols by high-performance liquid chromatography with U. V. detection, *J. Chromatogr.*, 153, 265, 1978.

41. **Munger, D., Sternson, L. A., Repta, A. J., and Higuchi, T.**, High-performance liquid chromatographic analysis of dianhydrogalactitol in plasma by derivatization with sodium diethyldithiocarbamate, *J. Chromatogr.*, 143, 375, 1977.

42. **Sugiura, T., Hayashi, T., Kawai, S., and Ohno, T.**, High speed liquid chromatographic determination of putrescine, spermidine and spermine, *J. Chromatogr.*, 110, 385, 1975.

43. **Suzuki, Y. and Tani, K.**, High-speed liquid-chromatography of the aliphatic alcohols as their trityl ether derivatives, *Buneseki Kagaku.*, 28, 610, 1979.

DERIVATIZING REAGENTS FOR HPLC

Derivatizing reagent	Structure/formula	Notes	Ref.
N-(9-Acridinyl) maleimide		Used for the precolumn preparation of fluorescent derivatives of thiols	1
9-Anthryldiazomethane		Used for the precolumn preparation of fluorescent derivatives of carboxylic acids; reagent reacts well with fatty acids at room temperature to give intensely fluorescent esters	2
Benzoyl chloride		Used to introduce chromophores into alcohols, and amines, using pyridine as a solvent; efficient means for the isolation of carbohydrates in complex mixtures	3
Benzyl bromide		Used to introduce chromophores into carboxylic acids	4
4-Bromomethyl-6,7-dimethoxycoumarin		Used for the precolumn preparation of fluorescent derivatives of carboxylic acids using acetone as solvent and with crown ether and alkali as catalysts	5
4-Bromomethyl-7-methoxy-coumarin (Br-Mmc)		Used for the precolumn preparation of fluorescent derivatives of carboxylic acids, using a crown ether (18-crown-6) as a catalyst	6
p-Bromophenacyl bromide		Used to introduce chromophores into carboxylic acids; crown ethers are used as phase transfer agents (e.g., 18-crown-6, and dicyclohexyl-18-crown-6)	4
9-(Chloromethyl) anthracene (9-CIMA)		Used for the precolumn preparation of fluorescent derivatives of carboxylic acids, using cyclohexane as a solvent	7
4-Chloro-7-nitrobenzo-2-oxa-1,3-diazole (NBD-Cl)		Used for the precolumn preparation of fluorescent derivatives of primary and secondary amines, phenols, and thiols (4-chloro-7-nitrobenzofurazan)	8
N-Chloromethyl-4-nitrophthalimide		Used to introduce chromophores into carboxylic acids	9

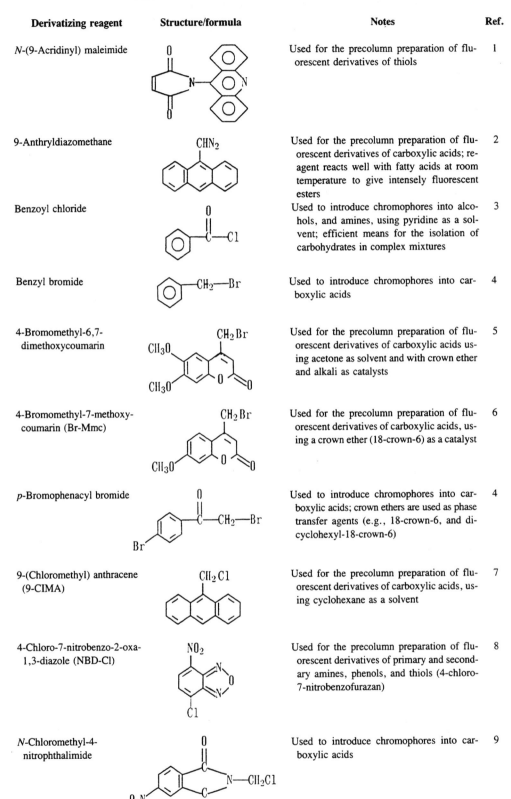

DERIVATIZING REAGENTS FOR HPLC (continued)

Derivatizing reagent	Structure/formula	Notes	Ref.
Dansyl chloride (DnS-Cl)		Used for the precolumn preparation of fluorescent derivatives of primary and secondary amines, phenols, amino acids, and imidazoles	10
Dansyl hydrazine (DnS-H)		Used for the precolumn preparation of fluorescent derivatives of aldehydes and ketones; optimal derivatization of glucose and other sugars occurs at pH 2 to 3	11
9,10-Diaminophenanthrene		Used for the precolumn preparation of fluorescent derivatives of carboxylic acids	12
Diazo-4-aminobenzonitrile		Used to introduce chromophores in phenols	13
2,5-Di-*n*-butylamino-naphthalene-1-sulfonyl chloride (BnS-Cl)		Used for the precolumn preparation of fluorescent derivatives of primary and secondary amines, phenols, amino acids, and imidazoles	10
4-Dimethylamino-1-naphthoyl nitrile		Used for the precolumn preparation of fluorescent derivatives of primary and secondary (but not tertiary) alcohols	14
p-Dimethylaminophenyl isocyanate		Used to introduce chromophores into alcohols; after reaction, excess reagent must be removed as it interferes with ensuing analysis	15
5-dimethylaminonaphthalene 1-sulfonylaziridine (dansylaziridine)		Used for the precolumn preparation of fluorescent derivatives of thiols; optimum derivatization conditions have been found to be pH 8.2 with a minimum of a 2.7-fold molar reagent excess using a reaction time of 1 h at 60 °C; under these conditions only free sulfhydryl groups are derivatized	16

DERIVATIZING REAGENTS FOR HPLC (continued)

Derivatizing reagent	Structure/formula	Notes	Ref.
5,5-Dimethyl-1,3-cyclohex-anedione (dimedone)		Used for the precolumn preparation of fluorescent derivatives of aldehydes, using isopropanol as a solvent in the presence of ammonium acetate	17
3,5-Dinitrobenzyl chloride		Used to introduce chromophores into amines (forming phenyl-substituted amines), alcohols, glycols, and phenols	18, 19
2,4-Dinitrophenylhydrazine		Used to introduce chromophores into aldehydes and ketones in a solution of carbonyl-free methanol; detection of the more common 17-keto steroids as their 2,4-dinitrophenyl derivatives from urine and plasma; suggested potential for clinical use	20
2-diphenylacetyl 1,3-indan-dione-1-hydrazone		Used for the precolumn preparation of fluorescent derivatives of aldehydes and ketones; reagent suggested to be especially useful because of its application on the micro level, for the analysis and identification of carbonyl compounds in smog, polluted air, and biochemical and pharmaceutical mixtures; reagent does not appear to be useful for analysis of sugars; derivatives are fluorescent in the UV_2 as solids and in solution	21, 22
1-Ethoxy-4-(dichloro-*s*-triazinyl)naphthalene or EDTN		Used for the precolumn preparation of fluorescent derivatives of primary and secondary alcohols and phenols	23
9-Fluorenylmethyl chloroformate (FMOCCl)		Used for the precolumn preparation of fluorescent derivatives of primary and secondary amines in acetone solvent; in the presence of sodium borate, derivation proceeds rapidly under alkaline conditions	24
Fluorescamine or fluram		Used for the precolumn preparation of fluorescent derivatives of primary amines and amino acids by HPLC	10

DERIVATIZING REAGENTS FOR HPLC (continued)

Derivatizing reagent	Structure/formula	Notes	Ref.
1-Fluoro-2,4-dinitrobenzene	NO_2 NO_2 F	Used to introduce chromophores into amines using Sanger's procedure	25
p-Iodobenzensulfonyl chloride	I SO_2Cl	Used to introduce chromophores into alcohols and phenols; aids in separation of estrogen derivatives	26
9-Isothiocyanatoacridine	NCS N	Used for the precolumn preparation of fluorescent derivatives of some primary and secondary amines using toluene as a solvent; only amines with $PK_a \geqq 9.33$ have been successfully determined	27
p-Methoxybenzoyl chloride	OCH_3 COCl	Used to introduce chromophores into amines using the reagent in tetrahydrofuran	28
2-Naphthacyl bromide (NPB)	O ‖ $C-CH_2-Br$	Used to introduce chromophores into amines in acetone as a solvent, with cesium bromide as a catalyst; it is suggested that elevated temperatures (up to 80 °C) are necessary for the complete derivatization of compounds containing diisopropylamines	29
1-Naphthyldiazomethane	CHN_2	Used to introduce chromophores into carboxylic acids; reagent is prepared from 1-naphthaldehyde hydrazone by oxidation with Hg(II) oxide, with diethyl ether as a solvent; acetic acid will destroy excess reagent	30
p-Nitrobenzenediazonium-tetrafluoroborate	$(O_2N-\langle\rangle-\overset{+}{N}{\equiv}N)BF_4^-$	Used to introduce chromophores into phenols, suggested derivation takes place in aqueous medium at pH 11.5	31
p-Nitrobenzoyl chloride (4-NBCl)	NO_2 COCl	Used to introduce chromophores into alcohols and amines, using pyridine as the solvent; with silica gel as the stationary phase, relatively low-viscosity, low-polarity solvents can be used for detection of digitalis glycosides by HPLC following derivatization with p-nitrobenzoyl chloride	32
p-Nitrobenzyl-N,N' diisopropylisourea	$CH_2-N-C=N-CH(CH_3)_2$ $\underset{CH(CH_3)_2}{OH}$ NO_2	Used to introduce chromophores into carboxylic acids, without the need for a base catalyst, and under mild conditions; picomolar concentrations are rendered detectable	33
p-Nitrobenzylhydroxylamine hydrochloride	$O_2N-\langle\rangle-CH_2-ONH_2 \cdot HCl$	Used to introduce chromophores into ketones and aldehydes	34

DERIVATIZING REAGENTS FOR HPLC (continued)

Derivatizing reagent	Structure/formula	Notes	Ref.

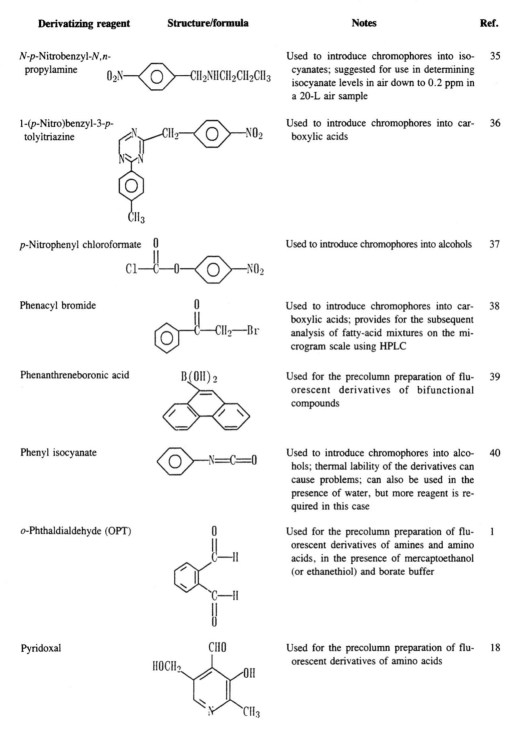

N-p-Nitrobenzyl-N,n-propylamine

Used to introduce chromophores into isocyanates; suggested for use in determining isocyanate levels in air down to 0.2 ppm in a 20-L air sample — 35

1-(p-Nitro)benzyl-3-p-tolyltriazine

Used to introduce chromophores into carboxylic acids — 36

p-Nitrophenyl chloroformate

Used to introduce chromophores into alcohols — 37

Phenacyl bromide

Used to introduce chromophores into carboxylic acids; provides for the subsequent analysis of fatty-acid mixtures on the microgram scale using HPLC — 38

Phenanthreneboronic acid

Used for the precolumn preparation of fluorescent derivatives of bifunctional compounds — 39

Phenyl isocyanate

Used to introduce chromophores into alcohols; thermal lability of the derivatives can cause problems; can also be used in the presence of water, but more reagent is required in this case — 40

o-Phthaldialdehyde (OPT)

Used for the precolumn preparation of fluorescent derivatives of amines and amino acids, in the presence of mercaptoethanol (or ethanethiol) and borate buffer — 1

Pyridoxal

Used for the precolumn preparation of fluorescent derivatives of amino acids — 18

DERIVATIZING REAGENTS FOR HPLC (continued)

Derivatizing reagent	Structure/formula	Notes	Ref.
Pyruvoyl chloride (2,4-dinitrophenyl hydrazone)	CH_3—C—C—Cl with N, O, NH, and dinitrophenyl (NO_2, NO_2) group	Used to introduce chromophores into alcohols, amines, ketones, aldehydes, mercaptans, and phenols; aids in separation of estrogen derivatives	26
Sodium diethyldithiocarbamate (DDTC)	$(CH_3CH_2)_2N$—C—S^- Na^+ with S below	Used to introduce chromophores into epoxides in the presence of phosphate buffer; dithiocarbamates retain high nucleophilicity, and are often water soluble	41
n-Succinimidyl-p-nitrophenylacetate	N—C—CH_2—⟨ring⟩—NO_2	Used to introduce chromophores into amines; reacts under mild conditions without the need for catalysis	34
p-Toluenesulfonyl chloride (TsCl)	CH_3—⟨ring⟩—SO_2Cl	Used to introduce chromophores into amines, aids in resolution of putrescine, spermidine, and spermine by HPLC; excess TsCl must be removed (by extraction with hexane, for example) before analysis	41
Trityl chloride	⟨triphenylmethyl⟩ C—Cl	Used to introduce chromophores into alcohols	42

III. Thin Layer Chromatography

STRENGTH OF COMMON TLC SOLVENTS

The following table lists the common solvents used in thin layer chromatography (TLC), with a measure of their "strengths" on silica gel and alumina. The solvent strength parameter, $\epsilon°$, is defined as a relative energy of adsorption per unit area of standard adsorbent.[1-3] It is defined as zero on alumina when pentane is used as the solvent. This series is what was called the eluotropic series in the older literature. For convenience, the solvent viscosity is also provided. Note that the viscosity is tabulated in centipoises (cP) for the convenience of most users. This is equivalent to milliPascals per second (mPa·s) in the SI convention. Additional data on these solvents may be found in the tables on HPLC.

REFERENCES

1. **Snyder, L. R.,** *Principles of Adsorption Chromatography,* Marcel Dekker, New York, 1968.
2. **Willard, H. H., Merritt, L. L., Dean, J. A., and Settle, F. A.,** *Instrumental Methods of Analysis,* 6th ed., Wadsworth Publishing, Belmont, CA, 1981.
3. **Hamilton, R. and Hamilton, S.,** *Thin Layer Chromatography,* John Wiley & Sons, Chichester, U.K., 1987.

STRENGTH OF COMMON TLC SOLVENTS

Solvent	$\epsilon°(Al_2O_3)$	Viscosity (cP, 20°C)	$\epsilon°(SiO_2)$
Fluoroalkanes	−0.25	—	
n-Hexane	0.00	0.23	0.00
n-Pentane	0.001	0.23	0.000
2,2,4-Trimethylpentane (isooctane)	0.01	0.54	
n-Heptane	0.01	0.41	
n-Decane	0.04	0.92	
Cyclohexane	0.04	1.00	−0.05
Cyclopentane	0.05	0.47	
Carbon disulfide	0.15	0.37	0.14
Tetrachloromethane (carbon tetrachloride)	0.18	0.97	
1-Chloropentane (*n*-pentylchloride)	0.26	0.43	
Diisopropyl ether	0.28	0.37	
2-Chloropropane (isopropyl chloride)	0.29	0.33	
Methylbenzene (toluene)	0.29	0.59	
1-Chloropropane (*n*-propyl chloride)	0.30	0.35	
Chlorobenzene	0.30	0.80	
Benzene	0.32	0.65	0.25
Bromoethane (ethyl bromide)	0.37	0.41	
Diethyl ether (ether)	0.38	0.23	0.38
Trichloromethane (chloroform)	0.40	0.57	
Dichloromethane (methylene chloride)	0.42	0.44	
Tetrahydrofuran	0.45	0.55	

STRENGTH OF COMMON TLC SOLVENTS (continued)

Solvent	$\epsilon°(Al_2O_3)$	Viscosity (cP, 20°C)	$\epsilon°(SiO_2)$
1,2-Dichloroethane	0.49	0.79	
Butanone (ethyl methyl ketone)	0.51	0.43	
Propanone (acetone)	0.56	0.32	0.47
1,4-Dioxane	0.56	1.54	0.49
Ethyl ethanoate (ethyl acetate)	0.58	0.45	0.38
Methyl ethanoate (methyl acetate)	0.60	0.37	
1-Pentanol (*n*-pentanol)	0.61	4.1	
Dimethyl sulfoxide (DMSO)	0.62	2.24	
Aminobenzene (aniline)	0.62	4.4	
Nitromethane	0.64	0.67	
Cyanomethane (acetonitrile)	0.65	0.37	0.50
Pyridine	0.71	0.94	
2-Propanol (isopropanol)	0.82	2.3	
Ethanol	0.88	1.20	
Methanol	0.95	0.60	
Ethylene glycol	1.11	19.9	
Ethanoic acid (acetic acid)	Large	1.26	
Water	Large	1.00	

MODIFICATION OF THE ACTIVITY OF ALUMINA BY ADDITION OF WATER

The following table describes five different activity grades of commercial alumina used in chromatography.[1-3] The activity grades are defined by the degree of adsorption of azobenzene (called *azobenzene number*) on the types of hydrated alumina. Those types are prepared by heating commercial alumina to redness, giving grade I, and then adding controlled amounts of water and allowing equilibration in a closed vessel. The azobenzene number decreases with the amount of water added. The R_f value is the ratio of distance traveled by the solute spot to that traveled by the solvent.

Water added (wt/wt%)	Activity grade	Azobenzene number (maximum adsorption of azobenzene [10^{-5} mol/g])	R_f (*p*-amino-azobenzene)
0	I	26	0.00
3	II	21	0.13
6	III	18	0.25
10	IV	13	0.45
15	V	0	0.55

REFERENCES

1. **Randerath, K.,** *Thin Layer Chromatography*, Academic Press, New York, 1968.
2. **Gordon, A. J. and Ford, R. A.,** *The Chemist's Companion: A Handbook of Practical Data, Techniques, and References*, John Wiley & Sons, New York, 1972.
3. **Brockmann, H. and Schodder, H.,** Aluminum oxide with buffered adsorptive properties for purposes of chromatography, *Ber. Dtsch. Chem. Ges.*, 74, 73, 1941.

STATIONARY AND MOBILE PHASES

The following table provides a comprehensive guide to the selection of thin layer chromatography media and solvents for a given chemical family. Mixed mobile phases are denoted with a slash, /, between components, and where available, the proportions are given. Among the references are several excellent texts,[1,2,3,60] review articles,[4-24] and original research papers and reports.[25-59,61-98] A table of abbreviations follows this section.

REFERENCES

1. **Krebs, K. G., Heusser, D., and Wimmer, H.,** in *Thin Layer Chromatography, A Laboratory Handbook,* Stahl, E., Ed., Springer-Verlag, New York, 1969.
2. **Bobbitt, J. B.,** *Thin Layer Chromatography,* Reinhold, New York, 1963.
3. **Touchstone, J. C.,** *Techniques and Application of Thin Layer Chromatography,* John Wiley & Sons, New York, 1985, 1972.
4. **Pataki, G.,** Paper, thin-layer, and electrochromatography of aminoacids in biological material, *Z. Klin. Chem.,* 2, 129, 1964; *Chem. Abstr.,* 64, 5425c, 1966.
5. **Padley, F. B.,** Thin-layer chromatography of lipids, *Thin-Layer Chromatography Proc. Symp., Rome 1963,* Marini-Bettolo, G. B., Ed., Elsevier, N. Y., 1964, 87.
6. **Honjo, M.,** Thin-layer chromatography of nucleic acid derivatives, *Kagaku No Ryoiki Zokan,* 64, 1, 1964.
7. **Kazumo, T.,** Thin-layer chromatography of bile acids, *Kagaku No Ryoiki Zokan,* 64, 19, 1964.
8. **Nakazawa, Y.,** Thin-layer chromatography of compound lipids, *Kagaku No Ryoiki Zokan,* 64, 31, 1964.
9. **Nishikaze, O.,** Separation and quantitative analysis of adrenocortical hormone and its metabolite (C_{21}) by thin-layer chromatography, *Kagaku No Ryoiki Zokan,* 64, 37, 1964.
10. **Shikita, M., Kakizazi, H., and Tamaoki, B.,** Thin-layer chromatography of radioactive substances, *Kagaku No Ryoiki Zokan,* 64, 45, 1964.
11. **Mo, I. and Hashimoto, Y.,** Method of thin-layer zone electrophoresis, *Kagaku No Ryoiki Zokan,* 64, 61, 1964.
12. **Kinoshita, S.,** Thin-layer chromatography of sugar esters, *Kagaku No Ryoiki Zokan,* 64, 79, 1964.
13. **Okada, M.,** Thin-layer chromatography of cardiotonic glycosides, *Kagaku No Ryoiki Zokan,* 64, 103, 1964.
14. **Omoto, T.,** Thin-layer chromatography of toad toxin, *Kagaku No Ryoiki Zokan,* 64, 115, 1964.
15. **Furuya, C. and Itokawa, H.,** Thin-layer chromatography of triterpenoids, *Kagaku No Ryoiki Zokan,* 64, 123, 1964.
16. **Zenda, H.,** Thin-layer chromatography of aconitine-type alkaloids, *Kagaku No Ryoiki Zokan,* 64, 133, 1964.
17. **Hara, S. and Tanaka, H.,** Thin-layer chromatography of mixed pharmaceutical preparations, *Kagaku No Ryoiki Zokan,* 64, 141, 1964.
18. **Katsui, G.,** Thin-layer chromatography of vitamins, *Kagaku No Ryoiki Zokan,* 64, 157, 1964.
19. **Fujii, S. and Kamikura, M.,** Thin-layer chromatography of pigments, *Kagaku No Ryoiki Zokan,* 64, 173, 1964.
20. **Hosogai, Y.,** Thin-layer chromatography of organic chlorine compounds, *Kagaku No Ryoiki Zokan,* 64, 185, 1964.
21. **Takeuchi, T.,** Thin-layer chromatography of metal complex salts, *Kagaku No Ryoiki Zokan,* 64, 197, 1964.
22. **Yamakawa, H. and Tanigawa, K.,** Thin-layer chromatography of organic metal compounds, *Kagaku No Ryoiki Zokan,* 64, 209, 1964.
23. **Takitani, S. and Kawanabe, K.,** Thin-layer chromatography of inorganic ions (anions), *Kagaku No Ryoiki Zokan,* 64, 221, 1964.
24. **Ibayashi, H.,** Thin-layer chromatography of steroid hormones and its clinical application, *Kagaku No Ryoiki Zokan,* 64, 227, 1964.
25. **Chilingarov, A. O. and Sobchinskaya, N. M.,** Quantitative ultramicroanalysis of monoamine dansyl derivatives in biological material, *Lab. Delo,* 1980, 333; *Chem. Abstr.,* 93, 109910t, 1980.
26. **Heacock, R. A., Nerenberg, C., and Payza, A. N.,** The chemistry of the "aminochromes". I. The preparation and paper chromatography of pure adrenochrome, *Can. J. Chem.,* 36, 853, 1958.
27. **Heacock, R. A. and Powell, W. S.,** Adrenochrome and related compounds, *Progr. Med. Chem.,* 9, 275, 1972.
28. **Heacock, R. A. and Scott, B. D.,** The chemistry of the "aminochromes". IV. Some new aminochromes and their derivatives, *Can, J. Chem.,* 38, 516, 1960.
29. **Heacock, R. A.,** The chemistry of adrenochrome and related compounds, *Chem. Rev.,* 59, 181, 1959.

30. **Suryaraman, M. G. and Cave, W. T.**, Detection of some aliphatic saturated long chain hydrocarbon derivatives by thin-layer chromatography, *Anal. Chim. Acta*, 30, 96, 1964; *Chem. Abstr.*, 60, 7463e, 1964.

31. **Knappe, E., Peteri, D., and Rohdewald, I.**, Thin-layer chromatographic identification of technically important polyhydric alcohols, *Z. Anal. Chem.*, 199, 270, 1964; *Chem. Abstr.*, 60, 7464f, 1964.

32. **Horak, V. and Klein, R. F. X.**, Microscale group test for carbonyl compounds, *J. Chem. Ed.*, 62, 806, 1985.

33. **Jaminet, F.**, Paper microchromatography in phytochemical analysis. Application to Congolian strychnos, *J. Pharm. Belg.*, 8, 339 and 449, 1953; *Chem. Abstr.*, 48, 8482c, 1954.

34. **Neu, R.**, A new color method for determining alkaloids and organic bases with sodium tetraphenylborate, *J. Chromatogr.*, 11, 364, 1963; *Chem. Abstr.*, 59, 12181d, 1963.

35. **Marini-Bettolo, B. G. and Caggiano, E.**, Paper chromatography and electrophoresis of tertiary bases, *Liblice Czech.*, 91, 1961; *Chem. Abstr.*, 60, 838d, 1964.

36. **Knappe, E. and Rohdewald, I.**, Impregnation of chromatographic thin layers with polyesters. III. Thin-layer chromatographic identification of acetoacetic acid amides, *Z. Anal. Chem.*, 208, 195, 1965; *Chem. Abstr.*, 62, 12424f, 1965.

37. **Lane, E. S.**, Thin-layer chromatography of long-chain tertiary amines and related compounds, *J. Chromatogr.*, 18, 426, 1965; *Chem. Abstr.*, 63, 7630f, 1965.

38. **Ashworth, M. R. F. and Bohnstedt, G.**, Reagent for the detection and determination of N-active hydrogen, *Talanta*, 13, 1631, 1966; *Chem. Abstr.*, 66, 25889x, 1967.

39. **Heacock, R. A.**, The aminochromes, in *Advances in Heterocyclic Chemistry*, Katrinsky, A. R., Ed., Academic Press, 1965, 205; *Chem. Abstr.*, 65, 5423d, 1966.

40. **Knappe, E., Peteri, D., and Rohdewald, I.**, Impregnation of chromatographic thin layers with polyesters for the separation and identification of substituted 2-hydroxybenzophenones and other ultraviolet absorbers, *Z. Anal. Chem.*, 197, 364, 1963; *Chem. Abstr.*, 60, 762g, 1964.

41. **Hara, S. and Takeuchi, M.**, Systematic analysis of bile acids and their derivatives by thin-layer chromatography, *J. Chromatogr.*, 11, 565, 1963; *Chem. Abstr.*, 60, 838f, 1964.

42. **Hauck, A.**, Detection of caffeine by paper chromatography, *Dtsche. Z. Gesamte Gerichtl. Med.*, 54, 98, 1963; *Chem. Abstr.*, 60, 838b, 1964.

43. **Knappe, E. and Rohdewald, I.**, Thin-layer chromatography of dicarboxylic acids. IV. Combination of thin-layer chromatographic systems for the identification of individual components in dicarboxylic acid mixtures, *Z. Anal. Chem.*, 210, 183, 1965; *Chem. Abstr.*, 63, 3600f, 1965.

44. **Passera, C., Pedrotti, A., and Ferrari, G.**, Thin-layer chromatography of carboxylic acids and ketoacids of biological interest, *J. Chromatogr.*, 14, 289, 1964; *Chem. Abstr.*, 60, 16191f, 1964.

45. **Knappe, E. and Peteri, D.**, Thin-layer chromatography of dicarboxylic acids. I. Separations in the homologous series oxalic to sebacic acids, *Z. Anal. Chem.*, 188, 184, 1962; *Chem. Abstr.*, 57, 11836a, 1962.

46. **Peteri, D.**, Thin-layer chromatography of dicarboxylic acids. II. Separation of carbocyclic dicarboxylic acids, *Z. Anal. Chem.*, 158, 352, 1962; *Chem. Abstr.*, 57, 11836b, 1962.

47. **Dutta, S. P. and Baruta, A. K.**, Separation of cis- and trans-isomers of α,β-unsaturated acids by thin-layer chromatography, *J. Chromatogr.*, 29, 263, 1967; *Chem. Abstr.*, 67, 96616n, 1967.

48. **Dalmaz, Y. and Peyrin, L.**, Rapid procedure for chromatographic isolation of DOPA, DOPAC, epinephrine, norepinephrine and dopamine from a single urinary sample at endogenous levels, *J. Chromatogr.*, 145, 11, 1978; *Chem. Abstr.*, 88, 59809c, 1978.

49. **Baumgartner, H., Ridl, W., Klein, G., and Preindl, S.**, Improved radioenzymic assay for the determination of catecholamines in plasma, *Clin. Chim. Acta*, 132, 111, 1983; *Chem. Abstr.*, 99, 99459k, 1983.

50. **Hansson, C., Agrup, G., Rorsman, H., Rosengren, A. M., and Rosengren, E.**, Chromatographic separation of catecholic aminoacids and catecholamines on immobilized phenylboronic acid, *J. Chromatogr.*, 161, 352, 1978; *Chem. Abstr.*, 90, 50771d, 1979.

51. **Endo, Y. and Ogura, Y.**, Separation of catecholamines on the phosphocellulose column, *Jpn. J. Pharmacol.*, 23, 491, 1973; *Chem. Abstr.*, 80, 12002s, 1974.

52. **Wada, H., Yamatodani, A., and Seki, T.**, Systematic determination of amino acids, amines and some nucleotides using dansylchloride, *Kagaku No Ryoiki Zokan*, 114, 1, 1976; *Chem. Abstr.*, 87, 1904f, 1977.

53. **Head, R. J., Irvine, R. J., and Kennedy, J. A.**, The use of sodium borate impregnated silica gel plates for the separation of 3-O-methyl catecholamines from their corresponding catecholamines, *J. Chromatogr. Sci.*, 14, 578, 1976; *Chem. Abstr.*, 86, 39601x, 1977.

54. **Adamec, O., Matis, J., and Galvanek, M.**, Fractionation and quantitative determination of urinary 17-hydroxycorticosteroids by thin-layer chromatography on silica gel, *Steroids*, 1, 495, 1963.

55. **Adamec, O., Matis, J., and Galvanek, M.**, Chromatographic separation of corticoids on a thin-layer of silica gel, *Lancet*, 1962-I, 81: *Chem. Abstr.*, 56, 9034d, 1962.

56. **Knappe, E. and Rohdewald, I.**, Thin-layer chromatography of dicarboxylic acids. V. Separation and identification of hydroxy dicarboxylic acids, of di- and tricarboxylic acids of the citrate cycle, and some other dicarboxylic acids of plant origin, *Z. Anal. Chem.*, 211, 49, 1965; *Chem. Abstr.*, 63, 7333c, 1965.

57. **Snegotskii, V. I. and Snegotskaya, V. A.**, Thin-layer chromatography of sulfur compounds, *Zavod. Lab.*, 35, 429, 1969; *Chem. Abstr.*, 71, 23436b, 1969.

58. **Borecky, J., Gasparic, J., and Vecera, M.**, Identification of organic compounds. XXV. Identification and separation of aliphatic C_1—C_{18} alcohols by paper chromatography, *Chem. Listy*, 52, 1283, 1958; *Chem. Abstr.*, 53, 8039h, 1958.

59. **Hörhammer, L., Wagner, H., and Hein, H.**, Thin-layer chromatography of flavonoids on silica gel, *J. Chromatogr.*, 13, 235, 1964; *Chem. Abstr.*, 60, 13856c, 1964.

60. **Mikes, O., Ed.**, *Laboratory Handbook of Chromatographic Methods*, D Van Nostrand, London, 1966.

61. **Wright, J.**, Detection of humectants in tobacco by thin-layer chromatography, *Chem. Ind. (London)*, 1963, p. 1125.

62. **Korte, F. and Vogel, J.**, Thin-layer chromatography of lactones, lactams and thiolactones, *J. Chromatogr.*, 9, 381, 1962; *Chem. Abstr.*, 58, 9609c, 1963.

63. **Heacock, R. A. and Mahon, M. E.**, Paper chromatography of some indole derivatives on acetylated paper, *J. Chromatogr.*, 6, 91, 1961.

64. **Hackman, R. H. and Goldberg, M.**, Microchemical detection of melanins, *Anal. Biochem.*, 41, 279, 1971; *Chem. Abstr.*, 74, 136114a, 1971.

65. **Preussmann, R., Neurath, G., Wulf-Lorentzen, G., Daiber, D., and Hengy, H.**, Color formation and thin-layer chromatography for N-nitrosocompounds, *Z. Anal. Chem.*, 202, 187, 1964.

66. **Preussmann, R., Daiber, D., and Hengy, H.**, Sensitive color reaction for nitrosamines on thin-layer chromatography, *Nature*, 201, 502, 1964; *Chem. Abstr.*, 60, 12663e, 1964.

67. **Hranisavljevic-Jakovljevic, M., Pejkovic-Tadic, I., and Stojiljkovic, A.**, Thin-layer chromatography of isomeric oximes, *J. Chromatogr.*, 12, 70, 1963; *Chem. Abstr.*, 60, 7d, 1964.

68. **Abraham, M. H., Davies, A. G., Llewellyn, D. R., and Thain, E. M.**, The chromatographic analysis of organic peroxides, *Anal. Chim. Acta*, 17, 499, 1957; *Chem. Abstr.*, 53, 120b, 1959.

69. **Seeboth, H.**, Thin-layer chromatography analysis of phenols, *Monatsber. Dtsch. Akad. Wiss, Berlin*, 5, 693, 1963; *Chem. Abstr.*, 61, 2489c, 1964.

70. **Knappe, E. and Rohdewald, I.**, Thin-layer chromatographic identification of simple phenols using the coupling products with Fast Red Salt AL, *Z. Anal. Chem.*, 200, 9, 1964; *Chem. Abstr.*, 60, 9913g, 1964.

71. **Donner, R. and Lohs, K.**, Cobalt chloride in the detection of organic phosphate ester by paper and especially thin-layer chromatography, *J. Chromatogr.*, 17, 349, 1965; *Chem. Abstr.*, 62, 13842d, 1965.

72. **Engel, J. F. and Barney, J. E.**, Chromatographic separation of hydrogenation products of dibenz[a,h]anthracene, *J. Chromatogr.*, 29, 232, 1967; *Chem. Abstr.*, 67, 96617p, 1967.

73. **Kucharczyk, N., Fohl, J., and Vymetal, J.**, Thin-layer chromatography of aromatic hydrocarbons and some heterocyclic compounds, *J. Chromatogr.*, 11, 55, 1963; *Chem. Abstr.*, 59, 9295g, 1963.

74. **Perifoy, P. V., Slaymaker, S. C., and Nager, M.**, Tetracyanoethylene as a color-developing reagent for aromatic hydrocarbons, *Anal. Chem.*, 31, 1740, 1959; *Chem. Abstr.*, 54, 5343e, 1960.

75. **Kodicek, E. and Reddi, K. K.**, Chromatography of nicotinic acid derivatives, *Nature*, 168, 475, 1951; *Chem. Abstr.*, 46, 3601g, 1952.

76. **Heacock, R. A. and Mahon, M. E.**, The color reactions of the hydroxyskatoles, *J. Chrom.*, 17, 338, 1965; *Chem. Abstr.*, 62, 13824g, 1965.

77. **Martin, H. P.**, Reversed phase paper chromatography and detection of steroids of the cholesterol class, *Biochim. Biophys. Acta*, 25, 408, 1957.

78. **Lisboa, B. P.**, Application of thin-layer chromatography to the steroids of the androstane series, *J. Chromatogr.*, 13, 391, 1964; *Chem. Abstr.*, 60, 13890b, 1964.

79. **Lisboa, B. P.**, Separation and characterization of Δ^5-3-hydroxy-C_{19}-steroids by thin-layer chromatography, *J. Chromatogr.*, 19, 333, 1965; *Chem. Abstr.*, 63, 16403h, 1965.

80. **Lisboa, B. P.**, Thin-layer chromatography of Δ^4-3-oxosteroids of the androstane series, *J. Chromatogr.*, 19, 81, 1965; *Chem. Abstr.*, 63, 13619e, 1965.

81. **Lisboa, B. P.**, Thin-layer chromatography of steroids, *J. Pharm. Belg.*, 20, 435, 1965; *Chem. Abstr.*, 65, 570c, 1966.

82. **Partridge, S. M.**, Aniline hydrogen phthalate as a spraying reagent for chromatography of sugars, *Nature*, 164, 443, 1949.

83. **Grossert, J. S. and Langler, R. F.**, A new spray reagent for organosulfur compounds, *J. Chromatogr.*, 97, 83, 1974; *Chem. Abstr.*, 82, 25473n, 1976.

84. **Petranek, J. and Vecera, M.**, Identification of organic compounds. XXIV. Separation and identification of sulfides by paper chromatography, *Chem. Listy*, 52, 1279, 1958; *Chem. Abstr.*, 53, 8039d, 1958.

85. **Bican-Fister, T. and Kajganovic, V.**, Quantitative analysis of sulfonamide mixtures by thin-layer chromatography, *J. Chromatogr.*, 16, 503, 1964; *Chem. Abstr.*, 62, 8943d, 1965.

86. **Bican-Fister, T. and Kajganovic, V.**, Separation and identification of sulfonamides by thin-layer chromatography, *J. Chromatogr.*, 11, 492, 1963; *Chem. Abstr.*, 60, 372f, 1964.

87. **Reisch, J., Bornfleth, H., and Rheinbay, J.**, Thin-layer chromatography of some useful sulfonamides, *Pharm. Ztg., Ver. Apotheker Ztg.*, 107, 920, 1962; *Chem. Abstr.*, 60, 372e, 1964.

88. **Prinzler, H. W., Tauchmann, H., and Tzcharnke, C.,** Thin-layer chromatography separation of organic sulfoxides and dinitrothioethers. Some observations on reproducibility and structural influence. II. Separation of sulfoxide mixtures by one and two-dimensional thin-layer chromatography, *J. Chromatogr.*, 29, 151, 1967; *Chem. Abstr.*, 67, 96615m, 1967.

89. **Wolski, T.,** Color reactions for the detection of sulfoxides, *Chem. Anal. (Warsaw)*, 14, 1319, 1969; *Chem. Abstr.*, 72, 106867q, 1970.

90. **Bergstrom, G. and Lagercrantz, C.,** Diphenylpicrylhydrazyl as a reagent for terpenes and other substances in thin-layer chromatography, *Acta Chem. Scand.*, 18, 560, 1964; *Chem. Abstr.*, 61, 2491h, 1964.

91. **Dietz, W. and Soehring, K.,** Identification of thiobarbituric acids in urine by paper chromatography, *Arch. Pharm.*, 290, 80, 1957; *Chem. Abstr.*, 52, 4736d, 1958.

92. **Curtis, R. F. and Philips, G. T.,** Thin-layer chromatography of thiophene derivatives, *J. Chromatogr.*, 9, 366, 1962; *Chem. Abstr.*, 58, 10705c, 1963.

93. **Salame, M.,** Detection and separation of the most important organophosphorus pesticides by thin-layer chromatography, *J. Chromatogr.*, 16, 476, 1964; *Chem. Abstr.*, 62, 11090b, 1965.

94. **Knappe, E. and Rohdewald, I.,** Thin-layer chromatography of substituted ureas and simple urethanes, *Z. Anal. Chem.*, 217, 110, 1966; *Chem. Abstr.*, 64, 16601g, 1966.

95. **Fishbein, L. and Fawkes, J.,** Detection and thin-layer chromatography of sulfur compounds. I. Sulfoxides, sulfones and sulfides, *J. Chromatogr.*, 22, 323, 1966; *Chem. Abstr.*, 65, 6281e, 1966.

96. **Prinzler, H. W., Pape, D., Tauchmann, H., Teppke, M., and Tzcharnke, C.,** Thin-layer chromatography of organic sulfur compounds, *Ropa Uhlie*, 8, 13, 1966; *Chem. Abstr.*, 65, 9710h, 1966.

97. **Karaulova, E. N., Bobruiskaya, T. S., and Gal'pern, G. D.,** Thin-layer chromatography of sulfoxides, *Zh. Anal. Khim.*, 21, 893, 1966; *Chem. Abstr.*, 65, 16046f, 1966.

98. **Knappe, E., and Yekundi, K. G.,** Impregnation of chromatographic thin layers with polyesters. II. Separation and identification of lower and middle fatty acids via the hydroxamic acid, *Z. Anal. Chem.*, 203, 87, 1964; *Chem. Abstr.*, 61, 5915e, 1964.

TABLE OF ABBREVIATIONS FOR SOLVENTS

Abbreviation	Solvent name	Abbreviation	Solvent name
Ac	Acetone	Foram	Amylformate
Ace	Acetate	HCl	Hydrochloric acid
AcOH	Acetic acid	H_3BO_3	Boric acid
n-AmOH	n-Amyl alcohol	Hex	Hexane
t-AmOH	t-Amyl alcohol	HForm	Formic acid
$AmSO_4$	Ammonium sulfate	MeCl	Methylene chloride
i-BuAc	Isobutylacetate	MeCN	Acetonitrile
BuFor	n-Butylformate	MEK	Methylethylketone
i-BuOH	Isobutanol	MeOH	Methanol
n-BuOH	n-Butanol	NaAc	Sodium acetate
i-Bu$_2$O	Diisobutylether	NH_3	Ammonia
CCl_4	Carbon tetrachloride	Petet	Petroleum ether
C_2HCl_3	Trichloroethene	Ph	Phosphate
$CHCl_3$	Chloroform	PhOH	Phenol
$(CH_2)_6$	Cyclohexane	PrAc	Propylacetate
C_6H_6	Benzene	PrFor	Propylformate
$n\text{-}C_6H_{14}$	n-Hexane	Progl	Propylene glycol
$n\text{-}C_7H_{16}$	n-Heptane	i-PrOH	Isopropanol
$i\text{-}C_8H_{18}$	Isooctane	n-PrOH	n-Propanol
$(ClCH_2)_2$	Dichloroethane	i-Pr$_2$NH	Diisopropylamine
Diox	Dioxane	i-Pr$_2$O	Diisopropylether
DMF	Dimethylformamide	Py	Pyridine
EtFor	Ethylformate	THF	Tetrahydrofuran
EtOAc	Ethylacetate	Tol	Toluene
EtOH	Ethanol	W	Water
Et$_2$NH	Diethylamine	m-X	m-Xylene
Et$_2$O	Diethylether		

Note: Other abbreviations: DEAE, diethyl aminoethyl.

STATIONARY AND MOBILE PHASES

Family	Stationary phase	Mobile phase	Ref.
Adrenaline and derivatives	Alumina (two dimensional)	C_6H_6/EtOAc (60:40)	25
		$CHCl_3$/EtOH/Tol (90:6.5:3.1)	
Adrenochromes	Cellulose	AcOH (2%)/w	26,27
	Whatman #1 (descending)	AcOH (2%)/w	26—29
Alcohols	Silica gel (G-coated)	EtOAc/Hex	30
Alcohols, polyhydric	Alumina or kieselguhr (impreg-nated with polyamide) or silica gel	$CHCl_3$/Tol/HForm or *n*-BuOH/NH_3 or $CHCl_3$	31
Aldehydes, 2,4-dinitro-phenylhydrazones	Silica gel (G-coated)	EtOAc/Hex	30
	Alumina	C_6H_6 or $CHCl_3$ or Et_2O or C_6H_6/Hex	3
	Alumina IB	MeCl or Tol/THF (4:1)	32
	Silica gel	Hex/EtOAc (4:1 or 3:2)	3
	Silica gel IB	MeCl or Tol/THF (4:1)	32
Alkaloids	Alumina	*i*-BuOH/AcOH or *i*-BuOH/NH_3 or *i*-PrOH/AcOH	33
	Alumina	$CHCl_3$ or EtOH or $(CH_2)_6$/$CHCl_3$ (3:7)	3
	Cellulose (impregnated with formamide)	C_6H_6/*n*-C_7H_{16}/$CHCl_3$/ET_2NH (6:5:1:0.02)	3
	Paper (S&S #2043b)	*n*-BuOH/HCl (25%)/w (100:26:39)	34
	Paper electrophoresis	*i*-BuOH/AcOH or *i*-BuOH/NH_3 or *i*-PrOH/HOAc	35
	Silica gel	C_6H_6/EtOH (9:1) or $CHCl_3$/Ac/Et_2NH (5:4:1)	3
Amides	Kieselguhr (adipic acid impregnated)	*i*-Pr_2O/Petet/CCl_4/HForm/w	36
	Silica gel	*i*-Pr_2O/Petet/CCl_4/HForm/w	36
Amines	Alumina	Ac/*n*-C_7H_{16} (1:1)	3
	Alumina G	*i*-BuAc or *i*-BuAc/AcOH	37
	Kieselguhr G	Ac/w (99:1)	3
	Silica gel	EtOH (95%)/NH_3 (25%)(4:1)	3
	Silica gel (aromatic only)		38
Amino acids	Alumina	*n*-BuOH/AcOH/w (3:1:1) or Py/w	3
	Cellulose	*n*-BuOH/AcOH/w (4:1:1)	3
	Cellulose (two-dimensional)	*n*-BuOH/Ac/NH_3/w (10:10:5:2) fol-lowed by *i*-PrOH/HForm/w (20:1:5)	3
	Silica gel	*n*-BuOH/AcOH/w (3 or 4:1:1) or PhOH/w (3:1) or *n*-PrOH/NH_3 (34%) (2:1)	3
Aminochromes	Whatman #1 (acid washed)	w or AcOH/w or MeOH/w or EtOH/w or *n*-BuOH/AcOH/w or *i*-PrOH/w	26,39
Barbiturates	Silica gel	$CHCl_3$/*n*-BuOH/NH_3 (25%) (14:8:1)	3
Benzophenones, hydroxy	Alumina or cellulose or kiesel-guhr (impregnated with adipic acid triethylene glycol polyes-ter) or silica gel	HForm/*m*-X	40
Bile acids	Silica gel	C_6H_6/Et_2O(4:1) or Et_2O/AcOH (99.6:04) or $CHCl_3$/MeOH (9:1)	41
Caffeine	Chromatography paper	*n*-BuOH/NH_3 or *n*-BuOH/HForm	42
Carboxylic acids	Kieselguhr/polyethylene glycol	*i*-Bu_2O/HForm/w (90:7:3)	43
	Polyamide powder	*i*-Pr_2O/Petet/CCl_4/HForm/w (50:20:20:8:1) or MeCN/EtOAc/HForm or BuForm/EtOAc/HForm	43
	Silica gel (G-coated)	EtOH/NH_3/THF	30
	Silica gel ($CaSO_4$ impregnated)	*n*-PrOH/NH_3 or EtOH/$CHCl_3$/NH_3	44
	Silica gel/polyethylene glycol M-1000	*i*-Pr_2O/HForm/w (90:7:3)	45,46
			43

STATIONARY AND MOBILE PHASES (continued)

Family	Stationary phase	Mobile phase	Ref.
Carboxylic acids, unsaturated	Silica gel	CHCl₃/MeOH	47
Catecholamines	Alumina		48
	Boric acid gel (neutral pH)	HCl (0.025 *N*)	48
	Kieselguhr		49
	Phenylboronate	Dilute acids	50
	Phosphocellulose	Ph buffer (pH = 6.2)/EDTA	51
Dansyl derivatives	Alumina (two dimensional)	C₆H₆/EtAc (60:40) or CHCl₃/EtOH/Tol (90:6.5:3.5)	25
	Amberlite IRC50		52
O-Methyl derivatives	Silica gel (sodium borate impregnated)		53
Corticosteroids	Silica gel	EtOH (5%)/MeCl or EtOH/CHCl₃	54,55
Coumarins	Polyamide	MeOH/w (4:1 or 3:2)	3
	Silica gel G	Petet/EtAc (2:1)	3
	Silica gel G (impregnated with NaAc)	Tol/EtFor/HForm (5:4:1)	3
	Silicic acid (starch bound)	EtOAc/Skellysolve B	3
Dicarboxylic acids	Kieselguhr/polyethylene glycol	i-Pr₂O/HForm/w (90:7:3)	43
	Polyamide powder	i-Pr₂O/Petet/CCl₄/HForm/w (50:20:20:8:1) or MeCN/EtAc/HForm (9:1:1) or BuFor/EtAc/HForm (9:1:1)	43
	Polyamide Woelm DC powder	MeCN/PrFor/PrAc/HForm (45:45:10:10) or i-Pr₂O/Petet/CCl₄/HForm/w (50:20:20:8:1) or n-AmOh/CCl₄/HFor (3:2:1)	56
	Silica gel	i-Pr₂O/HForm/w (90:7:3)	43
	Silica gel (G-coated)	EtOH/NH₃/THF	30
Diols (see alcohols, polyhydric)			
Disulfides	Alumina	Hex	57
Disulfides, 3.5-dinitrobenzoates	Whatman #3 (impregnated with 10% paraffin oil in (CH₂)₆)	DMF/MeOH/w or Foram/MeOH/w	58
Flavonoids	Paper	n-BuOH/AcOH/w or EtOAc/w or AcOH/w or C₆H₆/AcOH/w	60
	Polyamide	MeOH/H₂O	3
	Silica gel	C₆H₆/Py/AcOH (36:9:5)	59
		Petet/EtAc (2:1)	3
	Silica gel (impregnated with NaAc)	Tol/EtForm/HForm (5:4:1)	3
	Silicic acid (starch bound)	EtOAc/Skellysolve B	3
Glycerides	Silica gel G	CHCl₃/C₆H₆ (7:3)	3
	Silica gel G (impregnated with silver nitrate)	CHCl₃/AcOH (99.5:0.5)	3
Glycolipids	Silica gel G	n-PrOH/NH₃(12%) (4:1)	3
Glycols, polyethylene	Paper	n-PrOH/EtOAc/w (7:1:2) or n-BuOH/AcOH/w (4:1:5) or t-AmOH/n-PrOH/w (8:2:3) or EtOAc/AcOH/w (9:2:2)	60
	Silica gel	Ac or n-BuOH/AcOH/w	61
Hydroxamates	Silica gel	i-Pr₂O or i-Pr₂O/EtOAc (1:4) or i-Pr₂O/i-C₈H₁₈	62
Hydroxamic acids	Kieselguhr G (impregnated with diethylene glycol or triethylene glycol adipate polyesters)	i-Pr₂O/Petet/CCl₄/HForm/w (50:20:20:8:1)	98

STATIONARY AND MOBILE PHASES (continued)

Family	Stationary phase	Mobile phase	Ref.
Indoles	Acetylated (ascending)	CHCl$_3$/MeOH/w (10:10:6)	63
	Cellulose (thin-layer)	w or HCl (0.005 N) or n-BuOH/ AcOH/w (12:3:5) or C$_6$H$_6$/AcOH/w (125:72:3)	64
α-Ketoacids	Silica gel (CaSO$_4$ impregnated)	EtOH/CHCl$_3$/NH$_3$	44
Ketones, 2,4-dinitrophenyl	Alumina IB	MeCl or Tol/THF (4:1)	32
hydrazones	Silica Gel IB	MeCl or Tol/THF (4:1)	32
Lactams	Silica gel	i-Pr$_2$O or i-Pr$_2$O/EtOAc (1:4) or i-Bu$_2$O/i-C$_8$H$_{18}$	62
Lactones	Silica gel	i-Pr$_2$O or i-Pr$_2$O/EtOAc (1:4) or i-Bu$_2$O/i-C$_8$C$_{18}$	62
Lipids	Alumina	Petet/Et$_2$O (95:5)	3
	Silica gel G	Petet/Et$_2$O/AcOH (90:10:1)	3
	Silicic acid	CHCl$_3$/MeOH/w (80:25:3)	3
Mercaptans (see Thiols)			
Nitrosamines	Silica gel	Hex/Et$_2$O/MeCl	65
	Kieselgel	MeCl/Hex/Et$_2$O (2:3:4); (aliphatic, aromatic); MeCl/Hex/Et$_2$O (5:7:10); (cyclic)	66
Nucleotides	Cellulose	AmSo$_4$(sat'd.)/NaAc(1 M)/i-PrOH (80:18:2)	3
	Cellulose (on DEAE)	HCl (aq.)	3
Oximes	Silica gel G	C$_6$H$_6$/EtOAc or C$_6$H$_6$/MeOH(abs.)	67
Peroxides	Silicone filter paper	w/EtOH/CHCl$_3$	68
Phenols	Alumina	Et$_2$O	3
	Alumina/AcOH	C$_6$H$_6$	3
	Silica gel A	CHCl$_3$/AcOH (5:1) or CHCl$_3$/Ac/ AcOH (10:2:1) or C$_6$H$_6$/AcOH (5:1) or Petet (80°)/CCl$_4$/AcOH (4:6:1) or CHCl$_3$/Ac/Et$_2$NH (4:2:0.2)	69
	Silica gel G	C$_6$H$_6$/Diox/AcOH (90:25:4)	3
	Silica gel/oxalic acid	C$_6$H$_6$	70
	Silica gel/potassium carbonate	MeCl/EtoAc/Et$_2$NH (92:5:3 or 93:5:2)	70
Phosphates, esters	Alumina	Hex/C$_6$H$_6$/MeOH (2:1:1) or Hex/ MeOH/Et$_2$O	71
	Kieselgel	Hex/C$_6$H$_6$/MeOH (2:1:1) or Hex/ MeOH/Et$_2$O	71
Phospholipids	Silica gel G	CHCl$_3$/MeOH/w	3
Polynuclear aromatics	Alumina	CCl$_4$	3
	Alumina	C$_6$H$_6$/(CH$_2$)$_6$ (15:85)	72
	Silica gel	Hex or CH$_3$CHCl$_2$ or C$_2$HCl$_3$ or CCl$_4$	73,74
Polypeptides	Sephadex G-25	w or NH$_3$(0.05 M)	3
	Silica gel G	CHCl$_3$/MeOH (9:1) or CHCl$_3$/Ac (9:1)	3
Pyridines	Whatman #1 (descending)	n-BuOH/w or n-BuOH/w/NH$_3$ or Ac or i-BuOH/w or MEK/AcOH/w	75
Quaternary salts (descending)	Whatman #1	Ac/w or AmSO$_4$/Ph buffer (pH = 6.8)/n-PrOH (2%) or n-PrOH	75
Purines	Silica gel	Ac/CHCl$_3$/n-BuOH/NH$_3$ (25%) (3:3:4:1)	3
Pyrrole, tricarboxylic acid	Silica gel	n-BuOH/EtOH/NH$_3$/w (10:10:1:1)	64
Skatoles, hydroxy	Silica gel G	i-Pr$_2$O or (ClCH$_2$)$_2$/i-Pr$_2$NH (6:1)	76
Steroids	Alumina	CHCl$_3$/EtOH (96:4)	3
	Paper	Petet/Tol/MeOH/w or Petet/C$_6$H$_6$/ MeOH/w	60

STATIONARY AND MOBILE PHASES (continued)

Family	Stationary phase	Mobile phase	Ref.
	Paper (impregnated with kerosene)	n-PrOH/w	77
	Silica gel G	EtOAc/$(CH_2)_6$/EtOH (abs.) or EtOAc/$(CH_2)_6$ or CHCl$_3$/EtOH (abs.) or C$_6$H$_6$/EtOH or n-C$_6$H$_{14}$/ EtOAc or EtOAc/n-C$_6$H$_{14}$/EtOH (abs.)/AcOH or EtOAc/n-C$_6$H$_{14}$/ AcOH	78,79 80,81
Sugars	Cellulose	n-BuOH/Py/w (6:4:3) or EtOAc/Py/ w (2:1:2)	3
	Kieselguhr G (buffered with 0.02 N NaAc)	EtOAc/i-PrOH/w	3
	Silica gel (buffered with H$_3$BO$_3$)	C$_6$H$_6$/AcOH/MeOH (1:1:3)	3
	Silica gel (impregnated with sodium bisulfite)	EtOAc/AcOH/MeOH/w (6:1:5:1:5:1) or n-PrOH/w (85:15) or i-PrOH/ EtOAc/w (7:1:2) or MEK/AcOH/w (6:1:3)	3
	Silica gel G	n-PrOH/conc NH$_3$/w (6:2:1)	3
	Whatman #1 (descending — two-dimensional)	PhOH or n-BuOH/AcOH	82
Sugars, aldoses	Paper	EtOAc/Py/w (2:1:2) or n-n-BuOH/ AcOH/w (4:1:5) or n-BuOH/EtOH/ H$_2$O (5:1:4) or EtOAc/AcOH/w (9:2:2) or EtOAc/AcOH/HForm/w or EtOAc/Py/NaAc (sat'd.)	60
	Whatman #1	PhOH or n-BuOH/AcOH	82
Sugars, carbamates	Silica gel	n-BuOH/H$_3$BO$_3$ (0.03 M) (9:1)	
Sugars, deoxy	Whatman #1	PhOH or n-BuOH/AcOH	82
Sugars, ketoses	Paper	EtOAc/Py/w (2:1:2) or n-BuOH/ AcOH/w (4:1:5) or n-BuOH/EtOH/ H$_2$O (5:1:4) or EtOAc/AcOH/w (9:2:2) or w/PhOH (pH = 5.5)	60
	Whatman #1	PhOH or n-BuOH/AcOH	82
Sulfides	Alumina	Hex	57
	Alumina	CHCl$_3$/MeOH	96
	Silica gel	CCl$_4$ or C$_6$H$_6$	83
	Silica gel DF-5	Ac/C$_6$H$_6$ or Tol/EtOAc	95
Sulfilimines, p-nitrosobenzene sulfonyl	Whatman #4 (impregnated with formamide)	C$_6$H$_6$ or C$_6$H$_6$/$(CH_2)_6$	84
Sulfonamides	Kieselguhr	CHCl$_3$/MeOH (9:1) or CHCl$_3$/ MeOH/NH$_3$	85
	Silica gel	Et$_2$O or CHCl$_3$/MeOH (10:1)	86
	Silica gel (neutral)	n-BuOH/MeOH/Ac/Et$_2$NH (9:1:1:1)	87
	Silica gel (G)	CHCl$_3$/EtOH/n-C$_7$H$_{16}$	3
Sulfones	Alumina	Et$_2$O or Hex/Ac (1:1)	57
	Silica gel DF-5	Ac/C$_6$H$_6$ or Tol/EtOAc	95
Esters	Alumina	Et$_2$O or Hex/Ac (1:1)	57
Hydroxyethyl	Alumina	Hex/w (1:3)	57
Sulfoxides	Alumina	C$_6$H$_6$/Py (20:1) and Diox	88
	Alumina	Ac/CCl$_4$ (1:4)	97
	Silica gel	Ac or EtOAc or CHCl$_3$/Et$_2$O	83
	Silica gel DF-5	Ac/C$_6$H$_6$ or Tol/EtOAc	95
	Whatman #1	PhOH/w (8:3) or n-BuOH/AcOH/w (9:1:2.5)	89
Sulfoxides, hydroxyethyl	Alumina	Et$_2$O or Hex/Ac (1:1) or Hex/Et$_2$O (1:3)	57
Terpenes	Alumina	C$_6$H$_6$ or C$_6$H$_6$/Petet or C$_6$H$_6$/EtOH	3
	Silica gel G	i-Pr$_2$O or i-Pr$_2$O/Ac	3

STATIONARY AND MOBILE PHASES (continued)

Family	Stationary phase	Mobile phase	Ref.
	Silica gel/gypsum	$CHCl_3/C_6H_6$ (1:1)	90
	Silicic acid (starch bound)	$n\text{-}C_6H_{14}$/EtOAc (85:15)	3
Thiobarbiturates	Paper	n-AmOH/n-BuOH/25% NH_3 (2:2:1)	91
Thiolactones	Silica gel	i-Pr_2O or i-Pr_2O/EtOAc (1:4) or i-Bu_2O/i-C_8H_{18}	62
Thiols	Alumina	Hex	57
	Alumina (activated)	AcOH/MeCN (3:1)	96
	Alumina (5% cetane impregnated)	AcOH/MeCN (3:1)	96
	Silica gel	EtOAc or $CHCl_3$	83
Thiophenes	Alumina G	Petet (40—60 °C)	92
	Silica gel	MeOH or $C_6H_6/CHCl_3$ (9:1)	92
Thiophosphate, esters		Petet or $C_6H_6/CHCl_3$ or Ac or EtOH or EtOAc or MeOH	93
Ureas	Acetylated plates	CCl_4/EtOAc/EtOH (100:5:2)	94
	Silica gel	CCl_4/MeCl/EtOAc/HOAc (70:50:15:10)	94
Urethanes (see ureas)			

SPRAY REAGENTS IN THIN LAYER CHROMATOGRAPHY

The following table lists the most popular spray reagents needed to identify organic compounds on chromatographic plates. These reagents have been thoroughly covered in several books,[1-3] catalogs,[4] and reviews.[5-24] Due to the aerosol nature of the spray and the chemical hazards associated with several of those chemicals, the use of a fume hood is highly recommended. The original references of the spray reagents are given in order to provide information about their results with individual compounds.[25-139] A list and description of some complicated protocols follows this section of the chapter. (Note: 1 γ = 1 $\mu g/cm^2$ on a TLC plate.)

REFERENCES

1. **Krebs, K. G., Heusser, D., and Wimmer, H.,** Spray reagents, in *Thin Layer Chromatography, a Laboratory Handbook,* Stahl, E., Ed., Springer-Verlag, New York, 1969, 854.
2. **Bobbitt, J. B.,** Visualization, in *Thin Layer Chromatography,* Van Nostrand Reinhold, New York, 1963, chap. 7.
3. **Touchstone, J. C.,** Visualization procedures, in *Techniques and Application of Thin Layer Chromatography,* John Wiley & Sons, New York, 1985, 172.
4. Kodak Products for Chromatography, a Kodak data service catalog, 1979.
5. **Pataki, G.,** Paper, thin-layer and electrochromatography of amino acids in biological material, *Z. Klin. Chem.,* 2, 129, 1964; *Chem. Abstr.,* 64, 5425c, 1966.
6. **Padley, F. B.,** Thin-layer chromatography of lipids, *Thin-Layer Chromatography Proc. Symp., Rome, 1963,* Marini-BeHob, G. B., Elsevier, New York, 1964, 87.
7. **Honjo, M.,** Thin-layer chromatography of nucleic acid derivatives, *Kagaku No Ryoiki Zokan,* 64, 1, 1964.
8. **Kazumo, T.,** Thin-layer chromatography of bile acids, *Kagaku No Ryoiki Zokan,* 64, 19, 1964.
9. **Nakazawa, Y.,** Thin-layer chromatography of compound lipids, *Kagaku No Ryoiki Zokan,* 64, 31, 1964.
10. **Nishikaze, O.,** Separation and quantitative analysis of adrenocortical hormone and its metabolite (C_{21}) by thin-layer chromatography, *Kagaku No Ryoiki Zokan,* 64, 37, 1964.
11. **Shikita, M., Kazikazi, H., and Tamaoki, B.,** Thin-layer chromatography of radioactive substances, *Kagaku No Ryoiki Zokan,* 64, 45, 1964.
12. **Mo, I. and Hashimoto, Y.,** Method of thin-layer zone electrophoresis, *Kagaku No Ryoiki Zokan,* 64, 61, 1964.
13. **Kinoshita, S.,** Thin-layer chromatography of sugar esters, *Kagaku No Ryoiki Zokan,* 64, 79, 1964.
14. **Okada, M.,** Thin-layer chromatography of cardiotonic glycosides, *Kagaku No Ryoiki Zokan,* 64, 103, 1964.
15. **Omoto, T.,** Thin-layer chromatography of toad toxin, *Kagaku No Ryoiki Zokan,* 64, 115, 1964.
16. **Furnya, C. and Itokawa, H.,** Thin-layer chromatography of triterpenoids, *Kagaku No Ryoiki Zokan,* 64, 123, 1964.
17. **Zenda, H.,** Thin-layer chromatography of aconitine-type alkaloids, *Kagaku No Ryoiki Zokan,* 64, 133, 1964.
18. **Hara, S. and Tanaka, H.,** Thin-layer chromatography of mixed pharmaceutical preparations, *Kagaku No Ryoiki Zokan,* 64, 141, 1964.
19. **Katsui, G.,** Thin-layer chromatography of vitamins, *Kagaku No Ryoiki Zokan,* 64, 157, 1964.
20. **Fujii, S. and Kamikura, M.,** Thin-layer chromatography of pigments, *Kagaku No Ryoiki Zokan,* 64, 173, 1964.
21. **Hosogai, Y.,** Thin-layer chromatography of organic chlorine compounds, *Kagaku No Ryoiki Zokan,* 64, 185, 1964.
22. **Takeuchi, T.,** Thin-layer chromatography of metal complex salts, *Kagaku No Ryoiki Zokan,* 64, 197, 1964.
23. **Yamakawa, H. and Tanigawa, K.,** Thin-layer chromatography of organic metal compounds, *Kagaku No Ryoiki Zokan,* 64, 209, 1964.
24. **Ibayashi, H.,** Thin-layer chromatography of steroid hormones and its clinical application, *Kagaku No Ryoiki Zokan,* 64, 227, 1964.
25. **Beckett, A. H., Beavan, M. A., and Robinson, A. E.,** Paper chromatography: multiple spot formation by sympathomimetic amines in the presence of acids, *J. Pharm. Pharmacol.,* 12, 203T, 1960; *Chem. Abstr.,* 55, 9785c, 1961.
26. **Heacock, R. A. and Scott, B. D.,** The chemistry of the "aminochromes". IV. Some new aminochromes and their derivatives, *Can. J. Chem.,* 38, 516, 1960.

27. **Matthews, J. S.**, Steroids (CCXXIII) color reagent for steroids in thin-layer chromatography, *Biochim. Biophys. Acta*, 69, 163, 1963; *Chem. Abstr.*, 58, 14043d, 1963.

28. **Wasicky, R. and Frehden, O.**, Spot-plate tests in the examination of drugs. I. Aldehyde and amine tests for the recognition of ethereal oils, *Mikrochim. Acta*, 1, 55, 1937; *Chem. Abstr.*, 31, 5944, 1937.

29. **Lane, E. S.**, Thin-layer chromatography of long-chain tertiary amines and related compounds, *J. Chromatogr.*, 189, 426, 1965; *Chem. Abstr.*, 63, 7630f, 1965.

30. **Neu, R.**, A new color method for determining alkaloids and organic bases with sodium tetraphenylborate, *J. Chromatogr.*, 11, 364, 1963; *Chem. Abstr.*, 59, 12181d, 1963.

31. **Zinser, M. and Baumgärtel, C.**, Thin-layer chromatography of ergot alkaloids, *Arch. Pharm.*, 297, 158, 1964; *Chem. Abstr.*, 60, 13095f, 1964.

32. **Ashworth, M. R. F. and Bohnstedt, G.**, Reagent for the detection and determination of N-active hydrogen, *Talanta*, 13, 1631, 1966.

33. **Whittaker, V. P. and Wijesundera, S.**, Separation of esters of choline, *Biochem. J.*, 51, 348, 1952; *Chem. Abstr.*, 46, 7940g, 1952.

34. **Heacock, R. A. and Mahon, M. E.**, The color reactions of the hydroxyskatoles, *J. Chromatogr.*, 17, 338, 1965; *Chem. Abstr.*, 62, 13824g, 1965.

35. **Micheel, F. and Schweppe, H.**, Paper chromatographic separation of hydrophobic compounds with acetylated cellulose paper, *Mikrochim, Acta*, 1954, 53; *Chem. Abstr.*, 48, 4354i, 1954.

36. **Smyth, R. B. and Mckeown, G. G.**, Analysis of arylamines and phenols in oxidation-type hair dyes by paper chromatography, *J. Chromatogr.*, 16, 454, 1964; *Chem. Abstr.*, 62, 8930e, 1963.

37. **Kawerau, E. and Wieland, T.**, Aminoacids chromatograms, *Nature*, 168, 77, 1951; *Chem. Abstr.*, 46, 382h, 1952.

38. **Sturm, A. and Scheja, H. W.**, Separation of phenolic acids by high voltage electrophoresis, *J. Chromatogr.*, 16, 194, 1964; *Chem. Abstr.*, 62, 6788b, 1965.

39. **Feigl, F.**, *Spot Tests in Organic Analysis*, 7th ed., Elsevier, Amsterdam, 1966, 251.

40. **Curzon, G. and Giltrow, J.**, A chromatographic color reagent for a group of aminoacids, *Nature*, 172, 356, 1953.

41. **Heacock, R. A., Nerenberg, C., and Payza, A. N.**, The chemistry of the "aminochromes". I. The preparation and paper chromatography of pure adrenochrome, *Can. J. Chem.*, 36, 853, 1958.

42. **Heacock, R. A.**, The aminochromes, in *Advances in Heterocyclic Chemistry*, Katrinsky, A. R., Ed., Academic Press, New York, 1965. 205; *Chem. Abstr.*, 65, 5432d, 1966.

43. **Wieland, T. and Bauer, L.**, Separation of purines and aminoacids, *Angew. Chem.*, 63, 511, 1951; *Chem. Abstr.*, 46, 1082h, 1952.

44. **Hara, S. and Takeuchi, M.**, Systematic analysis of bile acids and their derivatives by thin layer chromatography, *J. Chromatogr.*, 11, 565, 1963; *Chem. Abstr.*, 60, 838f, 1964.

45. **Anthony, W. L. and Beher, W. T.**, Color detection of bile acids using thin layer chromatography, *J. Chromatogr.*, 13, 570, 1964; *Chem. Abstr.*, 60, 13546c, 1964.

46. **Hauck, A.**, Detection of caffeine by paper chromatography, *Dtsch. Z. Gesamte Gerichtl. Med.*, 54, 98, 1963; *Chem. Abstr.*, 60, 838b, 1964.

47. **Suryaraman, M. G. and Cave, W. T.**, Detection of some aliphatic saturated long chain hydrocarbon derivatives by thin-layer chromatography, *Anal. Chim. Acta*, 30, 96, 1964; *Chem. Abstr.*, 60, 7463e, 1964.

48. **Passera, C., Pedrotti, A., and Ferrari, G.**, Thin-layer chromatography of carboxylic acids and ketoacids of biological interest, *J. Chromatogr.*, 14, 289, 1964; *Chem. Abstr.*, 60, 16191f, 1964.

49. **Grant, D. W.**, Detection of some aromatic acids, *J. Chromatogr.*, 10, 511, 1963; *Chem. Abstr.*, 59, 5772a, 1963.

50. **Roux, D. G.**, Some recent advances in the identification of leucoanthocyanins and the chemistry of condensed tanins, *Nature*, 180, 973, 1957; *Chem. Abstr.*, 52, 5212f, 1958.

51. **Abbott, D. C., Egan, H., and Thompson, J.**, Thin-layer chromatography of organochlorine pesticides, *J. Chromatogr.*, 16, 481, 1964; *Chem. Abstr.*, 62, 11090c, 1965.

52. **Adamec, O., Matis, J., and Galvanek, M.**, Fractionation and quantitative determination of urinary 17-hydroxycorticosteroids by thin layer chromatography on silica gel, *Steroids*, 1, 495, 1963.

53. **French, D., Levine, M. L., Pazur, J. H., and Norberg, E.**, Studies on the Schardinger dextrins. The preparation and solubility characteristics of alpha, beta and gamma dextrins, *J. Am. Chem. Soc.*, 71, 353, 1949.

54. **Knappe, E. and Rohdewald, I.**, Thin-layer chromatography of dicarboxylic acids. V. Separation and identification of hydroxy dicarboxylic acids, of di- and tricarboxylic acids of the citrate cycle, and some other dicarboxylic acids of plant origin, *Z. Anal. Chem.*, 211, 49, 1965; *Chem. Abstr.*, 63, 7333c, 1965.

55. **Wright, J.**, Detection of humectants in tobacco by thin layer chromatography, *Chem. Ind. (London)*, 1963, p. 1125.

56. **Toennies, G. and Kolb, J. J.**, Techniques and reagents for paper chromatography, *Anal. Chem.*, 23, 823, 1951; *Chem. Abstr.*, 45, 8392i, 1951.

57. **Kaufmann, H. P. and Sen Gupta, A. K.**, Terpenes as constituents of the unsaponifiables of fats, *Chem. Ber.*, 97, 2652, 1964; *Chem. Abstr.*, 61, 14723b, 1964.

58. **Gage, T. B., Douglass, C. D., and Wender, S. H.**, Identification of flavonoid compounds by filter paper chromatography, *Anal. Chem.*, 23, 1582, 1951; *Chem. Abstr.*, 46, 2449c, 1952.

59. **Hörhammer, L., Wagner, H., and Hein, K.**, Thin layer chromatography of flavonoids on silica gel, *J. Chromatogr.*, 13, 235, 1964; *Chem. Abstr.*, 60, 13856c, 1964.

60. **Nakamura, H. and Pisano, J. J.**, Specific detection of primary catecholamines and their 3-0-methyl derivatives on thin-layer plates using a fluorigenic reaction with fluorescamine, *J. Chromatogr.*, 154, 51, 1978; *Chem. Abstr.*, 89, 117958x, 1978.

61. **Neu, R.**, Analyses of washing and cleaning agents. XVIII. A new test for polyethylene glycols and their esters, *Chem. Abstr.*, 49, 16475c, 1955; *Chem. Abstr.*, 54, 2665e, 1960.

62. **Korte, F. and Vogel, J.**, Thin-layer chromatography of lactones, lactams and thiolactones, *J. Chromatogr.*, 9, 381, 1962; *Chem. Abstr.*, 58, 9609c, 1963.

63. **Harley-Mason, J. and Archer, A. A. P. G.**, *p*-Dimethylamino-cinnamaldehyde as a spray reagent for indole derivatives on paper chromatograms, *Biochem. J.*, 69, 60, 1958; *Chem. Abstr.*, 52, 18600g, 1958.

64. **Heacock, R. A. and Mahon, M. E.**, Paper chromatography of some indole derivatives on acetylated paper, *J. Chromatogr.*, 6, 91, 1961.

65. **Adams, G. W. M.**, A perchloric acid-naphthoquinone method for the histochemical localization of cholesterol, *Nature*, 192, 331, 1961.

66. **Bennet-Clark, T. A., Tamblah, M. S., and Kefford, N. P.**, Estimation of plant growth substances by partition chromatography, *Nature*, 169, 452, 1951; *Chem. Abstr.*, 46, 6181c, 1952.

67. **Gordon, S. A. and Weber, R. P.**, Estimation of indole acetic acid, *Plant Physiol.*, 26, 192, 1951; *Chem. Abstr.*, 45, 4605c, 1951.

68. **Dickmann, S. R. and Crockett, A. L.**, Reactions of xanthydrol. IV. Determination of tryptophan in blood plasma and proteins, *J. Biol. Chem.*, 220, 957, 1956; *Chem. Abstr.*, 49, 7028h, 1956.

69. **Mangold, H. K., Lamp, B. G., and Schlenk, H.**, Indicators for the paper chromatography of lipides, *J. Am. Chem. Soc.*, 77, 6070, 1953; *Chem. Abstr.*, 50, 5074f, 1956.

70. **Witter, R. F., Marinetti, G. V., Morrison, A., and Heicklin, L.**, Paper chromatography of phospholipides with solvent mixtures of ketones and acetic acid, *Arch. Biochem. Biophys.*, 68, 15, 1957; *Chem. Abstr.*, 51, 12200a, 1957.

71. **Martin, H. P.**, Reversed phase paper chromatography and detection of steroids of the cholesterol class, *Biochim. Biophys. Acta*, 25, 408, 1957.

72. **Preussmann, R., Daiber, D., and Hengy, H.**, Sensitive color reaction for nitrosamines on thin-layer chromatography, *Nature*, 201, 502, 1964; *Chem. Abstr.*, 60, 12663e, 1964.

73. **Preussmann, R., Neurath, G., Wulf-Lorentzen, G., Daiber, D., and Hengy, H.**, Color formation and thin-layer chromatography of N-nitrosocompounds, *Z. Anal. Chem.*, 202, 187, 1964.

74. **Hranisavljevic-Jakovljevic, M., Pejkovic-Tadic, I., and Stojiljkovic, A.**, Thin-layer chromatography of isomeric oximes, *J. Chromatogr.*, 12, 70, 1963; *Chem. Abstr.*, 60, 7d, 1964.

75. **Abraham, M. H., Davies, A. G., Llewellyn, D. R., and Thain, E. M.**, Chromatographic analysis of organic peroxides, *Anal. Chem. Acta*, 17, 499, 1957; *Chem. Abstr.*, 53, 120b, 1959.

76. **Knappe, E. and Peteri, D.**, Thin-layer chromatographic identification of organic peroxides, *Z. Anal. Chem.*, 190, 386, 1962; *Chem. Abstr.*, 58, 5021a, 1963.

77. **Servigne, Y. and Duval, C.**, Paper chromatographic separation of mineral anions containing sulfur, *Compt. Rend. Acad. Sci.*, 245, 1803, 1957; *Chem. Abstr.*, 52, 5207b, 1958.

78. **Lisboa, B. P.**, Characterization of Δ^4-3-oxo-C_{21}-steroids on thin-layer chromatography, *J. Chromatogr.*, 16, 136, 1964; *Chem. Abstr.*, 62, 3409, 1965.

79. **Sherma, J. and Hood, L. V. S.**, Thin-layer solubilization chromatography. I. Phenols, *J. Chromatogr.*, 17, 307, 1965; *Chem. Abstr.*, 62, 13819b, 1965.

80. **Gumprecht, D. L.**, Paper chromatography of some isomeric monosubstituted phenols, *J. Chromatogr.*, 18, 336, 1965; *Chem. Abstr.*, 63, 7630h, 1965.

81. **Barton, G. M.**, α, α'-Dipyridyl as a phenol-detecting reagent, *J. Chromatogr.*, 20, 189, 1965; *Chem. Abstr.*, 64, 2724a, 1966.

82. **Sajid, H.**, Separation of chlorinated cresols and chlorinated xylenols by thin-layer chromatography, *J. Chromatogr.*, 18, 419, 1965; *Chem. Abstr.*, 63, 7630d, 1965.

83. **Seeboth, H.**, Thin-layer chromatography analysis of phenols, *Monatsber. Dtsch. Akad. Wiss. Berlin*, 5, 693, 1963; *Chem. Abstr.*, 61, 2489c, 1964.

84. **Burke, W. J., Potter, A. D., and Parkhurst, R. M.**, Neutral silver nitrate as a reagent in the chromatographic characterization of phenolic compounds, *Anal. Chem.*, 32, 727, 1960; *Chem. Abstr.*, 54, 13990d, 1960.

85. **Perifoy, P. V., Slaymaker, S. C., and Nager, M.**, Tetracyanoethylene as a color-developing reagent for aromatic hydrocarbons, *Anal. Chem.*, 31, 1740, 1959; *Chem. Abstr.*, 54, 5343e, 1960.

86. **Bate-Smith, E. C. and Westall, R. G.**, Chromatographic behavior and chemical structure. I. Naturally occurring phenolic substances, *Biochem. Biophys. Acta*, 4, 427, 1950; *Chem. Abstr.*, 44, 5677a, 1950.

87. **Noirfalise, A. and Grosjean, M. H.**, Detection of phenothiazine derivatives by thin-layer chromatography, *J. Chromatogr.*, 16, 236, 1964; *Chem. Abstr.*, 62, 10295f, 1965.

88. **Schreiber, K., Aurich, O., and Osske, G.**, Solanum alkaloids. XVIII. Thin-layer chromatography of Solanum steroid alkaloids and steroidal sapogenins, *J. Chromatogr.*, 12, 63, 1963; *Chem. Abstr.*, 60, 4442h, 1964.

89. **Clarke, E. G. C.**, Identification of solanine, *Nature*, 181, 1152, 1958; *Chem. Abstr.*, 53, 7298h, 1959.

90. **Donner, R. and Lohs, K.**, Cobalt chloride in the detection of organic phosphate ester by paper and especially thin-layer chromatography, *J. Chromatogr.*, 17, 349, 1965; *Chem. Abstr.*, 62, 13842d, 1965.

91. **Kucharczyk, N., Fohl, J., and Vymetal, J.**, Thin-layer chromatography of aromatic hydrocarbons and some heterocyclic compounds, *J. Chromatogr.*, 11, 55, 1963; *Chem. Abstr.*, 59, 9295g, 1963.

92. **Kodicek, E. and Reddi, K. K.**, Chromatography of nicotinic acid derivatives, *Nature*, 168, 475, 1951; *Chem. Abstr.*, 46, 3601g, 1952.

93. **Hodgson, E., Smith, E., and Guthrie, F. E.**, Two-dimensional thin-layer chromatography of tobacco alkaloids and related compounds, *J. Chromatogr.*, 20, 176, 1965; *Chem. Abstr.*, 64, 3960b, 1966.

94. **Stevens, P. J.**, Thin-layer chromatography of steroids. Specificity of two location reagents, *J. Chromatogr.*, 14, 269, 1964; *Chem. Abstr.*, 61, 2491b, 1964.

95. **Lisboa, B. P.**, Application of thin-layer chromatography to the steroids of the androstane series, *J. Chromatogr.*, 13, 391, 1964; *Chem. Abstr.*, 60, 13890b, 1964.

96. **Lisboa, B. P.**, Separation and characterization of Δ^5-3-hydroxy-C_{19}-steroids by thin-layer chromatography, *J. Chromatogr.*, 19, 333, 1965; *Chem. Abstr.*, 63, 16403h, 1965.

97. **Lisboa, B. P.**, Thin-layer chromatography of Δ^4-3-oxosteroids of the androstane series, *J. Chromatogr.*, 19, 81, 1965; *Chem. Abstr.*, 63, 13619e, 1965.

98. **Neher, R. and Wettstein, A.**, Steroids (CVII) color reactions: corticosteroids in the paper chromatogram, *Helv. Chim. Acta*, 34, 2278, 1951; *Chem. Abstr.*, 46, 3110d, 1952.

99. **Michalec, C.**, Paper chromatography of cholesterol and cholesterol esters, *Naturwissenschaften*, 42, 509, 1955; *Chem. Abstr.*, 51, 5884a, 1957.

100. **Scheidegger, J. J. and Cherbuliez, E.**, Hederacoside A, a heteroside extracted from english ivy, *Helv. Chim. Acta*, 38, 547, 1955; *Chem. Abstr.*, 50, 1685g, 1956.

101. **Richter, E.**, Detection of sterols with naphthoquinone-perchloric acid on silica gel layers, *J. Chromatogr.*, 18, 164, 1965; *Chem. Abstr.*, 63, 7653a, 1965.

102. **Lisboa, B. P.**, Thin-layer chromatography of steroids, *J. Pharm. Belg.*, 20, 435, 1965; *Chem. Abstr.*, 65, 570c, 1966.

103. **Adachi, S.**, Thin-layer chromatography of carbohydrates in the presence of bisulfite, *J. Chromatogr.*, 17, 295, 1965; *Chem. Abstr.*, 62, 13818g, 1965.

104. **Bryson, J. L. and Mitchell, T. J.**, Spraying reagents for the detection of sugar, *Nature*, 167, 864, 1951; *Chem. Abstr.*, 45, 8408b, 1951.

105. **Sattler, L. and Zerban, F. W.**, Limitations of the anthrone test for carbohydrates, *J. Am. Chem. Soc.*, 72, 3814, 1950; *Chem. Abstr.*, 45, 1039b, 1951.

106. **Bacon, J. S. D. and Edelmann, J.**, Carbohydrates of the Jerusalem artichoke and other Compositae, *Biochem. J.*, 48, 114, 1951; *Chem. Abstr.*, 45, 5242b, 1951.

107. **Timell, T. E., Glaudemans, C. P. J., and Currie, A. L.**, Spectrophotometric method for determination of sugars, *Anal. Chem.*, 28, 1916, 1956.

108. **Hay, G. W., Lewis, B. A., and Smith, F.**, Thin-film chromatography in the study of carbohydrates, *J. Chromatogr.*, 11, 479, 1963; *Chem. Abstr.*, 60, 839b, 1964.

109. **Edward, J. T. and Waldron, D. M.**, Detection of deoxy sugars, glycols and methyl pentoses, *J. Chem. Soc.*, 1952, 3631; *Chem. Abstr.*, 47, 1009h, 1953.

110. **Johanson, R.**, New specific reagent for keto-sugars, *Nature*, 172, 956, 1953.

111. **Adachi, S.**, Use of dimedon for the detection of keto sugars by paper chromatography, *Anal. Biochem.*, 9, 224, 1964; *Chem. Abstr.*, 61, 13616g, 1964.

112. **Sattler, L. and Zerban, F. W.**, New spray reagents for paper chromatography of reducing sugars, *Anal. Chem.*, 24, 1862, 1952; *Chem. Abstr.*, 47, 1543d, 1953.

113. **Bailey, R. W. and Bourne, E. J.**, Color reactions given by sugars and diphenylamine-aniline spray reagents on paper chromatograms, *J. Chromatogr.*, 4, 206, 1960; *Chem. Abstr.*, 55, 4251c, 1961.

114. **Buchan, J. L. and Savage, R. J.**, Paper chromatography of starch-conversion products, *Analyst*, 77, 401, 1952; *Chem. Abstr.*, 48, 8568c, 1954.

115. **Schwimmer, S. and Bevenue, A.**, Reagent for differentiation of 1,4- and 1,6-linked glucosaccharides, *Science*, 123, 543, 1956; *Chem. Abstr.*, 50, 8376a, 1956.

116. **Partridge, S. M.**, Aniline hydrogen phthalate as a spraying reagent for chromatography of sugars, *Nature*, 164, 443, 1949.

117. **Grossert, J. S. and Langler, R. F.**, A new spray reagent for organosulfur compounds, *J. Chromatogr.*, 97, 83, 1974; *Chem. Abstr.*, 82, 25473n, 1976.
118. **Snegotskii, V. I. and Snegotskaya, V. A.**, Thin-layer chromatography of sulfur compounds, *Zavod. Lab.*, 35, 429, 1969; *Chem. Abstr.*, 71, 23436b, 1969.
119. **Fishbein, L. and Fawkes, J.**, Detection and thin-layer chromatography of sulfur compounds. I. Sulfoxides, sulfones and sulfides, *J. Chromatogr.*, 22, 323, 1966; *Chem. Abstr.*, 65, 6281e, 1966.
120. **Svoronos, P. D. N.**, On the Synethesis and Characteristics of Sulfonyl Sulfilimines Derived from Aromatic Sulfides, Dissertation, Georgetown University, Washington, D. C., 1980; (Available from University Microfilms, Order No. 8021272.)
121. **Petranek, J. and Vecera, M.**, Identification of organic compounds. XXIV. Separation and identification of sulfides by paper chromatography, *Chem. Listy*, 52, 1279, 1958; *Chem. Abstr.*, 53, 8039d, 1958.
122. **Bican-Fister, T. and Kajganovic, V.**, Quantitative analysis of sulfonamide mixtures by thin-layer chromatography, *J. Chromatogr.*, 16, 503, 1964; *Chem. Abstr.*, 62, 8943d, 1965.
123. **Bratton, A. C. and Marshall, E. K., Jr.**, A new coupling component for sulfanilamide determination, *J. Biol. Chem.*, 128, 537, 1939.
124. **Borecky, J.**, Pinakryptol yellow, reagent for the identification of arenesulfonic acids, *J. Chromatogr.*, 2, 612, 1959; *Chem. Abstr.*, 54, 16255a, 1960.
125. **Pollard, F. H., Nickless, G., and Burton, K. W. C.**, A spraying reagent for anions, *J. Chromatogr.*, 8, 507, 1962; *Chem. Abstr.*, 58, 3873b, 1963.
126. **Coyne, C. M. and Maw, G. A.**, Paper chromatography for aliphatic sulfonates, *J. Chromatogr.*, 14, 552, 1964; *Chem. Abstr.*, 61, 7679d, 1964.
127. **Wolski, T.**, Color reactions for the detection of sulfoxides, *Ann. Chem. (Warsaw)*, 14, 1319, 1969; *Chem. Abstr.*, 72, 106867q, 1970.
128. **Suchomelova, L., Horak, V., and Zyka, J.**, The detection of sulfoxides, *Microchem. J.*, 9, 196, 1965; *Chem. Abstr.*, 63, 9062a, 1965.
129. **Thompson, J. F., Arnold, W. N., and Morris, C. J.**, A sensitive qualitative test for sulfoxides on paper chromatograms, *Nature*, 197, 380, 1963; *Chem. Abstr.*, 58, 7351d, 1963.
130. **Karaulova, E. N., Bobruiskaya, T. S., and Gal'pern, G. D.**, Thin-layer chromatography of sulfoxides, *Zh. Anal. Khim.*, 21, 893, 1966; *Chem. Abstr.*, 65, 16046f, 1966.
131. **Bergstrom, G. and Lagercrantz, C.**, Diphenylpicrylhydrazyl as a reagent for terpenes and other substances in thin-layer chromatography, *Acta Chem. Scand.*, 18, 560, 1964; *Chem. Abstr.*, 61, 2491h, 1964.
132. **Urx, M., Vondrackova, J., Kovarik, L., Horsky, O., and Herold, M.**, Paper chromatography of tetracyclines, *J. Chromatogr.*, 11, 62, 1963; *Chem. Abstr.*, 59, 9736g, 1963.
133. **Dietz, W. and Soehring, K.**, Identification of thiobarbituric acids in urine by paper chromatography, *Arch. Pharm.*, 290, 80, 1957; *Chem. Abstr.*, 52, 4736d, 1958.
134. **Prinzler, H. W., Pape, D., Tauchmann, H., Teppke, M., and Tzcharnke, C.**, Thin-layer chromatography of organic sulfur compound, *Ropa Uhlie*, 8, 13, 1966; *Chem. Abstr.*, 65, 9710h, 1966.
135. **Curtis, R. F. and Philips, G. T.**, Thin-layer chromatography of thiophene derivatives, *J. Chromatogr.*, 9, 366, 1962; *Chem. Abstr.*, 58, 10705c, 1963.
136. **Salame, M.**, Detection and separation of the most important organophosphorus pesticides by thin-layer chromatography, *J. Chromatogr.*, 16, 476, 1964; *Chem. Abstr.*, 62, 11090b, 1965.
137. **Siliprandi, D. and Siliprandi, M.**, Separation and determination of phosphate esters of thiamine, *Biochem. Biophys. Acta*, 14, 52, 1954; *Chem. Abstr.*, 49, 6036f, 1955.
138. **Nuernberg, E.**, Thin-layer chromatography of vitamins, *Dtsch. Apotheker Ztg.*, 101, 268, 1961; *Chem. Abstr.*, 60, 372, 1964.
139. **Mariani, A. and Vicari, C.**, Determination of vitamin D in the presence of interfering substances, *Chem. Abstr.*, 60, 373a, 1964.

SPRAY REAGENTS IN THIN LAYER CHROMATOGRAPHY

Family/functional group	Test	Result	Ref.
Adrenaline (and derivatives)	2,6-Dichloroquinonechloroimide (0.5% in absolute ethanol)	Variety of colors	1
	Potassium ferricyanide (0.6% in 0.5% sodium hydroxide)	Red spots	25
Adrenochromes	4-*N,N*-Dimethylaminocinnamaldehyde	Blue-green to grey-green spots	26
	Ehrlich reagent	Blue-violet to red-violet spots	26
	Zinc acetate (20%)	Blue or yellow fluorescent spots	26
Alcohols	Ceric ammonium sulfate (or nitrate)	Yellow/green spots on red background	1,3
	2,2-Diphenylpicrylhydrazyl (0.06% in chloroform)	Yellow spots on purple background after heating (110 °C, 5 min)	3
	Vanillin (1% in conc sulfuric acid)	Variety of spots after heating (120 °C) — good only for higher alcohols	27
Aldehydes	*o*-Dianisidine (saturated solution in acetic acid)	Variety of spots	28
	2,4-Dinitrophenylhydrazine	Blue colors (saturated ketones); olive-green colors (saturated aldehydes); slow developing colors (unsaturated carbonyl compounds)	1
	2,4-Diphenylpicrylhydrazyl (0.06% in chloroform)	Yellow spots on a purple background after heating (110 °C, 5 min)	3
	Hydrazine sulfate (1% in 1 *N* hydrochloric acid)	Spots under UV (especially after heating)	3
	Tollens reagent	Dark spots	1
Aldehydes, carotenoids	Rhodamine (1—5% in ethanol)	Variety of spots after treatment with strong alkali (sensitivity 0.03 γ)	1
Alkaloids	Bromcresol green (0.05% in ethanol)	Green spots, especially after exposure to ammonia	3
	Chloramine-T (10% aqueous)	Rose spots after exposure to hydrochloric acid and heat	4
	Cobalt(II) thiocyanate	Blue spots on a light-pink background	29
	4-*N-N*-Dimethylaminobenzaldehyde (4% in 1:3 hydrochloric acid/methanol)	Characteristic spots for individual alkaloids	4
	Iodine/potassium iodide (in 2 *N* acetic acid)	Variety of spots	3
	Kalignost test	Orange/red spots fluorescing under long-wave UV	30
	Sonnenschein test	Variety of spots	1
Ergot	4-*N,N*-Dimethylaminobenzaldehyde/sulfuric acid	Blue spots	31
Amides	Chlorine/pyrazolinone/cyanide	Red spots turning blue (detection limit 0.5 μg)	32
	Hydroxylamine/ferric chloride	Variety of spots	33
Amines (all types unless specified)	Alizarin (0.1% in ethanol)	Violet spots on yellow background	3
	Chlorine/pyrazolinone/cyanide	Red spots turning blue (aromatic only)	32
	Cobalt(II) thiocyanate	Blue spots on white/pink background	29
	Diazotization and α-naphthol coupling	Variety of spots (1° aromatic amines only)	1
	Ehrlich reagent	Yellow spots for aromatic amines	34

SPRAY REAGENTS IN THIN LAYER CHROMATOGRAPHY (continued)

Family/functional group	Test	Result	Ref.
	Fast Blue B Salt	Variety of spots (only for amines that can couple)	1
	Glucose/phosphoric acid (4%)	Variety of spots (aromatic amines only), especially after heating	35
	Malonic acid (0.2%)/salicylaldehyde (0.1%; in ethanol)	Yellow spots after heating (120 °C, 15 min)	3,4
	1,2-Naphthoquinone-4-sulfonic acid, sodium salt (0.5% in 1 *N* acetic acid)	Variety of colors after 30 min. (aromatic amines only)	36
	Ninhydrin	Red colors when exposed to ammonium hydroxide	37
	p-Nitroaniline, diazotized	Variety of colored spots	38
	Nitroprusside (2.5%)/acetaldehyde (5%)/sodium carbonate (1%)	Variety of spots (2° aliphatic only)	39
	Picric acid (3% in ethanol)/sodium hydroxide (10%); (5:1)	Orange spots	4
	Potassium iodate (1%)	Variety of spots for phenylethylamines (after heating)	3
	Vanillin-potassium hydroxide	Variety of colors	40
Amino acids	Dehydroascorbic acid (0.1% in 95% *n*-butanol)	Variety of colored spots	3
	2,4-Dinitrofluorobenzene	Variety of spots	1
	Isatin-zinc acetate	Variety of colors	1
	Folin reagent	Variety of colors	1
	Ninhydrin	Red colors when exposed to ammonium hydroxide	36
	Vanillin/potassium hydroxide	Variety of colors	40
Aminoalcohols	Alizarin (0.1% in ethanol)	Violet on yellow background	3
Aminochromes	*p*-*N,N*-Dimethylaminocinnamaldehyde	Variety of colors	41
	Ehrlich reagent	Violet spots	26,41
	Ferric chloride (3%)	Gray-brown spots	41
	p-Nitroaniline, diazotized	Red/brown spots	26,41
	Sodium bisulfite, aqueous	Yellow fluorescence under UV	41,42
Aminosugars	Ninhydrin	Red colors when exposed to ammonium hydroxide	37
Ammonium salts, quaternary	Cobalt(II) thiocyanate	Variety of spots	29
Anhydrides	Hydroxylamine/ferric chloride	Variety of spots	33
Arginine	Sakaguchi reagent	Orange/red spots	1
Azulenes	EP-reagent	Blue spots (room temp) that fade to green/yellow shades and can be regenerated with steam	1
Barbiturates	Cobalt(II) nitrate (2%)/lithium hydroxide (0.5%)	Variety of colors	1
	Cupric sulfate/quinine/pyridine	Variety of colors (white, yellow, violet)	1
	s-Diphenylcarbazone (0.1% in ethanol)	Purple spots	3
	Ferrocyanide/hydrogen peroxide	Yellow/red colors	1
	Fluorescein (0.005% in 0.5 *M* ammonia)	Variety of spots under long- or short-wave UV	43
	Mercurous nitrate (1%)	Variety of spots	1
	Zwikker reagent	Variety of spots	1
Bile acids	Anisaldehyde/sulfuric acid	Variety of spots	4
	Antimony trichloride (in chloroform)	Variety of spots	44

SPRAY REAGENTS IN THIN LAYER CHROMATOGRAPHY (continued)

Family/functional group	Test	Result	Ref.
	Perchloric acid (60%)	Fluorescent spots (long-wave UV) after heating (150 °C, 10 min)	44
	Sulfuric acid	Variety of spots	44,45
Bromides	Fluorescein/hydrogen peroxide	Nonfluorescent spots	1
Caffeine	Chloramine-T	Pink-red spots	1
	Silver nitrate (2% in 10% sulfuric acid)	Carmine-red spots (limit 2γ)	46
Carboxylic acids	Bromcresol Blue (0.5% in 0.2% citric acid)	Yellow spots on blue background	3
	Bromothymol blue (0.2% in ethanol, pH = 7)	Yellow spots upon exposure to ammonia	47
	2,6-Dichlorophenol/indophenol (0.1% in ethanol)	Red spots on blue background after heating	48
	Hydrogen peroxide (0.3%)	Blue fluorescence under long-wave UV	49
Ammonium salts	Schweppe reagent	Dark brown spots	1
Catechins	*p*-Toluenesulfonic acid (20% in chloroform)	Fluorescent spots under long-wave UV	50
Catecholamines	Ethylenediamine (50%)	Spots under short- and long-wave UV after heating (50 °C, 20 min)	1
Chlorides, alkyl	2,6-Dichlorophenol indophenol (0.2%)/silver nitrate (3%) in ethanol	Variety of spots	1
	Silver nitrate (0.5% in ethanol)	Dark spots upon UV irradiation	51
	Silver nitrate/formaldehyde	Dark gray spots	1
	Silver nitrate/hydrogen peroxide	Dark spots	1
Chlorinated insecticides and pesticides	Diphenylamine (0.5%)/zinc chloride (0.5%) in acetone	Variety of colors upon heating (200 °C)	1
	2-Phenoxyethanol (5%) in 0.05% silver nitrate	Variety of spots	4
	Silver nitrate/formaldehyde	Dark gray spots	1
	o-Toluidine (0.5%) in ethanol	Green spots under UV (sensitivity 0.5 μg)	4
Choline derivatives	Dipicrylamine (0.2% in 50% aqueous acetone)	Red spots on yellow background	1
Corticosteroids	Blue tetrazolium (0.05%)/sodium hydroxide (2.5 *M*)	Violet spots (limit 1γ/cm²)	1,52
	2,3,5-Triphenyl-*H*-tetrazolium chloride (2% in 0.5 NaOH)	Red spots after heating (100 °C, 5 min)	1
Coumarins	Benedict reagent	Fluorescent spots under long-wave UV	1
	Potassium hydroxide (5% in methanol)	Variety of spots under long-wave UV	1
Dextrins	Iodine/potassium iodide	Blue-black spots (α-dextrins); brown-yellow spots (β- or γ-dextrins)	53
Dicarboxylic acids	Bromcresol purple (0.04% in basic 50% ethanol, pH = 10)	Yellow spots on blue background	4,54
Diols (1,2-)	Lead tetraacetate (1% in benzene)	White spots after heating (110 °C, 5 min) (limit 2 μg)	55
Disulfides	Iodine (1.3% in ethanol)/sodium azide (3.3% in ethanol)	White spots on brown iodine background	3
	Nitroprusside (sodium)	Red spots	56
Diterpenes	Antimony (III) chloride/acetic acid	Reddish yellow to blue-violet	57

SPRAY REAGENTS IN THIN LAYER CHROMATOGRAPHY (continued)

Family/functional group	Test	Result	Ref.
Esters	Hydroxylamine/ferric chloride	Variety of spots	33
Flavonoids	Aluminum chloride	Yellow fluorescence on long-wave UV	58
	Antimony (III) chloride (10% in chloroform)	Fluorescence on long-wave UV	59
	Benedict's reagent	Fluorescence on long-wave UV (only for *o*-dihydroxy compounds)	
	Lead acetate (basic, 25%)	Fluorescent spots under long-wave UV	59
	p-Toluenesulfonic acid (20% in chloroform)	Fluorescent spots under long-wave UV after heating (100 °C, 10 min)	4,50
Fluorescamines	Perchloric acid (70%)	Blue fluorescent spots	60
Glycols, polyethylene	Quercetin/sodium tetraphenylborate	Orange-red spots	61
Glycolipids	Diphenylamine (5% in ethanol) dissolved in 1:1 hydrochloric acid/acetic acid	Blue-gray spots	1
Glycosides, triterpene	Liebermann-Burchard	Fluorescence under long-wave UV	1
Hydroxamates	Ferric chloride (10% in acetic acid)	Brown spots	62
Hydroxamic acids	Ferric chloride (1—5% in 0.5 *N* hydrochloric acid)	Red spots	1
Imidazoles	*p*-Anisidine/amyl nitrite	Red/brown spots	3
Indoles	Chlorine/pyrazolinone/cyanide	Red spots turning blue after a few minutes (limit 0.5 μg)	32
	Cinnamaldehyde/hydrochloric acid	Red spots	1
	4-*N*,*N*-Dimethylamino-cinnamaldehyde	Variety of colored spots	63
	Ehrlich reagent	Purple for indoles; blue for hydroxy indoles	9,34,64
	Ferric chloride (0.001 *M*) in 5% perchloric acid	Red spots	3
	Naphthoquinone/perchloric acid	Orange spots	65
	Perchloric acid (5%)/ferric chloride (0.001 *M*)	Variety of colored spots	66
	Prochazka reagent	Fluorescent (yellow/orange/green) spots under long-wave UV	1
	Salkowski reagent	Variety of colored spots	67
	van Urk (or Stahl) reagent	Variety of colored spots	1
	Xanthydrol (0.1% in acidified ethanol)	Variety of colored spots after heating (100 °C)	68
Iodides	Sonnenschein test	Variety of spots	1
α-Ketoacids	2,6-Dichlorophenol/indophenol (0.1% in ethanol)	Pink spots upon heating	4,48
	o-Phenylenediamine (0.05% in 10% trichloroacetic acid or 0.2% in 0.1 *N* H_2SO_4/ethanol)	Green fluorescence under long-wave UV after heating (100 °C, 2 min)	1
Ketones	*o*-Dianisidine (saturated solution in acetic acid)	Characteristic spots	28
	2,4-Dinitrophenylhydrazine	Yellow-red spots	3
Lactones	Hydroxylamine/ferric chloride	Variety of colors	33
Lipids	α-Cyclodextrin	Variety of spots (for straight-chain lipids)	53
	2',7'-Dichlorofluorescein (0.2%) in ethanol	Spots under long-wave UV	1,69

SPRAY REAGENTS IN THIN LAYER CHROMATOGRAPHY (continued)

Family/functional group	Test	Result	Ref.
	Fluorescein	Spots after treatment with steam	1
	Rhodamine 6G (1% in acetone)	Spots under long-wave UV	70
	Tungstophosphoric acid (20% in ethanol)	Variety of colored spots after heating	71
Mercaptans (see Thiols)			
Nitrocompounds	4-*N*-*N*-Dimethylaminobenzalde-hyde/stannous chloride/hydro-chloric acid	Yellow spots	3
Nitrosamines	Diphenylamine/palladium chloride	Violet spots after exposure to short-wave UV (limit 0.5 γ)	1,72
	Sulfanilic acid (0.5%)/α-na-phthylamine (0.05%) in 30% acetic acid	Spraying is preceded by short-wave UV irradiation (3 min). Aliphatic nitrosamines yield red/violet spots, while aromatic ones green/blue spots (limit 0.2—0.5 γ)	1,72,73
Oximes	Cupric chloride (0.5%)	Immediate green spots (β-oximes); green-brown spots after 10 min (α-oximes)	74
Peroxides	Ammonium thiocyanate (1.2%)/ferrous sulfate (4%)	Brown-red spots	74
	N-*N*-Dimethyl-*p*-phenylene diammonium dichloride	Purple spots	76
	Ferrous thiocyanate	Red-brown spots	1,75
	Iodide (potassium)/starch	Blue spots	1
Persulfates	Benzidine (0.05% in 1 *N* acetic acid)	Blue spots	77
Phenols	Anisaldehyde/sulfuric acid	Variety of colors	1,78
	p-Anisidine/ammonium vanadate	Variety of spots on pink background	3
	Benzidine, diazotized	Variety of colors	79
	Ceric ammonium nitrate (46% in 2 *M* nitric acid)	Variety of spots	80
	α,α'-Dipyridyl (0.5%)/ferric chloride (0.5%) in ethanol	Variety of spots	4,81
	Emerson	Red-orange to pink spots	1
	Fast Blue B Salt	Variety of spots	1
	Ferric chloride (1—5% in 0.5 *N* HCl)	Blue-greenish spots	1
	Folin-Denis	Variety of spots	82
	Gibbs reagent	Variety of colors	1
	Millon reagent	Variety of colors after heating	1
	Naphthoquinone/perchloric acid	Yellow spots (phenol, catechol); dark-blue spot (resorcinol)	65
	p-Nitroaniline, diazotized	Variety of colored spots	38
	p-Nitrobenzenediazonium fluoroborate	Variety of spots	84
	Silver nitrate (saturated in acetone)	Pink to deep green colors	84
	Stannic chloride (5%) in equal volumes of chloroform/acetic acid	Variety of spots after heating (100 °C, 5 min)	1
	Tetracyanoethylene (10% in benzene)	Variety of colors	85
	Tollen's (or Zaffaroni) reagent	Dark spots	86
	Vanillin (1% in sulfuric acid)	Variety of colors after heating	27
Phenols, chlorinated	Folin-Denis reagent	Variety of spots	82

SPRAY REAGENTS IN THIN LAYER CHROMATOGRAPHY (continued)

Family/functional group	Test	Result	Ref.
Phenothiazines	Ferric chloride (5%)/perchloric acid (20%)/nitric acid (50%) (1:9:10)	Variety of colors	4,87
	Formaldehyde (0.03% in phosphoric acid)	Variety of spots	88,89
	Palladium (II) chloride (0.5% pH <7)	Variety of spots	1
Phosphates, esters	Cobalt(II) chloride (1% in acetone or acetic acid)	Blue spots upon warming the plate at 40 °C	90
Polynuclear aromatics	Formaldehyde (2%) in conc sulfuric acid	Variety of colors	91
	Tetracyanoethylene (10% in benzene)	Variety of colors	85
Purines	Fluorescein (0.005% in 0.5 M ammonia)	Variety of spots under long- or short-wave UV	43
Pyrazolones	Ferric chloride (5%)/acetic acid (2 N) (1:11)	Variety of colors	4
Pyridines	König reagent	Variety of spots (for free α-position pyridines)	92,93
Pyridines, quaternary	König reagent	Blue-white fluorescence under UV	93
Pyrimidines	Fluorescein (0.005% in 0.5 M ammonia)	Variety of spots under long- or short-wave UV	43
Pyrones (α- and γ-)	Neu reagent	Fluorescent spots under long-wave UV	1
Quinine derivatives	Formic acid vapors	Fluorescent blue spots	3
Sapogenins	Komarowsky reagent	Yellow/pink spots	94
	Paraformaldehyde (0.03% in 85% phosphoric acid)	Variety of spots	88
	Zinc chloride (30% in methanol)	Fluorescent spots after heating (105 °C, 1 h) in a moisture-free atmosphere	94
Steroids	Anisaldehyde/sulfuric acid	Variety of colors	95—97
	Antimony (III) chloride (in acetic acid)	Variety of colors	57,96
	Carr-Price	Variety of colors	1
	Chlorosulfonic acid/acetic acid	Fluorescence under long-wave UV	
	Dragendorff	Variety of spots	78,95
	Formaldehyde (0.03% in phosphoric acid)	Variety of spots	88,89
	Hanes and Isherwood	Variety of spots (only for 3-hydroxy-Δ^5-steroids)	96
	Liebermann-Burchard	Fluorescence under long-wave UV	1
	Perchloric acid (20%)	Fluorescent spots (long-wave UV) after heating (150 °C, 10 min)	1,44
	Phosphomolybdic acid	Blue color	96
	Phosphoric acid (50%)	Fluorescent spots after heating (120 °C) (limit 0.005 γ)	95,96,98
	Phosphotungstic acid (10% in ethanol)	Variety of spots	99
	Stannic chloride (5%) in equal volumes of chloroform/acetic acid (1:1)	Variety of spots after heating (100 °C, 5 min)	1,100
	Sulfuric acid	Variety of spots	1
	p-Toluenesulfonic acid (20% in chloroform)	Fluorescent spots under long-wave UV	50
	Trichloroacetic acid (50% aqueous)	Variety of colors	96

SPRAY REAGENTS IN THIN LAYER CHROMATOGRAPHY (continued)

Family/functional group	Test	Result	Ref.
	Zimmermann	Variety of colors	95,96
Sterols	Antimony(III) chloride (50% in acetic acid)	Variety of spots	99
	Bismuth(III) chloride	Fluorescence under long-wave UV	1
	Chlorosulfonic acid/acetic acid	Fluorescence under long-wave UV	1
	Liebermann-Burchard	Fluorescence under long-wave UV	1
	1,2-Naphthoquinone-4-sulfonic acid/perchloric acid	Pink spots that change to blue upon prolonged heating (cholesterol limit 0.03 γ)	65,101
	Phosphoric acid (50%)	Fluorescent spots after heating (120 °C, 15 min)	98,102
	Phosphotungstic acid (10% in ethanol)	Variety of spots	99
	Stannic chloride (5%) in equal volumes of chloroform/acetic acid	Variety of spots after heating (100 °C, 5 min)	1
	Sulfuric acid	Variety of spots	102
Sugars	o-Aminodiphenyl (0.3%)/ortho-phosphoric acid (5%)	Brown spots after heating	103
	Aniline/phosphoric acid	Variety of colors	104
	Anisaldehyde/sulfuric acid	Variety of colors	1,78
	Anthrone test	Yellow spot	105
	Benzidine/trichloroacetic acid	Red-brown/dark spots	106
	Carbazole/sulfuric acid	Violet spots on blue background	103
	Lewis-Smith	Brown spots	107
	Naphthoquinone/perchloric acid	Pink-brown spots (glucose, mannose, lactose, sucrose)	65
	Naphthoresorcinol (0.2% in ethanol)/phosphoric acid (10:1)	Variety of spots (100 °C, 5—10 min)	1
	Naphthoresorcinol (0.1%)/sulfuric acid (10%)	Variety of spots after heating (100 °C, 5—10 min)	1
	Orcinol reagent	Variety of spots	1
	Permanganate, potassium (0.5% in 1 N sodium hydroxide)	Variety of spots after heating (100 °C)	108
	Phenol (3%)/sulfuric acid (5% in ethanol)	Brown spots after heating (100 °C, 10 min)	103
	Silver nitrate (0.2% in methanol)/ammonia (saturated)/sodium methoxide (2% in methanol)	Variety of spots after heating (110 °C, 10 min)	1
	Silver nitrate/sodium hydroxide	Variety of spots	1
	Sulfuric acid	Variety of spots	108
	Thymol (0.5%) in sulfuric acid (5%)	Pink spots after heating (120 °C, 20 min)	103
Sugars, deoxy	Metaperiodate/p-nitroaniline	Fluorescent (long-wave UV) yellow spots	109
Sugars, ketoses	Anthrone	Bright purple (pentoses); orange-yellow (heptoses); blue fluorescence (aldoses)	110
	Dimedone (0.3%)/phosphoric acid (10% in ethanol)	Dark-gray spots (white light); dark pink fluorescing spots (UV) after heating (110 °C, 15 min)	1,111
Sugars, reducing	4-Aminohippuric acid	Fluorescence under long-wave UV	112
	Aniline/diphenylamine/phosphoric acid	Variety of colors	113—115
	Aniline hydrogen phthalate	Variety of colors (limit 1 μg)	116

SPRAY REAGENTS IN THIN LAYER CHROMATOGRAPHY (continued)

Family/functional group	Test	Result	Ref.
	p-Anisidine phthalate	Variety of colors	1
	3,5-Dinitrosalicylic acid (0.5% in 4% sodium hydroxide)	Brown spots (sensitivity 1 μg)	3
Sulfides	Ceric ammonium nitrate (in 2 M HNO₃)	Colorless spots (limit <100 μg/ spot)	117
	Chloranil (1%) in benzene	Yellow-brown spots	119
	2,3-Dichloro-5,6-dicyano-1,4-benzoquinone (2%) in benzene	Purple blue spots changing to orange upon ammonia exposure	119
	Gibbs	Yellow-brown spots changing to blue-orange upon exposure to ammonia	119
	Iodine vapors	Brown spots	118
	Tetracyanoethylene (2%) in benzene	Orange spots	119
	N-2,6-Trichloro *p*-benzoquinoneimine (2%) in ethanol	Brown spots	119
Sulfilimines	Potassium permanganate	Colorless spots	120
p-Nitrobenzenesulfonyl	Tin chloride/4-*N*,*N*-dimethylaminobenzaldehyde	Yellow spots	121
Sulfites	Malachite green oxalate	White spots on blue background	4
Sulfonamides	Chlorine/pyrazolinone/cyanide	Red spots changing to blue	32
	Diazotization and coupling	Variety of spots (limit 0.25 γ)	122,123
	Ehrlich	Variety of colors	124
	Chloranil (1%) in benzene	Pink turning to violet or green after heating	119
	2,3-Dichloro-5,6-dicyano-1,4-benzoquinone (2%) in benzene	Lilac-violet turning to yellow-green upon ammonia exposure	119
	Gibbs	Violet turning to tan upon exposure to ammonia and heat	119
Sulfones	Iodine vapors	Brown spots	118
	Tetracyanoethylene (2%) in benzene	Pink to yellow upon exposure to ammonia and heat	119
Sulfonic acids	Pinacryptol yellow (0.1%)	Yellow-orange spots under long-wave UV	124
	Silver nitrate/fluorescein	Yellow spots under long-wave UV	125,126
Sulfoxides	Acetyl bromide	Yellow-orange spots	127
	Ceric ammonium nitrate (40%) in 2 *M* nitric acid	Brown spots after heating (especially good for α-polychlorosulfoxides); limit 80 μg/spot	117
	Chloranil (1%) in benzene	Yellow-blue spots	119
	2,3-Dichloro-5,6-dicyano-1,4-benzoquinone (2%) in benzene	Orange-crimson spots	119
	Dragendorff	Orange-brown-red spots (limit 30—150 γ)	128
	Gibbs	Yellow turning to brown upon ammonia exposure	168
	Iodide (sodium)/starch	Brown spots (limits 0.01 (μmol/20 μL solution)	129
	Iodine vapors	Brown spots	118,130
	Tetracyanoethylene (2%) in benzene	Yellow or crimson turning to white or tan upon exposure to ammonia	119
	N-2,6-Trichloro-*p*-benzoquinoneimine (2%) in ethanol	Yellow spots	119
Terpenes	Anisaldehyde/sulfuric acid	Variety of colors	1
	Antimony(V) chloride	Variety of colors	1
	Carr-Price	Variety of colors	1

SPRAY REAGENTS IN THIN LAYER CHROMATOGRAPHY (continued)

Family/functional group	Test	Result	Ref.
	Diphenylpicryldrazyl in chloroform	Yellow spots on purple background after heating (110 °C) (limit 1 γ/ 0.5 cm diameter	131
	Phenol (50% in carbon tetrachloride)	Variety of spots upon exposure to bromine vapors	3
	Vanillin (1% in 50% H_3PO_4)	Variety of spots after heating (120 °C, 20 min)	4
Tetracyclines	Ammonium hydroxide	Yellow fluorescence under long-wave UV	132
Thioacids	Silver nitrate/ammonium hydroxide/sodium chloride	Yellow-brown spots	1
Thiobarbiturates	Cupric sulfate (0.5%)/diethylamine (3% in methanol)	Green spots (limit 15 γ)	3,133
Thiolactones	Nitroprusside (sodium), basic	Red spots	62
Thiols (mercaptans)	Ceric ammonium nitrate (in 2 M nitric acid)	Colorless spots on yellow background (limit <100 μg/spot)	117
	Iodine (1.3% in ethanol/ethanol)	White spots in brown iodine background	3
	Nitroprusside (sodium)(3%)	Red spots	134
Thiophenes	Isatin (0.4% in conc sulfuric acid)	Variety of colors	135
Thiophosphates, esters	Ferric chloride/sulfosalicylic acid	White spots on violet background	136
	Palladium(II) chloride (0.5% in acidified water)	Variety of spots	1,136
	Periodic acid (10% in 70% perchloric acid)	Variety of spots	3
Unsaturated compounds	Fluorescein (0.1% in ethanol)/ bromine	Yellow spots on a pink background upon exposure to bromine vapors	1
	Osmium tetroxide vapors	Brown/black spots	3,95
Ureas	p-N,N-Dimethylaminobenzaldehyde (1% in ethanol)	Characteristic spots after exposure to hydrochloric acid	4
Vitamin A	Antimony (V) chloride	Variety of colors	1
	Carr-Price	Variety of colors	1
	Sulfuric (50% in methanol) followed by heating	Blue spots that turn brown	1
Vitamin B1	Dipicrylamine	Characteristic spots	3
	Thiochrome	Variety of spots under long-wave UV	137
Vitamin B6	N,2,6-Trichloro-p-benzoquinoneimine (0.1% in ethanol)	Blue spots after exposure to ammonia	3
Vitamin B_6, acetal	2,6-Dibromo-p-benzoquinone 4-chlorimine (0.4% in methanol)	Characteristic spot	138
Vitamin C	Cacotheline (2% aqueous)	Purple spot after heating (100 °C)	3
	Iodine (0.005%) in starch (0.4%)	White spot on blue background	3
	Methoxynitroaniline/sodium nitrite	Blue spots on orange background	3
Vitamin D	Antimony(V) chloride	Variety of colors	139
	Carr-Price	Variety of colors	1
	Trichloroacetic (1% in chloroform)	Variety of spots after heating (120 °C, 5 min)	1
Vitamin E	2′,7′-Dichlorofluorescein (0.01% in ethanol)	Spots under long-wave UV light	1
	α,α′-Dipyridyl (0.5%)/ferric chloride (0.5% in ethanol)	Variety of colors	1

Note: 1 γ = 1 μg/cm² on a TLC plate.

PROTOCOL FOR REAGENT PREPARATION

The following section gives a summary for the preparation of the major spray reagents listed in the previous section (Spray Reagents in Thin Layer Chromatography). Reference to the original literature is recommended for any reagents not listed here.[1-4]

REFERENCES

1. **Krebs, K. G., Heusser, D., and Wimmer, H.,** *Thin-Layer Chromatography, a Laboratory Handbook,* E. Shahl, Ed., Springer-Verlag, New York, 1969.
2. **Bobbitt, J. B.,** *Thin-Layer Chromatography,* Van Nostrand Reinhold, New York, 1963.
3. **Touchstone, J. C. and Dobbins, M. F,,** *Practice of Thin-Layer Chromatography,* John Wiley & Sons, New York, 1983.
4. **Randerath, K.,** *Thin-Layer Chromatography,* 2nd ed., Academic Press, New York, 1968.

Acetic anhydride-sulfuric acid

See Liebermann-Burchard.

Alizarin

A saturated solution of alizarin in ethanol is sprayed on the moist plate, which is then placed in a chamber containing 25% ammonium hydroxide solution to yield a variety of colors.

Aluminum chloride

A 1% aluminum chloride solution in ethanol is sprayed on the plate, which is then observed under long-wave UV light.

4-Aminoantipyrine-potassium ferricyanide

See Emerson.

4-Aminobiphenyl-phosphoric acid

See Lewis-Smith.

4-Aminohippuric acid

A 0.3% 4-aminohippuric acid solution in ethanol is sprayed on the plate, which is then heated at 140 °C (8 min) and observed under long-wave UV light.

Ammonium hydroxide

The chromatogram is placed in a chamber containing 25% ammonium hydroxide, dried, and then observed under long-wave UV light.

Aniline-diphenylamine-phosphoric acid

An aniline (1 g)/diphenylamine (1 g)/phosphoric acid (5 mL) solution in acetone (50 mL) is sprayed on the plate, which is then heated at 85 °C (10 min), yielding a variety of colors.

Aniline-phosphoric acid

A 20% aniline solution in *n*-butanol, saturated with an aqueous (2 *N*) orthophosphoric acid solution is sprayed on the plate, which is then heated at 105 °C (10 min), yielding a variety of colors.

Aniline phthalate

An aniline (1 g)/*o*-phthalic acid (1.5 g) in *n*-butanol (100 mL) (saturated with water) is sprayed on the plate, which is then heated at 105 °C (10 min), yielding a variety of colors.

Anisaldehyde-sulfuric acid

A 1% anisaldehyde solution in acetic acid (acidified by conc sulfuric acid) is sprayed on the plate, which is then heated at 105 °C to yield a variety of colors.

***p*-Anisidine phthalate**

A 0.1 *M* solution of *p*-anisidine and phthalic acid in ethanol is sprayed on the plate, which is then heated at 100 °C (10 min) to yield a variety of colors.

Anthrone

A 1% anthrone solution in 60% aqueous ethanol solution acidified with 10 ml 60% phosphoric acid is sprayed on the plate, which is then heated at 110 °C (5 min) to yield yellow spots.

Antimony(III) chloride

See Carr-Price.

Antimony(III) chloride-acetic acid

A 20% antimony(III) chloride solution in 75% chloroform-acetic acid solution is sprayed on the plate, which upon heating at 100 °C (5 min) yields a variety of colors.

Antimony(V) chloride

A 20% antimony (V) chloride solution in chloroform or carbon tetrachloride is sprayed on the plate yielding a variety of colors upon heating.

Benedict's

A solution that is 0.1 *M* in cupric sulfate, 1.0 *M* in sodium citrate, and 1.0 *M* in sodium carbonate is sprayed on the plate which is then observed under long-wave UV light.

Benzidine diazotized

A 0.5% benzidine solution in 0.005% hydrochloric acid is mixed with an equal volume of 10% sodium nitrite solution in water; the mixture is sprayed on the plate to yield a variety of colors.

Benzidine-trichloroacetic acid

A 0.5% benzidine in (1:1:8) acetic acid/trichloroacetic acid/ethanol is sprayed on the plate to yield red-brown spots upon heating (110 °C) or exposure to unfiltered UV light (15 min).

Bismuth(III) chloride

A 33% ethanol solution of bismuth(III) chloride is sprayed on the plate, which upon heating (110 °C) yields fluorescent spots under long-wave UV light.

Carbazole-sulfuric acid

A 0.5% carbazole in ethanol/sulfuric acid (95:5) is sprayed on the plate which yields violet spots (on blue background) after heating at 120 °C (10 min).

Carr-Price

A 25% antimony(III) chloride solution in chloroform or carbon tetrachloride is sprayed on the plate which is heated at 100 °C (10 min) to yield a variety of colors.

Ceric ammonium sulfate

A 1% solution of ceric ammonium sulfate in strong acids (phosphoric, nitric) is sprayed on the plate to yield yellow/green spots on a red background, after heating at 105 °C (10 min).

Chloramine-T

A 10% chloramine-T solution is sprayed on the plate, followed by 1 *N* hydrochloric acid. The chromatogram is dried and exposed to 25% ammonium hydroxide and warmed.

Chlorine-pyrazolinone-cyanide

An equal volume mixture of 0.2 *M* 1-phenyl-3-methyl-2-pyrazolin-5-one solution (in pyridine) and 1 *M* aqueous potassium cyanide solution is sprayed on the plate that has been previously exposed to chlorine vapors. The resulting red spots turn blue after a few minutes.

Chlorosulfonic acid-acetic acid

A 35% chlorosulfonic acid solution in acetic acid is sprayed on the plate which is then heated at 130 °C (5 min) to produce fluorescence under long-wave UV.

Cinnamaldehyde-hydrochloric acid

A 5% cinnamaldehyde solution in ethanol (acidified with hydrochloric acid) is sprayed on the plate, which is then placed in a hydrochloric acid chamber to yield red spots.

Cobalt(II) thiocyanate

An ammonium thiocyanate (15%)/cobalt(II) chloride (5%) solution in water is sprayed on the plate, yielding blue spots.

Cupric sulfate-quinine-pyridine

A solution that is 0.4% in cupric sulfate, 0.04% in quinine hydrochloride, and 4% in pyridine in water is sprayed on the plate followed by a 0.5% aqueous potassium permanganate solution. A variety of colors (white, yellow, violet) is detected on the chromatogram.

α-Cyclodextrin

A 30% α-cyclodextrin solution in ethanol is sprayed on the plate, which is further developed in an iodine chamber.

Diazonium

See Fast Blue B salt.

Diazotization and coupling

A 1% sodium nitrite solution (in 1 *M* hydrochloric acid) is sprayed on the plate, followed by a 0.2% α-naphthol solution in 1 *M* potassium hydroxide and drying.

4-*N*,*N*-Dimethylaminobenzaldehyde-sulfuric acid

A 0.125% solution of 4-*N*,*N*-dimethylaminobenzaldehyde in 65% sulfuric acid mixed with 5% ferric chloride (0.05 mL/100 mL solution) is sprayed on the plate giving a variety of spots.

4-*N*,*N*-Dimethylaminocinnamaldehyde

A 0.2% solution of 4-*N*,*N*-dimethylaminocinnamaldehyde in 6 *N* HCl/ethanol (1:4) is sprayed on the plate, which is then heated at 105 °C (5 min) revealing a variety of colored spots. Vapors of aqua regia tend to intensify the spots.

2,4-Dinitrofluorobenzene

A 1% sodium bicarbonate solution in 0.025 *M* sodium hydroxide is sprayed on the plate followed by a 2,4-dinitrofluorobenzene (10%) solution in methanol. Heating the plate in the dark (40 °C, 1 h) and further spraying it with diethyl ether yields a variety of spots.

2,4-Dinitrophenylhydrazine

A 0.4% solution of 2,4-dinitrophenylhydrazine in 2 *N* hydrochloric acid is sprayed on the plate followed by a 0.2% solution of potassium ferricyanide in 2 *N* hydrochloric acid yielding orange/yellow spots.

Dragendorff

A 1.7% aqueous solution of basic bismuth nitrate in weak acids (tartaric, acetic) mixed with an aqueous potassium iodide or barium chloride solution is sprayed on the plate to yield a variety of spots.

Ehrlich

A 1% 4-*N*,*N*-dimethylaminobenzaldehyde solution in ethanol is sprayed on the plate, which is dried and then placed in a hydrochloric acid chamber to yield various spots.

Emerson

A 2% 4-aminotipyrine solution in ethanol is sprayed on the plate, followed by an 8% aqueous potassium ferricyanide solution. The chromatogram is then placed in a chamber containing 25% ammonium hydroxide.

EP

A 0.3% solution of 4-*N*,*N*-dimethylaminobenzaldehyde in acetic acid/phosphoric acid/water (10:1:4) is sprayed on the plate to yield a variety of spots.

Fast Blue B Salt (diazonium)

A 0.5% aqueous solution of Fast Blue B Salt is sprayed on the plate followed by a 0.1 *M* sodium hydroxide.

Ferric chloride-perchloric acid

A solution made out of 5 mL 5% aqueous ferric chloride, 45 mL 20% perchloric acid, and 50 mL 50% nitric acid is sprayed on the plate to yield a variety of spots.

Ferric chloride-sulfosalicylic acids

The plate is first exposed to a bromine atmosphere then sprayed with a 0.1% ethanolic solution of ferric chloride. After air drying (15 min) the chromatogram is sprayed with a 1% ethanolic solution of sulfosalicylic acid to yield a variety of spots.

Ferrocyanide-hydrogen peroxide

Ammonium chloride (0.5 g) is added to a 0.1% potassium ferrocyanide solution in 0.2% hydrochloric acid and the resulting solution is sprayed on the plate, which is then dried (100 °C). The chromatogram is further sprayed with 30% hydrogen peroxide, heated (150 °C, 30 min) and sprayed with 10% potassium carbonate to yield yellow/red spots.

Ferrous thiocyanate

A 2:3 mixture of a 4% aqueous ferrous sulfate and 1.3% acetone solution of ammonium thiocyanate is sprayed on the plate yielding red-brown spots.

Fluorescein-hydrogen peroxide

A 0.1% fluorescein solution in 50% aqueous ethanol is sprayed on the plate followed by a 15% hydrogen peroxide in glacial acetic acid and heating (90 °C, 20 min), yielding nonfluorescent spots.

Folin

A 0.02% sodium 1,2-naphthoquinone-4-sulfonate in 5% sodium carbonate is sprayed on the plate, which is then dried to yield a variety of colors.

Folin-Denis

A tungstomolybdophosphoric acid solution is sprayed on the plate, which is then exposed to ammonia vapors.

Gibbs

A 0.4% methanolic solution of 2,6-dibromoquinonechloroimide is sprayed on the plate followed by a 10% aqueous sodium carbonate yielding a variety of spots.

Glucose-phosphoric acid

A 2% glucose solution in phosphoric acid/water/ethanol/*n*-butanol (1:4:3:3) is sprayed on the plate followed by heating (115 °C, 10 min) to yield a variety of spots.

Hydroxylamine-ferric chloride

A 1:2 mixture of 10% hydroxylammonium chloride/10% potassium hydroxide in aqueous ethanol is sprayed on the plate followed by drying. The chromatogram is then sprayed with an ether solution of ferric chloride in hydrochloric acid to yield a variety of spots.

Iodide (potassium)-starch

A 1% potassium iodide solution in 80% aqueous acetic acid is sprayed on the plate followed by a 1% aqueous starch solution. A pinch of zinc dust is recommended as an addition to the potassium iodide solution.

Iodide (sodium)-starch

A solution made by mixing a 5% starch/0.5% sodium iodide solution with an equal volume of concentrated hydrochloric acid is sprayed on the plate, which is then exposed to dry sodium hydroxide (dessicator) and evacuated (30—60 min) to yield brown spots.

Isatin-zinc acetate

An isatin (1%)/zinc acetate (1.5%) solution in isopropanol acidified with acetic acid is sprayed on the plate which is then heated to yield a variety of spots.

Kalignost

A 1% solution of sodium tetraphenylborate in aqueous butanone is sprayed on the plate, followed by a 0.015% methanolic solution of fischtin or quercetin to yield orange-red spots that fluoresce under long-wave UV.

König

A 2% *p*-aminobenzoic acid in ethanolic hydrochloric acid (0.6 *M*) is sprayed on the plate that has been exposed (1 h) to vapors of cyanogen bromide.

Komarowski

A 2% methanolic solution of *p*-hydroxybenzaldehyde that is 5% in sulfuric acid is sprayed on the plate which is then heated (105 °C, 3 min) to yield yellow or pink spots.

Lewis-Smith

o-Aminobiphenyl (0.3 g dissolved in 100 mL of a 19:1 ethanol/phosphoric acid mixture) is sprayed on the plate, which is then heated at 110 °C (15 min).

Liebermann-Burchard

A freshly prepared mixture of 5 mL acetic anhydride/5 mL conc sulfuric acid in 50 mL cold absolute ethanol is sprayed on the plate, which is heated at 100 °C (10 min) and observed under long-wave UV light.

Malachite green oxalate

A 1% ethenolic potassium hydroxide solution is sprayed, the plate heated (150 °C, 5 min) and further sprayed with a buffered (pH = 7) water/acetone solution of malachite green oxalate to yield white spots on blue background.

Metaperiodate (sodium)-*p*-nitroaniline

A 35% saturated solution of sodium metaperiodate is sprayed on the plate which is let to dry (10 min). The chromatogram is then sprayed with a 0.2% *p*-nitroaniline solution in ethanol/hydrochloric acid (4:1) to yield fluorescing (long-wave UV) yellow spots.

Methoxynitroaniline-sodium nitrite

A 0.02 *M* 4-methoxy-2-nitroaniline solution in 50% aqueous acetic acid/5 *N* sulfuric acid is sprayed on the plate, which is dried and resprayed with 0.2% sodium nitrite to yield blue spots on an orange background.

Millon reagent

A solution of mercury (5 g) in fuming nitric acid (10 mL) diluted with water (10 mL) is sprayed on the plate to yield yellow/orange spots that are intensified by heat (100 °C).

1,2-Naphthoquinone-4-sulfonic acid/perchloric acid

A 0.1% 1,2-napthoquinone-4-sulfonic acid solution in ethanol/perchloric acid/40% formaldehyde/water (20:10:1:9) is sprayed on the plate which is then heated (70 °C) to yield pink spots that turn to blue on prolonged heating.

Neu reagent

A 1% methanolic solution of the β-aminoethylester of diphenylboric acid is sprayed on the plate to yield fluorescent spots under long-wave UV light.

Ninhydrin

A ninhydrin solution (0.3% in acidified *n*-butanol or 0.2% in ethanol) is sprayed on the plate which is then heated (110 °C). The resulting spots are stabilized by spraying with a solution made of 1 mL saturated aqueous cupric nitrate, 0.2 mL 10% nitric acid, and 100 mL 95% ethanol, to yield red spots when exposed to ammonium hydroxide (25%).

p-Nitroaniline, diazotized

A solution made by mixing 0.1% aqueous *p*-nitroaniline/0.2% aqueous sodium nitrite/10% aqueous potassium carbonate (1:1:2) is sprayed on the plate to yield colored spots.

p-Nitroaniline, diazotized (buffered)

A solution of 0.5% *p*-nitroaniline (in 2 *N* hydrochloric acid), 5% aqueous sodium nitrite and 20% aqueous sodium acetate (10:1:30) is sprayed on the plate to yield a variety of colored spots.

Nitroprusside (sodium)

A solution made by mixing sodium nitroprusside (1.5 g), 2 *N* hydrochloric acid (5 mL), methanol (95 mL), and 25% ammonium hydroxide (10 mL) is sprayed on the plate to yield a variety of colors.

Nitroprusside (sodium), basic

A 2% sodium nitroprusside solution in 75% ethanol is sprayed on the plate which has already been treated with 1 *N* sodium hydroxide to yield red spots.

Orcinol

A mixture consisting of 0.6% ethanolic orcinol and 1% ferric chloride in dilute sulfuric acid is sprayed on the plate which is further heated (100 °C, 10 min) to yield characteristic spots.

Prochazka

A 10% formaldehyde solution in 5% hydrochoric acid solution in ethanol is sprayed in the plate which is then heated to yield fluorescent spots (yellow/orange/green) under long-wave UV.

Quercetin-sodium tetraphenylborate

A mixture of quercetin (0.015% in methanol) and sodium tetraphenylborate (1% in *n*-butanol saturated with water) is sprayed on the plate to yield orange/red spots.

Quinaldine

A 1 to 1.5% solution of 3,5-diaminobenzoic acid dihydrochloride in 30% phosphoric acid is sprayed on the plate which is then heated (100 °C, 15 min) to yield fluorescent (green/yellow) spots under long-wave UV or (in case of high concentrations) brown spots in daylight.

Sakaguchi

A 0.1% acetone solution of 8-hydroxyquinoline is sprayed on the plate followed by a 0.2% 0.5 *N* sodium hydroxide solution to yield orange/red spots.

Salkowski

A 0.01 *M* aqueous ferric chloride/35% perchloric acid solution is sprayed on the plate which is then heated (60 °C, 5 min) to yield a variety of colors intensified when exposed to aqua regia.

Schweppe

A mixture of 2% aqueous glucose/2% ethanolic aniline in *n*-butanol is sprayed on the plate which is heated (125 °C, 5 min) to yield a variety of spots.

Silver nitrate-ammonium hydroxide-sodium chloride

A mixture of silver nitrate (0.05 *M*)/ammonium hydroxide (5%) is sprayed on the plate, followed by drying and further spraying with 10% aqueous sodium chloride to yield yellow/brown spots.

Silver nitrate-fluorescein

A mixture of silver nitrate (2%)/sodium-fluorescein (0.2%) in 80% ethanol is sprayed on the plate to yield yellow spots on pink background.

Silver nitrate-formaldehyde

The plate is consecutively sprayed with 0.05 *M* ethanolic silver nitrate, 35% aqueous formaldehyde, 2 *M* potassium hydroxide, and, finally, a solution made of equal volumes of hydrogen peroxide (30%) and nitric acid (65%). Each spraying is preceded by a 30-min drying and at the end, the plate is kept in the dark for 12 h before exposing to sunlight to yield dark grey spots.

Silver nitrate-hydrogen peroxide

A 0.05% silver nitrate solution in water/cellosolve/acetone (1:10:190) (to which a drop of 30% hydrogen peroxide has been added) is sprayed on the plate, which is then treated under unfiltered UV to yield dark spots.

Silver nitrate-sodium hydroxide

A saturated silver nitrate solution is sprayed on the plate followed by a 0.5 *M* aqueous/methanol solution. Subsequent drying (100 °C, 2 min) yields a variety of spots.

Sonnenschein

A 2% ceric sulfate solution in 20% aqueous trichloroacetic acid (that has been acidified with sulfuric acid) is sprayed on the plate. A variety of colors appears upon heating (110 °C, 5 min).

Stahl

See van Urk.

Sulfanilic acid-1-naphthylamine

A mixture of sulfanilic acid/1-naphthylamine in 30% acetic acid is sprayed on the plate to yield a variety (violet/green/blue) of colors.

Thiochrome

A 0.3 *M* aqueous potassium ferricyanide solution that is 15% in sodium hydroxide is sprayed on the plate yielding a variety of spots under long-wave UV.

Tollen's

See Zaffaroni reagent.

Vanillin-potassium hydroxide

A 2% solution of vanillin in *n*-propanol is sprayed on the plate which is heated (100 °C, 10 min) and sprayed again with 1% ethanolic potassium hydroxide. Reheating yields a variety of colors observed under daylight.

van Urk (Stahl)

A 0.5% solution of 4-*N*,*N*-dimethylaminobenzaldehyde in concentrated hydrochloric acid/ethanol (1:1) is sprayed on the preheated plate, which is then subjected to aqua regia vapors to yield a variety of colors.

Zaffaroni (Tollen's)

A mixture of silver nitrate (0.02 *M*)/ammonium hydroxide (5 *M*) is sprayed on the plate which is then heated (105 °C, 10 min) to yield black spots.

Zwikker

A 1% cobaltous nitrate in absolute ethanol is sprayed on the plate, which is dried (at room temperature) and exposed to a wet chamber containing 25% ammonium hydroxide.

IV. Supercritical Fluid Chromatography

SOME USEFUL FLUIDS FOR SUPERCRITICAL FLUID CHROMATOGRAPHY

The following table lists some useful carrier fluids for supercritical fluid chromatography, along with relevant critical properties.[1,2] These fluids have either been used or proposed for use in supercritical fluid chromatography or supercritical extraction. The reader should note that some of these fluids (for example, methanol, benzene, and toluene) will undergo serious chemical degradation under near critical conditions while in contact with stainless steels and other common materials.[3,4]

REFERENCES

1. **Bruno, T. J.**, Instrumental aspects of supercritical fluid chromatography Proc. 2nd Symp. Energy Engineering Sciences (CONF-8404123), Argonne, IL, 1984, p.81.
2. **Bruno, T. J. and Svoronos, P. D. N.**, Basic Tables for Chemical Analysis, NBS Tech. Note 1096, U.S. Department of Commerce, National Bureau of Standards, Washington, D.C., 1986.
3. **Bruno, T. J. and Straty, G. C.**, Thermophysical property measurement on chemically reactive systems — a case study, *J. Res. Natl. Bur. Stand.*, 91(3), 135, 1986.
4. **Straty, G. C., Ball, M. J., and Bruno, T. J.**, Experimental determination of the PVT surface for benzene, *J. Chem. Eng. Data*, 32, 163, 1987.

SOME USEFUL FLUIDS FOR SUPERCRITICAL FLUID CHROMATOGRAPHY

Fluid	T_c(°C)	ρ_c (g/mL)	P_c (MPa)
Carbon dioxide	31.413	0.460	7.4
Ammonia	132.5	0.235	11.4
Nitrous oxide	36.5	0.450	7.3
Sulfur dioxide	157.8	0.520	7.9
Water	374.1	0.40	22.1
Methanol	239.5	0.272	8.0
Isopropanol	235.3	0.273	4.8
n-Pentane	196.6	0.232	3.4
n-Hexane	234.2	0.234	3.0
Dichlorofluoromethane	178.5	0.522	5.2
Trichlorofluoromethane	198	0.554	4.4
Chlorotrifluoromethane	28.9	0.578	3.9
Dichlorotetrafluoroethane	145.7	0.582	3.6
Dichlorodifluoromethane	111.5	0.558	4.0
Benzene	288.9	0.304	4.9
Xenon	16.6	1.155	5.9
Toluene	320.8	0.29	4.2

P-ρ-T TABLE FOR CARBON DIOXIDE

The following table provides a numerical listing of the P-ρ-T surface for carbon dioxide in the region of interest for supercritical fluid chromatography. These data were calculated using an empirical equation of state (the Schmidt-Wagner equation),[1-3] the parameters of which were determined from a fit of experimental P-V-T measurements.[4-9] Note that the pressures are tabulated in bars for convenience. The appropriate SI unit of pressure is the megapascal (1 bar = 0.1 MPa).

REFERENCES

1. **Reister, R., Bruno, T. J. and Martire, D. E.**, Generalized treatment of spatial and temporal column parameters applicable to gas, liquid, and supercritical fluid chromatography: applications to supercritical CO_2, *J. Chromatogr.*, in press.
2. **Schmidt, R. and Wagner, W.**, A new form of the equation of state for pure substances and its application to oxygen, *Fluid Phase Equilibria*, 19, 175, 1985.
3. **Ely, J. F.**, National Bureau of Standards, Boulder, CO, Coefficients for Carbon Dioxide, private communication, 1987.
4. **Reid, R. C., Prausnitz, J. M., and Sherwood, T. K.**, *The Properties of Gases and Liquids*, 3rd ed., McGraw-Hill, New York, 1977.
5. **Prausnitz, J. M.**, *Molecular Thermodynamics of Fluid Phase Equilibria*, Prentice-Hall, Englewood Cliffs, N.J., 1969.
6. **Chao, K. C. and Greenkorn, R. A.**, *The Thermodynamics of Fluids*, Marcel Dekker, New York, 1975.
7. **Jacobson, R. T. and Stewart, R. B.**, Thermodynamic properties of nitrogen including liquid and vapor phases from 63 K to 2000 K with pressures to 10,000 bar, *J. Phys. Chem. Ref. Data*, 2(4), 757, 1973.
8. **Ely, J. F.**, Proc. 63rd Gas Processors Assn. Annu. Conv., 1984, 9.
9. **Angus, S., Armstrong, B., and deReuck, K. M.**, *Carbon Dioxide, International Thermodynamic Tables of the Fluid State*, Pergamon Press, Oxford, 1976.

P-ρ-T TABLE FOR CARBON DIOXIDE

Temperature = 308.15 K

Density (ρ; mol/L)	Pressure (P; bar)	Density (ρ; mol/L)	Pressure (P; bar)	Density (ρ; mol/L)	Pressure (P; bar)
0.20	5.009	2.30	44.938	4.40	66.343
0.30	7.429	2.40	46.323	4.50	67.015
0.40	9.792	2.50	47.668	4.60	67.661
0.50	12.100	2.60	48.971	4.70	68.282
0.60	14.353	2.70	50.235	4.80	68.878
0.70	16.552	2.80	51.460	4.90	69.450
0.80	18.698	2.90	52.646	5.00	69.998
0.90	20.791	3.00	53.795	5.10	70.523
1.00	22.832	3.10	54.907	5.20	71.027
1.10	24.821	3.20	55.983	5.30	71.509
1.20	26.760	3.30	57.024	5.40	71.970
1.30	28.648	3.40	58.030	5.50	72.410
1.40	30.487	3.50	59.002	5.60	72.831
1.50	32.278	3.60	59.941	5.70	73.233
1.60	34.020	3.70	60.847	5.80	73.617
1.70	35.716	3.80	61.722	5.90	73.983
1.80	37.365	3.90	62.565	6.00	74.332
1.90	38.968	4.00	63.379	6.10	74.665
2.00	40.526	4.10	64.162	6.20	74.981
2.10	42.040	4.20	64.917	6.30	75.282
2.20	43.510	4.30	65.644	6.40	75.569

P-ρ-T TABLE FOR CARBON DIOXIDE

Temperature = 308.15 K (continued)

Density (ρ; mol/L)	Pressure (P; bar)	Density (ρ; mol/L)	Pressure (P; bar)	Density (ρ; mol/L)	Pressure (P; bar)
6.50	75.841	11.90	81.574	17.30	117.040
6.60	76.100	12.00	81.663	17.40	119.096
6.70	76.345	12.10	81.756	17.50	121.248
6.80	76.578	12.20	81.854	17.60	123.498
6.90	76.800	12.30	81.957	17.70	125.850
7.00	77.010	12.40	82.065	17.80	128.308
7.10	77.209	12.50	82.180	17.90	130.874
7.20	77.398	12.60	82.301	18.00	133.554
7.30	77.577	12.70	82.430	18.10	136.350
7.40	77.746	12.80	82.567	18.20	139.267
7.50	77.907	12.90	82.713	18.30	142.309
7.60	78.060	13.00	82.868	18.40	145.478
7.70	78.205	13.10	83.033	18.50	148.781
7.80	78.342	13.20	83.210	18.60	152.221
7.90	78.473	13.30	83.399	18.70	155.802
8.00	78.597	13.40	83.601	18.80	159.530
8.10	78.515	13.50	83.817	18.90	163.407
8.20	78.828	13.60	84.047	19.00	167.440
8.30	78.935	13.70	84.294	19.10	171.633
8.40	79.037	13.80	84.558	19.20	175.991
8.50	79.135	13.90	84.841	19.30	180.519
8.60	79.228	14.00	85.143	19.40	185.222
8.70	79.318	14.10	85.466	19.50	190.105
8.80	79.405	14.20	85.811	19.60	195.173
8.90	79.488	14.30	86.179	19.70	200.432
9.00	79.568	14.40	86.573	19.80	205.887
9.10	79.646	14.50	86.993	19.90	211.544
9.20	79.721	14.60	87.441	20.00	217.408
9.30	79.794	14.70	87.919	20.10	223.485
9.40	79.865	14.80	88.427	20.20	229.781
9.50	79.935	14.90	88.969	20.30	236.302
9.60	80.003	15.00	89.545	20.40	243.053
9.70	80.069	15.10	90.158	20.50	250.040
9.80	80.135	15.20	90.808	20.60	257.271
9.90	80.200	15.30	91.499	20.70	264.750
10.00	80.264	15.40	92.231	20.80	272.484
10.10	80.328	15.50	93.007	20.90	280.480
10.20	80.391	15.60	93.829	21.00	288.743
10.30	80.454	15.70	94.700	21.10	297.280
10.40	80.517	15.80	95.620	21.20	306.098
10.50	80.580	15.90	96.592	21.30	315.203
10.60	80.643	16.00	97.619	21.40	324.601
10.70	80.706	16.10	98.703	21.50	334.300
10.80	80.771	16.20	99.845	21.60	344.307
10.90	80.835	16.30	101.049	21.70	354.627
11.00	80.901	16.40	102.317	21.80	365.268
11.10	80.968	16.50	103.652	21.90	376.236
11.20	81.037	16.60	105.055	22.00	387.540
11.30	81.107	16.70	106.530	22.10	399.185
11.40	81.178	16.80	108.079	22.20	411.179
11.50	81.252	16.90	109.706	22.30	423.529
11.60	81.328	17.00	111.412	22.40	436.242
11.70	81.407	17.10	113.201	22.50	449.326
11.80	81.489	17.20	115.076	22.60	462.787

P-ρ-T TABLE FOR CARBON DIOXIDE

Temperature = 308.15 K (continued)

Density (ρ; mol/L)	Pressure (P; bar)	Density (ρ; mol/L)	Pressure (P; bar)	Density (ρ; mol/L)	Pressure (P; bar)
22.70	476.634	23.50	602.197	24.30	756.952
22.80	490.874	23.60	619.859	24.40	778.564
22.90	505.514	23.70	637.984	24.50	800.709
23.00	520.562	23.80	656.582	24.60	823.398
23.10	536.026	23.90	675.660	24.70	846.639
23.20	551.914	24.00	695.227	24.80	870.441
23.30	568.233	24.10	715.292	24.90	894.816
23.40	584.991	24.20	735.864	25.00	919.772

Temperature = 313.15 K

Density (ρ; mol/L)	Pressure (P; bar)	Density (ρ; mol/L)	Pressure (P; bar)	Density (ρ; mol/L)	Pressure (P; bar)
0.20	5.095	4.40	69.260	8.60	86.053
0.30	7.559	4.50	70.019	8.70	86.233
0.40	9.968	4.60	70.754	8.80	86.409
0.50	12.323	4.70	71.463	8.90	86.581
0.60	14.625	4.80	72.148	9.00	86.750
0.70	16.874	4.90	72.180	9.10	86.917
0.80	19.070	5.00	73.449	9.20	87.080
0.90	21.215	5.10	74.006	9.30	87.241
1.00	23.309	5.20	74.661	9.40	87.401
1.10	25.352	5.30	75.235	9.50	87.558
1.20	27.346	5.40	75.788	9.60	87.714
1.30	29.291	5.50	76.322	9.70	87.869
1.40	31.188	5.60	76.836	9.80	88.023
1.50	33.038	5.70	77.332	9.90	88.176
1.60	34.840	5.80	77.810	10.00	88.329
1.70	36.597	5.90	78.271	10.10	88.481
1.80	38.308	6.00	78.714	10.20	88.634
1.90	39.975	6.10	79.142	10.30	88.787
2.00	41.598	6.20	79.553	10.40	88.940
2.10	43.177	6.30	79.950	10.50	89.095
2.20	44.714	6.40	80.331	10.60	89.250
2.30	46.209	6.50	80.699	10.70	89.407
2.40	47.664	6.60	81.053	10.80	89.566
2.50	49.078	6.70	81.395	10.90	89.727
2.60	50.453	6.80	81.723	11.00	89.890
2.70	51.788	6.90	82.040	11.10	90.056
2.80	53.086	7.00	82.345	11.20	90.224
2.90	54.346	7.10	82.640	11.30	90.397
3.00	55.570	7.20	82.923	11.40	90.572
3.10	56.758	7.30	83.197	11.50	90.753
3.20	57.911	7.40	83.462	11.60	90.937
3.30	59.029	7.50	83.717	11.70	91.127
3.40	60.114	7.60	83.964	11.80	91.322
3.50	61.165	7.70	84.202	11.90	91.523
3.60	62.185	7.80	84.433	12.00	91.730
3.70	63.172	7.90	84.657	12.10	91.944
3.80	64.129	8.00	84.874	12.20	92.166
3.90	65.056	8.10	85.084	12.30	92.396
4.00	65.953	8.20	85.289	12.40	92.635
4.10	66.821	8.30	85.487	12.50	92.883
4.20	67.661	8.40	85.681	12.60	93.141
4.30	68.474	8.50	85.869	12.70	93.410

P-ρ-T TABLE FOR CARBON DIOXIDE

Temperature = 313.15 K (continued)

Density (ρ; mol/L)	Pressure (P; bar)	Density (ρ; mol/L)	Pressure (P; bar)	Density (ρ; mol/L)	Pressure (P; bar)
12.80	93.691	16.90	130.049	21.00	325.848
12.90	93.984	17.00	132.065	21.10	334.899
13.00	94.290	17.10	134.168	21.20	344.235
13.10	94.610	17.20	136.362	21.30	353.863
13.20	94.945	17.30	138.649	21.40	363.790
13.30	95.296	17.40	141.033	21.50	374.021
13.40	95.664	17.50	143.517	21.60	384.564
13.50	96.049	17.60	146.104	21.70	395.426
13.60	96.453	17.70	148.798	21.80	406.613
13.70	96.878	17.80	151.603	21.90	418.132
13.80	97.323	17.90	154.522	22.00	429.990
13.90	97.791	18.00	157.558	21.10	442.195
14.00	98.283	18.10	160.717	22.20	454.753
14.10	98.799	18.20	164.001	22.30	467.672
14.20	99.341	18.30	167.414	22.40	480.958
14.30	99.911	18.40	170.962	22.50	494.620
14.40	100.510	18.50	174.647	22.60	508.664
14.50	101.139	18.60	178.474	22.70	523.099
14.60	101.800	18.70	182.449	22.80	537.931
14.70	102.495	18.80	186.574	22.90	553.168
14.80	103.224	18.90	190.855	23.00	568.818
14.90	103.991	19.00	195.297	23.10	584.899
15.00	104.796	19.10	199.904	23.20	601.389
15.10	105.640	19.20	204.681	23.30	618.325
15.20	106.527	19.30	209.633	23.40	635.705
15.30	107.458	19.40	214.766	23.50	653.539
15.40	108.435	19.50	220.084	23.60	671.833
15.50	109.459	19.60	225.592	23.70	690.597
15.60	110.533	19.70	231.296	23.80	709.839
15.70	111.659	19.80	237.201	23.90	729.567
15.80	112.839	19.90	243.314	24.00	749.790
15.90	114.075	20.00	249.638	24.10	770.517
16.00	115.369	20.10	256.181	24.20	791.756
16.10	116.725	20.20	262.947	24.30	813.518
16.20	118.143	20.30	269.943	24.40	835.810
16.30	119.627	20.40	277.175	24.50	858.643
16.40	121.179	20.50	284.648	24.60	882.025
16.50	122.801	20.60	292.368	24.70	905.967
16.60	124.496	20.70	300.342	24.80	930.477
16.70	126.268	20.80	308.576	24.90	955.567
16.80	128.118	20.90	317.976	25.00	981.245

Temperature = 318.15 K

Density (ρ; mol/L)	Pressure (P; bar)	Density (ρ; mol/L)	Pressure (P; bar)	Density (ρ; mol/L)	Pressure (P; bar)
0.20	5.180	1.20	27.930	2.30	47.470
0.30	7.689	1.30	29.931	2.40	48.992
0.40	10.144	1.40	31.885	2.50	50.475
0.50	12.546	1.50	33.793	2.60	51.919
0.60	14.896	1.60	35.655	2.70	53.326
0.70	17.194	1.70	37.472	2.80	54.695
0.80	19.440	1.80	39.245	2.90	56.028
0.90	21.637	1.90	40.974	3.00	57.325
1.00	23.783	2.10	44.305	3.10	58.588
1.10	25.881	2.20	45.908	3.20	59.816

P-ρ-T TABLE FOR CARBON DIOXIDE

Temperature = 318.15 K (continued)

Density (ρ; mol/L)	Pressure (P; bar)	Density (ρ; mol/L)	Pressure (P; bar)	Density (ρ; mol/L)	Pressure (P; bar)
3.30	61.110	8.90	93.596	14.30	113.980
3.40	62.172	9.00	93.859	14.40	114.783
3.50	63.301	9.10	94.118	14.50	115.619
3.60	64.399	9.20	94.375	14.60	116.492
3.70	65.466	9.30	94.631	14.70	117.401
3.80	66.503	9.40	94.884	14.80	118.348
3.90	67.511	9.50	95.136	14.90	119.336
4.00	68.490	9.60	95.386	15.00	120.366
4.10	69.441	9.70	95.636	15.10	121.440
4.20	70.364	9.80	95.886	15.20	122.559
4.30	71.261	9.90	96.435	15.30	123.725
4.40	72.131	10.00	96.385	15.40	124.942
4.50	72.976	10.10	96.635	15.50	126.209
4.60	73.797	10.20	96.886	15.60	127.530
4.70	74.593	10.30	97.138	15.70	128.907
4.80	75.365	10.40	97.392	15.80	130.341
4.90	76.115	10.50	97.648	15.90	131.836
5.00	76.842	10.60	97.906	16.00	133.392
5.10	77.547	10.70	98.167	16.10	135.014
5.20	78.231	10.80	98.431	16.20	136.702
5.30	78.895	10.90	98.669	16.30	138.461
5.50	80.163	11.00	98.970	16.40	140.291
5.60	80.769	11.10	99.246	16.50	142.196
5.70	81.356	11.20	99.527	16.60	144.178
5.80	81.926	11.30	99.813	16.70	146.241
5.90	82.479	11.40	100.104	16.80	148.387
6.00	83.015	11.50	100.402	16.90	150.618
6.10	83.535	11.60	100.707	17.00	152.938
6.20	84.039	11.70	101.018	17.10	155.351
6.30	84.529	11.80	101.338	17.20	157.858
6.40	85.004	11.90	101.666	17.30	160.463
6.50	85.466	12.00	102.003	17.40	163.170
6.60	85.914	12.10	102.349	17.50	165.982
6.70	86.349	12.20	102.705	17.60	168.902
6.80	86.771	12.30	103.073	17.70	171.933
6.90	87.182	12.40	103.452	17.80	175.080
7.00	87.582	12.50	103.843	17.90	178.346
7.10	87.970	12.60	104.247	18.00	181.735
7.20	88.348	12.70	104.665	18.10	185.251
7.30	88.716	12.80	105.097	18.20	188.897
7.40	89.074	12.90	105.545	18.30	192.678
7.50	89.424	13.00	106.009	18.40	196.598
7.60	89.764	13.10	106.490	18.50	200.661
7.70	90.097	13.20	106.990	18.60	204.871
7.80	90.422	13.30	107.508	18.70	209.233
7.90	90.739	13.40	108.046	18.80	213.751
8.00	91.409	13.50	108.606	18.90	218.430
8.10	91.953	13.60	109.187	19.00	223.275
8.20	91.651	13.70	109.792	19.10	228.291
8.30	91.943	13.80	110.422	19.20	233.481
8.40	92.230	13.90	111.077	19.30	238.852
8.50	92.512	14.00	111.759	19.40	244.408
8.60	92.789	14.10	112.469	19.50	250.154
8.70	93.062	14.20	113.209	19.60	256.096

P-ρ-T TABLE FOR CARBON DIOXIDE

Temperature = 318.15 K (continued)

Density (ρ; mol/L)	Pressure (P; bar)	Density (ρ; mol/L)	Pressure (P; bar)	Density (ρ; mol/L)	Pressure (P; bar)
19.70	262.239	21.60	424.764	23.50	704.704
19.80	268.588	21.70	436.159	23.60	723.627
19.90	275.149	21.80	447.885	23.70	743.025
20.00	281.927	21.90	459.948	23.80	762.907
20.10	288.928	22.00	472.354	23.90	783.280
20.20	296.158	22.10	485.111	23.90	804.155
20.30	303.623	22.20	498.227	24.00	804.155
20.40	311.327	22.30	511.708	24.10	825.540
20.50	319.278	22.40	525.561	24.20	847.444
20.60	327.482	22.50	539.794	24.30	869.877
20.70	335.943	22.60	554.415	24.40	892.847
20.80	344.669	22.70	569.431	24.50	916.364
20.90	353.666	22.80	584.849	24.60	940.437
21.00	362.940	22.90	600.677	24.70	965.077
21.10	372.497	23.00	616.924	24.80	990.294
21.20	382.345	23.10	633.596	24.90	1016.096
21.30	392.488	23.20	650.702	25.00	1042.495
21.40	402.935	23.30	668.250		
21.50	413.691	23.40	686.248		

Temperature = 323.15 K

Density (ρ; mol/L)	Pressure (P; bar)	Density (ρ; mol/L)	Pressure (P; bar)	Density (ρ; mol/L)	Pressure (P; bar)
0.20	5.266	3.40	64.207	6.60	90.702
0.30	7.819	3.50	65.413	6.70	91.230
0.40	10.320	3.60	66.588	6.80	91.746
0.50	12.768	3.70	67.733	6.90	92.250
0.60	15.166	3.80	68.849	7.00	92.743
0.70	17.513	3.90	69.936	7.10	93.225
0.80	19.810	4.00	70.996	7.20	93.697
0.90	22.057	4.10	72.027	7.30	94.159
1.00	24.256	4.20	73.032	7.40	94.612
1.10	26.407	4.30	74.011	7.50	95.055
1.20	28.511	4.40	74.965	7.60	95.491
1.30	30.568	4.50	75.894	7.70	95.918
1.40	32.579	4.60	76.798	7.80	96.337
1.50	34.544	4.70	77.679	7.90	96.749
1.60	36.465	4.80	78.537	8.00	97.155
1.70	38.342	4.90	79.372	8.10	97.554
1.80	40.176	5.00	80.186	8.20	97.947
1.90	41.967	5.10	80.978	8.30	98.334
2.00	43.716	5.20	81.750	8.40	98.716
2.10	45.424	5.30	82.502	8.50	99.094
2.20	47.092	5.40	83.235	8.60	99.467
2.30	48.720	5.50	83.948	8.70	99.836
2.40	50.309	5.60	84.643	8.80	100.201
2.50	51.860	5.70	85.320	8.90	100.564
2.60	53.373	5.80	85.980	9.00	100.923
2.70	54.849	5.90	86.624	9.10	101.280
2.80	56.289	6.00	87.251	9.20	101.635
2.90	57.693	6.10	87.862	9.30	101.988
3.00	59.063	6.20	88.458	9.40	102.340
3.10	60.398	6.30	89.040	9.50	102.691
3.20	61.700	6.40	89.607	9.60	103.042
3.30	62.970	6.50	90.161	9.70	103.392

P-ρ-T TABLE FOR CARBON DIOXIDE

Temperature = 323.15 K (continued)

Density (ρ; mol/L)	Pressure (P; bar)	Density (ρ; mol/L)	Pressure (P; bar)	Density (ρ; mol/L)	Pressure (P; bar)
9.80	103.743	14.90	134.894	20.00	314.253
9.90	104.094	15.00	136.148	20.10	321.706
10.00	104.447	15.10	137.448	20.20	329.393
10.10	104.801	15.20	138.798	20.30	337.320
10.20	105.157	15.30	140.199	20.40	345.491
10.30	105.515	15.40	141.652	20.50	353.914
10.40	105.876	15.50	143.161	20.60	362.593
10.50	106.240	15.60	144.727	20.70	371.536
10.60	106.608	15.70	146.352	20.80	380.747
10.70	106.980	15.80	148.039	20.90	390.235
10.80	107.357	15.90	149.789	21.00	400.004
10.90	107.738	16.00	151.606	21.10	410.061
11.00	108.125	16.10	153.492	21.20	420.412
11.10	108.519	16.20	155.448	21.30	431.065
11.20	108.918	16.30	157.478	21.40	442.025
11.30	109.325	16.40	159.584	21.50	453.300
11.40	109.740	16.50	161.770	21.60	464.895
11.50	110.162	16.60	164.036	21.70	476.819
11.60	110.593	16.70	166.388	21.80	489.077
11.70	111.034	16.80	168.826	21.90	501.676
11.80	111.485	16.90	171.355	22.00	514.624
11.90	111.946	17.00	173.977	22.10	527.928
12.00	112.419	17.10	176.696	22.20	541.595
12.10	112.903	17.20	179.514	22.30	555.632
12.20	113.400	17.30	182.435	22.40	570.046
12.30	113.911	17.40	185.461	22.50	584.846
12.40	114.436	17.50	188.598	22.60	600.037
12.50	114.975	17.60	191.847	22.70	615.629
12.60	115.530	17.70	195.213	22.80	631.628
12.70	116.101	17.80	198.699	22.90	648.043
12.80	116.690	17.90	202.309	23.00	664.880
12.90	117.297	18.00	206.046	23.10	682.149
13.00	117.923	18.10	209.916	23.20	699.857
13.10	118.569	18.20	213.920	23.30	718.013
13.20	119.236	18.30	218.065	23.40	736.623
13.30	119.924	18.40	222.353	23.50	755.698
13.40	120.636	18.50	226.789	23.60	775.246
13.50	121.372	18.60	231.378	23.70	795.274
13.60	122.133	18.70	236.123	23.80	815.791
13.70	122.920	18.80	241.030	23.90	836.808
13.80	123.735	18.90	246.102	24.00	858.331
13.90	124.578	19.00	251.345	24.10	880.371
14.00	125.452	19.10	256.764	24.20	902.937
14.10	126.356	19.20	262.363	24.30	926.037
14.20	127.294	19.30	268.147	24.40	949.682
14.30	128.265	19.40	274.121	24.50	973.881
14.40	129.272	19.50	280.290	24.60	998.644
14.50	130.316	19.60	286.660	24.70	1023.980
14.60	131.399	19.70	293.236	24.80	1049.901
14.70	132.521	19.80	300.023	24.90	1076.415
14.80	133.686	19.90	307.027	25.00	1103.533

P-ρ-T TABLE FOR CARBON DIOXIDE

Temperature = 328.15 K

Density (ρ; mol/L)	Pressure (P; bar)	Density (ρ; mol/L)	Pressure (P; bar)	Density (ρ; mol/L)	Pressure (P; bar)
0.20	5.351	5.60	88.468	11.00	117.339
0.30	7.949	5.70	89.234	11.10	117.854
0.40	10.495	5.80	89.984	11.20	118.378
0.50	12.990	5.90	90.716	11.30	118.910
0.60	15.435	6.00	91.434	11.40	119.452
0.70	17.831	6.10	92.135	11.50	120.004
0.80	20.178	6.20	92.822	11.60	120.566
0.90	22.477	6.30	93.495	11.70	121.141
1.00	24.728	6.40	94.154	11.80	121.727
1.10	26.932	6.50	94.800	11.90	122.326
1.20	29.089	6.60	95.433	12.00	122.938
1.30	31.201	6.70	96.053	12.10	123.564
1.40	33.269	6.80	96.662	12.20	124.206
1.50	35.291	6.90	97.259	12.30	124.863
1.60	37.271	7.00	97.846	12.40	125.536
1.70	39.207	7.10	98.421	12.50	126.227
1.80	41.101	7.20	98.987	12.60	126.396
1.90	42.953	7.30	99.544	12.70	127.664
2.00	44.765	7.40	100.091	12.80	128.411
2.10	46.536	7.50	100.629	12.90	129.180
2.20	48.268	7.60	101.160	13.00	129.970
2.30	49.961	7.70	101.682	13.10	130.782
2.40	51.616	7.80	102.197	13.20	131.619
2.50	53.234	7.90	102.706	13.30	132.480
2.60	54.815	8.00	103.207	13.40	133.366
2.70	56.360	8.10	103.703	13.50	134.280
2.80	57.869	8.20	104.193	13.60	135.221
2.90	59.344	8.30	104.677	13.70	136.192
3.00	60.785	8.40	105.157	13.80	137.193
3.10	62.192	8.50	105.633	13.90	138.225
3.20	63.567	8.60	106.104	14.00	139.291
3.30	64.911	8.70	106.572	14.10	140.390
3.40	66.222	8.80	107.036	14.20	141.526
3.50	67.504	8.90	107.498	14.30	142.698
3.60	68.755	9.00	107.958	14.40	143.909
3.70	69.977	9.10	108.415	14.50	145.160
3.80	71.170	9.20	108.872	14.60	146.453
3.90	72.336	9.30	109.327	14.70	147.789
4.00	73.474	9.40	109.781	14.80	149.171
4.10	74.585	9.50	110.235	14.90	150.599
4.20	75.671	9.60	110.690	15.00	152.076
4.30	76.731	9.70	111.145	15.10	153.603
4.40	77.766	9.80	111.602	15.20	155.183
4.50	78.777	9.90	112.059	15.30	156.817
4.60	79.764	10.00	112.519	15.40	158.508
4.70	80.728	10.10	112.982	15.50	160.257
4.80	81.670	10.20	113.447	15.60	162.067
4.90	82.590	10.30	113.916	15.70	163.940
5.00	83.489	10.40	114.389	15.80	165.878
5.10	84.368	10.50	114.867	15.90	167.884
5.20	85.226	10.60	115.349	16.00	169.960
5.30	86.064	10.70	115.837	16.10	172.108
5.40	86.884	10.80	116.331	16.20	174.332
5.50	87.685	10.90	116.831	16.30	176.633

P-ρ-T TABLE FOR CARBON DIOXIDE

Temperature = 328.15 K (continued)

Density (ρ; mol/L)	Pressure (P; bar)	Density (ρ; mol/L)	Pressure (P; bar)	Density (ρ; mol/L)	Pressure (P; bar)
16.40	179.014	19.40	303.883	22.30	599.442
16.50	181.478	19.50	310.471	22.40	614.413
16.60	184.028	19.60	317.264	22.50	629.773
16.70	186.666	19.70	324.267	22.60	645.530
16.80	189.396	19.80	331.487	22.70	661.693
16.90	192.221	19.90	338.928	22.80	678.268
17.00	195.143	20.00	346.596	22.90	695.264
17.10	198.166	20.10	354.496	23.00	712.689
17.20	201.293	20.20	362.635	23.10	730.550
17.30	204.527	20.30	371.017	23.20	748.856
17.40	207.872	20.40	379.650	23.30	767.615
17.50	211.331	20.50	388.538	23.40	786.835
17.60	214.907	20.60	397.688	23.50	806.525
17.70	218.605	20.70	407.106	23.60	826.693
17.80	222.428	20.80	416.797	23.70	847.348
17.90	226.379	20.90	426.769	23.80	868.499
18.00	230.462	21.00	437.027	23.90	890.155
18.10	234.682	21.10	447.578	24.00	912.324
18.20	239.042	21.20	458.428	24.10	935.017
18.30	243.546	21.30	469.584	24.20	958.241
18.40	248.199	21.40	481.052	24.30	982.008
18.50	253.005	21.50	492.839	24.40	1006.326
18.60	257.969	21.60	504.951	24.50	1031.205
18.70	263.093	21.70	517.397	24.60	1056.655
18.80	268.384	21.80	530.181	24.70	1082.686
18.90	273.846	21.90	543.312	24.80	1109.308
19.00	279.483	22.00	556.797	24.90	1136.533
19.10	285.301	22.10	570.642	25.00	1164.369
19.20	291.303	22.20	584.855		
19.30	297.496				

Temperature = 333.15 K

Density (ρ; mol/L)	Pressure (P; bar)	Density (ρ; mol/L)	Pressure (P; bar)	Density (ρ; mol/L)	Pressure (P; bar)
0.20	5.437	2.30	51.194	4.40	80.538
0.30	8.078	2.40	52.914	4.50	81.630
0.40	10.670	2.50	54.598	4.60	82.699
0.50	13.212	2.60	56.246	4.70	83.746
0.60	15.704	2.70	57.859	4.80	84.771
0.70	18.149	2.80	59.437	4.90	85.775
0.80	20.545	2.90	60.982	5.00	86.758
0.90	22.895	3.00	62.493	5.10	87.720
1.00	25.197	3.10	63.972	5.20	88.664
1.10	27.454	3.20	65.419	5.30	89.588
1.20	29.666	3.30	66.835	5.40	90.494
1.30	31.832	3.40	68.220	5.50	91.382
1.40	33.955	3.50	69.575	5.60	92.253
1.50	36.035	3.60	70.902	5.70	93.106
1.60	38.072	3.70	72.200	5.80	93.944
1.70	40.067	3.80	73.469	5.90	94.765
1.80	42.021	3.90	74.712	6.00	95.572
1.90	43.934	4.00	75.928	6.10	96.363
2.00	45.807	4.10	77.118	6.20	97.140
2.10	47.641	4.20	78.283	6.30	97.903
2.20	49.437	4.30	79.422	6.40	98.653

P-ρ-T TABLE FOR CARBON DIOXIDE

Temperature = 333.15 K (continued)

Density (ρ; mol/L)	Pressure (P; bar)	Density (ρ; mol/L)	Pressure (P; bar)	Density (ρ; mol/L)	Pressure (P; bar)
6.50	99.391	11.90	132.777	17.30	226.715
6.60	100.115	12.00	133.532	17.40	230.376
6.70	100.828	12.10	134.303	17.50	234.156
6.80	101.529	12.20	135.091	17.60	238.058
6.90	102.220	12.30	135.897	17.70	242.085
7.00	102.900	12.40	136.721	17.80	246.242
7.10	103.569	12.50	137.566	17.90	250.531
7.20	104.229	12.60	138.430	18.00	254.958
7.30	104.880	12.70	139.316	18.10	259.526
7.40	105.523	12.80	140.224	18.20	264.238
7.50	106.157	12.90	141.156	18.30	269.100
7.60	106.783	13.00	142.111	18.40	274.115
7.70	107.402	13.10	143.092	18.50	279.288
7.80	108.013	13.20	144.099	18.60	284.622
7.90	108.619	13.30	145.133	18.70	290.123
8.00	109.218	13.40	146.196	18.80	295.795
8.10	109.811	13.50	147.288	18.90	301.642
8.20	110.400	13.60	148.411	19.00	307.669
8.30	110.983	13.70	149.566	19.10	313.881
8.40	111.562	13.80	150.753	19.20	320.283
8.50	112.137	13.90	151.976	19.30	326.879
8.60	112.709	14.00	153.234	19.40	333.676
8.70	113.278	14.10	154.529	19.50	340.677
8.80	113.844	14.20	155.863	19.60	347.888
8.90	114.408	14.30	157.237	19.70	355.315
9.00	114.970	14.40	158.652	19.80	362.962
9.10	115.531	14.50	160.111	19.90	370.836
9.20	116.091	14.60	161.614	20.00	378.941
9.30	116.651	14.70	163.164	20.10	387.283
9.40	117.211	14.80	164.763	20.20	395.868
9.50	117.772	14.90	166.411	20.30	404.702
9.60	118.333	15.00	168.111	20.40	413.790
9.70	118.897	15.10	169.866	20.50	423.139
9.80	119.462	15.20	171.676	20.60	432.754
9.90	120.030	15.30	173.543	20.70	442.641
10.00	120.601	15.40	175.471	20.80	452.807
10.10	121.175	15.50	177.461	20.90	463.258
10.20	121.754	15.60	179.515	21.00	474.000
10.30	122.337	15.70	181.636	21.10	485.039
10.40	122.925	15.80	183.825	21.20	496.382
10.50	123.519	15.90	186.086	21.30	508.036
10.60	124.120	16.00	188.421	21.40	520.006
10.70	124.727	16.10	190.832	21.50	532.301
10.80	125.342	16.20	193.321	21.60	544.926
10.90	125.965	16.30	195.892	21.70	557.888
11.00	126.596	16.40	198.548	21.80	571.194
11.10	127.237	16.50	201.290	21.90	584.851
11.20	127.888	16.60	204.122	22.00	598.867
11.30	128.549	16.70	207.047	22.10	613.248
11.40	129.222	16.80	210.068	22.20	628.002
11.50	129.906	16.90	213.187	22.30	643.136
11.60	130.603	17.00	216.409	22.40	658.658
11.70	131.314	17.10	219.735	22.50	674.574
11.80	132.038	17.20	223.169	22.60	690.894

P-ρ-T TABLE FOR CARBON DIOXIDE

Temperature = 333.15 K (continued)

Density (ρ; mol/L)	Pressure (P; bar)	Density (ρ; mol/L)	Pressure (P; bar)	Density (ρ; mol/L)	Pressure (P; bar)
22.70	707.623	23.50	857.189	24.30	1037.796
22.80	724.771	23.60	877.975	24.40	1062.785
22.90	742.344	23.70	899.254	24.50	1088.342
23.00	760.352	23.80	921.036	24.60	1114.478
23.10	778.802	23.90	943.328	24.70	1141.202
23.20	797.702	24.00	966.141	24.80	1168.526
23.30	817.061	24.10	989.484	24.90	1196.459
23.40	836.887	24.20	1013.365	25.00	1225.013

Temperature = 338.15 K

Density (ρ; mol/L)	Pressure (P; bar)	Density (ρ; mol/L)	Pressure (P; bar)	Density (ρ; mol/L)	Pressure (P; bar)
0.20	5.522	4.40	83.285	8.60	119.287
0.30	8.208	4.50	84.458	8.70	119.958
0.40	10.845	4.60	85.607	8.80	120.628
0.50	13.433	4.70	86.736	8.90	121.296
0.60	15.973	4.80	87.842	9.00	121.963
0.70	18.466	4.90	88.929	9.10	122.629
0.80	20.912	5.00	89.995	9.20	123.296
0.90	23.311	5.10	91.042	9.30	123.963
1.00	25.665	5.20	92.069	9.40	124.630
1.10	27.975	5.30	93.078	9.50	125.300
1.20	30.240	5.40	94.070	9.60	125.971
1.30	32.461	5.50	95.044	9.70	126.645
1.40	34.639	5.60	96.001	9.80	127.322
1.50	36.775	5.70	96.942	9.90	128.002
1.60	38.870	5.80	97.867	10.00	128.687
1.70	40.923	5.90	98.776	10.10	129,376
1.80	42.936	6.00	99.671	10.20	130.070
1.90	44.910	6.10	100.552	10.30	130.771
2.00	46.844	6.20	101.418	10.40	131.477
2.10	48.740	6.30	102.272	10.50	132.191
2.20	50.598	6.40	103.112	10.60	132.912
2.30	52.419	6.50	103.941	10.70	133.642
2.40	54.204	6.60	104.757	10.80	134.380
2.50	55.954	6.70	105.562	10.90	135.128
2.60	57.668	6.80	106.356	11.00	135.886
2.70	59.348	6.90	107.139	11.10	136.655
2.80	60.994	7.00	107.912	11.20	137.436
2.90	62.607	7.10	108.676	11.30	138.229
3.00	64.188	7.20	109.431	11.40	139.034
3.10	65.737	7.30	110.177	11.50	139.854
3.20	67.256	7.40	110.914	11.60	140.687
3.30	68.743	7.50	111.644	11.70	141.536
3.40	70.201	7.60	112.367	11.80	142.401
3.50	71.630	7.70	113.082	11.90	143.282
3.60	73.031	7.80	113.792	12.00	144.181
3.70	74.404	7.90	114.495	12.10	145.099
3.80	75.749	8.00	115.193	12.20	146.035
3.90	77.068	8.10	115.885	12.30	146.992
4.00	78.361	8.20	116.573	12.40	147.969
4.10	79.628	8.30	117.257	12.50	148.968
4.20	80.871	8.40	117.937	12.60	149.990
4.30	82.090	8.50	118.613	12.70	151.036

P-ρ-T TABLE FOR CARBON DIOXIDE

Temperature = 338.15 K (continued)

Density (ρ; mol/L)	Pressure (P; bar)	Density (ρ; mol/L)	Pressure (P; bar)	Density (ρ; mol/L)	Pressure (P; bar)
12.80	152.106	16.90	234.233	21.00	510.914
12.90	153.201	17.00	237.752	21.10	522.437
13.00	154.324	17.10	241.380	21.20	534.269
13.10	155.473	17.20	245.121	21.30	546.416
13.20	156.652	17.30	248.978	21.40	558.884
13.30	157.860	17.40	252.954	21.50	571.681
13.40	159.100	17.50	257.052	21.60	584.813
13.50	160.371	17.60	261.277	21.70	598.287
13.60	161.676	17.70	265.632	21.80	612.111
13.70	163.016	17.80	270.121	21.90	626.290
13.80	164.391	17.90	274.748	22.00	640.833
13.90	165.804	18.00	279.515	22.10	655.746
14.00	167.255	18.10	284.429	22.20	671.037
14.10	168.746	18.20	289.491	22.30	686.713
14.20	170.279	18.30	294.707	22.40	702.782
14.30	171.855	18.40	300.082	22.50	719.251
14.40	173.476	18.50	305.618	22.60	736.128
14.50	175.142	18.60	311.320	22.70	753.421
14.60	176.857	18.70	317.194	22.80	771.137
14.70	178.621	18.80	323.242	22.90	189.285
14.80	180.437	18.90	329.471	23.00	807.872
14.90	182.306	19.00	335.885	23.10	826.907
15.00	184.230	19.10	342.488	23.20	846.399
15.10	186.211	19.20	349.286	23.30	866.355
15.20	188.252	19.30	356.283	23.40	886.784
15.30	190.353	19.40	363.484	23.50	907.695
15.40	192.518	19.50	370.895	23.60	929.096
15.50	194.749	19.60	378.520	23.70	950.997
15.60	197.047	19.70	386.365	23.80	973.407
15.70	199.415	19.80	394.436	23.90	996.334
15.80	201.856	19.90	402.737	24.00	1019.788
15.90	204.372	20.00	411.275	24.10	1043.779
16.00	206.965	20.10	420.054	24.20	1068.315
16.10	209.639	20.20	429.081	24.30	1093.408
16.20	212.394	20.30	438.362	24.40	1119.066
16.30	215.236	20.40	447.901	24.50	1145.301
16.40	218.165	20.50	457.705	24.60	1172.121
16.50	221.185	20.60	467.781	24.70	1199.537
16.60	224.299	20.70	478.133	24.80	1227.561
16.70	227.509	20.80	488.769	24.90	1256.202
16.80	230.820	20.90	499.694	25.00	1285.472

Temperature = 343.15 K

Density (ρ; mol/L)	Pressure (P; bar)	Density (ρ; mol/L)	Pressure (P; bar)	Density (ρ; mol/L)	Pressure (P; bar)
0.20	5.607	1.10	28.494	2.00	47.875
0.30	8.337	1.20	30.812	2.10	49.833
0.40	11.019	1.30	33.087	2.20	51.753
0.50	13.654	1.40	35.321	2.30	53.638
0.60	16.241	1.50	37.513	2.40	55.487
0.70	18.782	1.60	39.664	2.50	57.301
0.80	21.277	1.70	41.775	2.60	59.081
0.90	23.727	1.80	43.847	2.70	60.827
1.00	26.132	1.90	45.880	2.80	62.541

P-ρ-T TABLE FOR CARBON DIOXIDE

Temperature = 343.15 K (continued)

Density (ρ; mol/L)	Pressure (P; bar)	Density (ρ; mol/L)	Pressure (P; bar)	Density (ρ; mol/L)	Pressure (P; bar)
2.90	64.222	8.30	123.502	13.70	176.525
3.00	65.872	8.40	124.284	13.80	178.089
3.10	67.491	8.50	125.063	13.90	179.693
3.20	69.080	8.60	125.840	14.00	181.338
3.30	70.639	8.70	126.616	14.10	183.026
3.40	72.169	8.80	127.390	14.20	184.758
3.50	73.670	8.90	128.163	14.30	186.536
3.60	75.144	9.00	128.937	14.40	188.362
3.70	76.591	9.10	129.710	14.50	190.238
3.80	78.011	9.20	130.485	14.60	192.164
3.90	79.405	9.30	131.261	14.70	194.143
4.00	80.774	9.40	132.038	14.80	196.176
4.10	82.119	9.50	132.819	14.90	198.266
4.20	83.439	9.60	133.602	15.00	200.414
4.30	84.736	9.70	134.388	15.10	202.623
4.40	86.010	9.80	135.179	15.20	204.894
4.50	87.261	9.90	135.974	15.30	207.229
4.60	88.491	10.00	136.774	15.40	209.632
4.70	89.700	10.10	137.580	15.50	212.103
4.80	90.888	10.20	138.393	15.60	214.646
4.90	92.056	10.30	139.212	15.70	217.262
5.00	93.205	10.40	140.039	15.80	219.954
5.10	94.335	10.50	140.875	15.90	222.725
5.20	95.446	10.60	141.719	16.00	225.577
5.30	96.539	10.70	142.573	16.10	228.512
5.40	97.615	10.80	143.437	16.20	231.534
5.50	98.675	10.90	144.313	16.30	234.645
5.60	99.718	11.00	145.199	16.40	237.848
5.70	100.745	11.10	146.099	16.50	241.145
5.80	101.757	11.20	147.011	16.60	244.540
5.90	102.754	11.30	147.937	16.70	248.036
6.00	103.737	11.40	148.878	16.80	251.635
6.10	104.705	11.50	149.834	16.90	255.341
6.20	105.662	11.60	150.806	17.00	259.157
6.30	106.605	11.70	151.795	17.10	263.086
6.40	107.536	11.80	152.802	17.20	267.132
6.50	108.455	11.90	153.828	17.30	271.298
6.60	109.363	12.00	154.873	17.40	275.587
6.70	110.259	12.10	155.938	17.50	280.003
6.80	111.146	12.20	157.024	17.60	284.550
6.90	112.022	12.30	158.133	17.70	289.231
7.00	112.889	12.40	159.264	17.80	294.050
7.10	113.747	12.50	160.419	17.90	299.011
7.20	114.596	12.60	161.600	18.00	304.118
7.30	115.437	12.70	162.806	18.10	309.374
7.40	116.271	12.80	164.039	18.20	314.784
7.50	117.097	12.90	165.300	18.30	320.352
7.60	117.916	13.00	166.589	18.40	326.083
7.70	118.730	13.10	167.909	18.50	331.980
7.80	119.537	13.20	169.261	18.60	338.047
7.90	120.339	13.30	170.644	18.70	344.290
8.00	121.136	13.40	172.061	18.80	350.713
8.10	121.928	13.50	173.513	18.90	357.321
8.20	122.717	13.60	175.000	19.00	364.118

P-ρ-T TABLE FOR CARBON DIOXIDE

Temperature = 343.15 K (continued)

Density (ρ; mol/L)	Pressure (P; bar)	Density (ρ; mol/L)	Pressure (P; bar)	Density (ρ; mol/L)	Pressure (P; bar)
19.10	371.108	21.10	559.767	23.10	874.869
19.20	378.298	21.20	572.082	23.20	894.949
19.30	385.692	21.30	584.718	23.30	915.500
19.40	393.294	21.40	597.680	23.40	936.529
19.50	401.110	21.50	610.975	23.50	958.047
19.60	409.146	21.60	624.611	23.60	980.061
19.70	417.406	21.70	638.593	23.70	1002.582
19.80	425.896	21.80	652.929	23.80	1025.617
19.90	434.621	21.90	667.627	23.90	1049.177
20.00	443.587	22.00	682.693	24.00	1073.271
20.10	452.800	22.10	698.134	24.10	1097.908
20.20	462.264	22.20	713.958	24.20	1123.098
20.30	471.987	22.30	730.173	24.30	1148.851
20.40	481.973	22.40	746.785	24.40	1175.178
20.50	492.229	22.50	763.803	24.50	1202.088
20.60	502.760	22.60	781.235	24.60	1229.591
20.70	513.573	22.70	799.087	24.70	1257.699
20.80	524.674	22.80	817.369	24.80	1286.421
20.90	536.069	22.90	836.088	24.90	1315.770
21.00	547.764	23.00	855.252	25.00	1345.755

Temperature = 348.15 K

Density (ρ; mol/L)	Pressure (P; bar)	Density (ρ; mol/L)	Pressure (P; bar)	Density (ρ; mol/L)	Pressure (P; bar)
0.20	5.692	3.20	70.892	6.20	109.875
0.30	8.467	3.30	72.521	6.30	110.907
0.40	11.194	3.40	74.122	6.40	111.928
0.50	13.874	3.50	75.696	6.50	112.938
0.60	16.508	3.60	77.243	6.60	113.936
0.70	19.097	3.70	78.763	6.70	114.925
0.80	21.642	3.80	80.257	6.80	115.904
0.90	24.142	3.90	81.726	6.90	116.873
1.00	26.598	4.00	83.171	7.00	117.833
1.10	29.011	4.10	84.591	7.10	118.785
1.20	31.382	4.20	85.988	7.20	119.729
1.30	33.712	4.30	87.362	7.30	120.666
1.40	36.000	4.40	88.714	7.40	121.595
1.50	38.247	4.50	90.045	7.50	122.518
1.60	40.455	4.60	91.354	7.60	123.435
1.70	42.624	4.70	92.642	7.70	124.346
1.80	44.754	4.80	93.911	7.80	125.252
1.90	46.847	4.90	95.160	7.90	126.154
2.00	48.902	5.00	96.391	8.00	127.051
2.10	50.920	5.10	97.603	8.10	127.944
2.20	52.902	5.20	98.797	8.20	128.834
2.30	54.850	5.30	99.974	8.30	129.721
2.40	56.762	5.40	101.134	8.40	130.606
2.50	58.641	5.50	102.278	8.50	131.489
2.60	60.486	5.60	103.406	8.60	132.370
2.70	62.298	5.70	104.519	8.70	133.251
2.80	64.078	5.80	105.618	8.80	134.131
2.90	65.827	5.90	106.702	8.90	135.011
3.00	67.545	6.00	107.773	9.00	135.893
3.10	69.233	6.10	108.830	9.10	136.775

P-ρ-T TABLE FOR CARBON DIOXIDE

Temperature = 348.15 K (continued)

Density (ρ; mol/L)	Pressure (P; bar)	Density (ρ; mol/L)	Pressure (P; bar)	Density (ρ; mol/L)	Pressure (P; bar)
9.20	137.659	14.50	205.384	19.80	457.333
9.30	138.545	14.60	207.522	19.90	466.478
9.40	139.434	14.70	209.716	20.00	475.869
9.50	140.326	14.80	211.968	20.10	485.511
9.60	141.223	14.90	214.279	20.20	495.409
9.70	142.123	15.00	216.652	20.30	505.570
9.80	143.029	15.10	219.088	20.40	515.999
9.90	143.941	15.20	221.590	20.50	526.702
10.00	144.859	15.30	224.160	20.60	537.685
10.10	145.784	15.40	226.800	20.70	548.955
10.20	146.716	15.50	229.512	20.80	560.517
10.30	147.657	15.60	232.299	20.90	572.378
10.40	148.606	15.70	235.163	21.00	584.545
10.50	149.565	15.80	238.106	21.10	597.022
10.60	150.534	15.90	241.132	21.20	609.818
10.70	151.514	16.00	244.242	21.30	622.938
10.80	152.506	16.10	247.439	21.40	636.390
10.90	153.510	16.20	250.727	21.50	650.180
11.00	154.528	16.30	254.107	21.60	664.315
11.10	155.559	16.40	257.582	21.70	678.802
11.20	156.605	16.50	261.156	21.80	693.648
11.30	157.666	16.60	264.832	21.90	708.860
11.40	158.743	16.70	268.612	22.00	724.445
11.50	159.838	16.80	272.499	21.10	740.411
11.60	160.950	16.90	276.497	22.20	756.765
11.70	162.081	17.00	280.609	22.30	773.515
11.80	163.231	17.10	284.838	22.40	790.668
11.90	164.402	17.20	289.188	22.50	808.232
12.00	165.595	17.30	293.662	22.60	826.215
12.10	166.809	17.40	298.263	22.70	844.624
12.20	168.047	17.50	302.995	22.80	863.468
12.30	169.308	17.60	307.862	22.90	882.755
12.40	170.595	17.70	312.867	23.00	902.493
12.50	171.908	17.80	318.014	23.10	922.691
12.60	173.247	17.90	323.307	23.20	943.357
12.70	174.615	18.00	328.751	23.30	964.499
12.80	176.012	18.10	334.348	23.40	986.127
12.90	177.439	18.20	340.104	23.50	1008.249
13.00	178.898	18.30	346.021	23.60	1030.875
13.10	180.389	18.40	352.105	23.70	1054.012
13.20	181.913	18.50	358.360	23.80	1077.672
13.30	183.472	18.60	364.791	23.90	1101.863
13.40	185.068	18.70	371.401	24.00	1126.595
13.50	186.700	18.80	378.195	24.10	1151.877
13.60	188.371	18.90	385.178	24.20	1177.719
13.70	190.083	19.00	392.355	24.30	1204.132
13.80	191.835	19.10	399.730	24.40	1231.125
13.90	193.630	19.20	407.308	24.50	1258.710
14.00	195.470	19.30	415.095	24.60	1286.896
14.10	197.355	19.40	423.095	24.70	1315.693
14.20	199.287	19.50	431.314	24.80	1345.114
14.30	201.269	19.60	439.756	24.90	1375.169
14.40	203.300	19.70	448.427	25.00	1405.870

P-ρ-T TABLE FOR CARBON DIOXIDE

Temperature = 353.15 K

Density (ρ; mol/L)	Pressure (P; bar)	Density (ρ; mol/L)	Pressure (P; bar)	Density (ρ; mol/L)	Pressure (P; bar)
0.20	5.777	5.60	107.070	11.00	163.865
0.30	8.596	5.70	108.268	11.10	165.030
0.40	11.368	5.80	109.453	11.20	166.211
0.50	14.094	5.90	110.624	11.30	167.408
0.60	16.776	6.00	111.781	11.40	168.624
0.70	19.412	6.10	112.927	11.50	169.858
0.80	22.005	6.20	114.060	11.60	171.112
0.90	24.555	6.30	115.182	11.70	172.386
1.00	27.062	6.40	115.292	11.80	173.681
1.10	29.527	6.50	117.392	11.90	174.999
1.20	31.951	6.60	118.482	12.00	176.340
1.30	34.334	6.70	119.562	12.10	177.704
1.40	36.676	6.80	120.633	12.20	179.094
1.50	38.979	6.90	121.695	12.30	180.510
1.60	41.244	7.00	122.749	12.40	181.953
1.70	43.469	7.10	123.795	12.50	183.424
1.80	45.658	7.20	124.834	12.60	184.925
1.90	47.809	7.30	125.866	12.70	186.455
2.00	49.923	7.40	126.892	12.80	188.017
2.10	52.002	7.50	127.911	12.90	189.611
2.20	54.046	7.60	128.926	13.00	191.239
2.30	56.056	7.70	129.935	13.00	192.902
2.40	58.031	7.80	130.940	13.10	194.600
2.50	59.973	7.90	131.941	13.30	196.336
2.60	61.883	8.00	132.939	13.40	198.111
2.70	63.761	8.10	133.933	13.50	199.925
2.80	65.607	8.20	134.925	13.60	201.780
2.90	67.423	8.30	135.915	13.70	203.678
3.00	69.209	8.40	136.904	13.80	205.620
3.10	70.966	8.50	137.891	13.90	207.607
3.20	72.693	8.60	138.878	14.00	209.642
3.30	74.393	8.70	139.864	14.10	211.724
3.40	76.064	8.80	140.852	14.20	213.857
3.50	77.709	8.90	141.840	14.30	216.042
3.60	79.328	9.00	142.830	14.40	218.280
3.70	80.921	9.10	143.822	14.50	220.573
3.80	82.488	9.20	144.816	14.60	222.924
3.90	84.032	9.30	145.814	14.70	225.333
4.00	85.551	9.40	146.816	14.80	227.803
4.10	87.047	9.50	147.822	14.90	230.335
4.20	88.520	9.60	148.832	15.00	232.933
4.30	89.971	9.70	149.849	15.10	235.597
4.40	91.400	9.80	150.872	15.20	238.330
4.50	92.809	9.90	151.901	15.30	241.134
4.60	94.196	10.00	152.938	15.40	244.011
4.70	95.564	10.10	153.983	15.50	246.964
4.80	96.913	10.20	155.036	15.60	249.996
4.90	98.243	10.30	156.100	15.70	253.107
5.00	99.554	10.40	157.716	15.80	256.302
5.10	100.848	10.50	158.257	15.90	259.582
5.20	102.125	10.60	159.353	16.00	262.950
5.30	103.385	10.70	160.460	16.10	266.409
5.40	104.628	10.80	161.581	16.20	269.961
5.50	105.857	10.90	162.716	16.30	273.610

P-ρ-T TABLE FOR CARBON DIOXIDE

Temperature = 353.15 K (continued)

Density (ρ; mol/L)	Pressure (P; bar)	Density (ρ; mol/L)	Pressure (P; bar)	Density (ρ; mol/L)	Pressure (P; bar)
16.40	277.358	19.30	444.483	22.20	799.459
16.50	281.208	19.40	452.878	22.30	816.741
16.60	285.163	19.50	461.496	22.40	834.431
16.70	289.277	19.60	470.341	22.50	852.538
16.80	293.401	19.70	479.420	22.60	871.070
16.90	297.690	19.80	488.738	22.70	890.033
17.00	302.097	19.90	498.300	22.80	909.437
17.10	306.625	20.00	508.112	22.90	929.290
17.20	311.277	20.10	518.180	23.00	949.600
17.30	316.057	20.20	528.508	23.10	970.375
17.40	320.968	20.30	539.104	23.20	991.625
17.50	326.015	20.40	549.972	23.30	1013.357
17.60	331.200	20.50	561.119	23.40	1035.581
17.70	336.528	20.60	572.551	23.50	1058.306
17.80	342.001	20.70	584.274	23.60	1081.541
17.90	347.625	20.80	596.294	23.70	1105.295
18.00	353.403	20.90	608.617	23.80	1129.577
18.10	359.340	21.00	621.251	23.90	1154.397
18.20	365.438	21.10	634.200	24.00	1179.765
18.30	371.703	21.20	647.473	24.10	1205.691
18.40	378.139	21.30	661.075	24.20	1323.185
18.50	384.749	21.40	675.013	24.30	1259.256
18.60	391.539	21.50	689.294	24.40	1286.915
18.70	398.513	21.60	703.925	24.50	1315.173
18.80	405.676	21.70	718.913	24.60	1344.040
18.90	413.032	21.80	734.265	24.70	1373.528
19.00	420.586	21.90	749.989	24.80	1403.636
19.10	428.342	22.00	766.090	24.90	1434.407
19.20	436.306	22.10	782.578	25.00	1465.822

Temperature = 363.15 K

Density (ρ; mol/L)	Pressure (P; bar)	Density (ρ; mol/L)	Pressure (P; bar)	Density (ρ; mol/L)	Pressure (P; bar)
0.20	5.947	2.30	58.452	4.40	96.723
0.30	8.854	2.40	60.552	4.50	98.286
0.40	11.716	2.50	62.620	4.60	99.830
0.50	14.534	2.60	64.658	4.70	101.355
0.60	17.309	2.70	66.665	4.80	102.862
0.70	20.041	2.80	68.643	4.90	104.351
0.80	22.731	2.90	70.591	5.00	105.823
0.90	25.380	3.00	72.510	5.10	107.279
1.00	27.988	3.10	74.402	5.20	108.719
1.10	30.556	3.20	76.267	5.30	110.144
1.20	33.084	3.30	78.104	5.40	111.553
1.30	35.573	3.40	79.916	5.50	112.949
1.40	38.024	3.50	81.702	5.60	114.330
1.50	40.436	3.60	83.463	5.70	115.698
1.60	42.812	3.70	85.200	5.80	117.054
1.70	45.151	3.80	86.912	5.90	118.397
1.80	47.454	3.90	88.602	6.00	119.728
1.90	49.722	4.00	90.269	6.10	121.048
2.00	51.955	4.10	91.914	6.20	122.357
2.10	54.154	4.20	93.538	6.30	123.656
2.20	56.319	4.30	95.141	6.40	124.945

P-ρ-T TABLE FOR CARBON DIOXIDE

Temperature = 363.15 K (continued)

Density (ρ; mol/L)	Pressure (P; bar)	Density (ρ; mol/L)	Pressure (P; bar)	Density (ρ; mol/L)	Pressure (P; bar)
6.50	126.225	11.90	196.233	17.30	360.903
6.60	127.497	12.00	197.874	17.40	366.431
6.70	128.759	12.10	199.542	17.50	372.101
6.80	130.014	12.20	201.240	17.60	377.918
6.90	131.262	12.30	202.968	17.70	383.886
7.00	132.503	12.40	204.727	17.80	390.007
7.10	133.737	12.50	206.517	17.90	396.287
7.20	134.966	12.60	208.341	18.00	402.728
7.30	136.189	12.70	210.200	18.10	409.336
7.40	137.408	12.80	212.094	18.20	416.114
7.50	138.622	12.90	214.025	18.30	423.067
7.60	139.832	13.00	215.994	18.40	430.199
7.70	141.038	13.10	218.003	18.50	437.513
7.80	142.242	13.20	220.052	18.60	445.016
7.90	143.444	13.30	222.143	18.70	452.711
8.00	144.643	13.40	224.278	18.80	460.603
8.10	145.841	13.50	226.457	18.90	468.697
8.20	147.039	13.60	228.682	19.00	476.997
8.30	148.236	13.70	230.956	19.10	485.508
8.40	149.433	13.80	233.278	19.20	494.236
8.50	150.631	13.90	235.651	19.30	503.184
8.60	151.830	14.00	238.077	19.40	512.360
8.70	153.030	14.10	240.556	19.50	521.766
8.80	154.234	14.20	243.091	19.60	531.410
8.90	155.440	14.30	245.684	19.70	541.295
9.00	156.649	14.40	248.335	19.80	551.429
9.10	157.863	14.50	251.048	19.90	561.815
9.20	159.081	14.60	253.823	20.00	572.460
9.30	160.305	14.70	256.664	20.10	583.370
9.40	161.534	14.80	259.571	20.20	594.549
9.50	162.770	14.90	262.546	20.30	606.005
9.60	164.013	15.00	265.593	20.40	617.742
9.70	165.264	15.10	268.713	20.50	629.768
9.80	166.523	15.20	271.908	20.60	642.087
9.90	167.792	15.30	275.180	20.70	654.707
10.00	169.070	15.40	278.532	20.80	667.633
10.10	170.358	15.50	281.967	20.90	680.872
10.20	171.658	15.60	285.486	21.00	694.430
10.30	172.970	15.70	289.092	21.10	708.314
10.40	174.294	15.80	292.787	21.20	722.531
10.50	175.632	15.90	296.575	21.30	737.086
10.60	176.984	16.00	300.458	21.40	751.988
10.70	178.351	16.10	304.438	21.50	767.243
10.80	179.734	16.20	308.519	21.60	782.857
10.90	181.133	16.30	312.703	21.70	798.839
11.00	182.550	16.40	316.994	21.80	815.195
11.10	183.985	16.50	321.393	21.90	831.932
11.20	185.439	16.60	325.905	22.00	849.059
11.30	186.913	16.70	330.533	22.10	866.581
11.40	188.409	16.80	335.279	22.20	884.508
11.50	189.926	16.90	340.146	22.30	902.847
11.60	191.466	17.00	345.139	22.40	921.605
11.70	193.030	17.10	350.261	22.50	940.791
11.80	194.618	17.20	355.514	22.60	960.413

P-ρ-T TABLE FOR CARBON DIOXIDE

Temperature = 363.15 K (continued)

Density (ρ; mol/L)	Pressure (P; bar)	Density (ρ; mol/L)	Pressure (P; bar)	Density (ρ; mol/L)	Pressure (P; bar)
22.70	980.478	23.50	1158.001	24.30	1369.054
22.80	1000.995	23.60	1182.449	24.40	1398.042
22.90	1021.973	23.70	1207.430	24.50	1427.645
23.00	1043.421	23.80	1232.954	24.60	1457.873
23.10	1065.345	23.90	1259.029	24.70	1488.738
23.20	1087.757	24.00	1285.667	24.80	1520.250
23.30	1110.664	24.10	1312.876	24.90	1552.421
23.40	1134.075	24.20	1340.669	25.00	1585.264

Temperature = 373.15 K

Density (ρ; mol/L)	Pressure (P; bar)	Density (ρ; mol/L)	Pressure (P; bar)	Density (ρ; mol/L)	Pressure (P; bar)
0.20	6.117	4.40	101.990	8.60	164.702
0.30	9.111	4.50	103.705	8.70	166.119
0.40	12.063	4.60	105.403	8.80	167.540
0.50	14.972	4.70	107.083	8.90	168.966
0.60	17.840	4.80	108.746	9.00	170.398
0.70	20.667	4.90	110.393	9.10	171.835
0.80	23.454	5.00	112.025	9.20	173.280
0.90	26.202	5.10	113.641	9.30	174.731
1.00	28.910	5.20	115.243	9.40	176.191
1.10	31.579	5.30	116.830	9.50	177.660
1.20	34.211	5.40	118.404	9.60	179.138
1.30	36.806	5.50	119.965	9.70	180.626
1.40	39.363	5.60	121.514	9.80	182.125
1.50	41.885	5.70	123.050	9.90	183.635
1.60	44.371	5.80	124.575	10.00	185.157
1.70	46.822	5.90	126.089	10.10	186.693
1.80	49.239	6.00	127.593	10.20	188.242
1.90	51.622	6.10	129.086	10.30	189.806
2.00	53.971	6.20	130.571	10.40	191.385
2.10	56.289	6.30	132.046	10.50	192.980
2.20	58.574	6.40	133.513	10.60	194.592
2.30	60.828	6.50	134.972	10.70	196.221
2.40	63.052	6.60	136.424	10.80	197.869
2.50	65.245	6.70	137.869	10.90	199.537
2.60	67.409	6.80	139.307	11.00	201.225
2.70	69.544	6.90	140.740	11.10	202.935
2.80	71.651	7.00	142.167	11.20	204.666
2.90	73.729	7.10	143.590	11.30	206.421
3.00	75.781	7.20	145.008	11.40	208.200
3.10	77.807	7.30	146.423	11.50	210.004
3.20	79.806	7.40	147.834	11.60	211.834
3.30	81.780	7.50	149.242	11.70	213.691
3.40	83.729	7.60	150.648	11.80	215.577
3.50	85.655	7.70	152.053	11.90	217.491
3.60	87.556	7.80	153.456	12.00	219.436
3.70	89.435	7.90	154.858	12.10	221.413
3.80	91.291	8.00	156.261	12.20	223.422
3.90	93.126	8.10	157.664	12.30	225.465
4.00	94.939	8.20	159.067	12.40	227.543
4.10	96.731	8.30	160.472	12.50	229.656
4.20	98.503	8.40	161.880	12.60	231.807
4.30	100.256	8.50	163.289	12.70	233.997

P-ρ-T TABLE FOR CARBON DIOXIDE

Temperature = 373.15 K (continued)

Density (ρ; mol/L)	Pressure (P; bar)	Density (ρ; mol/L)	Pressure (P; bar)	Density (ρ; mol/L)	Pressure (P; bar)
12.80	236.227	16.90	382.643	21.00	767.285
12.90	238.497	17.00	388.217	21.10	782.093
13.00	240.811	17.10	393.927	21.20	797.243
13.10	243.168	17.20	399.776	21.30	812.742
13.20	245.570	17.30	405.768	21.40	828.597
13.30	248.019	17.40	411.907	21.50	844.815
13.40	250.516	17.50	418.196	21.60	861.404
13.50	253.062	17.60	424.638	21.70	878.369
13.60	255.660	17.70	431.239	21.80	895.719
13.70	258.310	17.80	438.001	21.90	913.461
13.80	261.015	17.90	444.929	22.00	931.602
13.90	263.775	18.00	452.027	22.10	950.151
14.00	266.593	18.10	459.299	22.20	969.115
14.10	269.471	18.20	466.749	22.30	988.502
14.20	272.409	18.30	474.382	22.40	1008.320
14.30	275.410	18.40	482.201	22.50	1028.576
14.40	278.476	18.50	490.212	22.60	1049.280
14.50	281.609	18.60	498.419	22.70	1070.440
14.60	284.810	18.70	506.826	22.80	1092.063
14.70	288.081	18.80	515.438	22.90	1114.159
14.80	291.426	18.90	524.260	23.00	1136.737
14.90	294.845	19.00	533.296	23.10	1159.805
15.00	298.340	19.10	542.553	23.20	1183.372
15.10	301.915	19.20	552.034	23.30	1207.447
15.20	305.571	19.30	561.744	23.40	1232.041
15.30	309.311	19.40	571.690	23.50	1257.161
15.40	313.137	19.50	581.875	23.60	1282.818
15.50	317.051	19.60	592.306	23.70	1309.022
15.60	321.056	19.70	602.988	23.80	1335.782
15.70	325.155	19.80	613.926	23.90	1363.108
15.80	329.350	19.90	625.126	24.00	1391.012
15.90	333.643	20.00	636.593	24.10	1419.502
16.00	338.038	20.10	648.334	24.20	1448.589
16.10	342.537	20.20	660.354	24.30	1478.285
16.20	347.144	20.30	672.658	24.40	1508.600
16.30	351.861	20.40	685.254	24.50	1539.545
16.40	356.690	20.50	698.147	24.60	1571.132
16.50	361.636	20.60	711.343	24.70	1603.371
16.60	366.701	20.70	724.848	24.80	1636.275
16.70	371.888	20.80	738.669	24.90	1669.856
16.80	377.201	20.90	752.813	25.00	1704.124

Temperature = 383.15 K

Density (ρ; mol/L)	Pressure (P; bar)	Density (ρ; mol/L)	Pressure (P; bar)	Density (ρ; mol/L)	Pressure (P; bar)
0.20	6.287	1.10	32.599	2.00	55.975
0.30	9.369	1.20	35.334	2.10	58.410
0.40	12.409	1.30	38.033	2.20	60.814
0.50	15.410	1.40	40.696	2.30	63.189
0.60	18.370	1.50	43.326	2.40	65.534
0.70	21.292	1.60	45.921	2.50	67.851
0.80	24.175	1.70	48.483	2.60	70.139
0.90	27.020	1.80	51.013	2.70	72.401
1.00	29.828	1.90	53.510	2.80	74.635

P-ρ-T TABLE FOR CARBON DIOXIDE

Temperature = 383.15 K (continued)

Density (ρ; mol/L)	Pressure (P; bar)	Density (ρ; mol/L)	Pressure (P; bar)	Density (ρ; mol/L)	Pressure (P; bar)
2.90	76.843	8.30	172.634	13.70	285.705
3.00	79.025	8.40	174.252	13.80	288.793
3.10	81.183	8.50	175.875	13.90	291.941
3.20	83.316	8.60	177.502	14.00	295.153
3.30	85.425	8.70	179.136	14.10	298.429
3.40	87.510	8.80	180.776	14.20	301.771
3.50	89.573	8.90	182.423	14.30	305.181
3.60	91.614	9.00	184.077	14.40	308.662
3.70	93.633	9.10	185.740	14.50	312.214
3.80	95.631	9.20	187.412	14.60	315.841
3.90	97.608	9.30	189.094	14.70	319.544
4.00	99.566	9.40	190.786	14.80	323.324
4.10	101.504	9.50	192.489	14.90	327.186
4.20	103.423	9.60	194.203	15.00	331.130
4.30	105.325	9.70	195.931	15.10	335.159
4.40	107.208	9.80	197.671	15.20	339.275
4.50	109.074	9.90	199.425	15.30	343.480
4.60	110.924	10.00	201.194	15.40	347.778
4.70	112.758	10.10	202.979	15.50	352.171
4.80	114.577	10.20	204.780	15.60	356.660
4.90	116.380	10.30	206.598	15.70	361.249
5.00	118.169	10.40	208.434	15.80	365.941
5.10	119.944	10.50	210.288	15.90	370.737
5.20	121.706	10.60	212.163	16.00	375.642
5.30	123.455	10.70	214.058	16.10	380.657
5.40	125.192	10.80	215.974	16.20	385.786
5.50	126.918	10.90	217.913	16.30	391.032
5.60	128.632	11.00	219.875	16.40	396.397
5.70	130.335	11.10	221.862	16.50	401.885
5.80	132.029	11.20	223.873	16.60	407.499
5.90	133.713	11.30	225.911	16.70	413.242
6.00	135.388	11.40	227.977	16.80	419.118
6.10	137.054	11.50	230.070	16.90	425.129
6.20	138.712	11.60	232.193	17.00	431.280
6.30	140.363	11.70	234.347	17.10	437.574
6.40	142.007	11.80	236.532	17.20	444.014
6.50	143.645	11.90	238.750	17.30	450.604
6.60	145.276	12.00	241.002	17.40	457.347
6.70	146.093	12.10	243.289	17.50	464.249
6.80	148.524	12.20	245.612	17.60	471.311
6.90	150.142	12.30	247.972	17.70	478.539
7.00	151.755	12.40	250.372	17.80	485.935
7.10	153.365	12.50	252.811	17.90	493.505
7.20	154.973	12.60	255.292	18.00	501.253
7.30	156.578	12.70	257.815	18.10	509.182
7.40	158.182	12.80	260.382	18.20	517.297
7.50	159.785	12.90	262.995	18.30	525.602
7.60	161.387	13.00	265.655	18.40	534.102
7.70	162.989	13.10	268.362	18.50	542.801
7.80	164.592	13.20	271.119	18.60	551.704
7.90	166.196	13.30	273.928	18.70	560.815
8.00	167.802	13.40	276.789	18.80	570.139
8.10	169.409	13.50	279.704	18.90	579.681
8.20	171.020	13.60	282.676	19.00	589.446

P-ρ-T TABLE FOR CARBON DIOXIDE

Temperature = 383.15 K (continued)

Density (ρ; mol/L)	Pressure (P; bar)	Density (ρ; mol/L)	Pressure (P; bar)	Density (ρ; mol/L)	Pressure (P; bar)
19.10	599.439	21.10	855.532	23.10	1253.778
19.20	609.665	21.20	871.607	23.20	1278.495
19.30	620.128	21.30	888.041	23.30	1303.735
19.40	630.835	21.40	904.840	23.40	1329.505
19.50	641.791	21.50	922.013	23.50	1355.816
19.60	653.000	21.60	939.566	23.60	1382.678
19.70	664.469	21.70	957.507	23.70	1410.101
19.80	676.202	21.80	975.843	23.80	1438.094
19.90	688.207	21.90	994.582	23.90	1466.668
20.00	700.487	22.00	1013.731	24.00	1495.833
20.10	713.050	22.10	1033.299	24.10	1525.601
20.20	725.900	22.20	1053.292	24.20	1555.981
20.30	739.045	22.30	1073.720	24.30	1586.985
20.40	752.489	22.40	1094.590	24.40	1618.624
20.50	766.240	22.50	1115.911	24.50	1650.910
20.60	780.303	22.60	1137.690	24.60	1683.853
20.70	794.685	22.70	1159.938	24.70	1717.466
20.80	809.393	22.80	1182.661	24.80	1751.761
20.90	824.432	22.90	1205.870	24.90	1786.749
21.00	839.809	23.00	1229.572	25.00	1822.442

Temperature = 393.15 K

Density (ρ; mol/L)	Pressure (P; bar)	Density (ρ; mol/L)	Pressure (P; bar)	Density (ρ; mol/L)	Pressure (P; bar)
0.20	6.457	3.20	86.799	6.20	146.792
0.30	9.626	3.30	89.042	6.30	148.617
0.40	12.755	3.40	91.262	6.40	150.438
0.50	15.847	3.50	93.462	6.50	152.253
0.60	18.900	3.60	95.640	6.60	154.064
0.70	21.915	3.70	97.798	6.70	155.871
0.80	24.894	3.80	99.936	6.80	157.675
0.90	27.837	3.90	102.055	6.90	159.477
1.00	30.743	4.00	104.156	7.00	161.276
1.10	33.615	4.10	106.239	7.10	163.073
1.20	36.452	4.20	108.304	7.20	164.870
1.30	39.254	4.30	110.352	7.30	166.666
1.40	42.024	4.40	112.384	7.40	168.462
1.50	44.760	4.50	114.400	7.50	170.259
1.60	47.464	4.60	116.401	7.60	172.057
1.70	50.136	4.70	118.387	7.70	173.857
1.80	52.777	4.80	120.359	7.80	175.660
1.90	55.388	4.90	122.318	7.90	177.465
2.00	57.968	5.00	124.263	8.00	179.274
2.10	60.519	5.10	126.196	8.10	181.087
2.20	63.041	5.20	128.118	8.20	182.904
2.30	65.534	5.30	130.027	8.30	184.728
2.40	68.000	5.40	131.926	8.40	186.557
2.50	70.439	5.50	133.815	8.50	188.393
2.60	72.851	5.60	135.693	8.60	190.236
2.70	75.238	5.70	137.563	8.70	192.086
2.80	77.599	5.80	139.424	8.80	193.946
2.90	79.935	5.90	141.277	8.90	195.815
3.00	82.246	6.00	143.122	9.00	197.693
3.10	84.534	6.10	144.960	9.10	199.582

P-ρ-T TABLE FOR CARBON DIOXIDE

Temperature = 393.15 K (continued)

Density (ρ; mol/L)	Pressure (P; bar)	Density (ρ; mol/L)	Pressure (P; bar)	Density (ρ; mol/L)	Pressure (P; bar)
9.20	201.483	14.60	346.882	19.90	751.041
9.30	203.395	14.70	351.015	20.00	764.126
9.40	205.320	14.80	355.232	20.10	777.503
9.50	207.258	14.90	359.534	20.20	791.176
9.60	209.211	15.00	363.925	20.30	805.153
9.70	211.178	15.10	368.407	20.40	819.438
9.80	213.162	15.20	372.981	20.50	834.039
9.90	215.161	15.30	377.651	20.60	848.962
10.00	217.178	15.40	382.419	20.70	864.213
10.10	219.213	15.50	387.288	20.80	879.798
10.20	221.267	15.60	392.259	20.90	895.725
10.30	223.341	15.70	397.336	21.00	912.000
10.40	225.436	15.80	402.522	21.10	928.630
10.50	227.552	15.90	407.819	21.20	945.622
10.60	229.690	16.00	413.230	21.30	962.983
10.70	231.852	16.10	418.758	21.40	980.720
10.80	234.039	16.20	424.406	21.50	998.841
10.90	236.250	16.30	430.177	21.60	1017.352
11.00	238.489	16.40	436.075	21.70	1036.262
11.10	240.754	16.50	442.101	21.80	1055.577
11.20	243.048	16.60	448.261	21.90	1075.306
11.30	245.371	16.70	454.556	22.00	1095.456
11.40	247.725	16.80	460.990	22.10	1116.036
11.50	250.110	16.90	467.567	22.20	1137.053
11.60	252.528	17.00	474.290	22.30	1158.516
11.70	254.980	17.10	481.162	22.40	1180.433
11.80	257.467	17.20	488.188	22.50	1202.812
11.90	259.990	17.30	495.371	22.60	1225.662
12.00	262.551	17.40	502.714	22.70	1248.922
12.10	265.150	17.50	510.222	22.80	1272.810
12.20	267.789	17.60	517.899	22.90	1297.127
12.30	270.470	17.70	525.748	23.00	1321.949
12.40	273.193	17.80	533.774	23.10	1347.288
12.50	275.959	17.90	541.980	23.20	1373.152
12.60	278.771	18.00	550.371	23.30	1399.551
12.70	281.630	18.10	558.950	23.40	1426.495
12.80	284.537	18.20	567.724	23.50	1453.994
12.90	287.493	18.30	576.695	23.60	1482.057
13.00	290.500	18.40	585.869	23.70	1510.695
13.10	293.560	18.50	595.249	23.80	1539.918
13.20	296.674	18.60	604.841	23.90	1569.737
13.30	299.843	18.70	614.649	24.00	1600.162
13.40	303.069	18.80	624.678	24.10	1631.204
13.50	306.355	18.90	634.933	24.20	1662.875
13.60	309.700	19.00	645.419	24.30	1695.186
13.70	313.108	19.10	656.141	24.40	1728.147
13.80	316.580	19.20	667.104	24.50	1761.772
13.90	320.118	19.30	678.313	24.60	1796.070
14.00	323.723	19.40	689.774	24.70	1831.055
14.10	327.398	19.50	701.491	24.80	1866.739
14.20	331.144	19.60	713.471	24.90	1903.133
14.30	334.964	19.70	725.719	25.00	1940.252
14.40	338.858	19.80	738.240		
14.50	342.830				

P-ρ-T TABLE FOR CARBON DIOXIDE

Temperature = 403.15 K

Density (ρ; mol/L)	Pressure (P; bar)	Density (ρ; mol/L)	Pressure (P; bar)	Density (ρ; mol/L)	Pressure (P; bar)
0.20	6.627	5.60	142.705	11.00	257.059
0.30	9.883	5.70	144.740	11.10	259.605
0.40	13.101	5.80	146.768	11.20	262.182
0.50	16.283	5.90	148.789	11.30	264.792
0.60	19.428	6.00	150.803	11.40	267.435
0.70	22.537	6.10	152.813	11.50	270.113
0.80	25.612	6.20	154.817	11.60	272.828
0.90	28.651	6.30	156.816	11.70	275.579
1.00	31.656	6.40	158.812	11.80	278.370
1.10	34.628	6.50	160.805	11.90	281.199
1.20	37.566	6.60	162.794	12.00	284.070
1.30	40.472	6.70	164.782	12.10	286.983
1.40	43.346	6.80	166.768	12.20	289.939
1.50	46.189	6.90	168.753	12.30	292.941
1.60	49.000	7.00	170.737	12.40	295.989
1.70	51.782	7.10	172.722	12.50	299.084
1.80	54.534	7.20	174.707	12.60	302.229
1.90	57.256	7.30	176.693	12.70	305.424
2.00	59.591	7.40	178.682	12.80	308.671
2.10	62.617	7.50	180.672	12.90	311.972
2.20	65.255	7.60	182.666	13.00	315.327
2.30	67.867	7.70	184.664	13.10	318.740
2.40	70.453	7.80	186.665	13.20	322.211
2.50	73.013	7.90	188.672	13.30	325.742
2.60	75.548	8.00	190.684	13.40	329.334
2.70	78.058	8.10	192.702	13.50	332.990
2.80	80.544	8.20	194.727	13.60	336.710
2.90	83.007	8.30	196.760	13.70	340.498
3.00	85.447	8.40	198.800	13.80	344.354
3.10	87.864	8.50	200.849	13.90	348.281
3.20	90.260	8.60	202.908	14.00	352.280
3.30	92.635	8.70	204.976	14.10	356.353
3.40	94.989	8.80	207.056	14.20	360.503
3.50	97.324	8.90	209.146	14.30	364.731
3.60	99.638	9.00	211.249	14.40	369.039
3.70	101.934	9.10	213.365	14.50	373.430
3.80	104.212	9.20	215.494	14.60	377.905
3.90	106.471	9.30	217.638	14.70	382.468
4.00	108.714	9.40	219.796	14.80	387.119
4.10	110.940	9.50	221.971	14.90	391.862
4.20	113.149	9.60	224.162	15.00	396.698
4.30	115.343	9.70	226.370	15.10	401.631
4.40	117.523	9.80	228.597	15.20	406.662
4.50	119.687	9.90	230.843	15.30	411.794
4.60	121.838	10.00	233.109	15.40	417.030
4.70	123.976	10.10	235.395	15.50	422.372
4.80	126.101	10.20	237.703	15.60	427.823
4.90	128.213	10.30	240.034	15.70	433.386
5.00	130.314	10.40	242.388	15.80	439.063
5.10	132.404	10.50	244.767	15.90	444.858
5.20	134.483	10.60	247.170	16.00	450.773
5.30	136.552	10.70	249.601	16.10	456.811
5.40	138.612	10.80	252.058	16.20	462.975
5.50	140.663	10.90	254.544	16.30	469.268

P-ρ-T TABLE FOR CARBON DIOXIDE

Temperature = 403.15 K (continued)

Density (ρ; mol/L)	Pressure (P; bar)	Density (ρ; mol/L)	Pressure (P; bar)	Density (ρ; mol/L)	Pressure (P; bar)
16.40	475.694	19.30	736.283	22.20	1220.413
16.50	482.256	19.40	748.491	22.30	1242.906
16.60	488.956	19.50	760.964	22.40	1265.865
16.70	495.799	19.60	773.707	22.50	1289.298
16.80	502.788	19.70	786.727	22.60	1313.214
16.90	509.926	19.80	800.029	22.70	1337.622
17.00	517.216	19.90	813.619	22.80	1362.531
17.10	524.663	20.00	827.503	22.90	1387.951
17.20	532.270	20.10	841.687	23.00	1413.890
17.30	540.041	20.20	856.176	23.10	1440.359
17.40	547.980	20.30	870.977	23.20	1467.366
17.50	556.090	20.40	886.097	23.30	1494.921
17.60	564.375	20.50	901.541	23.40	1523.035
17.70	572.841	20.60	917.316	23.50	1551.718
17.80	581.489	20.70	933.429	23.60	1580.979
17.90	590.326	20.80	949.886	23.70	1610.830
18.00	599.355	20.90	966.695	23.80	1641.281
18.10	608.580	21.00	983.861	23.90	1672.342
18.20	618.006	21.10	1001.392	24.00	1704.025
18.30	627.637	21.20	1019.294	24.10	1736.340
18.40	637.479	21.30	1037.576	24.20	1769.300
18.50	647.534	21.40	1056.245	24.30	1802.915
18.60	657.809	21.50	1075.307	24.40	1837.198
18.70	668.308	21.60	1094.770	24.50	1872.159
18.80	679.036	21.70	1114.642	24.60	1907.812
18.90	689.997	21.80	1134.931	24.70	1944.168
19.00	701.198	21.90	1155.645	24.80	1981.239
19.10	712.642	22.00	1176.791	24.90	1019.040
19.20	724.335	22.10	1198.378	25.00	2057.581

Temperature = 413.15 K

Density (ρ; mol/L)	Pressure (P; bar)	Density (ρ; mol/L)	Pressure (P; bar)	Density (ρ; mol/L)	Pressure (P; bar)
0.20	6.796	2.30	70.189	4.40	122.628
0.30	10.139	2.40	72.893	4.50	124.941
0.40	13.446	2.50	75.573	4.60	127.241
0.50	16.718	2.60	78.229	4.70	129.528
0.60	19.955	2.70	80.863	4.80	131.805
0.70	23.158	2.80	83.473	4.90	134.070
0.80	26.327	2.90	86.062	5.00	136.326
0:90	29.463	3.00	88.629	5.10	138.571
1.00	32.566	3.10	91.175	5.20	140.808
1.10	35.638	3.20	93.701	5.30	143.036
1.20	38,677	3.30	96.207	5.40	145.256
1.30	41.686	3.40	98.694	5.50	147.468
1.40	44.664	3.50	101.162	5.60	149.673
1.50	47.612	3.60	103.612	5.70	151.873
1.60	50.531	3.70	106.045	5.80	154.066
1.70	53.421	3.80	108.460	5.90	156.254
1.80	56.283	3.90	110.860	6.00	158.438
1.90	59.117	4.00	113.243	6.10	160.617
2.00	61.924	4.10	115.611	6.20	162.793
2.10	64.705	4.20	117.964	6.30	164.966
2.20	67.459	4.30	120.303	6.40	167.137

P-ρ-T TABLE FOR CARBON DIOXIDE

Temperature = 413.15 K (continued)

Density (ρ; mol/L)	Pressure (P; bar)	Density (ρ; mol/L)	Pressure (P; bar)	Density (ρ; mol/L)	Pressure (P; bar)
6.50	169.306	11.90	302.368	17.30	584.594
6.60	171.474	12.00	305.550	17.40	593.124
6.70	173.641	12.10	308.777	17.50	601.831
6.80	175.809	12.20	312.052	17.60	610.721
6.90	177.977	12.30	315.375	17.70	619.798
7.00	180.146	12.40	318.748	17.80	629.065
7.10	182.317	12.50	322.173	17.90	638.527
7.20	184.490	12.60	325.650	18.00	648.189
7.30	186.667	12.70	329.182	18.10	658.054
7.40	188.847	12.80	332.770	18.20	668.127
7.50	191.031	12.90	336.416	18.30	678.413
7.60	193.220	13.00	340.121	18.40	688.917
7.70	195.415	13.10	343.887	18.50	699.642
7.80	197.616	13.20	347.715	18.60	710.594
7.90	199.824	13.30	351.607	18.70	721.778
8.00	202.039	13.40	355.566	18.80	733.199
8.10	204.262	13.50	359.592	18.90	744.861
8.20	206.494	13.60	363.687	19.00	756.770
8.30	208.736	13.70	367.854	19.10	768.931
8.40	210.987	13.80	372.094	19.20	781.349
8.50	213.250	13.90	376.409	19.30	794.030
8.60	215.524	14.00	380.801	19.40	806.978
8.70	217.810	14.10	385.273	19.50	820.200
8.80	220.109	14.20	398.825	19.60	833.702
8.90	222.422	14.30	394.461	19.70	847.488
9.00	224.749	14.40	399.182	19.80	861.565
9.10	227.091	14.50	403.991	19.90	875.938
9.20	229.450	14.60	408.889	20.00	890.614
9.30	231.825	14.70	413.880	20.10	905.598
9.40	234.217	14.80	418.964	20.20	920.898
9.50	236.628	14.90	424.145	20.30	936.518
9.60	239.058	15.00	429.426	20.40	952.466
9.70	241.508	15.10	434.808	20.50	968.747
9.80	243.979	15.20	440.294	20.60	985.369
9.90	246.471	15.30	445.886	20.70	1002.338
10.00	248.986	15.40	451.588	20.80	1019.661
10.10	251.524	15.50	457.402	20.90	1037.345
10.20	254.087	15.60	463.330	21.00	1055.396
10.30	256.675	15.70	469.376	21.10	1073.822
10.40	259.290	15.80	475.542	21.20	1092.631
10.50	261.931	15.90	481.831	21.30	1111.828
10.60	264.601	16.00	488.247	21.40	1131.422
10.70	267.300	16.10	494.791	21.50	1151.421
10.80	270.029	16.20	501.468	21.60	1171.831
10.90	272.790	16.30	508.281	21.70	1192.661
11.00	275.583	16.40	515.232	21.80	1213.919
11.10	278.410	16.50	522.325	21.90	1235.613
11.20	281.271	16.60	529.563	22.00	1257.750
11.30	284.168	16.70	536.950	22.10	1280.339
11.40	287.102	16.80	544.489	22.20	1303.389
11.50	290.074	16.90	552.184	22.30	1326.907
11.60	293.085	17.00	560.038	22.40	1350.903
11.70	296.137	17.10	568.056	22.50	1375.386
11.80	299.231	17.20	576.240	22.60	1400.365

P-ρ-T TABLE FOR CARBON DIOXIDE

Temperature = 413.15 K (continued)

Density (ρ; mol/L)	Pressure (P; bar)	Density (ρ; mol/L)	Pressure (P; bar)	Density (ρ; mol/L)	Pressure (P; bar)
22.70	1425.847	23.50	1649.012	24.30	1910.199
22.80	1451.844	23.60	1679.470	24.40	1945.801
22.90	1478.363	23.70	1710.531	24.50	1982.099
23.00	1505.416	23.80	1742.207	24.60	2019.104
23.10	1533.011	23.90	1774.509	24.70	2056.830
23.20	1561.157	24.00	1807.447	24.80	2095.289
23.30	1589.867	24.10	1841.034	24.90	2134.494
23.40	1619.148	24.20	1875.281	25.00	2174.459

Temperature = 423.15 K

Density (ρ; mol/L)	Pressure (P; bar)	Density (ρ; mol/L)	Pressure (P; bar)	Density (ρ; mol/L)	Pressure (P; bar)
0.20	6.965	4.40	127.705	8.60	228.088
0.30	10.396	4.50	130.164	8.70	230.591
0.40	13.791	4.60	132.612	8.80	233.110
0.50	17.153	4.70	135.049	8.90	235.645
0.60	20.482	4.80	137.476	9.00	238.197
0.70	23.778	4.90	139.894	9.10	240.766
0.80	27.042	5.00	142.303	9.20	243.353
0.90	30.274	5.10	144.704	9.30	245.960
1.00	33.475	5.20	147.096	9.40	248.586
1.10	36.645	5.30	149.482	9.50	251.233
1.20	39.785	5.40	151.861	9.60	253.902
1.30	42.896	5.50	154.234	9.70	256.594
1.40	45.977	5.60	156.602	9.80	259.308
1.50	49.031	5.70	158.965	9.90	262.048
1.60	52.056	5.80	161.323	10.00	264.812
1.70	55.054	5.90	163.678	10.10	267.602
1.80	58.025	6.00	166.030	10.20	270.420
1.90	60.971	6.10	168.379	10.30	273.265
2.00	63.890	6.20	170.726	10.40	276.140
2.10	66.784	6.30	173.072	10.50	279.045
2.20	69.654	6.40	175.417	10.60	281.981
2.30	72.500	6.50	177.762	10.70	284.949
2.40	75.322	6.60	180.108	10.80	287.950
2.50	78.121	6.70	182.454	10.90	290.986
2.60	80.899	6.80	184.802	11.00	294.057
2.70	83.654	6.90	187.153	11.10	297.165
2.80	86.388	7.00	189.506	11.20	300.311
2.90	89.101	7.10	191.864	11.30	303.495
3.00	91.795	7.20	194.225	11.40	306.720
3.10	94.469	7.30	196.591	11.50	309.986
3.20	97.124	7.40	198.962	11.60	313.295
3.30	99.760	7.50	201.340	11.70	316.648
3.40	102.379	7.60	203.724	11.80	320.046
3.50	104.980	7.70	206.116	11.90	323.490
3.60	107.564	7.80	208.516	12.00	326.983
3.70	110.133	7.90	210.924	12.10	330.525
3.80	112.685	8.00	213.342	12.20	334.118
3.90	115.223	8.10	215.770	12.30	337.763
4.00	117.746	8.20	218.209	12.40	341.462
4.10	120.255	8.30	220.660	12.50	345.216
4.20	122.751	8.40	223.123	12.60	349.027
4.30	125.234	8.50	225.598	12.70	352.896

P-ρ-T TABLE FOR CARBON DIOXIDE

Temperature = 423.15 K (continued)

Density (ρ; mol/L)	Pressure (P; bar)	Density (ρ; mol/L)	Pressure (P; bar)	Density (ρ; mol/L)	Pressure (P; bar)
12.80	356.825	16.90	594.327	21.00	1126.615
12.90	360.815	17.00	602.741	21.10	1145.932
13.00	364.869	17.10	611.325	21.20	1165.640
13.10	368.988	17.20	620.082	21.30	1185.749
13.20	373.173	17.30	629.016	21.40	1206.264
13.30	377.427	17.40	638.131	21.50	1227.195
13.40	381.751	17.50	647.432	21.60	1248.548
13.50	386.147	17.60	656.922	21.70	1270.331
13.60	390.617	17.70	666.606	21.80	1292.553
13.70	395.163	17.80	676.487	21.90	1315.222
13.80	399.786	17.90	686.570	22.00	1338.347
13.90	404.489	18.00	696.860	22.10	1361.934
14.00	409.274	18.10	707.360	22.20	1385.994
14.10	414.143	18.20	718.076	22.30	1410.535
14.20	419.097	18.30	729.012	22.40	1435.565
14.30	424.139	18.40	740.173	22.50	1461.095
14.40	429.272	18.50	751.563	22.60	1487.132
14.50	434.497	18.60	763.188	22.70	1513.686
14.60	439.817	18.70	775.052	22.80	1540.766
14.70	445.234	18.80	787.160	22.90	1568.383
14.80	450.750	18.90	799.517	23.00	1596.546
14.90	456.368	19.00	812.130	23.10	1625.264
15.00	462.091	19.10	825.002	23.20	1654.548
15.10	467.920	19.20	838.139	23.30	1684.409
15.20	473.859	19.30	851.548	23.40	1714.855
15.30	479.910	19.40	865.232	23.50	1745.899
15.40	486.075	19.50	879.198	23.60	1777.550
15.50	492.358	19.60	893.452	23.70	1809.820
15.60	498.761	19.70	907.999	23.80	1842.720
15.70	505.288	19.80	922.845	23.90	1876.260
15.80	511.940	19.90	937.996	24.00	1910.453
15.90	518.721	20.00	953.459	24.10	1945.309
16.00	525.635	20.10	969.239	24.20	1980.842
16.10	532.683	20.20	985.343	24.30	2017.062
16.20	539.870	20.30	1001.777	24.40	2053.982
16.30	547.198	20.40	1018.548	24.50	2091.614
16.40	554.671	20.50	1035.662	24.60	2129.972
16.50	562.292	20.60	1053.125	24.70	2169.066
16.60	570.065	20.70	1070.945	24.80	2208.912
16.70	577.993	20.80	1089.129	24.90	2249.522
16.80	586.079	20.90	1107.683	25.00	2290.909

Temperature = 433.15 K

Density (ρ; mol/L)	Pressure (P; bar)	Density (ρ; mol/L)	Pressure (P; bar)	Density (ρ; mol/L)	Pressure (P; bar)
0.20	7.135	1.10	37.650	2.00	65.848
0.30	10.562	1.20	40.890	2.10	68.856
0.40	14.136	1.30	44,103	2.20	71.840
0.50	17.588	1.40	47.287	2.30	74.802
0.60	21.008	1.50	56.445	2.40	77.741
0.70	24.397	1.60	53.577	2.50	80.659
0.80	27.755	1.70	56.682	2.60	83.556
0.90	31.083	1.80	59.762	2.70	86.433
1.00	34.381	1.90	62.817	2.80	89.290

P-ρ-T TABLE FOR CARBON DIOXIDE

Temperature = 433.15 K (continued)

Density (ρ; mol/L)	Pressure (P; bar)	Density (ρ; mol/L)	Pressure (P; bar)	Density (ρ; mol/L)	Pressure (P; bar)
2.90	92.127	8.30	232.536	13.70	422.414
3.00	94.946	8.40	235.210	13.80	427.419
3.10	97.747	8.50	237.898	13.90	432.509
3.20	100.530	8.60	240.603	14.00	437.686
3.30	103.296	8.70	243.324	14.10	442.950
3.40	106.045	8.80	246.062	14.20	448.306
3.50	108.779	8.90	248.819	14.30	453.754
3.60	111.497	9.00	251.595	14.40	459.296
3.70	114.200	9.10	254.391	14.50	464.937
3.80	116.889	9.20	257.207	14.60	470.677
3.90	119.564	9.30	260.045	14.70	476.519
4.00	122.227	9.40	262.905	14.80	482.465
4.10	124.876	9.50	265.789	14.90	488.518
4.20	127.514	9.60	268.696	15.00	494.681
4.30	130.140	9.70	271.629	15.10	500.956
4.40	132.755	9.80	274.588	15.20	507.345
4.50	135.360	9.90	277.573	15.30	513.852
4.60	137.955	10.00	280.587	15.40	520.479
4.70	140.541	10.10	283.629	15.50	527.229
4.80	143.118	10.20	286.802	15.60	534.105
4.90	145.687	10.30	289.805	15.70	541.109
5.00	148.249	10.40	292.940	15.80	548.245
5.10	150.804	10.50	296.108	15.90	555.516
5.20	153.352	10.60	299.310	16.00	562.924
5.30	155.895	10.70	302.547	16.10	570.474
5.40	158.433	10.80	305.821	16.20	578.168
5.50	160.966	10.90	309.131	16.30	586.009
5.60	163.495	11.00	312.481	16.40	594.001
5.70	166.021	11.10	315.870	16.50	602.147
5.80	168.544	11.20	319.300	16.60	610.450
5.90	171.064	11.30	322.772	16.70	618.915
6.00	173.584	11.40	326.287	16.80	627.545
6.10	176.102	11.50	329.847	16.90	636.343
6.20	178.620	11.60	333.453	17.00	645.313
6.30	181.138	11.70	337.107	17.10	654.460
6.40	183.657	11.80	340.809	17.20	663.786
6.50	186.177	11.90	344.561	17.30	673.296
6.60	188.700	12.00	348.365	17.40	682.994
6.70	191.225	12.10	352.221	17.50	692.884
6.80	193.754	12.20	356.132	17.60	702.970
6.90	196.286	12.30	360.099	17.70	713.256
7.00	198.824	12.40	364.124	17.80	723.747
7.10	201.366	12.50	368.207	17.90	734.447
7.20	203.915	12.60	372.350	18.00	745.360
7.30	206.470	12.70	376.556	18.10	756.492
7.40	209.033	12.80	380.826	18.20	767.846
7.50	211.603	12.90	385.161	18.30	779.428
7.60	214.182	13.00	389.564	18.40	791.241
7.70	216.771	13.10	394.035	18.50	803.292
7.80	219.369	13.20	398.577	18.60	815.584
7.90	221.978	13.30	403.192	18.70	828.123
8.00	224.599	13.40	407.881	18.80	840.915
8.10	227.231	13.50	412.646	18.90	853.963
8.20	229.877	13.60	417.490	19.00	867.274

P-ρ-T TABLE FOR CARBON DIOXIDE

Temperature = 433.15 K (continued)

Density (ρ; mol/L)	Pressure (P; bar)	Density (ρ; mol/L)	Pressure (P; bar)	Density (ρ; mol/L)	Pressure (P; bar)
19.10	880.853	21.10	1217.728	23.10	1717.139
19.20	894.705	21.20	1238.333	23.20	1747.558
19.30	908.836	21.30	1259.349	23.30	1778.567
19.40	923.251	21.40	1280.781	23.40	1810.177
19.50	937.957	21.50	1302.640	23.50	1842.398
19.60	952.958	21.60	1324.931	23.60	1875.241
19.70	968.261	21.70	1347.665	23.70	1908.718
19.80	983.872	21.80	1370.848	23.80	1942.839
19.90	999.797	21.90	1394.489	23.90	1977.617
20.00	1016.042	22.00	1418.597	24.00	2013.062
20.10	1032.613	22.10	1443.179	24.10	2049.187
20.20	1049.517	22.20	1468.246	24.20	2086.004
20.30	1066.760	22.30	1493.806	24.30	2123.524
20.40	1084.349	22.40	1519.867	24.40	2161.762
20.50	1102.290	22.50	1546.440	24.50	2200.728
20.60	1120.591	22.60	1573.533	24.60	2240.436
20.70	1139.257	22.70	1601.155	24.70	2280.899
20.80	1158.297	22.80	1629.317	24.80	2322.131
20.90	1177.718	22.90	1658.028	24.90	2364.144
21.00	1197.526	23.00	1687.299	25.00	2406.952

Temperature = 443.15 K

Density (ρ; mol/L)	Pressure (P; bar)	Density (ρ; mol/L)	Pressure (P; bar)	Density (ρ; mol/L)	Pressure (P; bar)
0.20	7.304	3.20	103.921	6.20	186.477
0.30	10.908	3.30	106.816	6.30	189.167
0.40	14.480	3.40	109.695	6.40	191.859
0.50	18.022	3.50	112.560	6.50	194.555
0.60	21.533	3.60	115.411	6.60	197.254
0.70	25.015	3.70	118.249	6.70	199.957
0.80	28.467	3.80	121.073	6.80	202.666
0.90	31.890	3.90	123.886	6.90	205.380
1.00	35.285	4.00	126.686	7.00	208.101
1.10	38.653	4.10	129.475	7.10	210.829
1.20	41.993	4.20	132.254	7.20	213.564
1.30	45.307	4.30	135.022	7.30	216.308
1.40	48.594	4.40	137.781	7.40	219.061
1.50	51.856	4.50	140.531	7.50	221.824
1.60	55.093	4.60	143.272	7.60	224.598
1.70	58.305	4.70	146.006	7.70	227.382
1.80	61.494	4.80	148.733	7.80	230.179
1.90	64.659	4.90	151.453	7.90	232.988
2.00	67.801	5.00	154.166	8.00	235.811
2.10	70.920	5.10	156.875	8.10	238.648
2.20	74.018	5.20	159.578	8.20	241.500
2.30	77.095	5.30	162.277	8.30	244.367
2.40	80.151	5.40	164.973	8.40	247.252
2.50	83.187	5.50	167.665	8.50	250.153
2.60	86.203	5.60	170.355	8.60	253.072
2.70	89.201	5.70	173.044	8.70	256.010
2.80	92.180	5.80	175.730	8.80	258.968
2.90	95.141	5.90	178.417	8.90	261.947
3.00	98.084	6.00	181.103	9.00	264.946
3.10	101.011	6.10	183.789	9.10	267.968

P-ρ-T TABLE FOR CARBON DIOXIDE

Temperature = 443.15 K (continued)

Density (ρ; mol/L)	Pressure (P; bar)	Density (ρ; mol/L)	Pressure (P; bar)	Density (ρ; mol/L)	Pressure (P; bar)
9.20	271.013	14.60	501.459	19.90	1061.343
9.30	274.082	14.70	507.724	20.00	1078.366
9.40	277.176	14.80	514.099	20.10	1095.724
9.50	280.296	14.90	520.585	20.20	1113.424
9.60	283.442	15.00	527.187	20.30	1131.472
9.70	286.616	15.10	533.905	20.40	1149.875
9.80	289.818	15.20	540.743	20.50	1168.640
9.90	293.050	15.30	547.704	20.60	1187.773
10.00	296.313	15.40	554.790	20.70	1207.283
10.10	299.607	15.50	562.005	20.80	1227.175
10.20	302.934	15.60	569.351	20.90	1247.458
10.30	306.294	15.70	576.831	21.00	1268.138
10.40	309.690	15.80	584.448	21.10	1289.223
10.50	313.121	15.90	592.206	21.20	1310.721
10.60	316.589	16.00	600.107	21.30	1332.639
10.70	320.094	16.10	608.155	21.40	1354.986
10.80	323.640	16.20	616.353	21.50	1377.768
10.90	327.225	16.30	624.704	21.60	1400.995
11.00	330.852	16.40	633.211	21.70	1424.675
11.10	334.522	16.50	641.879	21.80	1448.816
11.20	338.236	16.60	650.711	21.90	1473.426
11.30	341.996	16.70	659.710	22.00	1498.514
11.40	345.802	16.80	668.880	22.10	1524.089
11.50	349.655	16.90	678.225	22.20	1550.159
11.60	353.559	17.00	687.748	22.30	1576.735
11.70	357.512	17.10	697.454	22.40	1603.825
11.80	361.518	17.20	707.346	22.50	1631.438
11.90	365.578	17.30	717.428	22.60	1659.584
12.00	369.692	17.40	727.705	22.70	1688.273
12.10	373.863	17.50	738.180	22.80	1717.514
12.20	378.091	17.60	748.859	22.90	1747.317
12.30	382.380	17.70	759.744	23.00	1777.693
12.40	386.729	17.80	770.841	23.10	1808.652
12.50	391.141	17.90	782.154	23.20	1840.204
12.60	395.617	18.00	793.687	23.30	1872.360
12.70	400.159	18.10	805.446	23.40	1905.131
12.80	404.769	18.20	817.434	23.50	1938.528
12.90	409.448	18.30	829.657	23.60	1972.561
13.00	414.199	18.40	842.119	23.70	2007.243
13.10	419.022	18.50	854.826	23.80	2042.585
13.20	423.920	18.60	867.782	23.90	2078.598
13.30	428.895	18.70	880.993	24.00	2115.295
13.40	433.948	18.80	894.463	24.10	2152.687
13.50	439.082	18.90	908.193	24.20	2190.787
13.60	444.298	19.00	922.204	24.30	2229.607
13.70	449.599	19.10	936.485	24.30	2269.161
13.80	454.986	19.20	951.047	24.50	2309.460
13.90	460.462	19.30	965.897	24.60	2350.518
14.00	466.028	19.40	981.039	24.70	2392.349
14.10	471.688	19.50	996.479	24.80	2434.965
14.20	477.443	19.60	1012.223	24.90	2478.381
14.30	483.295	19.70	1028.278	25.00	2522.610
14.40	489.247	19.80	1044.650		
14.50	495.300				

P-ρ-T TABLE FOR CARBON DIOXIDE

Temperature = 453.15 K

Density (ρ; mol/L)	Pressure (P; bar)	Density (ρ; mol/L)	Pressure (P; bar)	Density (ρ; mol/L)	Pressure (P; bar)
0.20	7.473	5.60	177.186	11.00	349.172
0.30	11.164	5.70	180.036	11.10	353.122
0.40	14.824	5.80	182.886	11.20	357.120
0.50	18.456	5.90	185.737	11.30	361.166
0.60	22.058	6.00	188.590	11.40	365.262
0.70	25.632	6.10	191.444	11.50	369.409
0.80	29.178	6.20	194.302	11.60	373.609
0.90	32.697	6.30	197.163	11.70	377.863
1.00	36.188	6.40	200.028	11.80	382.172
1.10	39.654	6.50	202.898	11.90	386.538
1.20	43.094	6.60	205.773	12.00	390.962
1.30	46.508	6.70	208.654	12.10	395.446
1.40	49.898	6.80	211.542	12.20	399.992
1.50	53.263	6.90	214.438	12.30	404.601
1.60	56.605	7.00	217.341	12.40	409.275
1.70	59.924	7.10	220.254	12.50	414.015
1.80	63.220	7.20	223.176	12.60	418.823
1.90	66.494	7.30	226.108	12.70	423.700
2.00	69.747	7.40	229.051	12.80	428.649
2.10	72.978	7.50	232.006	12.90	433.672
2.20	76.190	7.60	234.974	13.00	438.769
2.30	79.381	7.70	237.954	13.10	443.944
2.40	82.553	7.80	240.949	13.20	449.197
2.50	85.706	7.90	243.958	13.30	454.531
2.60	88.841	8.00	246.982	13.40	459.947
2.70	91.059	8.10	250.023	13.50	465.448
2.80	95.059	8.20	253.081	13.60	471.036
2.90	98.142	8.30	253.157	13.70	476.713
3.00	101.210	8.40	259.251	13.80	482.480
3.10	104.262	8.50	262.364	13.90	488.341
3.20	107.299	8.60	265.498	14.00	494.296
3.30	110.322	8.70	268.653	14.10	500.349
3.40	113.331	8.80	271.830	14.20	506.502
3.50	116.327	8.90	275.030	14.30	512.757
3.60	119.310	9.00	278.253	14.40	519.116
3.70	122.281	9.10	281.501	14.50	525.582
3.80	125.240	9.20	284.774	14.60	532.157
3.90	128.188	8.30	288.074	14.70	538.844
4.00	131.126	9.40	291.041	14.80	545.646
4.10	134.054	9.50	294.756	14.90	552.564
4.20	136.973	9.60	298.141	15.00	559.602
4.30	139.883	9.70	301.556	15.10	566.762
4.40	142.785	9.80	305.001	15.20	574.047
4.50	145.679	9.90	308.479	15.30	581.459
4.60	148.567	10.00	311.991	15.40	589.003
4.70	151.448	10.10	215.536	15.50	596.680
4.80	154.323	10.20	319.117	15.60	604.494
4.90	157.193	10.30	322.735	15.70	612.447
5.00	160.058	10.40	326.390	15.80	620.543
5.10	162.919	10.50	330.083	15.90	628.785
5.20	165.777	10.60	333.817	16.00	637.176
5.30	168.632	10.70	337.591	16.10	645.720
5.40	171.485	10.80	341.407	16.20	654.419
5.50	174.336	10.90	345.267	16.30	663.277

P-ρ-T TABLE FOR CARBON DIOXIDE

Temperature = 453.15 K (continued)

Density (ρ; mol/L)	Pressure (P; bar)	Density (ρ; mol/L)	Pressure (P; bar)	Density (ρ; mol/L)	Pressure (P; bar)
16.40	672.298	19.30	1022.732	22.20	1631.749
16.50	681.485	19.40	1038.597	22.30	1659.338
16.60	690.842	19.50	1054.768	22.40	1687.453
16.70	700.372	19.60	1071.252	22.50	1716.104
16.80	710.080	19.70	1088.055	22.60	1745.301
16.90	719.968	19.80	1105.183	22.70	1775.054
17.00	730.041	19.90	1122.642	22.80	1805.372
17.10	740.303	20.00	1140.439	22.90	1836.265
17.20	750.757	20.10	1158.580	23.00	1867.745
17.30	761.409	20.20	1177.071	12.10	1899.821
17.40	772.261	20.30	1195.921	23.20	1932.504
17.50	783.319	20.40	1215.134	23.30	1965.806
17.60	794.586	20.50	1234.719	23.40	1999.736
17.70	806.067	20.60	1254.682	23.50	2034.307
17.80	817.766	20.70	1275.031	23.60	2069.529
17.90	829.688	20.80	1295.772	23.70	2105.415
18.00	841.838	20.90	1316.913	23.80	2141.976
18.10	854.220	21.00	1338.462	23.90	2179.223
18.20	866.839	21.10	1360.427	24.00	2217.170
18.30	879.699	21.20	1382.814	24.10	2225.828
18.40	892.807	21.30	1405.632	24.20	2295.210
18.50	906.166	21.40	1428.889	24.30	2335.329
18.60	919.782	21.50	1452.593	24.40	2376.197
18.70	933.660	21.60	1476.752	24.50	2417.829
18.80	947.805	21.70	1501.375	24.60	2460.236
18.90	962.223	21.80	1526.470	24.70	2503.433
19.00	976.919	21.90	1552.046	24.80	2547.434
19.10	991.899	22.00	1578.112	24.90	2592.252
19.20	1007.168	22.10	1604.677	25.00	2637.902

Temperature = 463.15 K

Density (ρ; mol/L)	Pressure (P; bar)	Density (ρ; mol/L)	Pressure (P; bar)	Density (ρ; mol/L)	Pressure (P; bar)
0.20	7.642	2.30	81.660	4.40	147.769
0.30	11.419	2.40	84.948	4.50	150.807
0.40	15.168	2.50	88.218	4.60	153.840
0.50	18.889	2.60	91.471	4.70	156.867
0.60	22.582	2.70	94.707	4.80	159.890
0.70	26.248	2.80	97.928	4.90	162.910
0.80	29.888	2.90	101.134	5.00	165.926
0.90	33.502	3.00	104.325	5.10	168.939
1.00	37.090	3.10	107.502	5.20	171.951
1.10	40.653	3.20	110.665	5.30	174.961
1.20	44.192	3.30	113.815	5.40	177.971
1.30	47.707	3.40	116.953	5.50	180.980
1.40	51.199	3.50	120.079	5.60	183.990
1.50	54.668	3.60	123.193	5.70	187.001
1.60	58.114	3.70	126.297	5.80	190.014
1.70	61.539	3.80	129.391	5.90	193.029
1.80	64.942	3.90	132.475	6.00	196.047
1.90	68.325	4.00	135.549	6.10	199.069
2.00	71.688	4.10	138.616	6.20	202.096
2.10	75.031	4.20	141.674	6.30	205.127
2.20	78.355	4.30	144.725	6.40	208.165

P-ρ-T TABLE FOR CARBON DIOXIDE

Temperature = 463.15 K (continued)

Density (ρ; mol/L)	Pressure (P; bar)	Density (ρ; mol/L)	Pressure (P; bar)	Density (ρ; mol/L)	Pressure (P; bar)
6.50	211.208	11.90	407.440	17.30	805.236
6.60	214.259	12.00	412.174	17.40	816.661
6.70	217.318	12.10	416.971	17.50	828.297
6.80	220.385	12.20	421.833	17.60	840.150
6.90	223.461	12.30	426.762	17.70	852.224
7.00	226.547	12.40	431.759	17.80	864.522
7.10	229.644	12.50	436.826	17.90	877.050
7.20	232.752	12.60	441.965	18.00	889.813
7.30	235.872	12.70	447.177	18.10	902.815
7.40	239.005	12.80	452.465	18.20	916.061
7.50	242.152	12.90	457.830	18.30	929.556
7.60	245.313	13.00	463.273	18.40	943.305
7.70	248.489	13.10	468.798	18.50	957.313
7.80	251.681	13.20	474.405	18.60	971.586
7.90	254.889	13.30	480.097	18.70	986.128
8.00	258.115	13.40	485.875	18.80	1000.945
8.10	261.359	13.50	491.743	18.90	1016.042
8.20	264.623	13.60	497.701	19.00	1031.425
8.30	267.906	13.70	503.752	19.10	1047.100
8.40	271.210	13.80	509.899	19.20	1063.072
8.50	274.535	13.90	516.142	19.30	1079.347
8.60	277.883	14.00	522.486	19.40	1095.931
8.70	281.255	14.10	528.931	19.50	1112.830
8.80	284.650	14.20	535.480	19.60	1130.050
8.90	288.071	14.30	542.136	19.70	1147.597
9.00	291.517	14.40	548.901	19.80	1165.478
9.10	294.991	14.50	555.778	19.90	1183.699
9.20	298.492	14.60	562.768	20.00	1202.267
9.30	302.022	14.70	569.875	20.10	1221.187
9.40	305.582	14.80	577.101	20.20	1240.467
9.50	309.172	14.90	584.449	20.30	1260.115
9.60	312.795	15.00	591.921	20.40	1280.135
9.70	316.450	15.10	599.521	20.50	1300.537
9.80	320.138	15.20	607.251	20.60	1321.326
9.90	323.862	15.30	615.114	20.70	1342.510
10.00	327.622	15.40	623.112	20.80	1364.097
10.10	331.418	15.50	631.250	20.90	1386.095
10.20	335.253	15.60	639.529	21.00	1408.509
10.30	339.127	15.70	647.954	21.10	1431.350
10.40	343.041	15.80	656.526	21.20	1454.623
10.50	346.996	15.90	665.250	21.30	1478.338
10.60	350.995	16.00	674.129	21.40	1502.503
10.70	355.037	16.10	683.166	21.50	1527.126
10.80	359.124	16.20	692.364	21.60	1552.214
10.90	363.258	16.30	701.727	21.70	1577.778
11.00	367.440	16.40	711.258	21.80	1603.825
11.10	371.670	16.50	720.962	21.90	1630.365
11.20	375.951	16.60	730.841	22.00	1657.406
11.30	380.283	16.70	740.899	22.10	1684.957
11.40	384.668	16.80	751.141	22.20	1713.028
11.50	389.108	16.90	761.570	22.30	1741.628
11.60	393.604	17.00	772.190	22.40	1770.767
11.70	398.157	17.10	783.005	22.50	1800.455
11.80	402.768	17.20	794.019	22.60	1830.700

P-ρ-T TABLE FOR CARBON DIOXIDE

Temperature = 463.15 K (continued)

Density (ρ; mol/L)	Pressure (P; bar)	Density (ρ; mol/L)	Pressure (P; bar)	Density (ρ; mol/L)	Pressure (P; bar)
22.70	1861.515	23.50	2129.752	24.30	2440.707
22.80	1892.907	23.60	2166.162	24.40	2482.890
22.90	1924.889	23.70	2203.250	24.50	2525.852
23.00	1957.471	23.80	2241.028	24.60	2569.608
23.10	1990.662	23.90	2279.509	24.70	2614.171
23.20	2024.475	24.00	2318.704	24.80	2659.555
23.30	2058.920	24.10	2358.627	24.90	2705.775
23.40	2094.008	24.20	2399.291	25.00	2752.844

Temperature = 473.15 K

Density (ρ; mol/L)	Pressure (P; bar)	Density (ρ; mol/L)	Pressure (P; bar)	Density (ρ; mol/L)	Pressure (P; bar)
0.20	7.811	4.40	152.735	8.60	290.229
0.30	11.675	4.50	155.916	8.70	293.816
0.40	15.512	4.60	159.093	8.80	297.430
0.50	19.322	4.70	162.267	8.90	301.071
0.60	23.106	4.80	165.437	9.00	304.740
0.70	26.864	4.90	168.605	9.10	308.439
0.80	30.597	5.00	171.772	9.20	312.167
0.90	34.306	5.10	174.937	9.30	315.927
1.00	37.990	5.20	178.102	9.40	319.720
1.10	41.651	5.30	181.267	9.50	323.545
1.20	45.289	5.40	184.432	9.60	327.405
1.30	48.904	5.50	187.599	9.70	331.299
1.40	52.497	5.60	190.768	9.80	335.231
1.50	56.069	5.70	193.940	9.90	339.199
1.60	59.620	5.80	197.115	10.00	343.207
1.70	63.150	5.90	200.294	10.10	347.254
1.80	66.660	6.00	203.477	10.20	351.341
1.90	70.152	6.10	206.666	10.30	355.471
2.00	73.624	6.20	209.861	10.40	359.644
2.10	77.078	6.30	213.063	10.50	363.861
2.20	80.514	6.40	216.272	10.60	368.124
2.30	83.933	6.50	219.489	10.70	372.433
2.40	87.335	6.60	222.715	10.80	376.791
2.50	90.721	6.70	225.950	10.90	381.198
2.60	94.092	6.80	229.196	11.00	385.656
2.70	97.448	6.90	232.452	11.10	390.165
2.80	100.789	7.00	235.721	11.20	394.728
2.90	104.116	7.10	239.001	11.30	399.346
3.00	107.430	7.20	242.295	11.40	404.020
3.10	110.731	7.30	245.603	11.50	408.752
3.20	114.020	7.40	248.925	11.60	413.543
3.30	117.297	7.50	252.263	11.70	418.394
3.40	120.563	7.60	255.617	11.80	423.307
3.50	123.818	7.70	258.988	11.90	428.284
3.60	127.064	7.80	262.377	12.00	433.327
3.70	130.299	7.90	265.784	12.10	438.436
3.80	133.527	8.00	269.211	12.20	443.614
3.90	136.745	8.10	272.658	12.30	448.861
4.00	139.956	8.20	276.127	12.40	454.181
4.10	143.160	8.30	279.617	12.50	459.574
4.20	146.358	8.40	283.131	12.60	465.043
4.30	149.549	8.50	286.668	12.70	470.589

P-ρ-T TABLE FOR CARBON DIOXIDE

Temperature = 473.15 K (continued)

Density (ρ; mol/L)	Pressure (P; bar)	Density (ρ; mol/L)	Pressure (P; bar)	Density (ρ; mol/L)	Pressure (P; bar)
12.80	476.214	16.90	803.030	21.00	1478.289
12.90	481.920	17.00	814.194	21.10	1502.003
13.00	487.708	17.10	825.560	21.20	1526.160
13.10	493.582	17.20	837.130	21.30	1550.770
13.20	499.542	17.30	848.911	21.40	1575.840
13.30	505.591	17.40	860.905	21.50	1601.378
13.40	511.730	17.50	873.118	21.60	1627.394
13.50	517.963	17.60	885.553	21.70	1653.896
13.60	524.290	17.70	898.216	21.80	1680.893
13.70	530.715	17.80	911.110	21.90	1708.393
13.80	537.239	17.90	924.242	22.00	1736.407
13.90	543.864	18.00	937.614	22.10	1764.943
14.00	550.594	18.10	951.233	22.20	1794.011
14.10	557.430	18.20	965.103	22.30	1823.620
14.20	564.374	18.30	979.230	22.40	1853.781
14.30	571.430	18.40	993.617	22.50	1884.502
14.40	578.599	18.50	1008.271	22.60	1915.795
14.50	585.885	18.60	1023.197	22.70	1947.669
14.60	593.288	18.70	1038.399	22.80	1980.135
14.70	600.814	18.80	1053.885	22.90	2013.204
14.80	608.463	18.90	1069.658	23.00	2046.885
14.90	616.329	19.00	1085.725	23.10	2081.191
15.00	624.144	19.10	1102.092	23.20	2116.131
15.10	632.181	19.20	1118.763	23.30	2151.718
15.20	640.354	19.30	1135.746	23.40	2187.963
15.30	648.664	19.40	1153.046	23.50	2224.878
15.40	657.116	19.50	1170.670	23.60	2262.474
15.50	665.712	19.60	1188.623	23.70	2300.763
15.60	674.455	19.70	1206.911	23.80	2339.758
15.70	683.348	19.80	1225.542	23.90	2379.471
15.80	692.395	19.90	1244.522	24.00	2419.914
15.90	701.599	20.00	1263.857	24.10	2461.101
16.00	710.963	20.10	1283.554	24.20	2503.044
16.10	720.490	20.20	1303.620	24.30	2545.757
16.20	730.185	20.30	1324.062	24.40	2589.254
16.30	740.050	20.40	1344.887	24.50	2633.546
16.40	750.090	20.50	1366.102	24.60	2678.650
16.50	760.307	20.60	1387.714	24.70	2724.578
16.60	770.706	20.70	1409.732	24.80	2771.345
16.70	781.290	20.80	1432.162	24.90	2818.965
16.80	792.064	20.90	1455.012	25.00	2867.454

Temperature = 493.15 K

Density (ρ; mol/L)	Pressure (P; bar)	Density (ρ; mol/L)	Pressure (P; bar)	Density (ρ; mol/L)	Pressure (P; bar)
0.20	8.149	1.10	43.643	2.00	77.483
0.30	12.186	1.20	47.477	2.10	81.157
0.40	16.199	1.30	51.292	2.20	84.817
0.50	20.187	1.40	55.088	2.30	88.462
0.60	24.152	1.50	58.864	2.40	92.092
0.70	28.094	1.60	62.622	2.50	95.710
0.90	32.014	1.70	66.363	2.60	99.314
0.90	35.911	1.80	70.086	2.70	102.906
1.00	39.787	1.90	73.793	2.80	106.486

P-ρ-T TABLE FOR CARBON DIOXIDE

Temperature = 493.15 K (continued)

Density (ρ; mol/L)	Pressure (P; bar)	Density (ρ; mol/L)	Pressure (P; bar)	Density (ρ; mol/L)	Pressure (P; bar)
2.90	110.055	8.30	302.933	13.70	584.405
3.00	113.613	8.40	306.864	13.80	591.680
3.10	117.161	8.50	310.822	13.90	599.065
3.20	120.699	8.60	314.810	14.00	606.563
3.30	124.228	8.70	318.827	14.10	614.176
3.40	127.749	8.80	322.876	14.20	621.906
3.50	131.262	8.90	326.956	14.30	629.757
3.60	134.767	9.00	331.069	14.40	637.729
3.70	138.266	9.10	335.217	14.50	645.827
3.80	141.759	9.20	339.399	14.60	654.053
3.90	145.245	9.30	343.617	14.70	662.410
4.00	148.727	9.40	347.872	14.80	670.900
4.10	152.205	9.50	352.166	14.90	679.526
4.20	155.679	9.60	356.498	15.00	688.292
4.30	159.149	9.70	360.871	15.10	697.199
4.40	167.617	9.80	365.286	15.20	706.252
4.50	166.083	9.90	369.743	15.30	715.453
4.60	169.547	10.00	374.244	15.40	724.805
4.70	173.011	10.10	378.790	15.50	734.311
4.80	176.475	10.20	383.382	15.60	743.975
4.90	179.939	10.30	388.022	15.70	753.800
5.00	183.404	10.40	392.710	15.80	763.789
5.10	186.871	10.50	397.448	15.90	773.946
5.20	190.341	10.60	402.238	16.00	784.273
5.30	193.813	10.70	407.080	16.10	794.776
5.40	197.289	10.80	411.976	16.20	805.457
5.50	200.770	10.90	416.927	16.30	816.319
5.60	204.256	11.00	421.935	16.40	827.368
5.70	207.747	11.10	427.001	16.50	838.606
5.80	211.245	11.20	432.127	16.60	850.037
5.90	214.750	11.30	437.313	16.70	861.665
6.00	218.262	11.40	442.562	16.80	873.495
6.10	221.784	11.50	447.875	16.90	885.530
6.20	225.314	11.60	453.254	17.00	897.774
6.30	228.855	11.70	458.699	17.10	910.232
6.40	232.406	11.80	464.213	17.20	922.908
6.50	235.968	11.90	469.798	17.30	935.807
6.60	239.543	12.00	475.455	17.40	948.932
6.70	243.130	12.10	481.185	17.50	962.289
6.80	246.732	12.20	486.991	17.60	975.882
6.90	250.347	12.30	492.874	17.70	989.715
7.00	253.978	12.40	498.836	17.80	1003.793
7.10	257.625	12.50	504.879	17.90	1018.122
7.20	261.289	12.60	511.004	18.00	1032.707
7.30	264.970	12.70	517.214	18.10	1047.551
7.40	268.669	12.80	523.511	18.20	1062.661
7.50	272.388	12.90	529.895	18.30	1078.041
7.60	276.127	13.00	536.371	18.40	1093.696
7.70	279.887	13.10	542.938	18.50	1109.633
7.80	283.668	13.20	549.601	18.60	1125.857
7.90	287.473	13.30	556.360	18.70	1142.373
8.00	291.300	13.40	563.217	18.80	1159.186
8.10	295.152	13.50	570.176	18.90	1176.303
8.20	299.029	13.60	577.238	19.00	1193.729

P-ρ-T TABLE FOR CARBON DIOXIDE

Temperature = 493.15 K (continued)

Density (ρ; mol/L)	Pressure (P; bar)	Density (ρ; mol/L)	Pressure (P; bar)	Density (ρ; mol/L)	Pressure (P; bar)
19.10	1211.471	21.10	1642.543	23.10	2261.365
19.20	1229.533	21.20	1668.461	23.20	2298.558
19.30	1247.924	21.30	1694.852	23.30	2336.425
19.40	1266.647	21.40	1721.725	23.40	2374.979
19.50	1285.711	21.50	1749.089	23.50	2414.233
19.60	1305.121	21.60	1776.952	23.60	2454.197
19.70	1324.884	21.70	1805.324	23.70	2494.885
19.80	1345.006	21.80	1834.214	23.80	2536.310
19.90	1365.495	21.90	1863.631	23.90	2578.484
20.00	1386.356	22.00	1893.584	24.00	2621.420
20.10	1407.598	22.10	1924.084	24.10	2665.131
20.20	1429.228	22.20	1955.140	24.20	2709.632
20.30	1451.251	22.30	1986.762	24.30	2754.936
20.40	1473.676	22.40	2018.960	24.40	2801.056
20.50	1496.511	22.50	2051.744	24.50	2848.007
20.60	1519.762	22.60	2085.126	24.60	2895.804
20.70	1543.438	22.70	2119.115	24.70	2944.460
20.80	1567.546	22.80	2153.722	24.80	2993.990
20.90	1592.094	22.90	2188.959	24.90	3044.411
21.00	1617.090	23.00	2224.836	25.00	3095.736

Temperature = 513.15 K

Density (ρ; mol/L)	Pressure (P; bar)	Density (ρ; mol/L)	Pressure (P; bar)	Density (ρ; mol/L)	Pressure (P; bar)
0.20	8.487	3.20	127.343	6.20	240.673
0.30	12.697	3.30	131.123	6.30	244.550
0.40	16.885	3.40	134.896	6.40	248.442
0.50	21.051	3.50	138.665	6.50	252.348
0.60	25.197	3.60	142.428	6.60	256.269
0.70	29.322	3.70	146.188	6.70	260.207
0.80	33.427	3.80	149.944	6.80	264.162
0.90	37.513	3.90	153.697	6.90	268.135
1.00	41.580	4.00	157.448	7.00	272.127
1.10	45.629	4.10	161.197	7.10	276.138
1.20	49.660	4.20	164.945	7.20	280.170
1.30	53.673	4.30	168.693	7.30	284.223
1.40	57.670	4.40	172.440	7.40	288.298
1.50	61.650	4.50	176.189	7.50	292.396
1.60	65.615	4.60	179.939	7.60	296.518
1.70	69.565	4.70	183.691	7.70	300.664
1.80	73.499	4.80	187.446	7.80	304.837
1.90	77.420	4.90	191.204	7.90	309.035
2.00	81.327	5.00	194.966	8.00	313.262
2.10	85.220	5.10	198.733	8.10	317.517
2.20	89.102	5.20	202.505	8.20	321.801
2.30	92.971	5.30	206.283	8.30	326.116
2.40	96.828	5.40	210.068	8.40	330.461
2.50	100.675	5.50	213.860	8.50	334.840
2.60	104.512	5.60	217.660	8.60	339.251
2.70	108.338	5.70	221.470	8.70	343.697
2.80	112.155	5.80	225.288	8.80	348.178
2.90	115.964	5.90	229.117	8.90	352.696
3.00	119.764	6.00	232.957	9.00	357.251
3.10	123.557	6.10	826.809	9.10	361.548

P-ρ-T TABLE FOR CARBON DIOXIDE

Temperature = 513.15 K (continued)

Density (ρ; mol/L)	Pressure (P; bar)	Density (ρ; mol/L)	Pressure (P; bar)	Density (ρ; mol/L)	Pressure (P; bar)
9.20	366.479	14.60	714.452	19.90	1485.622
9.30	371.153	14.70	723.633	20.00	1508.001
9.40	375.869	14.80	732.958	20.10	1530.778
9.50	380.628	14.90	742.428	20.20	1553.961
9.60	385.431	15.00	752.046	20.30	1577.557
9.70	390.280	15.10	761.817	20.40	1601.573
9.80	395.175	15.20	771.742	20.50	1626.018
9.90	400.119	15.30	781.826	20.60	1650.899
10.00	405.111	15.40	792.070	20.70	1676.224
10.10	410.153	15.50	802.480	20.80	1702.001
10.20	415.247	15.60	813.057	20.90	1728.239
10.30	420.394	15.70	823.805	21.00	1754.946
10.40	425.595	15.80	834.729	21.10	1782.130
10.50	430.852	15.90	845.830	21.20	1809.800
10.60	436.165	16.00	857.114	21.30	1837.964
10.70	441.537	16.10	868.583	21.40	1866.632
10.80	446.968	16.20	880.242	21.50	1895.814
10.90	452.461	16.30	892.093	21.60	1925.517
11.00	458.016	16.40	904.142	21.70	1955.751
11.10	463.635	16.50	916.392	21.80	1986.526
11.20	469.320	16.60	928.846	21.90	2017.852
11.30	475.071	16.70	941.510	22.00	2049.738
11.40	480.892	16.80	954.386	22.10	2082.195
11.50	486.783	16.90	967.480	22.20	2115.232
11.60	492.746	17.00	980.796	22.30	2148.860
11.70	498.782	17.10	994.338	22.40	2183.089
11.80	504.894	17.20	1008.110	22.50	2217.931
11.90	511.082	17.30	1022.117	22.60	2253.395
12.00	517.349	17.40	1036.364	22.70	2289.493
12.10	523.697	17.50	1050.855	22.80	2326.236
12.20	530.127	17.60	1065.595	22.90	2363.636
12.30	536.641	17.70	1080.589	23.00	2401.704
12.40	543.241	17.80	1095.842	23.10	2440.451
12.50	549.929	17.90	1111.358	23.20	2479.890
12.60	556.707	18.00	1127.144	23.30	2520.033
·12.70	563.576	18.10	1143.204	23.40	2560.892
12.80	570.540	18.20	1159.543	23.50	2602.479
12.90	577.599	18.30	1176.167	23.60	2644.808
13.00	584.756	18.40	1193.081	23.70	2687.891
13.10	592.013	18.50	1210.291	23.80	2371.741
13.20	599.373	18.60	1227.802	23.90	2776.372
13.30	606.837	18.70	1245.620	24.00	2821.797
13.40	614.407	18.80	1263.752	24.10	2868.030
13.50	622.087	18.90	1282.202	24.20	2915.085
13.60	629.878	18.90	1300.978	24.30	2962.976
13.70	637.783	19.00	1300.978	24.40	3011.717
13.80	645.803	19.10	1320.084	24.50	3061.323
13.90	653.942	19.20	1339.527	24.60	3111.810
14.00	662.203	19.30	1359.315	24.70	3163.191
14.10	670.587	19.40	1379.452	24.80	3215.483
14.20	679.097	19.50	1399.946	24.90	3268.701
14.30	687.735	19.60	1420.803	25.00	3322.860
14.40	696.506	19.70	1442.030		
14.50	705.410	19.80	1463.634		

P-ρ-T TABLE FOR CARBON DIOXIDE

Temperature = 533.15 K

Density (ρ; mol/L)	Pressure (P; bar)	Density (ρ; mol/L)	Pressure (P; bar)	Density (ρ; mol/L)	Pressure (P; bar)
0.20	8.825	5.60	230.992	11.00	493.905
0.30	13.207	5.70	235.117	11.10	500.074
0.40	17.570	5.80	239.255	11.20	506.315
0.50	21.914	5.90	243.406	11.30	512.629
0.60	26.240	6.00	247.572	11.40	519.017
0.70	30.548	6.10	251.752	11.50	525.482
0.80	34.839	6.20	255.948	11.60	532.026
0.90	39.112	6.30	260.160	11.70	538.026
1.00	43.370	6.40	264.390	11.80	545.354
1.10	47.611	6.50	268.638	11.90	552.143
1.20	51.837	6.60	272.904	12.00	559.016
1.30	56.049	6.70	277.191	12.10	565.977
1.40	60.246	6.80	281.498	12.20	573.027
1.50	64.429	6.90	285.826	12.30	580.168
1.60	68.599	7.00	290.177	12.40	587.402
1.70	72.756	7.10	294.551	12.50	594.731
1.80	76.902	7.20	298.949	12.60	602.157
1.90	81.035	7.30	303.371	12.70	609.681
2.00	85.157	7.40	307.820	12.80	617.307
2.10	89.269	7.50	312.295	12.90	625.036
2.20	93.371	7.60	316.799	13.00	632.871
2.30	97.463	7.70	321.330	13.10	640.813
2.40	101.546	7.80	325.891	13.20	648.865
2.50	105.621	7.90	330.483	13.30	657.029
2.60	109.688	8.00	335.106	13.40	665.307
2.70	113.747	8.10	339.762	13.50	673.703
2.80	117.800	8.20	344.451	13.60	682.217
2.90	121.847	8.30	349.175	13.70	690.854
3.00	125.889	8.40	353.934	13.80	699.615
3.10	129.925	8.50	358.730	13.90	708.503
3.20	133.957	8.60	363.564	14.00	717.520
3.30	137.985	8.70	368.436	14.10	726.669
3.40	142.010	8.80	373.348	14.20	735.953
3.50	146.032	8.90	378.301	14.30	745.374
3.60	150.052	9.00	383.296	14.40	754.936
3.70	154.071	9.10	388.334	14.50	764.641
3.80	158.088	9.20	393.416	14.60	774.492
3.90	162.106	9.30	398.544	14.70	784.492
4.00	166.124	9.40	403.719	14.80	794.645
4.10	170.143	9.50	408.941	14.90	804.952
4.20	174.164	9.60	414.213	15.00	815.417
4.30	178.186	9.70	419.535	15.10	826.044
4.40	182.212	9.80	424.909	15.20	836.836
4.50	186.242	9.90	430.335	15.30	847.795
4.60	190.275	10.00	435.816	15.40	858.926
4.70	194.314	10.10	441.352	15.50	870.231
4.80	198.358	10.20	446.945	15.60	881.714
4.90	202.408	10.30	452.596	15.70	893.379
5.00	206.465	10.40	458.307	15.80	905.230
5.10	210.530	10.50	464.079	15.90	917.269
5.20	214.603	10.60	469.913	16.00	929.501
5.30	218.685	10.70	475.812	16.10	941.929
5.40	222.777	10.80	481.775	16.20	954.558
5.50	226.879	10.90	487.806	16.30	967.390

P-ρ-T TABLE FOR CARBON DIOXIDE

Temperature = 533.15 K (continued)

Density (ρ; mol/L)	Pressure (P; bar)	Density (ρ; mol/L)	Pressure (P; bar)	Density (ρ; mol/L)	Pressure (P; bar)
16.40	980.432	19.30	1469.973	22.20	2274.376
16.50	993.685	19.40	1491.515	22.30	2310.004
16.60	1007.155	19.50	1513.431	22.40	2346.259
16.70	1020.845	19.60	1535.726	22.50	2383.152
16.80	1034.761	19.70	1558.410	22.60	2420.694
16.90	1048.906	19.80	1581.487	22.70	2458.896
17.00	1063.284	19.90	1604.966	22.80	2497.770
17.10	1077.901	20.00	1628.854	22.90	2537.328
17.20	1092.761	20.10	1653.158	23.00	2577.581
17.30	1107.868	20.20	1677.886	23.10	2618.542
17.40	1123.228	20.30	1703.046	23.20	2660.223
17.50	1138.845	20.40	1728.646	23.30	2702.637
17.60	1154.724	20.50	1754.693	23.40	2745.796
17.70	1170.870	20.60	1781.196	23.50	2789.714
17.80	1187.288	20.70	1808.163	23.60	2834.403
17.90	1203.984	20.80	1835.602	23.70	2879.877
18.00	1220.962	20.90	1863.522	23.80	2926.149
18.10	1238.229	21.00	1891.931	23.90	2973.233
18.20	1255.788	21.10	1920.839	24.00	3021.144
18.30	1273.647	21.20	1950.254	24.10	3069.895
18.40	1291.811	21.30	1980.185	24.20	3119.501
18.50	1310.285	21.40	2010.642	24.30	3169.976
18.60	1329.075	21.50	2041.633	24.40	3221.335
18.70	1348.188	21.60	2073.169	24.50	3273.594
18.80	1367.628	21.70	2105.260	24.60	3326.768
18.90	1387.403	21.80	2137.914	24.70	3380.872
19.00	1407.519	21.90	2171.143	24.80	3435.922
19.10	1427.981	22.00	2204.956	24.90	3491.935
19.20	1448.797	22.10	2239.363	25.00	3548.926

Temperature = 553.15 K

Density (ρ; mol/L)	Pressure (P; bar)	Density (ρ; mol/L)	Pressure (P; bar)	Density (ρ; mol/L)	Pressure (P; bar)
0.20	9.162	2.30	101.940	4.40	191.939
0.30	13.717	2.40	106.248	4.50	196.248
0.40	18.255	2.50	110.549	4.60	200.563
0.50	22.777	2.60	114.846	4.70	204.886
0.60	27.282	2.70	119.137	4.80	209.218
0.70	31.773	2.80	123.425	4.90	213.559
0.80	36.248	2.90	127.708	5.00	217.909
0.90	40.709	3.00	131.989	5.10	222.270
1.00	45.156	3.10	136.267	5.20	226.643
1.10	49.590	3.20	140.544	5.30	231.027
1.20	54.011	3.30	144.819	5.40	235.424
1.30	58.149	3.40	149.094	5.50	239.834
1.40	62.816	3.50	153.368	5.60	244.258
1.50	67.201	3.60	157.643	5.70	248.698
1.60	71.576	3.70	161.919	5.80	253.153
1.70	75.940	3.80	166.197	5.90	257.625
1.80	80.294	3.90	170.478	6.00	262.114
1.90	84.640	4.00	174.761	6.10	266.621
2.00	88.976	4.10	179.049	6.20	271.147
2.10	93.305	4.20	183.340	6.30	275.693
2.20	97.626	4.30	187.637	6.40	280.259

P-ρ-T TABLE FOR CARBON DIOXIDE

Temperature = 553.15 K (continued)

Density (ρ; mol/L)	Pressure (P; bar)	Density (ρ; mol/L)	Pressure (P; bar)	Density (ρ; mol/L)	Pressure (P; bar)
6.50	284.847	11.90	592.988	17.20	1176.892
6.60	289.457	12.00	600.465	17.30	1193.092
6.70	294.090	12.10	608.035	17.40	1209.557
6.80	298.747	12.20	615.701	17.50	1226.292
6.90	303.429	12.30	623.465	17.60	1243.302
7.00	308.137	12.40	631.328	17.70	1260.593
7.10	312.872	12.50	639.294	17.80	1278.169
7.20	317.634	12.60	647.363	17.90	1296.036
7.30	322.425	12.70	655.539	18.00	1314.199
7.40	327.245	12.80	663.823	18.10	1332.665
7.50	332.096	12.90	672.217	18.20	1351.438
7.60	336.979	13.00	680.724	18.30	1370.524
7.70	341.893	13.10	689.346	18.40	1389.929
7.80	346.842	13.20	698.086	18.50	1409.659
7.90	351.824	13.30	706.945	18.60	1429.721
8.00	356.842	13.40	715.927	18.70	1450.120
8.10	361.897	13.50	725.033	18.80	1470.862
8.20	366.989	13.60	734.267	18.90	1491.954
8.30	372.120	13.70	743.630	19.00	1513.402
8.40	377.291	13.80	753.126	19.10	1535.213
8.50	382.502	13.90	762.757	19.20	1557.393
8.60	387.756	14.00	772.526	19.30	1579.950
8.70	393.052	14.00	772.526	19.40	1602.890
8.80	398.393	14.10	782.435	19.50	1626.220
8.90	403.779	14.20	792.487	19.60	1649.947
9.00	409.211	14.30	802.686	19.70	1674.078
9.10	414.691	14.40	813.034	19.80	1698.622
9.20	420.220	14.50	823.534	19.90	1723.584
9.30	425.800	14.60	834.188	20.00	1748.974
9.40	431.431	14.70	845.001	20.10	1774.798
9.50	437.114	14.80	855.975	20.20	1801.065
9.60	442.851	14.90	867.114	20.30	1827.782
9.70	448.644	15.00	878.420	20.40	1854.958
9.80	454.494	15.10	889.897	20.50	1882.601
9.90	460.401	15.20	901.548	20.60	1920.719
10.00	466.367	15.30	913.377	20.70	1939.321
10.10	472.395	15.40	925.387	20.80	1968.415
10.20	478.484	15.50	937.582	20.90	1998.011
10.30	484.637	15.60	949.965	21.00	2028.116
10.40	490.855	15.70	962.540	21.10	2058.742
10.50	497.139	15.80	975.311	21.20	2089.895
10.60	503.492	15.90	988.281	21.30	2121.587
10.70	509.913	16.00	1001.454	21.40	2153.826
10.80	516.406	16.10	1014.835	21.50	2186.623
10.90	522.972	16.20	1028.426	21.60	2219.986
11.00	529.612	16.30	1042.233	21.70	2253.927
11.10	536.327	16.40	1056.260	21.80	2288.455
11.20	543.120	16.50	1070.510	21.90	2323.581
11.30	549.993	16.60	1084.988	22.00	2359.315
11.40	556.946	16.70	1099.698	22.10	2395.668
11.50	563.982	16.80	1114.645	22.20	2432.651
11.60	571.102	16.90	1129.833	22.30	2470.275
11.70	578.309	17.00	1145.268	22.40	2508.551
11.80	585.604	17.10	1160.952	22.50	2547.490

P-ρ-T TABLE FOR CARBON DIOXIDE

Temperature = 553.15 K (continued)

Density (ρ; mol/L)	Pressure (P; bar)	Density (ρ; mol/L)	Pressure (P; bar)	Density (ρ; mol/L)	Pressure (P; bar)
22.60	2587.105	23.50	2976.024	24.30	3376.025
22.70	2627.406	23.60	3023.070	24.40	3430.000
22.80	2668.407	23.70	3070.931	24.50	3484.909
22.90	2710.118	23.80	3119.622	24.60	3540.767
23.00	2752.553	23.90	3169.156	24.70	3597.592
23.10	2795.723	24.00	3219.549	24.80	3655.398
23.20	2839.643	24.10	3270.815	24.90	3714.203
23.30	2884.324	24.20	3222.969	25.00	3774.024
23.40	2929.780				

Temperature = 573.15 K

Density (ρ; mol/L)	Pressure (P; bar)	Density (ρ; mol/L)	Pressure (P; bar)	Density (ρ; mol/L)	Pressure (P; bar)
0.20	9.500	4.30	197.048	8.40	400.539
0.30	14.227	4.40	201.625	8.50	406.164
0.40	18.940	4.50	206.212	8.60	411.835
0.50	23.638	4.60	210.808	8.70	417.553
0.60	28.324	4.70	215.414	8.80	423.320
0.70	32.996	4.80	220.032	8.90	429.137
0.80	37.656	4.90	224.662	9.00	435.005
0.90	42.304	5.00	229.304	9.10	440.925
1.00	46.940	5.10	233.960	9.20	446.899
1.10	51.565	5.20	238.630	9.30	452.927
1.20	56.180	5.30	243.314	9.40	459.012
1.30	60.785	5.40	248.015	9.50	465.154
1.40	65.381	5.50	252.732	9.60	471.355
1.50	69.967	5.60	257.466	9.70	477.616
1.60	74.546	5.70	262.218	9.80	483.938
1.70	79.116	5.80	266.989	9.90	490.324
1.80	83.679	5.90	271.779	10.00	496.773
1.90	88.236	6.00	276.590	10.10	503.289
2.00	92.786	6.10	281.423	10.20	509.872
2.10	97.330	6.20	286.277	10.30	516.524
2.20	101.870	6.30	291.154	10.40	523.246
2.30	106.404	6.40	296.056	10.50	530.040
2.40	110.935	6.50	300.982	10.60	536.907
2.50	115.463	6.60	305.934	10.70	543.850
2.60	119.987	6.70	310.912	10.80	550.869
2.70	124.509	6.80	315.918	10.90	557.966
2.80	129.030	6.90	320.952	11.00	565.144
2.90	133.549	7.00	326.015	11.10	572.403
3.00	138.068	7.10	331.109	11.20	579.745
3.10	142.587	7.20	336.234	11.30	587.173
3.20	147.107	7.30	341.391	11.40	594.687
3.30	151.628	7.40	346.581	11.50	602.291
3.40	156.151	7.50	351.806	11.60	609.985
3.50	160.677	7.60	357.066	11.70	617.771
3.60	165.205	7.70	362.362	11.80	625.652
3.70	169.738	7.80	367.695	11.90	633.629
3.80	174.275	7.90	373.067	12.00	641.705
3.90	178.816	8.00	378.478	12.10	649.880
4.00	183.364	8.10	383.930	12.20	658.159
4.10	187.918	8.20	389.423	12.30	666.541
4.20	192.479	8.30	394.959	12.40	675.030

P-ρ-T TABLE FOR CARBON DIOXIDE

Temperature = 573.15 K (continued)

Density (ρ; mol/L)	Pressure (P; bar)	Density (ρ; mol/L)	Pressure (P; bar)	Density (ρ; mol/L)	Pressure (P; bar)
12.50	683.628	16.70	1178.096	20.90	2131.770
12.60	692.337	16.80	1194.068	21.00	2163.566
12.70	701.160	16.90	1210.293	21.10	2195.903
12.80	710.097	17.00	1226.776	21.20	2228.790
12.90	719.152	17.10	1243.522	21.30	2262.237
13.00	728.328	17.20	1260.535	21.40	2296.253
13.10	737.626	17.30	1277.821	21.50	2330.849
13.20	747.048	17.40	1295.385	21.60	2366.035
13.30	756.599	17.50	1313.231	21.70	2401.821
13.40	766.279	17.60	1331.366	21.80	2438.218
13.50	776.091	17.70	1349.794	21.90	2475.237
13.60	786.039	17.80	1368.521	22.00	2512.887
13.70	796.125	17.90	1387.552	22.10	2551.181
13.80	806.351	18.00	1406.894	22.20	2590.130
13.90	816.720	18.10	1426.551	22.30	2629.745
14.00	827.235	18.20	1446.530	22.40	2670.037
14.10	837.899	18.30	1466.837	22.50	2711.019
14.20	848.715	18.40	1487.477	22.60	2752.702
14.30	859.686	18.50	1508.457	22.70	2795.099
14.40	870.814	18.60	1529.783	22.80	2838.221
14.50	882.103	18.70	1551.461	22.90	2882.083
14.60	893.556	18.80	1573.498	23.00	2926.695
14.70	905.177	18.90	1595.900	23.10	2972.072
14.80	916.967	19.00	1618.674	23.20	3018.226
14.90	928.931	19.10	1641.827	23.30	3065.170
15.00	941.072	19.20	1665.365	23.40	3112.919
15.10	953.394	19.30	1689.296	23.50	3161.486
15.20	965.899	19.40	1713.627	23.60	3210.886
15.30	978.592	19.50	1738.364	23.70	3261.131
15.40	991.476	19.60	1763.516	23.80	3312.238
15.50	1004.554	19.70	1789.090	23.90	3364.220
15.60	1017.831	19.80	1815.093	24.00	3417.092
15.70	1031.309	19.90	1841.533	24.10	3470.870
15.80	1044.994	20.00	1868.418	24.20	3525.569
15.90	1058.889	20.10	1895.756	24.30	3581.204
16.00	1072.997	20.20	1923.555	24.40	3637.792
16.10	1087.324	20.30	1951.824	24.50	3695.348
16.20	1101.872	20.40	1980.570	24.60	3753.888
16.30	1116.647	20.50	2009.802	24.70	3813.431
16.40	1131.653	20.60	2039.530	24.80	3873.991
16.50	1146.893	20.70	2069.761	24.90	3935.587
16.60	1162.373	20.80	2100.504	25.00	3998.235

V. Ultraviolet Spectrophotometry

SOLVENTS FOR ULTRAVIOLET SPECTROPHOTOMETRY

The following table lists some solvents useful in UV spectrophotometry, along with their wavelength cutoffs and dielectric constants.[1-4]

REFERENCES

1. **Willard, H. H., Merritt, L. L., Dean, J. A., and Settle, F. A.,** *Instrumental Methods of Analysis,* 6th ed., Wadsworth Publishing, Belmont, CA, 1981.
2. **Weast, R. C., Ed.,** *Handbook of Chemistry and Physics,* 69th ed., CRC Press, Boca Raton, FL, 1987.
3. **Dreisbach, R. R.,** *Physical Properties of Chemical Compounds,* Advances in Chemistry Ser., No. 15, American Chemical Society, Washington, D.C., 1955.
4. **Dreisbach, R. R.,** *Physical Properties of Chemical Compounds,* Advances in Chemistry Ser., No. 22, American Chemical Society, Washington, D.C., 1959.

SOLVENTS FOR ULTRAVIOLET SPECTROPHOTOMETRY

Solvent	Wavelength cutoff (nm)	Dielectric constant (20°C)	
Acetic acid	260	6.15	
Acetone	330	20.7	(25°C)
Acetonitrile	190	37.5	
Benzene	280	2.284	
2-Butanol	260	15.8	(25°C)
n-Butyl acetate	254		
Carbon disulfide	380	2.641	
Carbon tetrachloride	265	2.238	
1-Chlorobutane	220	7.39	(25°C)
Chloroform[a]	245	4.806	
Cyclohexane	210	2.023	
1,2-Dichloroethane	226	10.19	(25°C)
1,2-Dimethoxyethane	240		
N,N-Dimethylacetamide	268	59	(83°C)
N,N-Dimethylformamide	270	36.7	
Dimethylsulfoxide	265	4.7	
1,4-Dioxane	215	2.209	(25°C)
Diethyl ether	218	4.335	
Ethanol	210	24.30	(25°C)
2-Ethoxyethanol	210		
Ethyl acetate	255	6.02	(25°C)
Glycerol	207	42.5	(25°C)
n-Hexadecane	200	2.06	(25°C)
n-Hexane	210	1.890	
Methanol	210	32.63	(25°C)
2-Methoxyethanol	210	16.9	
Methyl cyclohexane	210	2.02	(25°C)
Methyl ethyl ketone	330	18.5	
Methyl isobutyl ketone	335		
2-Methyl-1-propanol	230	1	
N-Methyl-2-pyrrolidone	285	32.0	
Pentane	210	1.844	
n-Pentyl acetate	212		
1-Propanol	210	20.1	(25°C)
2-Propanol	210	18.3	(25°C)
Pyridine	330	12.3	(25°C)
Tetrachloroethylene[b]	290		
Tetrahydrofuran	220	7.6	
Toluene	286	2.379	(25°C)
1,1,2-Trichloro-1,2,2-trifluoroethane	231		
2,2,4-Trimethylpentane	215	1.936	(25°C)
o-Xylene	290	2.568	
m-Xylene	290	2.374	
p-Xylene	290	2.270	
Water		78.54	(25°C)

[a] Stabilized with ethanol to avoid phosgene formation.

[b] Stabilized with thymol (isopropyl meta-cresol).

ULTRAVIOLET SPECTRA OF COMMON LIQUIDS

The following tables present the UV spectra of some common solvents and liquids used in chemical analysis. The data were obtained, using a 1.00-cm path-length cell, against a water reference.[1]

REFERENCES

1. **Krieger, P. A.,** *High Purity Solvent Guide,* Burdick and Jackson, McGaw Park, IL, 1984.

ACETONE

Wavelength (nm)	Max absorbance
330	1.000
340	0.060
350	0.010
375	0.005
400	0.005

BENZENE

Wavelength (nm)	Max absorbance
278	1.000
300	0.020
325	0.010
350	0.005
400	0.005

ACETONITRILE

Wavelength (nm)	Max absorbance
190	1.000
200	0.050
225	0.010
250	0.005
350	0.005

1-BUTANOL

Wavelength (nm)	Max absorbance
215	1.000
225	0.500
250	0.040
275	0.010
300	0.005

2-BUTANOL

Wavelength (nm)	Max absorbance
260	1.000
275	0.300
300	0.010
350	0.005
400	0.005

CARBON TETRACHLORIDE

Wavelength (nm)	Max absorbance
263	1.000
275	0.100
300	0.005
350	0.005
400	0.005

n-BUTYL ACETATE

Wavelength (nm)	Max absorbance
254	1.000
275	0.050
300	0.010
350	0.005
400	0.005

CHLOROBENZENE

Wavelength (nm)	Max absorbance
287	1.000
300	0.050
325	0.040
350	0.020
400	0.005

n-BUTYL CHLORIDE

Wavelength (nm)	Max absorbance
220	1.000
225	0.300
250	0.010
300	0.005
400	0.005

CHLOROFORM

Wavelength (nm)	Max absorbance
245	1.000
250	0.300
275	0.005
300	0.005
400	0.005

CYCLOHEXANE

Wavelength (nm)	Max absorbance
200	1.000
225	0.170
250	0.020
300	0.005
400	0.005

o-DICHLOROBENZENE

Wavelength (nm)	Max absorbance
295	1.000
300	0.300
325	0.100
350	0.050
400	0.005

CYCLOPENTANE

Wavelength (nm)	Max absorbance
200	1.000
215	0.300
225	0.020
300	0.005
400	0.005

DIETHYL CARBONATE

Wavelength (nm)	Max absorbance
256	1.000
265	0.150
275	0.050
300	0.040
400	0.010

DECAHYDRONAPHTHALENE

Wavelength (nm)	Max absorbance
200	1.000
225	0.500
250	0.050
300	0.005
400	0.005

DIMETHYL ACETAMIDE

Wavelength (nm)	Max absorbance
268	1.000
275	0.300
300	0.080
350	0.005
400	0.005

DIMETHYL FORMAMIDE

Wavelength (nm)	Max absorbance
268	1.000
275	0.300
300	0.050
350	0.005
400	0.005

2-ETHOXYETHANOL

Wavelength (nm)	Max absorbance
210	1.000
225	0.500
250	0.200
300	0.005
400	0.005

DIMETHYL SULFOXIDE

Wavelength (nm)	Max absorbance
268	1.000
275	0.500
300	0.200
350	0.020
400	0.005

ETHYL ACETATE

Wavelength (nm)	Max absorbance
256	1.000
275	0.050
300	0.030
325	0.005
350	0.005

1,4-DIOXANE

Wavelength (nm)	Max absorbance
215	1.000
250	0.300
300	0.020
350	0.005
400	0.005

DIETHYL ETHER

Wavelength (nm)	Max absorbance
215	1.000
250	0.080
275	0.010
300	0.005
400	0.005

ETHYLENE DICHLORIDE

Wavelength (nm)	Max absorbance
228	1.000
240	0.300
250	0.100
300	0.005
400	0.005

HEXADECANE

Wavelength (nm)	Max absorbance
190	1.000
200	0.500
250	0.020
300	0.005
400	0.005

ETHYLENE GLYCOL DIMETHYL ETHER (GLYME)

Wavelength (nm)	Max absorbance
220	1.000
250	0.250
300	0.050
350	0.010
400	0.005

HEXANE

Wavelength (nm)	Max absorbance
195	1.000
225	0.050
250	0.010
275	0.005
300	0.005

HEPTANE

Wavelength (nm)	Max absorbance
200	1.000
225	0.100
250	0.010
300	0.005
400	0.005

ISOBUTANOL

Wavelength (nm)	Max absorbance
220	1.000
250	0.050
275	0.030
300	0.020
400	0.010

METHANOL

Wavelength (nm)	Max absorbance
205	1.000
225	0.160
250	0.020
300	0.005
400	0.005

METHYL *t*-BUTYL ETHER

Wavelength (nm)	Max absorbance
210	1.000
225	0.500
250	0.100
300	0.005
400	0.005

2-METHOXYETHANOL

Wavelength (nm)	Max absorbance
210	1.000
250	0.130
275	0.030
300	0.005
400	0.005

METHYLENE CHLORIDE

Wavelength (nm)	Max absorbance
233	1.000
240	0.100
250	0.010
300	0.005
400	0.005

2-METHOXYETHYL ACETATE

Wavelength (nm)	Max absorbance
254	1.000
275	0.150
300	0.050
350	0.005
400	0.005

METHYL ETHYL KETONE

Wavelength (nm)	Max absorbance
329	1.000
340	0.100
350	0.020
375	0.010
400	0.005

METHYL ISOAMYL KETONE

Wavelength (nm)	Max absorbance
330	1.000
340	0.100
350	0.050
375	0.010
400	0.005

n-METHYLPYRROLIDONE

Wavelength (nm)	Max absorbance
285	1.000
300	0.500
325	0.100
350	0.030
400	0.010

METHYL ISOBUTYL KETONE

Wavelength (nm)	Max absorbance
334	1.000
340	0.500
350	0.250
375	0.050
400	0.005

PENTANE

Wavelength (nm)	Max absorbance
190	1.000
200	0.600
250	0.010
300	0.005
400	0.005

METHYL *n*-PROPYL KETONE

Wavelength (nm)	Max absorbance
331	1.000
340	0.150
350	0.020
375	0.005
400	0.005

β-PHENETHYLAMINE

Wavelength (nm)	Max absorbance
285	1.000
300	0.300
325	0.100
350	0.050
400	0.005

1-PROPANOL

Wavelength (nm)	Max absorbance
210	1.000
225	0.500
250	0.050
300	0.005
400	0.005

PYRIDINE

Wavelength (nm)	Max absorbance
330	1.000
340	0.100
350	0.010
375	0.010
400	0.005

2-PROPANOL

Wavelength (nm)	Max absorbance
205	1.000
225	0.160
250	0.020
300	0.005
400	0.010

TETRAHYDROFURAN

Wavelength (nm)	Max absorbance
212	1.000
250	0.180
300	0.020
350	0.005
400	0.005

PROPYLENE CARBONATE

Wavelength (nm)	Max absorbance
280	1.000
300	0.500
350	0.050
375	0.030
400	0.020

TOLUENE

Wavelength (nm)	Max absorbance
284	1.000
300	0.120
325	0.020
350	0.050
400	0.005

1,2,4-TRICHLOROBENZENE

Wavelength (nm)	Max absorbance
308	1.000
310	0.500
350	0.050
375	0.010
400	0.005

2,2,4-TRIMETHYLPENTANE

Wavelength (nm)	Max absorbance
215	1.000
225	0.100
250	0.020
300	0.005
400	0.005

TRICHLOROETHYLENE

Wavelength (nm)	Max absorbance
273	1.000
300	0.100
325	0.080
350	0.060
400	0.060

WATER

Wavelength (nm)	Max absorbance
190	0.010
200	0.010
250	0.005
300	0.005
400	0.005

1,1,2-TRICHLOROTRIFLUOROETHANE

Wavelength (nm)	Max absorbance
231	1.000
250	0.050
300	0.005
350	0.005
400	0.005

o-XYLENE

Wavelength (nm)	Max absorbance
288	1.000
300	0.200
325	0.050
350	0.010
400	0.005

TRANSMITTANCE-ABSORBANCE CONVERSION

The following is a conversion table for absorbance and transmittance, assuming no reflection. Included for each pair is the percent error propagated into a measured concentration (using the Beer-Lambert Law), assuming an uncertainty in transmitance of $\pm$ 0.005.[1] The value of transmittance which will give the lowest percent error in concentration is 0.368. Where possible, analyses should be designed for the low-error area.

REFERENCES

1. **Kennedy, J. H.**, *Analytical Chemistry Principles*, Harcourt, Brace and Jovanovich, San Diego, 1984.

TRANSMITTANCE-ABSORBANCE CONVERSION

Transmittance	Absorbance	Percent Error
0.990	0.004	50.227
0.980	0.009	25.242
0.970	0.013	16.915
0.960	0.018	12.752
0.950	0.022	10.256
0.940	0.027	8.592
0.930	0.032	7.405
0.920	0.036	6.515
0.910	0.041	5.823
0.900	0.046	5.270
0.890	0.051	4.818
0.880	0.056	4.442
0.870	0.060	4.125
0.860	0.065	3.853
0.850	0.071	3.618
0.840	0.076	3.412
0.830	0.081	3.231
0.820	0.086	3.071
0.810	0.091	2.928
0.800	0.097	2.799
0.790	0.102	2.684
0.780	0.108	2.579
0.770	0.113	2.483
0.760	0.119	2.396
0.750	0.125	2.316
0.740	0.131	2.243
0.730	0.137	2.175
0.720	0.143	2.113
0.710	0.149	2.055
0.700	0.155	2.002
0.690	0.161	1.952
0.680	0.167	1.906
0.670	0.174	1.863
0.660	0.180	1.822
0.650	0.187	1.785
0.640	0.194	1.750
0.630	0.201	1.717
0.620	0.208	1.686
0.610	0.215	1.657
0.600	0.222	1.631
0.590	0.229	1.605

TRANSMITTANCE-ABSORBANCE CONVERSION (continued)

Transmittance	Absorbance	Percent Error
0.580	0.237	1.582
0.570	0.244	1.560
0.560	0.252	1.539
0.550	0.260	1.520
0.540	0.268	1.502
0.530	0.276	1.485
0.520	0.284	1.470
0.510	0.292	1.455
0.500	0.301	1.442
0.490	0.310	1.430
0.480	0.319	1.419
0.470	0.328	1.408
0.460	0.337	1.399
0.450	0.347	1.391
0.440	0.356	1.383
0.430	0.366	1.377
0.420	0.377	1.372
0.410	0.387	1.367
0.400	0.398	1.364
0.390	0.409	1.361
0.380	0.420	1.359
0.370	0.432	1.358
0.360	0.444	1.359
0.350	0.456	1.360
0.340	0.468	1.362
0.330	0.481	1.366
0.320	0.495	1.371
0.310	0.509	1.376
0.300	0.523	1.384
0.290	0.538	1.392
0.280	0.553	1.402
0.270	0.569	1.414
0.260	0.585	1.427
0.250	0.602	1.442
0.240	0.620	1.459
0.230	0.638	1.478
0.220	0.657	1.500
0.210	0.678	1.525
0.200	0.699	1.553
0.190	0.721	1.584
0.180	0.745	1.619
0.170	0.769	1.659
0.160	0.796	1.704
0.150	0.824	1.756
0.140	0.854	1.816
0.130	0.886	1.884
0.120	0.921	1.964
0.110	0.958	2.058
0.100	1.000	2.170
0.090	1.046	2.306
0.080	1.097	2.473
0.070	1.155	2.685
0.060	1.222	2.961
0.050	1.301	3.336
0.040	1.398	3.881
0.030	1.523	4.751
0.020	1.699	6.387
0.010	2.000	10.852

CORRELATION TABLE FOR ULTRAVIOLET ACTIVE FUNCTIONALITIES

The following table presents a correlation between common chromophoric functional groups and the expected absorptions form UV spectrophotometry.[1] While not as informative as infrared correlations, UV can often provide valuable qualitative information.

REFERENCES

1. **Willard, H. H., Merritt, L. L., Dean, J. A., and Settle, F. A.,** *Instrumental Methods of Analysis,* 6th ed., Wadsworth Publishing, Belmont, CA, 1981.
2. **Silverstein, R. M., Bassler, G. C., and Morrill, T. C.,** *Spectrometric Identification of Organic Compounds,* 4th ed., John Wiley & Sons, New York, 1981.
3. **Lambert J. B., Shuruell, H. F., Verbit, L., Cooks, R. G., and Stout, G. H.,** *Organic Structural Analysis,* Macmillan, New York, 1976.

CORRELATION TABLE FOR ULTRAVIOLET ACTIVE FUNCTIONALITIES

Chromophore	System	λ_{max} (nm)	ϵ_{max}	λ_{max} (nm)	ϵ_{max}	λ_{max}	ϵ_{max}
Ether	–O–	185	1,000				
Thioether	–S–	194	4,600	215	1,600		
Amine	–NH$_2$–	195	2,800				
Amide	–CONH$_2$	<210	—				
Thiol (mercaptan)	–SH	195	1,400				
Disulfide	–S–S–	194	5,500	255	400		
Bromide	–Br	208	300				
Iodide	–I	260	400				
Nitrile	–C≡N	160	—				
Acetylide (alkyne)	–C≡C–	175—180	6,000				
Sulfone	–SO$_2$–	180	—				
Oxime	>C=NOH	190	5,000				
Azido	>C=N–	190	5,000				
Alkene	–C=C–	190	8,000				
Ketone	>C=O	195	1,000	270—285	18—30		
Thioketone	>C=S	205	Strong				
Ester	–COOR	205	50				
Aldehyde	–CHO	210	Strong	280—300	11—18		
Carboxyl	–COOH	200—210	50—70				
Sulfoxide	>S→O	210	1,500				
Nitro	–NO$_2$	210	Strong				
Nitrite	–ONO	220—230	1,000—2,000	300—4,000	10		
Azo	–N=N–	285—400	3—25				
Nitroso	–N=O	302	100				
Nitrate	–ONO$_2$	270 (shoulder)	12				
← Conjugated systems →	–(C=C)$_2$– (acyclic)	210—230	21,000				
	–(C=C)$_3$–	260	35,000				
	–(C=C)$_4$–	300	52,000				
	–(C=C)$_5$–	330	118,000				
	–(C=C)$_2$– (alicyclic)	230—260	3,000—8,000				
	>C=C–C=C<	219	6,500				
	>C=C–C=N–	220	23,000				
	>C=C–C=O	210—250	10,000—20,000			300—350	Weak
	>C=C–NO$_2$	229	9,500				

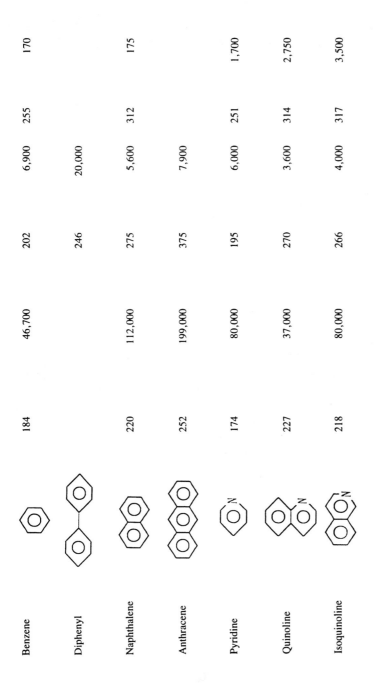

Compound	λ	ε	λ	ε	λ	ε
Benzene	184	46,700	202	6,900	255	170
Diphenyl			246	20,000		
Naphthalene	220	112,000	275	5,600	312	175
Anthracene	252	199,000	375	7,900		
Pyridine	174	80,000	195	6,000	251	1,700
Quinoline	227	37,000	270	3,600	314	2,750
Isoquinoline	218	80,000	266	4,000	317	3,500

WOODWARD'S RULES FOR BATHOCHROMIC SHIFTS

Conjugated systems show bathochromic shifts in their $\pi \rightarrow \pi^*$ transition bands. Empirical methods for predicting those shifts were originally formulated by Woodward, Fiesser, and Fieser[1,4] This section includes the most important conjugated system rules.[1-6] The reader should consult References 5 and 6 for more details on how to apply the wavelength-increment data.

REFERENCES

1. **Woodward, R. B.,** Structure and the absorption spectra of α,β-unsaturated ketones, *J. Am. Chem. Soc.,* 63, 1123, 1941.
2. **Woodward, R. B.,** Structure and absorption spectra. III. Normal conjugated dienes, *J. Am. Chem. Soc.,* 64, 72, 1942.
3. **Woodward, R. B.,** Structure and absorption spectra. IV. Further observations on α,β-unsaturated ketones, *J. Am. Chem. Soc.,* 64, 76, 1942.
4. **Fieser, L. F. and Fieser, M.,** *Natural Products Related to Phenanthrene,* Van Nostrand Reinhold, New York, 1949.
5. **Silverstein, R. M., Bassler, G. C., and Morrill, T. C.,** *Spectrometric Identification of Organic Compounds,* John Wiley & Sons, New York, 1981.
6. **Lambert, J. B., Shuzvell, H. F., Verbit, L., Cooks, R. G., and Stout, G. H.,** *Organic Structural Analysis,* Macmillan, New York, 1976.

RULES OF DIENE ABSORPTION

Base value for diene: 214 nm
Increments for (each) in nanometers:

Heteroannular diene	+0
Homoannular diene	+39
Extra double bond	+30
Alkyl substituent or ring residue	+5
Exocyclic double bond	+5
Polar groups:	
–OOCR	+0
–OR	+6
–S–R	+30
halogen	+5
–NR$_2$	+60
λ Calculated	= Total

RULES OF ENONE ABSORPTION[a]

$$
\begin{array}{cccc}
\delta & \gamma & \beta & \alpha \\
\end{array}
$$
$$
-C{=}C{-}C{=}C{-}C{-}
$$
$$
\begin{array}{ccccc}
| & | & | & | & \| \\
& & & & O
\end{array}
$$

Base value for acyclic (or six-membered) α-β-unsaturated ketone: 215 nm
Base value for five-membered α,β-unsaturated ketone: 202 nm
Base value for α,β-unsaturated esters or carboxylic acids: 195 nm
Increments for (each) in nanometers:

Heteroannular diene	+0
Homoannular diene	+39
Double bond	+30
Alkyl group	
α-	+10
β-	+12
γ and higher	+18
Polar groups	
−OH:	
α-	+35
β-	+30
δ-	+50
−OOCR α,β,δ	+6
−OR	
α-	+35
β-	+30
γ-	+17
δ-	+31
−SR β-	+85
−Cl	
α-	+15
β-	+12
−Br	
α-	+25
β-	+30
−NR$_2$ β	+95
Exocyclic double bond	+5

λ Calculated = Total

[a] Solvent corrections should be included. These are water (-8), chloroform ($+1$), dioxane ($+5$), ether ($+7$), hexane ($+11$), and cyclohexane ($+11$). No correction for methanol or ethanol.

RULES FOR MONOSUBSTITUTED BENZENE DERIVATIVES
C_6H_5–G;G = substituent

Parent chromophore (benzene): 250 nm

Substitutent	Increment
–R	– 4
–COR –	4
–CHO	0
–OH	– 16
–OR	– 16
–COOR	– 16

where R is an alkyl group, and the substitution is on C_6H_3–.

RULES FOR DISUBSTITUTED BENZENE DERIVATIVES
$C_6H_4G_2$;G = substituent

Parent Chromophore (benzene): 250 nm

Substituent	o-	m-	p-
–R	+ 3	+ 3	+ 10
–COR	+ 3	+ 3	+ 10
–OH	+ 7	+ 7	+ 25
–O⁻	+ 11	+ 20	+ 78
			(Variable)
–Cl	+ 0	+ 0	+ 10
–Br	+ 2	+ 2	+ 15
–NH_2	+ 13	+ 13	+ 58
–$NHCOCH_3$	+ 20	+ 20	+ 45
–$NHCH_3$	—	—	+ 73
–$N(CH_3)_2$	+ 20	+ 20	+ 85

R indicates an alkyl group.

VI. Infrared Spectrophotometry

INFRARED OPTICS MATERIALS

The following table lists the more common materials used for optical components (windows, prisms, etc.) in the infrared region of the electromagnetic spectrum. The properties listed are needed to choose the materials with optimal transmission characteristics.[1,2] The thermal properties are useful when designing experiments for operation at elevated temperatures.[3-6] This listing is far from exhaustive, but these are the most common materials used in instrumentation laboratories.

REFERENCES

1. **Gordon, A. J. and Ford, G. A.,** *The Chemist's Companion,* John Wiley & Sons, New York, 1972.
2. **Willard, H. H., Merritt, L. L., Dean, J. A., and Settle, F. A.,** *Instrumental Methods of Analysis,* 6th ed., Van Nostrand, New York, 1985.
3. **Touloukien, Y. S., Powell, R., W., Ho, C. Y., and Klemens, P. G.,** *Thermophysical Properties of Matter: Thermal Conductivity of Nonmetallic Solids,* Vol. 2, Plenum Data, New York, 1970.
4. **Touloukien, Y. S., Kirby, R. K., Taylor, R. E., and Lee, T.,** *Thermophysical Properties of Matter: Thermal Expansion of Nonmetallic Solids,* Vol. 13, Plenum Data, New York, 1977.
5. **Weast, R. C., Ed.,** *Handbook of Chemistry and Physics,* 69th, ed., CRC Press, Boca Raton, FL, 1988.
6. **Wolfe, W. L. and Zissis, G. J., Eds.,** The Infrared Handbook, Office of Naval Research, Department of the Navy, Washington, D.C., 1978.

INFRARED OPTICS MATERIALS

Material	Wavelength range (μm)	Wave number range (cm⁻¹)	Refractive index at 2 μm	Thermal conductivity (W/[m·K])(×10²)	Thermal expansion (ΔL/L, %)	Notes
Sodium chloride (NaCl)	0.25—16	40,000—625	1.52	7.61 (273 K) 6.61 (300 K) 4.85 (400 K)	0.448 (400 K) 0.896 (500 K)	Most common material; absorbs water; for aqueous solutions, use saturated NaCl solution as the solvent
Potassium bromide (KBr)	0.25—25	40,000—400	1.53	5.00 (275 K) 4.87 (301.5 K) 4.80 (372.2 K)	0.028 (400 K) 0.429 (500 K) 0.846 (600 K)	Useful for the study of C-Br stretch region; useful for solid sample pellets
Silver chloride (AgCl)	0.4—23	25,000—435	2.0	1.19 (269.8 K) 1.10 (313.0 K) 1.05 (372.5 K)	0.356 (400 K) 0.729 (500 K) 1.183 (600 K)	Not good for amines or liquids with basic nitrogen; light sensitive
Silver bromide (AgBr)	0.50—35	20,000—286	2.2	0.90 (308.2 K) 0.79 (353.2 K) 0.71 (413.2 K)	0.024 (300 K) 0.109 (325 K) 0.196 (350 K)	Not good for amines or liquids with basic nitrogen; light sensitive
Calcium fluoride (CaF$_2$)	0.15—9	66,700—1,110	1.40	10.40 (237 K) 9.60 (309 K) 4.14 (402 K)	0.214 (400 K) 0.431 (500 K) 0.670 (500 K)	Useful for obtaining high-resolution for OH, NH, and CH stretching frequencies
Barium fluoride (BaF$_2$)	0.20—11.5	50,000—870	1.46	11.7 (284 K) 10.9 (305 K) 10.5 (370 K)	0.233 (400 K) 0.461 (500 K) 0.698 (600 K)	Shock sensitive; should be handled with care
Cesium bromide (CsBr)	1—37	10,000—270	1.67	9.24 (269.4 K) 8.00 (337.5 K) 7.76 (367.5 K)	0.526 (400 K) 1.063 (500 K) 1.645 (600 K)	Useful for C-Br stretching frequencies
Cesium iodide (CsI)	1—50	10,000—200	1.74	1.15 (277.7 K) 1.05 (296.0 K) 0.950 (360.7 K)		Useful for C-Br stretching frequencies
Thalium bromide, thalium iodide; TlBr-TlI (KRS-5)	0.5—35	20,000—286	2.37		0.464 (373 K) 1.026 (473 K)	Highly toxic, handle with care; 42% TlBr, 58% TlI

Material					
Zinc selenide (ZnSe)	1—18	10,000—555	2.4		0.086 (400 K) 0.175 (500 K) 0.272 (600 K)	Vacuum deposited
Germanium (Ge)	0.5—11.5	20,000—870	4.0		0.033 (400 K) 0.066 (500 K) 0.102 (600 K)	
Silicon (Si)	0.20—6.2	50,000—1,613	3.5			
Aluminum oxide (saphire; Al$_2$O$_3$)	0.20—6.5	50,000—1,538	1.76	25.1 (393.2 K) 21.3 (323 K) 14.2 (432.2 K)	0.075 (400 K) 0.148 (500 K) 0.225 (600 K)	
Polyethylene	16—300	625—33	1.54			Not useful for many organic compounds
	200—425	50—23.5				

WAVELENGTH-WAVE NUMBER CONVERSION TABLE

The following table provides a conversion between wavelength and wave number units, used in infrared spectrophotometry.

Wavelength (μm)	Wave number (cm⁻¹)									
	0	1	2	3	4	5	6	7	8	9
2.0	5000	4975	4950	4926	4902	4878	4854	4831	4808	4785
2.1	4762	4739	4717	4695	4673	4651	4630	4608	4587	4566
2.2	4545	4525	4505	4484	4464	4444	4425	4405	4386	4367
2.3	4348	4329	4310	4292	4274	4255	4237	4219	4202	4184
2.4	4167	4149	4232	4115	4098	4082	4065	4049	4032	4016
2.5	4000	3984	3968	4953	3937	3922	3006	3891	3876	3861
2.6	3846	3831	3817	3802	3788	3774	3759	3745	3731	3717
2.7	3704	3690	3676	3663	3650	3636	3623	3610	3597	3584
2.8	3571	3559	3546	3534	3521	3509	3497	3484	3472	3460
2.9	3448	3436	3425	3413	3401	3390	3378	3367	3356	3344
3.0	3333	3322	3311	3300	3289	3279	3268	3257	3247	3236
3.1	3226	3215	3205	3195	3185	3175	3165	3155	3145	3135
3.2	3125	3115	3106	3096	3086	3077	3067	3058	3049	3040
3.3	3030	3021	3012	3003	2994	2985	2976	2967	2959	2950
3.4	2941	2933	2924	2915	2907	2899	2890	2882	2874	2865
3.5	2857	2849	2841	2833	2825	2817	2809	2801	2793	2786
3.6	2778	2770	2762	2755	2747	2740	2732	2725	2717	2710
3.7	2703	2695	2688	2681	2674	2667	2660	2653	2646	2639
3.8	2632	2625	2618	2611	2604	2597	2591	2584	2577	2571
3.9	2654	2558	2551	2545	2538	2532	2525	2519	2513	2506
4.0	2500	2494	2488	2481	2475	2469	2463	2457	2451	2445
4.1	2439	2433	2427	2421	2415	2410	2404	2398	2387	2387
4.2	2381	2375	2370	2364	2358	2353	2347	2342	2336	2331
4.3	2326	2320	2315	2309	2304	2299	2294	2288	2283	2278
4.4	2273	2268	2262	2257	2252	2247	2242	2237	2232	2227
4.5	2222	2217	2212	2208	2203	2198	2193	2188	2183	2179
4.6	2174	2169	2165	2160	2155	2151	2146	2141	2137	2132
4.7	2128	2123	2119	2114	2110	2105	2101	2096	2092	2088
4.8	2083	2079	2075	2070	2066	2062	2058	2053	2049	2045
4.9	2041	2037	2033	2028	2024	2020	2016	2012	2008	2004
5.0	2000	1996	1992	1988	1984	1980	1976	1972	1969	1965
5.1	1961	1957	1953	1949	1946	1942	1938	1934	1931	1927
5.2	1923	1919	1916	1912	1908	1905	1901	1898	1894	1890
5.3	1887	1883	1880	1876	1873	1869	1866	1862	1859	1855
5.4	1852	1848	1845	1842	1838	1835	1832	1828	1825	1821
5.5	1818	1815	1812	1808	1805	1802	1799	1795	1792	1788
5.6	1786	1783	1779	1776	1773	1770	1767	1764	1761	1757
5.7	1754	1751	1748	1745	1742	1739	1736	1733	1730	1727
5.8	1724	1721	1718	1715	1712	1709	1706	1704	1701	1698
5.9	1695	1692	1689	1686	1684	1681	1678	1675	1672	1669

WAVELENGTH-WAVE NUMBER CONVERSION TABLE (continued)

Wavelength (μm)	Wave number (cm⁻¹)									
	0	1	2	3	4	5	6	7	8	9
6.0	1667	1664	1661	1568	1656	1653	1650	1647	1645	1642
6.1	1639	1637	1634	1631	1629	1626	1623	1621	1618	1616
6.2	1613	1610	1608	1605	1603	1600	1597	1595	1592	1590
6.3	1587	1585	1582	1580	1577	1575	1572	1570	1567	1565
6.4	1563	1560	1558	1555	1553	1550	1548	1546	1543	1541
6.5	1538	1536	1534	1531	1529	1527	1524	1522	1520	1517
6.6	1515	1513	1511	1508	1506	1504	1502	1499	1497	1495
6.7	1493	1490	1488	1486	1484	1481	1479	1477	1475	1473
6.8	1471	1468	1466	1464	1462	1460	1458	1456	1453	1451
6.9	1449	1447	1445	1443	1441	1439	1437	1435	1433	1431
7.0	1429	1427	1425	1422	1420	1418	1416	1414	1412	1410
7.1	1408	1406	1404	1403	1401	1399	1397	1395	1393	1391
7.2	1389	1387	1385	1383	1381	1379	1377	1376	1374	1372
7.3	1370	1368	1366	1364	1362	1361	1359	1357	1355	1353
7.4	1351	1350	1348	1346	1344	1342	1340	1339	1337	1335
7.5	1333	1332	1330	1328	1326	1325	1323	1321	1319	1318
7.6	1316	1314	1312	1311	1309	1307	1305	1304	1302	1300
7.7	1299	1297	1295	1294	1292	1290	1289	1287	1285	1284
7.8	1282	1280	1279	1277	1276	1274	1272	1271	1269	1267
7.9	1266	1264	1263	1261	1259	1258	1256	1255	1253	1252
8.0	1250	1248	1247	1245	1244	1242	1241	1239	1238	1236
8.1	1235	1233	1232	1230	1229	1227	1225	1224	1222	1221
8.2	1220	1218	1217	1215	1214	1212	1211	1209	1208	1206
8.3	1205	1203	1202	1200	1199	1198	1196	1195	1193	1192
8.4	1190	1189	1188	1186	1185	1183	1182	1181	1179	1178
8.5	1176	1175	1174	1172	1171	1170	1168	1167	1166	1164
8.6	1163	1161	1160	1159	1157	1156	1155	1153	1152	1151
8.7	1149	1148	1147	1145	1144	1143	1142	1140	1139	1138
8.8	1136	1135	1134	1133	1131	1130	1129	1127	1126	1125
8.9	1124	1122	1121	1120	1119	1117	1116	1115	1114	1112
9.0	1111	1110	1109	1107	1106	1105	1104	1103	1101	1100
9.1	1099	1098	1096	1095	1094	1093	1092	1091	1089	1088
9.2	1087	1086	1085	1083	1082	1081	1080	1079	1078	1076
9.3	1075	1074	1073	1072	1071	1070	1068	1067	1066	1065
9.4	1064	1063	1062	1060	1059	1058	1057	1056	1055	1054
9.5	1053	1052	1050	1049	1048	1047	1046	1045	1044	1043
9.6	1042	1041	1040	1038	1037	1036	1035	1034	1033	1032
9.7	1031	1030	1029	1028	1027	1026	1025	1024	1022	1021
9.8	1020	1019	1018	1017	1016	1015	1014	1013	1012	1011
9.9	1010	1009	1008	1007	1006	1005	1004	1003	1002	1001
10.0	1000	999	998	997	996	995	994	993	992	991
10.1	990	989	988	987	986	985	984	983	982	981
10.2	980	979	978	978	977	976	975	974	973	972
10.3	971	970	969	968	967	966	965	964	963	962
10.4	962	961	960	959	958	957	956	955	954	953
10.5	952	951	951	950	949	948	947	946	945	944
10.6	943	943	942	041	940	939	938	937	936	935
10.7	935	934	933	932	931	930	929	929	928	927
10.8	926	925	924	923	923	922	921	920	919	918
10.9	917	917	916	915	914	913	912	912	911	910

WAVELENGTH-WAVE NUMBER CONVERSION TABLE (continued)

Wavelength (μm)	Wave number (cm⁻¹)									
	0	1	2	3	4	5	6	7	8	9
11.0	909	908	907	907	906	905	904	903	903	902
11.1	901	900	899	898	898	897	896	895	894	894
11.2	893	892	891	890	890	889	888	887	887	886
11.3	885	884	883	883	882	881	880	880	879	878
11.4	877	876	876	875	874	873	873	872	871	870
11.5	870	869	868	867	867	866	865	864	864	863
11.6	862	861	861	860	859	858	858	857	856	855
11.7	855	854	853	853	852	851	850	850	849	848
11.8	847	847	846	845	845	844	843	842	842	841
11.9	840	840	839	838	838	837	836	835	835	834
12.0	833	833	832	831	831	830	829	829	828	827
12.1	826	826	825	824	824	823	822	822	821	820
12.2	820	819	818	818	817	816	816	815	814	814
12.3	813	812	812	811	810	810	809	808	808	807
12.4	806	806	805	805	804	803	803	802	801	801
12.5	800	799	799	798	797	797	796	796	795	794
12.6	794	793	792	792	791	791	790	789	789	788
12.7	787	787	786	786	785	784	784	783	782	782
12.8	781	781	780	779	779	778	778	777	776	776
12.9	775	775	774	773	773	772	772	771	770	770
13.0	769	769	768	767	767	766	766	765	765	764
13.1	763	763	762	762	761	760	760	759	759	758
13.2	758	757	756	756	755	755	754	754	753	752
13.3	752	751	751	750	750	749	749	748	747	747
13.4	746	746	745	745	744	743	743	742	742	741
13.5	741	740	740	739	739	738	737	737	736	736
13.6	735	735	734	734	733	733	732	732	731	730
13.7	730	729	729	728	728	727	727	726	726	725
13.8	725	724	724	723	723	722	722	721	720	720
13.9	719	719	718	718	717	717	716	716	715	715
14.0	714	714	713	713	712	712	711	711	710	710
14.1	709	709	708	708	707	707	706	706	705	705
14.2	704	704	703	703	702	702	702	701	701	700
14.3	699	699	698	698	697	697	696	696	695	695
14.4	694	694	693	693	693	692	692	691	691	690
14.5	690	689	689	688	688	687	687	686	686	685
14.6	685	684	684	684	683	683	682	682	681	681
14.7	680	680	679	679	678	678	678	677	677	676
14.8	676	675	675	674	674	673	673	672	672	672
14.9	671	671	670	670	669	669	668	668	668	667

USEFUL SOLVENTS FOR INFRARED SPECTROPHOTOMETRY

The following tables provide the infrared absorption spectra of several useful solvents. The spectra were measured via a double-beam spectrophotometer using a neat sample against an air reference. They are presented in both wave number (cm^{-1}) and micron (μm) scales. The spectra were recorded under high concentration conditions (in terms of path length and attenuation) in order to emphasize the characteristics of each solvent. Thus, these spectra are not meant to be "textbook" examples of infrared spectra. The physical properties listed are those needed most often in designing spectrophotometric experiments.[1-7] The refractive indices are values measured with the sodium-D line. General information and suggestions on safe handling of the solvents are provided in the notes. The maximum exposure limits quoted are as of November 1986. The Chemical Abstract Service (CAS) registry numbers are also provided for each solvent, to allow the reader to easily obtain further information using computerized database services. Note that the heat of vaporization is presented in the commonly used calories per gram (cal/g) unit. To convert to the appropriate SI unit (J/g), multiply by 4.184.

REFERENCES

1. **Hawley, G. G. Ed.,** *The Condensed Chemical Dictionary,* Van Nostrand Reinhold, New York, 1971.
2. **Weast, R. C., Ed.,** *Handbook of Chemistry and Physics,* 69th ed., CRC Press, Boca Raton, FL, 1971.
3. **Dreisbach, R. R.,** *Physical Properties of Chemical Compounds,* Advances in Chemistry Ser., No. 22, American Chemical Society, Washington, D.C., 1959.
4. **Jamieson, D. T., Irving, J. B., and Tudhope, J. S.,** Liquid Thermal Conductivity — a Data Survey to 1973, Her Majesty's Stationery Office, Edinburgh, 1975.
5. **Sax, N. I.,** *Dangerous Properties of Industrial Materials,* Van Nostrand Reinhold, New York, 1984.
6. **Sedivec, V. and Flek, J.,** *Handbook of Analysis of Organic Solvents,* John Wiley & Sons (Halsted Press), New York, 1976.
7. **Epstein, W. W. and Sweat, F.W.,** Dimethyl Sulfoxide oxidations, *Chem. Rev.,* 67, 247, 1967.

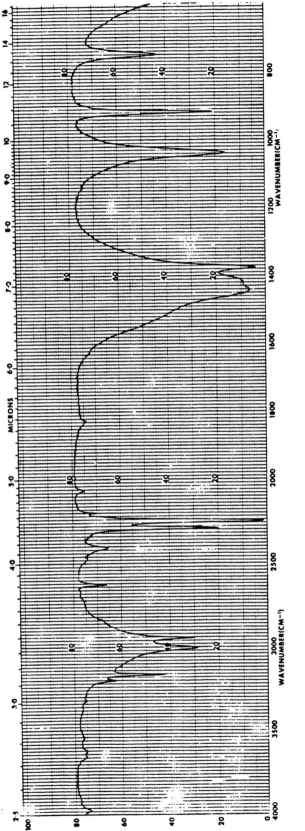

ACETONITRILE: CH₃CN

ACETONITRILE: CH₃CN

Physical Properties

Molecular mass	41.05
Melting point	−45.7 °C
Boiling point	81.6 °C
Refractive index (20 °C)	1.34423
Density (20 °C)	0.7857 g/mL
Viscosity (25 °C)	0.345 mPa·s
Surface tension (20 °C)	29.30 mN/m
Heat of vaporization (at boiling point)	199.1 cal/g
Thermal conductivity	0.1762 W/(m·K)
Dielectric constant (20 °C)	37.5
Relative vapor density (air = 1)	1.41
Vapor pressure (20 °C)	0.0097 MPa
Solubility in water[a]	∞ %, w/w
Flash point (open cup [oc])	6 °C
Autoignition temperature	509 °C
Explosive limits in air	4.4—16%, v/v
CAS registry number	75-05-8
Exposure limits	40 ppm, 8 h TWA
Solubility parameter, δ	11.9

[a] Forms azeotrope at 16% (w/w).

Notes: Highly polar solvent; sweet, ethereal odor; soluble in water; flammable; highly toxic. Incompatible with strong oxidants; hydrolyzes in the presence of aqueous bases and strong aqueous acids. Synonyms: methyl cyanide, acetic acid nitrile, cyanomethane, ethylnitrile.

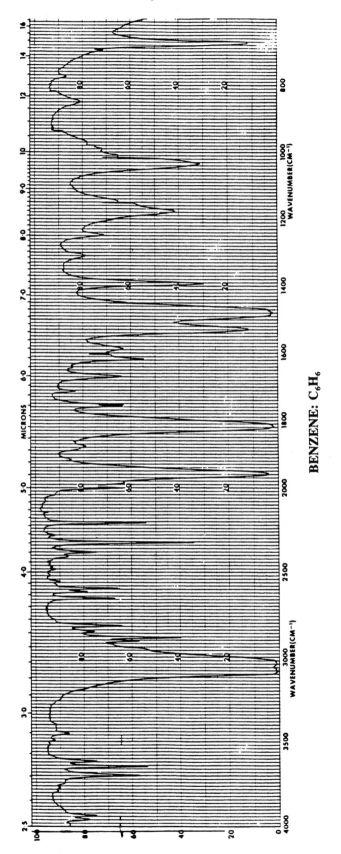

BENZENE: C$_6$H$_6$

BENZENE: C_6H_6

Physical Properties

Molecular mass	78.11
Melting point	5.5 °C
Boiling point	80.1 °C
Refractive index (20 °C)	1.50110
(25 °C)	1.4979
Density (20 °C)	0.8790 g/mL
(25 °C)	0.8737 g/mL
Viscosity (20 °C)	0.654 mPa·s
Surface tension (20 °C)	28.87 mN/m
Heat of vaporization (at boiling point)	94.325 cal/g
Thermal conductivity (25 °C)	0.1424 W/(m·K)
Dielectric constant (20 °C)	2.284
Relative vapor density (air = 1)	2.77
Vapor pressure (25 °C)	0.0097 MPa
Solubility in water[a]	0.07%, w/w
Flash point (OC)	−11 °C
Autoignition temperature	562 °C
Explosive limits in air	1.4—8.0%, v/v
CAS registry number	71-43-2
Exposure limits	10 ppm, 8 h TWA
Solubility parameter, δ	9.2
Hydrogen bond index, λ	2.2

[a]   Forms azeotrope (approximately 65 °C).

Notes: Nonpolar, aromatic solvent; sweet odor; very flammable and toxic; CONFIRMED HUMAN CARCINOGEN; soluble in alcohols, hydrocarbons (aliphatic and aromatic), ether, chloroform, carbon tetrachloride, carbon disulfide, slightly soluble in water. Incompatible with some strong acids and oxidants, chlorine trifluoride (zinc in the presence of steam); decomposes at high temperature to form biphenyl. Synonyms: cyclohexatriene, benzin, benzol, phenylhydride (these are the most common, although there are many other synonyms).

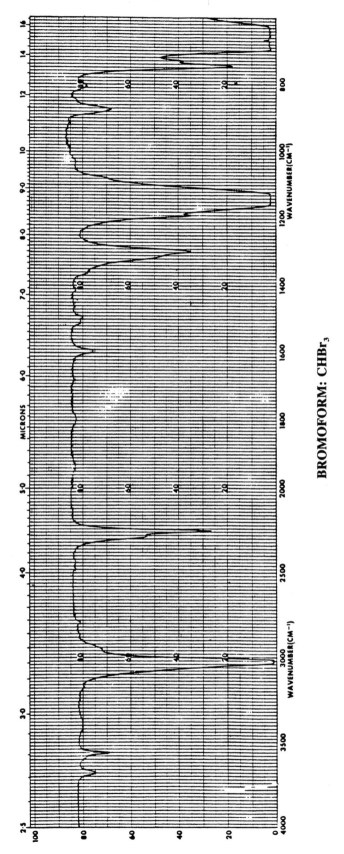

BROMOFORM: CHBr$_3$

BROMOFORM: CHBr$_3$

Physical Properties

Molecular mass	252.75
Melting point	8.3 °C
Boiling point	149.5 °C
Refractive index (20 °C)	1.6005
Density (20 °C)	2.8899 g/mL
Viscosity (25 °C)	1.89 mPa·s
Surface tension (20 °C)	41.53 mN/m
Heat of vaporization (at boiling point)	38.27 cal/g
Thermal conductivity	0.0961 W/(m·K)
Dielectric constant (20 °C)	4.39
Relative vapor density (air = 1)	2.77
Vapor pressure (25 °C)	0.0008 MPa
Solubility in water	(Slightly)
Flash point (OC)	None
Autoignition temperature	None
Explosive limits in air	Nonflammable
CAS registry number	75-25-2
Exposure limits	0.5 ppm (skin)

Notes: Moderately polar, weakly hydrogen bonding solvent; non-flammable; highly toxic, soluble in alcohols, organohalogen compounds, hydrocarbons, benzene and many oils. Incompatible with many alkali and alkaline earth metals. Synonym: tribromomethane.

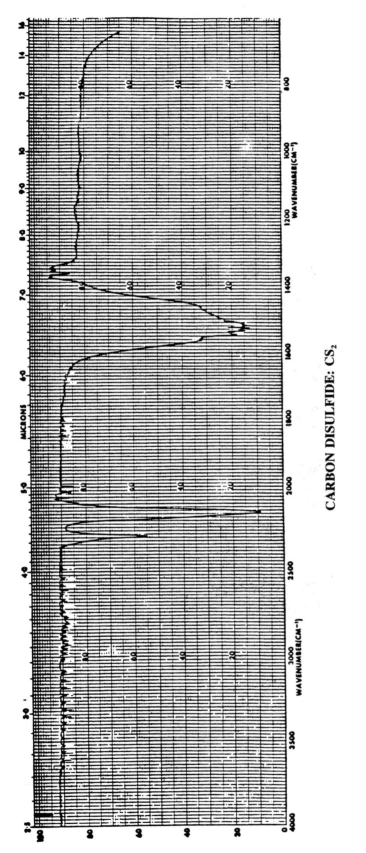

CARBON DISULFIDE: CS$_2$

CARBON DISULFIDE: CS₂

Physical Properties

Molecular mass	76.14
Melting point	-111 °C
Boiling point	46.3 °C
Refractive index (20 °C)	1.6280
(25 °C)	1.6232
Density (20 °C)	1.2631 g/mL
(25 °C)	1.2556 g/mL
Viscosity (20 °C)	0.363 mPa·s
Surface tension (20 °C)	32.25 mN/m
Heat of vaporization (at boiling point)	84.07 cal/g
Dielectric constant (20 °C)	2.641
Relative vapor density (air = 1)	2.64
Vapor pressure (25 °C)	0.0448 MPa
Solubility in water (20 °C)	0.29%, w/w
Flash point (OC)	-30 °C
Autoignition temperature	125 °C
Explosive limits in air	1.0—50%, v/v
CAS registry number	75-15-0
Exposure limits	20 ppm, 8 h TWA

Notes: Moderately polar solvent, soluble in alcohols, benzene, ethers, and chloroform; slightly soluble in water; very flammable; strong disagreeable odor when impure; incompatible with aluminum (powder), azides, chlorine, chlorine monoxide, ethylene diamine, ethyleneamine, fluorine, nitrogen, oxides, potassium, and zinc. Synonyms: carbon bisulfide, dithiocarbon anhydride.

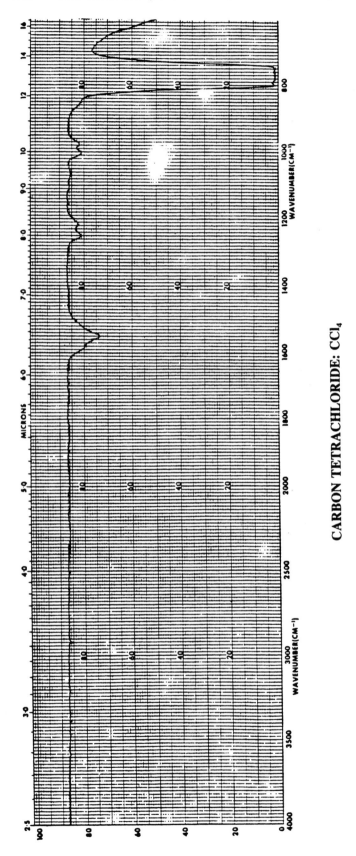

CARBON TETRACHLORIDE: CCl$_4$

CARBON TETRACHLORIDE: CCl_4

Physical Properties

Molecular mass	153.82
Melting point	-23.0 °C
Boiling point	76.74 °C
Refractive index (20 °C)	1.4607
(25 °C)	1.4570
Density (20 °C)	1.5940 g/mL
(25 °C)	1.5843 g/mL
Viscosity (20 °C)	0.969 mPa·s
Surface tension (20 °C)	26.75 mN/m
Heat of vaporization (at boiling point)	46.8 cal/g
Thermal conductivity (20 °C)	0.1070 W/(m·K)
Dielectric constant (20 °C)	2.238
Relative vapor density (air $=$ 1)	5.32
Vapor pressure (25 °C)	0.0122 MPa
Solubility in water (20 °C)	0.08%, w/w
Flash point (OC)	Incombustible
Autoignition temperature	Incombustible
Explosive limits in air	Nonexplosive %, v/v
CAS registry number	56-23-5
Exposure limits	5 ppm (skin)
Solubility parameter, δ	8.6
Hydrogen bond index, λ	2.2

Notes: Nonpolar solvent; soluble in alcohols, ethers, chloroform, benzene and most oils, insoluble in water; nonflammable; extremely toxic; carcinogenic; incompatible with allyl alcohol, silanes, triethyldialuminum, many metals (e.g., sodium). Synonyms: tetrachloromethane, perchloromethane, methane tetrachloride, Halon-104.

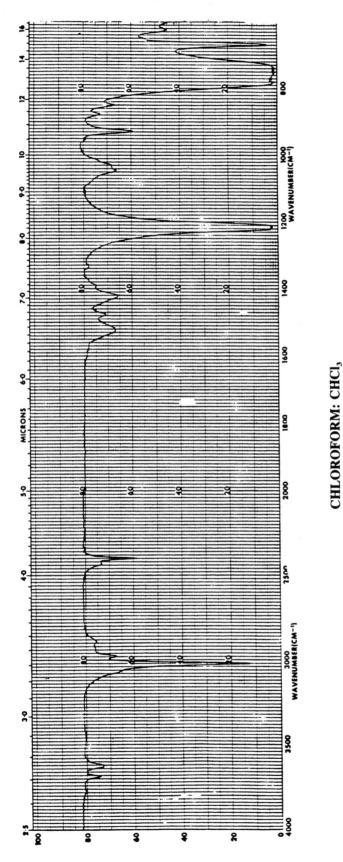

CHLOROFORM: CHCl₃

CHLOROFORM: CHCl$_3$

Physical Properties

Molecular mass	119.38
Melting point	−63.2 °C
Boiling point	61.2 °C
Refractive index (20 °C)	1.4458
(25 °C)	1.4422
Density (20 °C)	1.4892 g/mL
(25 °C)	1.4798 g/mL
Viscosity (20 °C)	0.566 mPa·s
Surface tension	27.2 mN/m
Heat of vaporization (at boiling point)	59.3 cal/g
Thermal conductivity (20 °C)	0.1164 W/(m·K)
Dielectric constant (20 °C)	4.806
Relative vapor density (air = 1)	4.13
Vapor pressure (25 °C)	0.0263 MPa
Solubility in water	0.815%, w/w
Flash point (OC)	Incombustible
Autoignition temperature	incombustible
Explosive limits in air	Nonexplosive
CAS registry number	67-66-3
Exposure limits	10 ppm, 8 h TWA
Solubility parameter, δ	9.3
Hydrogen bond index, λ	2.2

Notes: Polar solvent; soluble in alcohols, ether, benzene, and most oils; usually stabilized with methanol to prevent phosgene formation; flammable and highly toxic; carcinogenic, incompatible with caustics, active metals, aluminum powder, potassium, sodium, magnesium. Synonyms: trichloromethane, methane trichloride.

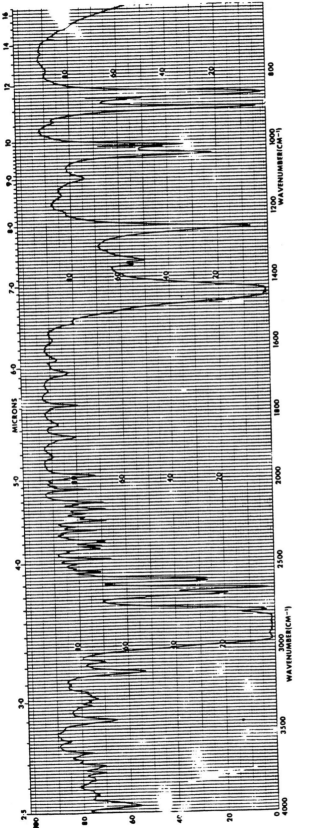

CYCLOHEXANE: C₆H₁₂

CYCLOHEXANE: C_6H_{12}

Physical Properties

Molecular mass	84.16
Melting point	6.3 °C
Boiling point	80.72 °C
Refractive index (20 °C)	1.4263
(25 °C)	1.4235
Density (20 °C)	0.7786 g/mL
(25 °C)	0.7739 g/mL
Viscosity (20 °C)	1.06 mPa·s
Surface tension (20 °C)	24.99 mN/m
Heat of vaporization (at boiling point)	85.62 cal/g
Thermal conductivity (20 °C)	0.122 W/(m·K)
Dielectric constant (20 °C)	2.023
Relative vapor density (air = 1)	2.90
Vapor pressure (25 °C)	0.0111 MPa
Solubility in water (20 °C)	<0.01%, w/w
Flash point (OC)	−17 °C
Autoignition temperature	245 °C
Explosive limits in air	1.31—8.35%, v/v
CAS registry number	110-82-7
Exposure limits	300 ppm, 8 h TWA

Notes: Nonpolar hydrocarbon solvent; mild, gasoline-like odor; soluble in hydrocarbons, alcohols, organic halides, acetone, benzene, flammable; moderately toxic, reacts with oxygen (air) at elevated temperatures, decomposes upon heating; incompatible with oxidants. Synonyms: benzene hexahydride, hexamethylene, hexanaphthene, hexahydrobenzene.

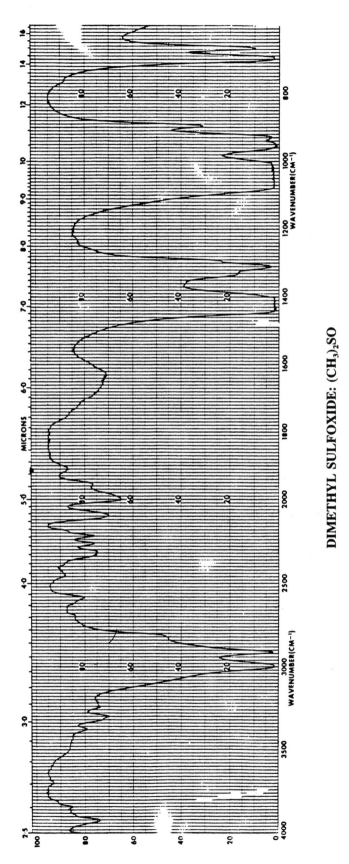

DIMETHYL SULFOXIDE: $(CH_3)_2SO$

DIMETHYL SULFOXIDE: $(CH_3)_2SO$

Physical Properties

Molecular mass	78.13
Melting point	18.5 °C
Boiling point	189 °C
Refractive index (20 °C)	1.4770
Density (20 °C)	1.1014 g/mL
Viscosity (25 °C)	1.98 mPa·s
Surface tension	43.5 mN/m
Heat of vaporization (at boiling point)	144 cal/g
Relative vapor density (air = 1)	2.7
Vapor pressure	5.3×10^{-5} MPa
Solubility in water	∞
Flash point (OC)	95 °C
Autoignition temperature	215 °C
Explosive limits in air	2.6—28.5%, v/v
CAS registry number	67-68-5
Exposure limits	None established
Solubility parameter, δ	13.0
Hydrogen bond index, λ	5.0

Notes: Colorless, odorless (when pure), hygroscopic liquid, powerful aprotic solvent; dissolves many inorganic salts, soluble in water; combustible; readily penetrates the skin; incompatible with strong oxidizers, and many halogenated compounds (e.g., alkyl halides, aryl halides), oxygen, peroxides, diborane, perchlorates. Synonyms: DMSO, methyl sulfoxide, sulfinylbismethane.

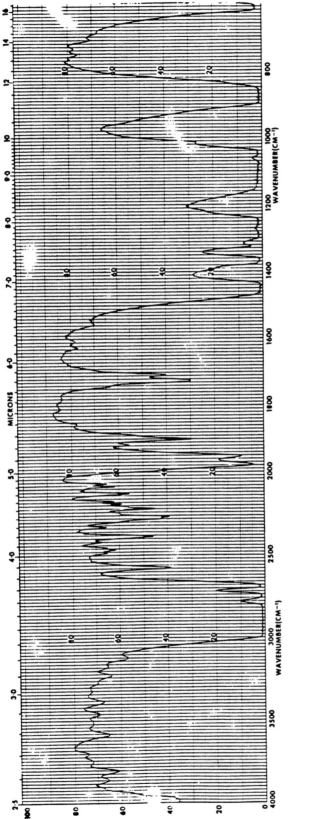

1,4-DIOXANE: OCH₂CH₂OCH₂CH₂

1,4-DIOXANE: OCH$_2$CH$_2$OCH$_2$CH$_2$

Physical Properties

Molecular mass	88.11
Melting point	11 °C
Boiling point	101.3 °C
Refractive index (20 °C)	1.4221
(25 °C)	1.4195
Density (20 °C)	1.0338 g/mL
(25 °C)	1.0282 g/mL
Viscosity (20 °C)	1.37 mPa·s
Surface tension (20 °C)	33.74 mN/m
Heat of vaporization (at boiling point)	98 cal/g
Dielectric constant (20 °C)	2.209
Relative vapor density (air = 1)	3.03
Vapor pressure (25 °C)	0.0053 MPa
Solubility in water	∞
Flash point (OC)	12 °C
Autoignition temperature	180 °C
Explosive limits in air	1.97—22.2%, v/v
CAS registry number	123-91-1
Exposure limits	100 ppm (skin)
Solubility parameter, δ	9.9
Hydrogen bond index, λ	5.7

Notes: Moderately polar solvent; soluble in water and most organic solvents; flammable; highly toxic by ingestion and inhalation; absorbed through the skin; incompatible with strong oxidizers. Synonyms: diethylene ether, 1,4-diethylene dioxide, diethylene dioxide, dioxyethylene ether.

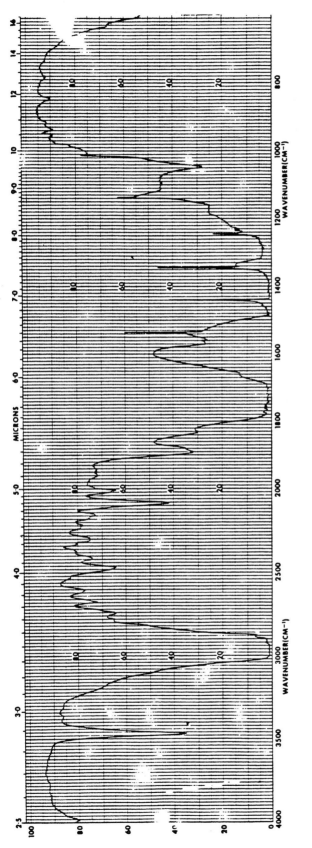

ETHYL ACETATE: CH₃COOC₂H₅

ETHYL ACETATE: $CH_3COOC_2H_5$

Physical Properties

Molecular mass	88.11
Melting point	-83.58 °C
Boiling point	77.06 °C
Refractive index (20 °C)	1.3723
(25 °C)	1.3698
Density (20 °C)	0.9006 g/mL
(25 °C)	0.8946 g/mL
Viscosity (20 °C)	0.452 mPa·s
Surface tension (20 °C)	23.95 mN/m
Heat of vaporization (at boiling point)	87.63 cal/g
Thermal conductivity (20 °C)	0.122 W/(m·K)
Dielectric constant (25 °C)	6.02
Relative vapor density (air $=$ 1)	3.04
Vapor pressure (20 °C)	0.0097 MPa
Solubility in water (20 °C)	3.3%, w/w
Flash point (OC)	-1 °C
Autoignition temperature	486 °C
Explosive limits in air	2.18—11.5%, v/v
CAS registry number	141-78-6
Exposure limits	400 ppm, 8 h, TWA
Solubility parameter, δ	9.1
Hydrogen bond index, λ	5.2

Notes: Polar solvent; soluble in water, alcohols (organic halides), ether, and many oils; flammable; moderately toxic by inhalation and skin absorption, incompatible with strong oxidizers, nitrates, strong alkalies, strong acids. Synonyms: acedin, acetic ether, acetic ester, vinegar naphtha, acetic acid, ethyl ester.

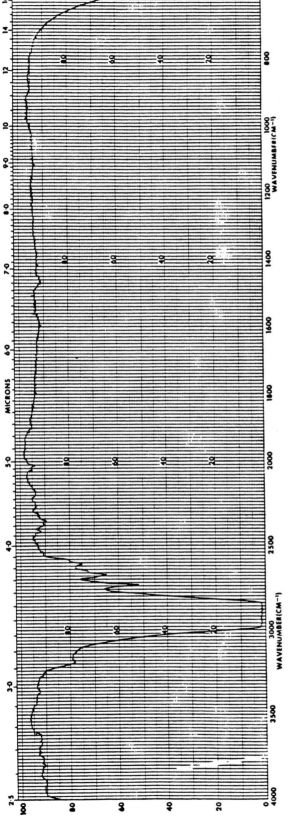

n-HEXANE: CH₃(CH₂)₄CH₃

n-HEXANE: CH$_3$(CH$_2$)$_4$CH$_3$

Physical Properties

Molecular mass	86.18
Melting point	−95 °C
Boiling point	68.742 °C
Refractive index (20 °C)	1.37486
(25 °C)	1.3723
Density (20 °C)	0.6594 g/mL
(25 °C)	0.6548 g/mL
Viscosity (20 °C)	0.31 mPa·s
Surface tension (20 °C)	18.42 mN/m
Heat of vaporization (at boiling point)	80.03 cal/g
Thermal conductivity (20 °C)	0.1217 W/(m·K)
Dielectric constant (20 °C)	1.890
Relative vapor density (air = 1)	2.97
Vapor pressure (25 °C)	0.0222 MPa
Solubility in water (20 °C)	0.011%, w/w
Flash point (OC)	−26 °C
Autoignition temperatures	247 °C
Explosive limits in air	1.25—6.90%, v/v
CAS registry number	110-54-3
Exposure limits	50 ppm, 8 h TWA
Solubility parameter, δ	9.3
Hydrogen bond index, λ	2.2

Notes: Nonpolar solvent; soluble in alcohols, hydrocarbons, organic halides, acetone and ethers; insoluble in water; flammable; moderately toxic by inhalation and ingestion; incompatible with strong oxidizers. Synonyms: hexane, hexyl hydride.

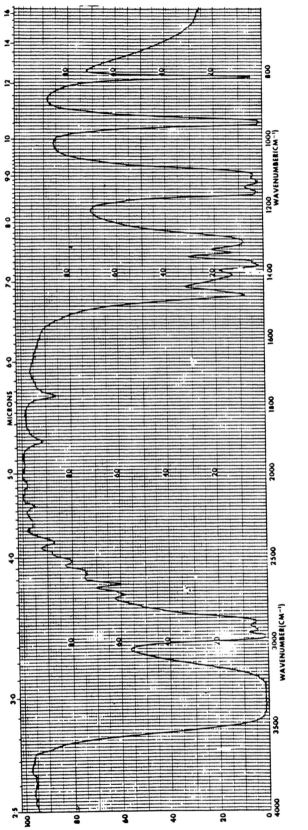

ISOPROPANOL: (CH₃)₂CHOH

ISOPROPANOL: $(CH_3)_2CHOH$

Physical Properties

Molecular mass	60.10
Melting point	89.8 °C
Boiling point	82.4 °C
Refractive index (20 °C)	1.3771
(25 °C)	1.3750
Density (20 °C)	0.7864 g/mL
(25 °C)	0.7812 g/mL
Viscosity (20 °C)	2.43 mPa·s
Surface tension (20 °C)	21.99 mN/m
Heat of vaporization (at boiling point)	160 cal/g
Dielectric constant (25 °C)	18.3
Relative vapor density (air = 1)	2.07
Vapor pressure	0.0044 MPa
Solubility in water (20 °C)	∞
Flash point (OC)	16 °C
Autoignition temperature	456 °C
Explosive limits in air	2.02—11.8%, v/v
CAS registry number	67-63-0
Exposure limits	400 ppm (skin)
Solubility parameter, δ	11.5
Hydrogen bond index, λ	8.9

Notes: Polar solvent; soluble in water, alcohols, ethers, many hydrocarbons and oils; flammable and moderately toxic by ingestion and inhalation; incompatible with strong oxidizers. Synonyms: dimethyl carbinol, *sec*-propyl alcohol, 2-propanol, *n*-propan-2-ol.

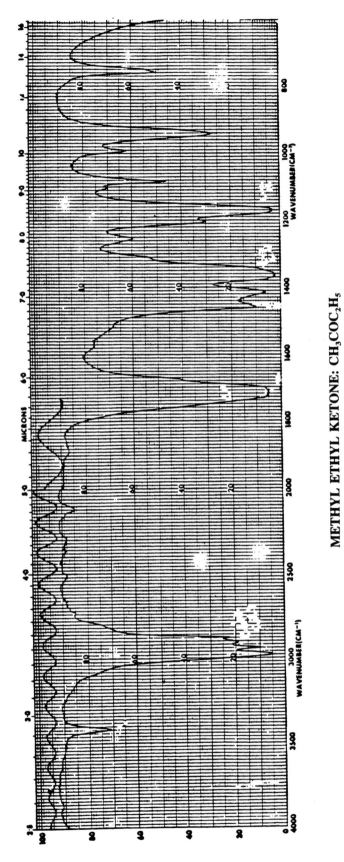

METHYL ETHYL KETONE: CH₃COC₂H₅

METHYL ETHYL KETONE: $CH_3COC_2H_5$

Physical Properties

Molecular mass	72.11
Melting point	-86.4 °C
Boiling point	79.6 °C
Refractive index (20 °C)	1.379
(25 °C)	1.3761
Density (20 °C)	0.8054 g/mL
(25 °C)	0.8002 g/mL
Viscosity (20 °C)	0.448 mPa·s
Surface tension (20 °C)	24.50 mN/m
Heat of vaporization (at boiling point)	108.74 cal/g
Thermal conductivity	0.1465 W/(m·K)
Dielectric constant (20 °C)	18.5
Relative vapor density (air = 1)	2.41
Vapor pressure (25 °C)	0.0129 MPa
Solubility in water (20 °C)	27.33%, w/w
Flash point (OC)	2 °C
Autoignition temperature	516 °C
Explosive limits in air	1.81—11.5%, v/v
CAS registry number	78-93-3
Exposure limits	200 ppm, 8 h, TWA
Solubility parameter, δ	9.3
Hydrogen bond index, λ	5.0

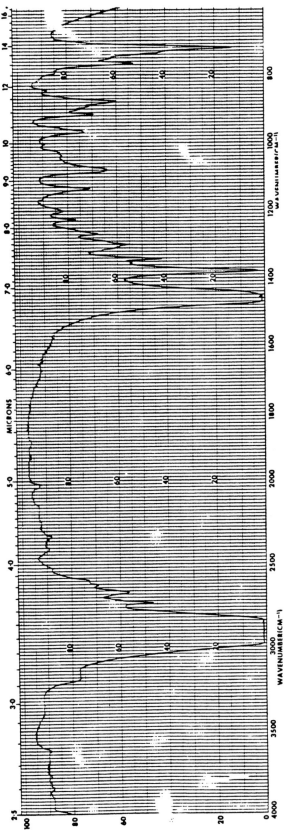

n-OCTANE: $CH_3(CH_2)_6CH_3$

n-OCTANE: $CH_3(CH_2)_6CH_3$

Physical Properties

Molecular mass	114.23
Melting point	$-56.66°C$
Boiling point	125.677°C
Refractive index (20°C)	1.39745
(25°C)	1.3951
Density (20°C)	0.7025 g/mL
(25°C)	0.6985 g/mL
Viscosity (20°C)	0.539 mPa·s
Surface tension (20°C)	21.75 mN/m
Heat of vaporization (at boiling point)	73.3 cal/g
Dielectric constant (20°C)	1.948
Relative vapor density (air = 1)	3.86
Vapor pressure (25°C)	0.0023 MPa
Solubility in water (20°C)	-0.002%, w/w
Flash point (CC)	13°C
Autoignition temperature	232°C
Explosive limits in air	0.84—3.2%, v/v
CAS registry number	111-65-9
Exposure limits	500 ppm, 8 hr TWA
Hydrogen bond index, λ	2.2

Notes: Nonpolar solvent; soluble in alcohol, acetone, and hydrocarbons, insoluble in water; flammable, incompatible with
strong oxidizers. Synonym: octane.

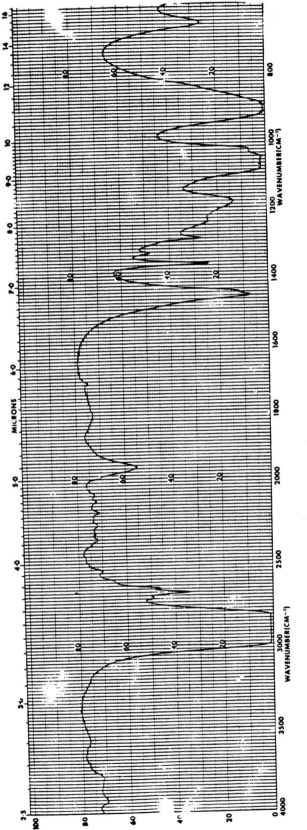

TETRAHYDROFURAN: $CH_2(CH_2)_2CH_2O$

TETRAHYDROFURAN: $CH_2(CH_2)_2CH_2O$

Physical Properties

Molecular mass	72.108
Melting point	$-65\ °C$
Boiling point	66 °C
Refractive (20 °C)	1.4070
(25 °C)	1.4040
Density (20 °C)	0.8880 g/mL
(25 °C)	0.8818 g/mL
Viscosity (20 °C)	0.55 mPa·s
Surface tension (20 °C)	26.4 mN/m
Heat of vaporization (at boiling point)	95 cal/g
Dielectric constant (20 °C)	7.54
Relative vapor density (air = 1)	2.5
Vapor pressure (20 °C)	0.0191 MPa, 143 mmHg
Solubility in water (20 °C)[a]	∞
Flash point (closed cup [CC])	260 °C
Explosive limits in air	1.8—11.8%, v/v
CAS registry number	109-99-9
Exposure limits	200 ppm, 8 h TWA
Solubility parameter, δ	9.1
Hydrogen bond index, λ	5.3

[a] pH of aqueous solution = 7.

Notes: Moderately polar solvent, ethereal odor; soluble in water and most organic solvents; flammable; moderately toxic. Incompatible with strong oxidizers; can form potentially explosive peroxides upon long standing in air; can polymerize in the presence of cationic initiators such as Lewis acids or strong proton acids. Synonyms: THF, tetramethylene oxide, diethylene oxide, 1,4-epoxybutane oxolane, oxacyclopentane.

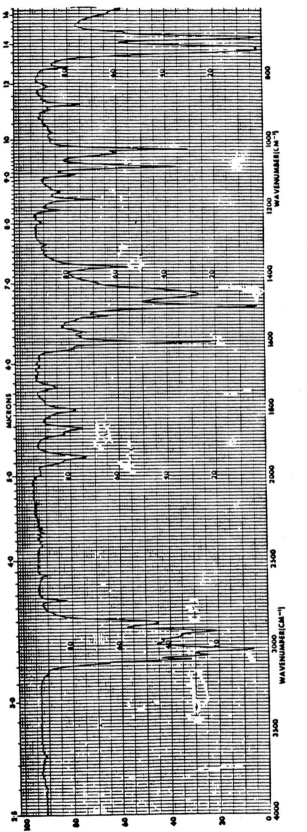

TOLUENE: $CH_3C_6H_5$

TOLUENE CH$_3$C$_6$H$_5$

Physical Properties

Molecular mass	92.14
Melting point	-94.5 °C
Boiling point	110.7 °C
Refractive index (20 °C)	1.4969
(25 °C)	1.4941
Density (20 °C)	0.8669 g/mL
(25 °C)	0.8623 g/mL
Viscosity (20 °C)	0.587 mPa·s
Surface tension (20 °C)	28.52 mN/m
Heat of vaporization (at boiling point)	86.50 cal/g
Thermal conductivity (20 °C)	0.1348 W/(m·K)
Dielectric constant (25 °C)	2.379
Relative vapor density (air = 1)	2.14
Vapor pressure (25 °C)	0.0036 MPa
Solubility in water	0.047%, w/w
Flash point (CC)	4 °C
Autoignition temperature	552 °C
Explosive limits in air	1.27—7.0%, v/v
CAS registry number	108-88-3
Exposure limits	200 ppm, 8 h TWA
Solubility parameter, δ	8.9
Hydrogen bond index, λ	3.8

Notes: Aromatic solvent; sweet pungent odor; soluble in benzene, alcohols, organic halides, ethers; insoluble in water; highly flammable; toxic by ingestion, inhalation and absorption through the skin. Incompatible with strong oxidants; decomposes under high heat to form (predominantly) dimethylbiphenyl. Synonyms: toluol, methylbenzene, methylbenzol, phenylmethane.

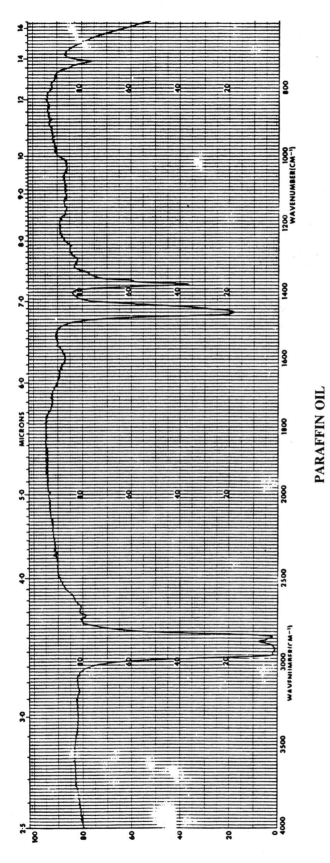

PARAFFIN OIL

PARAFFIN OIL

Physical Properties

Molecular mass	Variable
Melting point	-20 °C
	(approximate)
Boiling point	315 °C
	(approximate)
Refractive index (20 °C)	1.4720
(25 °C)	1.4697
Solubility in water	Insoluble
Flash point (OC)	229 °C
Explosive limits in air	0.6—6.5% v/v
CAS registry number	8012—95-1
Exposure limits	50 ppm 8 h TWA

Notes: Viscous, odorless, moderately combustible liquid used for mull preparation, low toxicity; soluble in benzene, chloroform, carbon disulfide, ethers; incompatible with oxidizing materials and amines. Synonyms: mineral oil, adepsine oil, lignite oil.

POLYSTYRENE WAVE NUMBER CALIBRATION

The following are wave number readings assigned to the peaks on the spectrum:

1	3027.1	6	1801.6	11	1069.1
2	2924.0	7	1601.4	12	1028.0
3	2850.7	8	1583.1	13	906.7
4	1944.0	9	1181.4	14	698.9
5	1871.0	10	1154.3		

Film thickness, 50 µm; molecular weight, 104.15 η.

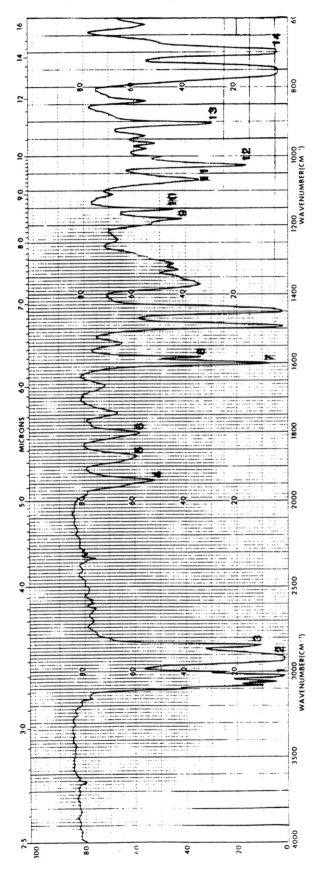

INFRARED ABSORPTION CORRELATION CHART

The following pages provide an infrared spectrophotometric correlation chart. This chart is based upon the work of Professor Charles F. Hammer of Georgetown University (copyright 1975) and is reproduced with permission, with slight modifications.

Key: AR = aromatic; b = broad; sd = solid; sn = solution; sp = sharp; ? = unreliable.

Peak intensity:
Strong   �■
Medium   ▨▨
Weak   ░░
Variable   ⧫⧫

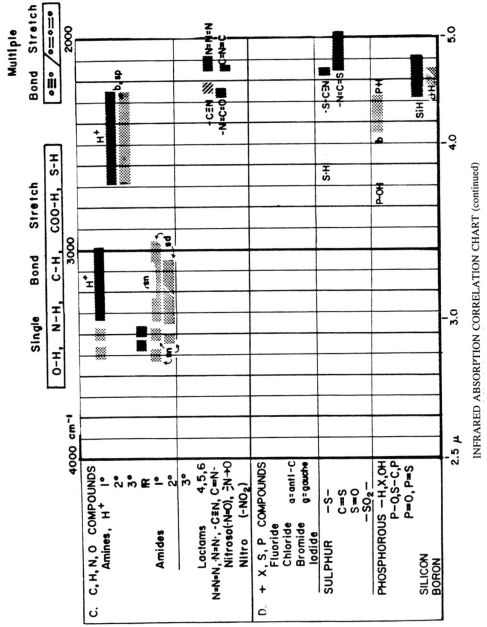

INFRARED ABSORPTION CORRELATION CHART (continued)

Reprinted by permission of Professor Charles F. Hammer, Chemistry Department, Georgetown University.

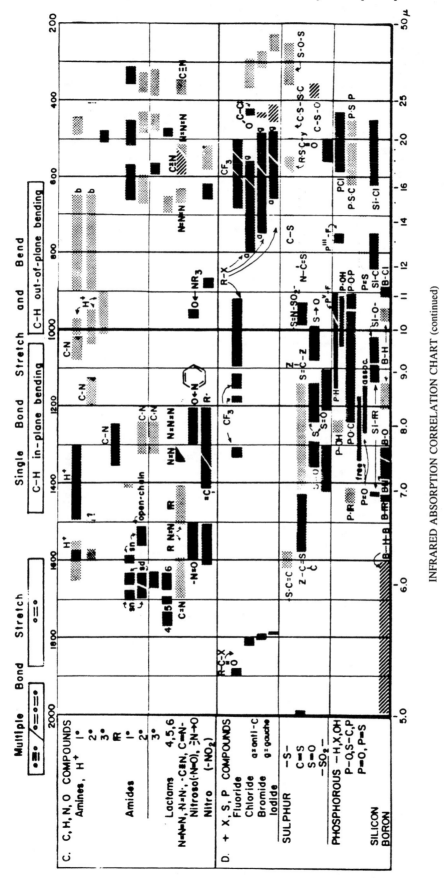

INFRARED ABSORPTION CORRELATION CHART (continued)

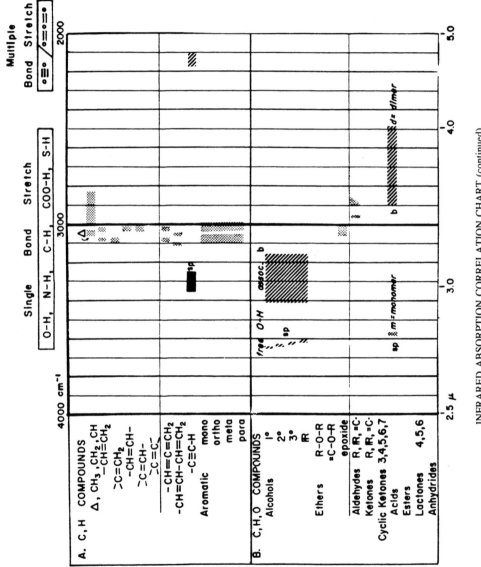

INFRARED ABSORPTION CORRELATION CHART (continued)

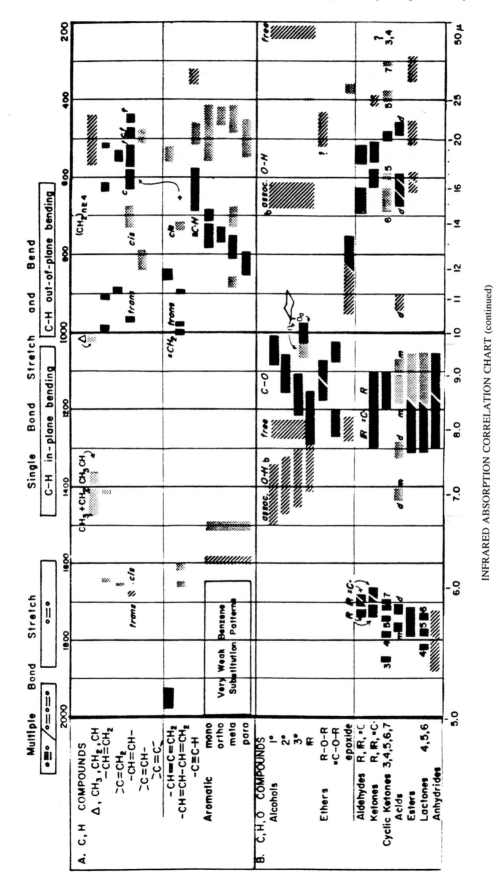

INFRARED ABSORPTION CORRELATION CHART (continued)

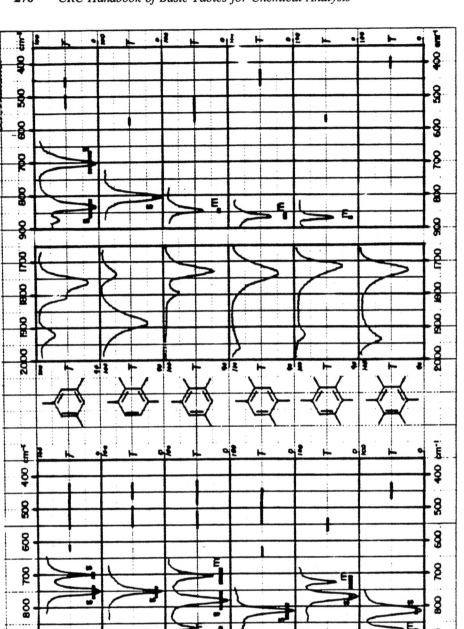

INFRARED ABSORPTION CORRELATION CHART (continued)

INFRARED ABSORPTIONS OF MAJOR CHEMICAL FAMILIES

The following tables provide expected IR absorptions of the major chemical families.[1-23] The ordering of these tables is: hydrocarbons, oxygen compounds, nitrogen compounds, sulfur compounds, silicon compounds, phosphorous compounds, and halogen compounds.

ABBREVIATIONS

s	Strong	1°	primary
m	Medium	2°	secondary
w	Weak	3°	tertiary
vs	Very strong		
vw	Very weak		
sym	Symmetrical		
asym	Asymmetrical		

REFERENCES

1. **Nakanishi, K. and Solomon, P. H.,** *Infrared Absorption Spectroscopy,* Holden-Day, San Francisco, 1977.
2. **Conley, R. T.,** *Infrared Spectroscopy,* 2nd ed., Allyn & Bacon, Boston, 1972.
3. **Silverstein, R. M., Bassler, G. C., and Morrill, T. C.,** *Spectrometric Identification of Organic Compounds,* John Wiley & Sons, New, York, 1981.
4. **Williams, D. H. and Fleming, I.,** *Spectroscopic Methods in Organic Chemistry,* McGraw-Hill, London, 1973.
5. **Lambert, J. B., Shuzvell, H. F., Verbit, L., Cooks, R. G., and Stout, G. H.,** *Organic Structural Analysis,* Macmillan, New York, 1976.
6. **Meyers, C. Y.,** Eighth Annual Report on Research, 1963 (Sponsored by the Petroleum Research Fund), American Chemical Society, Washington, D.C., 1964.
7. **Kucsman, A., Ruff, F., and Kapovits, I.,** Bond system of *N*-acylsulfilimines. I. Infrared spectroscopic investigation of *N*-sulfonylsulfilimines, *Tetrahedron,* 25, 1575, 1966.
8. **Kucsman, A., Kapovits, I., and Ruff, F.,** Infrared absorption of *N*-acylsulfilimines, *Acta Chim. Acad. Sci. Hung.,* 40, 75, 1964.
9. **Kucsman, A., Ruff, F., and Kapovits, I.,** Bond system of *N*-acylsulfilimines. V. IR spectroscopic study on *N*-(*p*-nitrophenylsulfonyl) sulfilimines, *Acta Chim. Acad. Sci. Hung.,* 54, 153, 1967.
10. **Shah, J. J.,** Iminosulfuranes (sulfilimines): infrared and ultraviolet spectroscopic studies, *Can. J. Chem.,* 53, 2381, 1975.
11. **Tsujihara, K., Furukawa, N., and Oae, S.,** Sulfilimine. II. IR, UV and NMR spectroscopic studies, *Bull. Chem. Soc. Jpn.,* 43, 2153, 1970.
12. **Fuson, N., Josien, M. L., and Shelton, E. M.,** An infrared spectroscopic study of the carbonyl stretching frequency in some groups of ketones and quinones, *J. Am. Chem. Soc.,* 76, 2526, 1954.
13. **Davis, F. A., Friedman, A. J., and Kluger, E. W.,** Chemistry of the sulfur-nitrogen bond. VIII. *N*-alkylidenesulfinamides, *J. Am. Chem. Soc.,* 96, 5000, 1974.
14. **Krueger, P. J. and Fulea, A. O.,** Rotation about the C-N bond in thioamides: influence of substituents on the potential function, *Tetrahedron,* 31, 1813, 1957.
15. **Baumgarten, H. E. and Petersen, J. M.,** Reactions of amines. V. Synthesis of α-aminoketones, *J. Am. Chem. Soc.,* 82, 459, 1960.
16. **Gaset, A., Lafaille, L., Verdier, A., and Lattes, A.,** Infrared spectra of α-aminoketones; configurational study and evidence of an enol form, *Bull. Soc. Chim. Fr.,* 10, 4108, 1958.
17. **Cagniant, D., Faller, P., and Cagniant, P.,** Contribution in the study of condensed sulfur heterocycles. XVII. Ultraviolet and infrared spectra of some alkyl derivatives of thianaphthene, *Bull. Soc. Chim. Fr.,* 2410, 1961.
18. **Tamres, M. and Searles, S. J.,** Hydrogen bonding abilities of cyclic sulfoxides and cyclic ketones, *J. Am. Chem. Soc.,* 81, 2100, 1959.
19. **George, W. O., Goodman, R. C. W., and Green, J. H. S.,** The infra-red spectra of alkyl mercapturic acids, their sulphoxides and sulphones, *Spectrochim. Acta,* 22, 1741, 1966.
20. **Cairns, T., Eglinton, G., and Gibson, D. T.,** Infra-red studies with sulphoxides. I. The S–O stretching absorptions of some simple sulphoxides, *Spectrochim. Acta,* 20, 31, 1964.

21. **Hadzi, D.,** Hydrogen bonding in some adducts of oxygen bases with acids. I. Infrared spectra and structure of crystalline adducts of some phosphine, arsine, and amine oxides, and sulphoxides with strong acids, *J. Chem. Soc.*, 1962, 5128.
22. **Currier, W. F. and Weber, J. H.,** Complexes of sulfoxides. I. Octahedral complexes of manganese (II), iron (II), cobalt (II), nickel (II), and zinc (II), *Inorg. Chem.*, 6, 1539, 1967.
23. **Kucsman, A., Ruff, F., and Tanacs, B.,** IR spectroscopic study of N^{15} labeled acylsulfilimines, *Int. J. Sulfur Chem.*, 8, 505, 1976.

HYDROCARBON COMPOUNDS

Family	General formula	Wave numbers (cm^{-1})			
		C–H stretch	C–H bend	C–C stretch	C–C bend
Alkanes					
Acyclic					
Straight chain	C_nH_{2n+2} $CH_3(CH_2)_nCH_3$	3000—2840 (s/m) CH$_3$– (asym): 3000—2960(s) CH$_3$– (sym): 2880—2870(s) >CH$_2$ (asym): 2930—2920(s) >CH$_2$ (sym): 2860—2840(s)	Below 1500 (w/m/s) CH$_3$– (asym): 1460—1440(s) CH$_3$– (sym): 1380—1370(s) >CH$_2$ (scissoring): ~1465(s) >CH$_2$ (rocking): ~720(s) >CH$_2$ (twisting and wagging): 1350—1150(w),	1200—800 (w) (not of practical value)	Below 500 (not of practical value)
Branched	R^1–CHR^3 | R^2	C–H (3°): ~2890 (vw)	Gem dimethyl [(CH$_3$)$_2$CH–]: 1380, 1370 (m, symmetrical doublet) tert-Butyl [(CH$_3$)$_3$C–]: 1390, 1370 (m, unsymmetrical doublet; latter more intense) CH$_3$– Rocking: 930—920 (w, not reliable).		
Cyclic	(CH$_2$)$_n$	Same as in acyclic alkanes; ring strain increases the wave numbers up to 3100 cm^{-1}	>CH$_2$ (scissoring): lower than in acyclic alkanes (10—15 cm^{-1}).		

HYDROCARBON COMPOUNDS (continued)

Family	General formula	Wave numbers (cm^{-1})				Notes
		>C=C< stretch	>C=C–H stretch	>C=C–H bend (in plane)	>C=C–H bend (out-of-plane)	
Alkenes (olefins) Acyclic Nonconjugated	C_nH_{2n}	1670—1600	Above 3000			
Monosubstituted (vinyl)	$R^1CH=CH_2$	1667—1640(m) 1658—1648(m)	3082—3000(m)	1420—1415(m) (scissoring)	~995(m) ~919(m)	
Disubstituted cis-	R^1 R^2 >C=C< H H	1662—1652(m)	3030—3015(m)	~1406(m)	715—675(s) (rocking)	C–H rocking not too dependable
trans-	R^1 H >C=C< H R^2	1678—1668(w)	3030—3020(m)	1325—1275(m) (deformation)	~965(s) (rocking)	
Vinylidene	R^1 H >C=C< R^2 H	1658—1648(m)	3090—3080(m) ~2980(m)	~1415(m)	~890(s) (rocking)	
Trisubstituted	R^1 H >C=C< R^2 R^3	1675—1665(w)	3090—3080(w)	~1415(w)	840—800(m) (deformation)	
Tetrasubstituted	R^1 R^3 >C=C< R^2 R^4	~1670(vw)	—	—	—	>C=V< stretch may be absent.
Conjugated	>C=C–C=C<	1610—1600(m) (frequently a doublet)	3050(vw)		~980 (rocking)	Conjugation of an olefinic >C=C< with an aromatic ring raises the frequency 20—25 cm^{-1}

					Notes
Cumulated	>C=C=C<	2000—1900(m)	3300(m)	2000—1900(s) 1800—1700(w)	880—850(s)	>C=C< stretch is coupled with C–C stretch of adjacent bonds. Alkyl substitution increases the >C=C< absorption frequency. >C=C< frequency increases with decreasing ring size
Cyclic		1640—1560 (variable)			697—625(w) (wagging)	
External exocyclic	=CH$_2$	1781—1650	3080, 2995(m)	~1300(w)		

Wave numbers (cm^{-1})

		—C≡C— stretch	—C≡C—H stretch	C–H bend	Notes
Alkynes **Nonconjugated**	C$_n$H$_{2n-2}$				—C≡C—H stretch peak is narrower than that of —OH or —NH stretch, which are broader due to hydrogen bonding
Terminal	R^1–C≡C–H	2150—2100(m)	3310—3200(m) (sharp)	700—610(s) 1370—1220(w) (overtone)	
Nonterminal	R^1–C≡C–R^2	2260—2190(vw)	—	700—610(s) 1370—1220(w) (overtone)	
Conjugated					
Terminal	R^1–C≡C–C≡C–H	2200, 2040 (doublet)	3310—3200(m) (sharp)	700—610(s) 1370—1220(w) (overtone)	
Nonterminal	R^1–C≡C–C≡C–R^2	2200, 2040 (doublet)	—	700—610(s) 1370—1220(w) (overtone)	

HYDROCARBON COMPOUNDS (continued)

Family	General formula	Wave numbers in (cm⁻¹)			Notes
		>C=C<^{H} stretch	>C=C< stretch	>C—H bend (out-of-plane)	
Aromatic compounds Monsubstituted		3100—3000	1600—1500	770—730(s) 710—690(s)	All show weak combination and overtone bands between 2000—16500 cm⁻¹; see aromatic substitution pattern chart
Disubstituted					
1,2-		3100—3000	1600—1500	770—735(s)	
1,3-		3100—3000	1600—1500	810—750(s) 710—690(s)	
1,4-		3100—3000	1600—1500	833—810(s)	
Trisubstituted					
1,2,3-		3100—3000	1600—1500	780—760(s) 745—705 (m)	

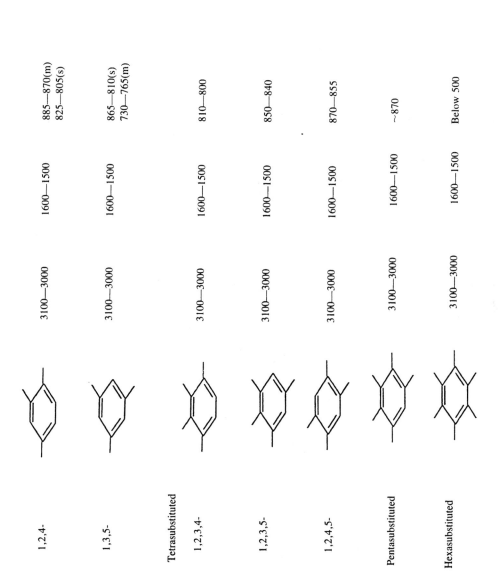

1,2,4-	3100—3000	1600—1500	885—870(m) 825—805(s)
1,3,5-	3100—3000	1600—1500	865—810(s) 730—765(m)
Tetrasubstituted			
1,2,3,4-	3100—3000	1600—1500	810—800
1,2,3,5-	3100—3000	1600—1500	850—840
1,2,4,5-	3100—3000	1600—1500	870—855
Pentasubstituted	3100—3000	1600—1500	~870
Hexasubstituted	3100—3000	1600—1500	Below 500

ORGANIC OXYGEN COMPOUNDS

Family	General formula	Wave numbers (cm⁻¹)			Notes
		O–H stretch	>C–O stretch	–O–H bend	
Acetals	OR² | R'–C–H | OR³		1195—1060(s) (three bands) 1055—1040(s) (sometimes obscured)		
Acyl halides	R–C–X ‖ O X=halogen				See Organic Halogen Compounds
Alcohols Primary	R–OH R–CH₂OH	3650—3584 (s, sharp)—for very dilute solutions or vapor phase spectra. 3550—3200 (s, broad)—for less dilute solutions where intermolecular hydrogen bonding is likely to occur. Intramolecular hydrogen bonding is responsible for a broad, shallow peak in the range of 3100—3050 cm⁻¹.	~1050	1420—1300(s) ~1420(m) and ~1330(m) (coupling of O–H inplane bending and C–H wagging)	α-Unsaturation decrease >C–O stretch by 30 cm⁻¹. Liquid spectra of alcohols show a broad out-of-plane bending band (769—650, s)
Secondary	R'–CHOH | R²		~1100	~1420(m) and ~1330 (m) (coupling of O–H inplane bending and C–H wagging)	
Tertiary	R² | R'–C–OH | R³		~1150	Only one band (1420—1330 cm⁻¹), position depending on the degree of hydrogen bonding)	

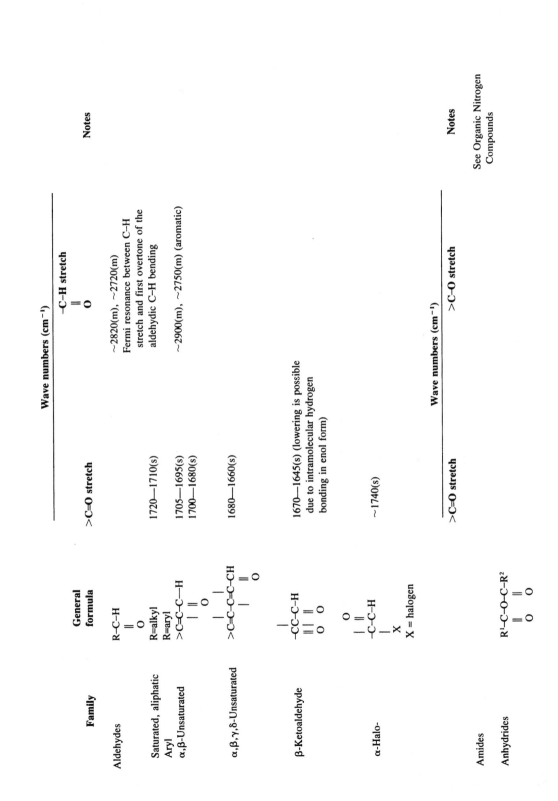

Wave numbers (cm⁻¹)

Family	General formula	>C=O stretch	—C—H stretch (=O)	Notes
Aldehydes	R—C—H (=O)			~2820(m), ~2720(m) Fermi resonance between C–H stretch and first overtone of the aldehydic C–H bending
Saturated, aliphatic		1720—1710(s)		
Aryl	R=alkyl R=aryl	1705—1695(s) 1700—1680(s)	~2900(m), ~2750(m) (aromatic)	
α,β-Unsaturated	>C=C—C—H (=O)	1680—1660(s)		
α,β,γ,δ-Unsaturated	>C=C—C=C—CH (=O)	1670—1645(s) (lowering is possible due to intramolecular hydrogen bonding in enol form)		
β-Ketoaldehyde	—CC—C—H (=O =O)			
α-Halo-	—C—C—H (O, X) X = halogen	~1740(s)		

Wave numbers (cm⁻¹)

Family	General formula	>C=O stretch	>C=O stretch	Notes
Amides				See Organic Nitrogen Compounds
Anhydrides	R¹—C—O—C—R² (=O =O)			

ORGANIC OXYGEN COMPOUNDS (continued)

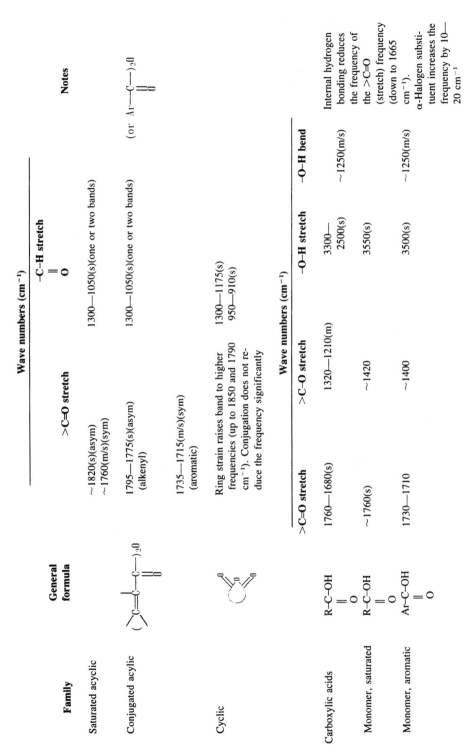

Wave numbers (cm⁻¹)

Family	General formula	>C=O stretch	–C–H stretch ‖ O	Notes
Saturated acyclic		~1820(s)(asym) ~1760(m/s)(sym)	1300—1050(s)(one or two bands)	
Conjugated acyclic		1795—1775(s)(asym) (alkenyl) 1735—1715(m/s)(sym) (aromatic)	1300—1050(s)(one or two bands) (or Ar—C—)₂O ‖ O	
Cyclic		Ring strain raises band to higher frequencies (up to 1850 and 1790 cm⁻¹). Conjugation does not reduce the frequency significantly	1300—1175(s) 950—910(s)	

Wave numbers (cm⁻¹)

Family	General formula	>C=O stretch	>C–O stretch	–O–H stretch	–O–H bend	Notes
Carboxylic acids	R–C–OH ‖ O	1760—1680(s)	1320—1210(m)	3300—2500(s)	~1250(m/s)	Internal hydrogen bonding reduces the frequency of the >C=O (stretch) frequency (down to 1665 cm⁻¹). α-Halogen substituent increases the frequency by 10—20 cm⁻¹
Monomer, saturated	R–C–OH ‖ O	~1760(s)	~1420	3550(s)		
Monomer, aromatic	Ar–C–OH ‖ O	1730—1710	~1400	3500(s)	~1250(m/s)	

		>C=O stretch		>C-O stretch	
Dimer, saturated	R = alkyl	1720—1706(s)	1315—1280(m) (sometimes doublet)	3300—2500(s, broad)	900—860(m, broad) (out-of-plane)
Dimer, α,β-unsaturated (or aromatic)	R = alkenyl	1700—1680(s)	1315—1280(m) (sometimes doublet)	3300—2500(s, broad)	900—860(m, broad) (out-of-plane)
Carboxylic acids salts	R–C–O–, =O	1610—1550(s)(asym) CO_2^-, ~1400(s)(sym) CO_2^-			
Epoxides	R¹, R² structure	~1250(s)(ring breathing, sym), 950—810(s)(asym), 840—810(s)(C–H bend), 3050—2990(m/s) (C–H stretch)			
Esters	R^1–C–OR^2, =O			(O–C–C) 1064—1031(s)(1° alcohol) ~1100(s)(2° alcohol)	
Saturated, aliphatic	R^1, R^2 = alkyl	1750—1735(s) α-Halogen substitution results in an increase in wave numbers (up to 30 cm⁻¹)	1210—1163(s) [acetates only: 1240(s)]		
Formates	R^1 = H, R^2 = alkyl	1730—1715(s)	~1180(s), ~1160(s)		
α,β-Unsaturated	>C=C–C–OR^2, =O, R^2 = alkyl	1730—1715(s)	1300—1250(s), 1200—1050(s)		

ORGANIC OXYGEN COMPOUNDS (continued)

Family		>C=O stretch	>C–O stretch	Notes
		Wave numbers (cm^{-1})		
Benzoate	C_6H_5–C(=O)–OR^2 R^2 = alkyl	1730—1715(s)	1310—1250(s) 1180—1100(s)	
Vinyl	R^1–C(=O)–O–CH=CH_2 R^1=alkyl	1775—1755(s)	1300—1250(s) ~1210(vs)	
Phenyl	R^1–C(=O)–O–C_6H_5 R^1=alkyl	~1770(s)	1300—1200(s) 1190—1140(s)	
α-Keto	–C(=O)–C(=O)–OR^2 R^2 = alkyl	1775—1740(s)	1300—1050(s) (two peaks)	
β-Ketoesters	–C(=O)–C–C(=O)–OR^2 R^2 = alkyl	~1735(s) ~1650(s) (due to enolization)	1300—1050(s) (two peaks)	

Aryl benzoates	$R^1-\underset{\underset{O}{\|}}{C}-OR^2$ $R^1, R^2 = aryl$	~1735(s)	1300—1050(s) (two peaks)

Wave numbers (cm⁻¹)

		>C–O–C< stretch asymmetrical	>C–O–C< stretch symmetrical	
Ethers				
Aliphatic	R^1-O-R^2 $R^1, R^2 = alkyl$	1150—1085(s) Branching off on the carbons adjacent to oxygen creates splitting	Very hard to trace	
Aryl alkyl	$R^1 = alkyl$ $R^2 = aryl$	1275—1200(s) (high due to resonance)	1075—1020(s)	
Vinyl	$R^1 = vinyl$ $R^2 = aryl$	1225—1200(s) (high due to resonance)	1075—1020(s)	1660—1610(m) (>C=CC) ~1000(m), 909(m) (>C=C–H) (wagging)
Imides	$(R-C=O)_2N$			See Organic Nitrogen Compounds
Isocyanates	$R-N=C=O$			See Organic Nitrogen Compounds
Ketals	$\underset{R^1}{\overset{OR^3}{\underset{\|}{\overset{\|}{-C-}}}}OR^4$	1190—1160(s) 1195—1125(s) 1098—1063(s) 1055—1035(s)		
Ketenes	>C=C=O			~2150(s) (>C=C=O)

ORGANIC OXYGEN COMPOUNDS (continued)

Family	General formula	Wave numbers (cm^{-1})		Notes
		>C=O stretch	>C=C< stretch	
Ketones	R^1C-R^2 $\parallel$ O			>C=O Overtone ~3400(w) Solid samples or solutions decrease >C=O stretch (10—20 cm^{-1}) α-Halogenation increases >C=O stretch (0—25 cm^{-1}) >C–H stretch is very weak (3100—2900 cm^{-1})
Aliphatic, saturated	R^1,R^2 = alkyl	1720—1710(s)		
α,β-Unsaturated	O $\parallel$ >C=C–C–R² R^2 = alkyl	~1690(s)(s-cis) ~1675(s)(s-trans)	1650—1600(m)	
α,β-α',β'-Unsaturated	(>C=C–)₂C=O	~1665(s)	~1640(m)	
α,β,γ,δ-Unsaturated	O $\parallel$ >C=C–C=C–C–R₂	~1665(s)	~1640(m)	
Aryl	R^1 = aryl R^2 = alkyl	~1690(s)	~1600, 1500(m/s) (aromatic)	
Diaryl	R^1,R^2 = aryl	~1665(s)	~1600, 1500(m/s) (aromatic)	
Cyclic	(cyclic C=O)	3-membered: 1850(s) 4-membered: 1780(s) 5-membered: 1745(s) 6-membered: 1715(s) Larger than 6-membered: 1705(s)		

α-Keto (s-*trans*)

$$R^1\text{-}C\text{-}C\text{-}R^2$$
(with two =O groups)

~1720(s) (aliphatic)
~1680(s) (aromatic)

β-Keto

$$R^1\text{-}C\text{-}CH_2\text{-}C\text{-}R^2$$
(with two =O groups)

~1720(s) (two bands)

1640—1580 (m, broad)
Due to enol from

$$R^1\text{-}C=CH\text{-}C\text{-}R^2$$
$$O\text{-}\ H\cdot O$$

Shows a shallow broad –OH band (enol form) at 3000—2700 cm⁻¹

>C=O decreases 10—15 cm⁻¹ with electron deactivating *p*-substituents

Strong bands at 3700—3600 cm⁻¹ (–OH) and 1700—1600 cm⁻¹ (>C=O) due to the presence of enolic forms

α-Amino ketone hydrochlorides

$$R^1\text{-}C\text{-}CH_2\text{-}NH_3^+ \quad Cl^-$$
(with =O)

α-Amino ketones

$$R\text{-}C\text{-}CH_2\text{-}NR_2$$
(with =O and NR₂)

Lactones

$$(CH_2)_x \quad C=O$$

| | **Wave numbers (cm⁻¹)** | | |
	>C=O stretch	>C=C< stretch	>C–O stretch
Saturated			
δ-	x = 4	~1735(s)		1300—1050(s, two peaks)
γ-	x = 3	~1770(s)		1300—1050(s, two peaks)
β-	x = 2	~1840(s)		1300—1050(s, two peaks)

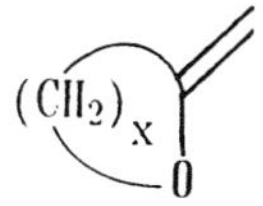

ORGANIC OXYGEN COMPOUNDS (continued)

		Wave numbers (cm^{-1})		
		>C=O stretch	>C=C< stretch	>C—O stretch
Unsaturated, α- to the carbonyl (>C=O)	x = 4	~1720(s)		
	x = 3	~1750(s) (doublet 1785—1755 cm^{-1} when α-hydrogen present)		
Unsaturated, α- to the oxygen	x = 4	~1760(s)	~1685(s)	1300—1050(s, two peaks)
	x = 3	~1790(s)	~1660(s)	1300—1050(s, two peaks)
Unsaturated, α- to the carbonyl and α- to the oxygen	x = 4 (α-pyrone; coumarin)	1775—1715 (s, doublet)	1650—1620(s) 1570—1540(s)	1300—1050(s, two peaks)

		Wave numbers (cm^{-1})			
		>C—O stretch	—O—H stretch	>C=O stretch	—O—H bend
Peroxides	R^1—O—O—R^2	—C—O—O—C—		(—C—O—O—C—) O= =O	
Aliphatic	R^1,R^2 = alkyl	890—820(vw)			
Aromatic	R^1,R^2 = aryl	~1000(vw)			
Acyl, aliphatic	R^1,R^2 = acyl (aliphatic)	890—820(vw)		1820—1810(s) 1800—1780(s)	
Acyl, aromatic	R^1,R^2 = acyl (aromatic)	~1000(vw)		1805—1780(s) 1785—1755(s)	
Peroxyacids	R^1-C-O-O-H, =O	~1260(s)	3300—3250(s, not as broad as in R—COOH)	1745—1735(s) (doublet)	~1400(m) ~850 cm^{-1}(m, —O—O— stretch)

Peroxyacids, anhydrides	$(R^1\text{-}C\text{-}O\text{-})_2O$ $\|$ O	$(\text{-}C\text{-}O\text{-}O\text{-}C\text{-})$ $\|$ $\|$ O O
Alkyl	R^1=alkyl	1815(s), 1790(s)
Aryl	R^1=aryl	1790(s), 1770(s)

	Wave numbers (cm^{-1})		
	>C—O stretch	—O—H stretch	—O—H bend
Phenols Ar—OH Ar = aryl	~1230(m)	~3610(m, sharp) (in CHCl$_3$ or CCl$_4$ solution) ~3100(m, broad) (in neat samples)	1410—1310(m, broad) (in-plane) ~650(m) (out-of-plane)

	Wave numbers (cm^{-1})	
	>C=O stretch	>C=C< stretch
Quinones 1,2- a) 1,2-	~1675(s)	~1600(s)
1,4- b) 1,4-	~1675(s)	~1600(s)

ORGANIC NITROGEN COMPOUNDS

Family	General formula	Wave numbers (cm^{-1})			Notes
		C–N	N–H	Others	
Amides Primary	O=C(R¹)–NH₂	1400(s) (stretch)	3520(m) (stretch) 3400(m) (stretch) 1655—1620(m) (bend) 860—666(m, broad) (wagging)	>C=O (1650)(s, solid state) (1690)(s, solution)	Lowering of N–H stretch occurs in solid samples due to hydrogen bonding; higher values arise in dilute samples
Secondary	O=C(R¹)–NHR²	1400(s) (stretch)	3500—3400(w) (stretch) 1570—1515(w) (bend) 860—666(m, broad) (wagging)	>C=O (1700—1670) (s, solution) (1680—1630)(s, solid state) Band due to interaction of N–H (bend) and (C–N) (stretch) (~1250) (m, broad)	Lowering of N–H stretch occurs in solid samples due to hydrogen bonding; higher values arise in dilute samples
Tertiary	O=C(R¹)–NR²R³	1400(s)(stretch)	—	>C=O (1680—1630)(s); higher values are obtained with electron-attracting groups attached to the nitrogen	
Amines Primary	R¹–NH₂	1250—1020(m) (for nonconjugated amines) 1342—1266(s) (for aromatic amines)	3500(w)(stretch) 3400(w)(stretch) 1650—1580(m) (scissoring) 909—666(m) (wagging)		
Secondary	R¹–NHR²	1250—1020(m) (for nonconjugated amines) 1342—1266(s) (for aromatic amines)	3350—3310(w) (stretch) 1515(vw) (scissoring) 909—666(m) (wagging)		
Tertiary	R¹–N(R²)–R³	1250—1020(m) (for nonconjugated amines) 1342—1266(s) (for aromatic amines)	—		

Amine salts		
Primary	$RNH_3^+X^-$	3000—2800(s) 2800—2200(m) (series of peaks) 1600—1575(m) 1550—1504(m)
Secondary	$R_2NH_2^+X^-$	3000—2700(s) 2700—2250(m) (series of peaks) 2000(w) 1620—1560(m) 2700—2250(s)
Tertiary Quaternary	R_3NHX^- R_4NX^-	—
Amino acids (alpha)	$R^1-CH-COO^-$ $\quad\;\; \vert$ $\quad\;\; NH_2$ $\quad\;\; H^+$	3100—2600(s, broad) 2222—2000(s, broad, overtone) $-\underset{\Vert\,O}{C}-O^-$ (1600—1590)(s), (1400)(w)
	$R^1-CH-COO^-$ $\quad\;\; \vert$ $\quad\;\; {}^+NH_3$ $\quad\;\; H^+$	1610(w)(bend) 1550—1485(s) (bend)
	$R^1-CH-COOH$ $\quad\;\; \vert$ $\quad\;\; {}^+NH_3$	$-\underset{\Vert\,O}{C}-OH$ (1755—1730)(s)

ORGANIC NITROGEN COMPOUNDS (continued)

Family	General formula	Wave numbers (cm⁻¹)			
		$-N=N-$	N–H	Others	>N–O stretch
Ammonium ion	NH_4^+		3300—3040(s) 2000—1709(m) 1429(s)		
Azides	$R–N_3$			2140(s) (asym stretch, N_3) 1295(s) (sym stretch, N_3)	
Azocompounds	$R^1–N=N–R^2$ (trans)	Forbidden in IR but allowed in Raman spectrum (1576)(w); peak is lowered down to 1429 cm⁻¹ in unsymmetrical p-electron-donating substituted azobenzenes			
Azoxy compounds	$R–N=N→O$				1310—1250(s)

Family	General formula	Wave numbers (cm⁻¹)		Others
		C–N multiple bond	Cumulated (–X=C=Y) double bond	
Cyanocompounds (nitriles)	$R–C≡N$	2260—2240(w) (aliphatic) 2240—2220(m) (aromatic, conjugated)		Electronegative elements α- to the C≡N group reduce the intensity of the absorption
Diazonium salts	$R–N≡N^+$			2280—2240(m) ($-N≡N^+$)
Imides	$R–C(=O)–NH–C(=O)–R$			1710, 1700 (>C=O six-membered ring) 1770, 1700 (>C=O five-membered ring)
Isocyanates	$R–N=C=O$		2273—2000(s) (broad) (asym) 1400—1350(w) (sym)	
Isocyanides (isonitriles)	$R–N≡C$	2400—2300(w) (aliphatic) 2300—2200(w) (aromatic)		

		Wave numbers (cm⁻¹)			
		>C–N	>N–O	Others	
Isothiocyanates	R–N=C=S			2140—2000(s) (stretch)	Add ~15 cm⁻¹ to every wave number in case of a >C=C< in conjugation; amide group is forced into the cis-conformation in rings of medium size
Ketene	$\begin{matrix} R^1 \\ R^2 \end{matrix} > C=C=O$			2150(stretch); 1120	
Ketenimine	$\begin{matrix} R^1 \\ R^2 \end{matrix} > C=C=N-$			2000(stretch)	
Lactams	$\underset{(CH_2)_n}{\overset{C=O}{\big|}}\ N-H$			>C=O (s) (stretch) 1670 (six-membered ring) 1700 (five-membered ring) 1745 (four-membered ring) N–H (out-of-plane wagging) (800—700) (broad)	
Nitramines	$R^1-\underset{R^2}{N}-NO_2$		1620—1580(s) (asym) 1320—1290(s) (asym)		
Nitrates	$RO-NO_2$			–N=O 1660—1625(s) (asym) 1300—1225(s) (asym) >N–O 870—833(s) (stretch) 763—690(s) (bend)	

ORGANIC NITROGEN COMPOUNDS (continued)

Family	General formula	Wave numbers (cm⁻¹)			Notes
		>C–N	>N–O	Others	
Nitrites	RO–N=O			–N=O stretch 1680—1650(vs) (*trans*) 1625—1610(vs) (*cis*) >N–O stretch 850—750(vs)	

Family	General formula	Wave numbers (cm⁻¹)			
		>C–N	–N=O (stretch)	>N–O (stretch)	Others
Nitrocompounds					
Aliphatic	R–NO₂ R = alkyl	870	1615—1540(vs) (asym) 1390—1320(vs) (sym)	1390—1320(vs)	~610(m) (CNO bend)
Aromatic	R = aryl	(difficult to assign)	1548—1508(s) (asym) 1356—1340(s) (sym) (Aromatics absorb at lower frequencies than aliphatics)	1356—1340(s)	
Nitrosamines	R¹ >N–N=O R²				>N–O stretch (1520—1500)(s) (vapor) (1500—1480)(s) (neat) N–N(1150—925)(m)

Wave numbers (cm⁻¹)

	C–N	N–O (asymmetric)	N–O (symmetric)	Others	
Nitrosocompounds	R–N=O			N=O stretch 1585—1539(s) (3°, aliphatic) 1511—1495(s) (3°, aromatic)	1° and 2° C-nitrosocompounds are unstable and rearrange or dimerize
Pyridines				N–H (3075, 3030)(s) C–H (out-of-plane) (920—720)(s) (2000—1650) (overtone) C=C ring stretch (1600, 1570, 1500, 1435)	Characteristic substitution pattern: α-Substitution: (795—780), (755—745) β-Substitution: (920—880), (840—770), 720

ORGANIC SULFUR COMPOUNDS

Family	General formula	Wave numbers (in cm^{-1})				Comments
		>S=O (asymmetric)	>S=O (symmetric)	>S=N-	Others	
Disulfides Mercaptans	R^1-S-S-R^2 R-S-H				-S-S- (<500)(w) -S-H (2600—2500)(w)	Only popular frequency around that region; lowering of 50—150 cm^{-1} due to hydrogen bonding
Mercapturic acids	R^{2}OCNH │ RSCH$_2$ CH │ HOOC	1295—1280(s) (for sulfones)	1135—1100(s) (for sulfones)		1025, 970 (>S→O) (for sulfoxides)	Reduction of all >S=O frequencies due to H- bonding with -NH
Sulfates Sulfides Sulfilimines	(RO)$_2$S(=O)$_2$ R^1-S-R^2 R$_2$S=N-R^1	1415—1380(s)	1200—1185(s)		R-S- (700—600)(w)	
N-Acyl	R$_2$S=N-C-R^1 ‖ O			800(s)	>C=O (1625—1600)(s)	
N-Alkyl N-Sulfonyl	R$_2$S=N-R^1 R$_2$S=N-SO$_2$R^1	1280—1200(s) 1095—1030(s)	1160—1135(s)	987—935(s) 980—901(s)		
Sulfinamides, N-alkylidene	RS(O)N=CR$_2$				1520 (amide II band) 1080 (s, S→O)	
Sulfonamides	R-SO$_2$NH$_2$	1370—1335(s)	1170—1155(s)		>N-H (1°) (3390—3330)(s) (3300—3247)(s) >N-H (2°) (3265)(s)	Solid phase spectra lower wave numbers by 10—20 cm^{-1}

Group	Formula		Remarks
Sulfonates	R¹–SO₂–OR²	1372—1335(s) 1195—1168(s)	Electron-donating groups on the aryl group cause higher frequency absorption
Sulfones	R¹–SO₂–R²	1350—1300(s) 1160—1120(s)	Hydrogen bonding reduces the frequency of absorption slightly
Sulfonic acids (anhydrous)	R–SO₃H	1350—1342(s) 1165—1150(s) –OH (3300—2500) (s, broad)	Hydrated sulfonic acids show broad bands at 1230—1150 cm⁻¹
Sulfonic acids, salts	R–SO₃⁻	ca. 1175(s) ca. 1055(s)	
Sulfonyl chlorides	R–SO₂Cl	1410—1380(s) 1204—1177(s)	
Sulfoxides	R₂S→O	>S→O (1070—1030)(s)	Hydrogen bonding reduces the frequency absorption slightly; electronegative substituents increase the >S→O frequency; inorganic complexation reduces the >S→O (up to 50 cm⁻¹)
Cyclic	(CH₂)ₓ, S→o	x = 3 1192(CCl₄) 1073(CHCl₃) x = 4 1035(CCl₄) 1020(CHCl₃) x = 5 1053(CCl₄) 1031(CHCl₃)	
Thiocarbonyls (not trimerized into cyclic sulfides)	R¹–C–R²(H) ‖ S	>C=S (1250—1020)(s)	
Thiocyanates	R–S–C≡N	–C≡N (2175—2140)(s); higher values for aryl thiocyanates	
Thiol esters	R¹–C–SR² ‖ O	>C=O (1690)(s) (S-alkyl thioester) (1710)(s) (S-aryl thioester)	
Thiols	R–SH	–S–H (2600—2500)(w)	The (+)mesomeric effect of sulfur is larger than its (–) inductive effect
Thiophenols	Ar–SH		See Mercaptans

ORGANIC SILICON COMPOUNDS

Family	General formula	Wave numbers (cm^{-1})					
		>Si–H stretch	>Si–H bend	>C–Si< stretch	>C–H bend	>Si–O– stretch	–OH stretch
Silanes	R$_x$SiH$_y$						
Monoalkyl	R–SiH$_3$	2130—2100(s)	890—860(s)	890—690(s)	~1260(s) (rocking)		
Dialkyl	R$_2$SiH$_2$	~2135(s)	890—860(s)	820—800(s)	~1260(s) (rocking)		
Trialkyl	R$_3$SiH	2360—2150(s)	890—860(s)	~840(s) ~755(s)	~1260(s) (rocking)		
Tetraalkyl	R$_4$Si			890—690(s)	~1260(s) (rocking)		
Alkoxy	R$_x^1$Si(OR^2)$_y$			890—690(s)	~1260(s) (rocking)	1090—1080(s) (doublet)	
Siloxanes	->Si–O–Si<--						
Disiloxanes						1110—1000(s) (Si–O–Si)	
Cyclic trimer						~1053(s)	
Cyclic tetramer						~1020(s) ~1082(s)	
Hydroxysilanes	R$_x$Si(OH)$_y$						~3680(s) (Confirmed by band at 870—820 cm^{-1})

ORGANIC PHOSPHORUS COMPOUNDS

Family	General formula	Wave numbers (cm^{-1})				Notes
		>P=O stretch	>P–H stretch	>P–O–C< stretch	–OH stretch	
Phosphates Alkyl Aryl	$O=P(OR)_3$	1300—1100(s) (doublet) 1285—1260(s) (doublet) 1315—1290(s) (doublet)		~1050(s) (alkyl) 950—875(s) (aryl)		->P=O stretch can shift up to 65 cm^{-1} due to change in solvent
Phosphinates	H_2P-OR $\parallel$ O	1220—1180(s)	~2380(m) ~2340(m) (sharp)	~1050(s) (alkyl) 950—875(s) (aryl)		
Phosphine oxides	$(R)H-PR_1R_2$ $\parallel$ O					->P=O decreases with complexation
Phosphonates Alkyl Aryl	$H-P(OR)_2$ $\parallel$ O	1185—1150(s) 1145—1095(s) 1265—1230(s)	2340—2280(m) 2340—2280(m) 2450—2420(m)	~1050(s) (alkyl) 950—875(s) (aryl)		
Phosphorus acids	$R^1P(=O)OH$ $\vert$ R^2	1240—1180(vs)			2700—2200(s, broad) (assoc)	
Phosphorus amides	$(RO)_2PNR^1R^2$ $\parallel$ O	1275—1200(s)				
Pyrophosphates	$R_2P-O-PR_2$ $\parallel \quad \parallel$ $O \quad\ O$	1310—1200(s, single band)				

ORGANIC HALOGEN COMPOUNDS

Wave numbers (cm^{-1})

Family	General formula	>C-X stretch	>CX$_2$ stretch	-CH$_3$ stretch	C-X stretch
Fluorides	X = F	1120—1010	1350—1200(asym) 1200—1080 (sym)	1350—1200(asym) 1200—1080 (sym)	1230—1100
Chlorides	X = Cl	830—500	845—795(asym) ~620(sym)		
Bromides	X = Br	1510—1480 (overtone) 667—290			
Iodides	X = I	500—200			

VII. Nuclear Magnetic Resonance Spectroscopy

PROPERTIES OF IMPORTANT NMR NUCLEI

The following table lists the magnetic properties required most often for choosing the nuclei to be used in NMR experiments.[1-12] The reader is referred to several excellent texts and the literature for guidelines in nucleus selection.

REFERENCES

1. **Silverstein, R. M., Bassler, G. C., and Morrill, T. C.,** *Spectrometric Identification of Organic Compounds,* 4th ed., John Wiley & Sons, New York, 1981.
2. **Yoder, C. H. and Shaeffer, C. D.,** *Introduction to Multinuclear NMR,* Benjamin/Cummings, Menlo Park, CA, 1987.
3. **Gordon, A. J. and Ford, R. A.,** *The Chemists Companion,* Interscience, New York, 1971.
4. **Bruno, T. J. and Svoronos, P. D. N.,** *Basic Tables for Chemical Analysis,* Technical Note #1096, National Bureau of Standards, Washington, D.C., 1986.
5. **Becker, E. D.,** *High Resolution NMR, Theory and Chemical Applications,* 2nd ed., Academic Press, New York, 1980.
6. **Gunther, H.,** *NMR Spectroscopy,* John Wiley & Sons, Chichester, 1980.
7. **Rahman, A.-U.,** *Nuclear Magnetic Resonance,* Springer-Verlag, New York, 1986.
8. **Harris, R. K.,** NMR and the periodic table, *Chem. Soc. Rev.,* 5, 1, 1976.
9. **Carrington, A. and McLaughlin, A.,** *Introduction to Magnetic Resonance,* Harper & Row, New York, 1967.
10. **Levine, I. M.,** *Molecular Spectroscopy,* John Wiley & Sons, New York, 1975.
11. **Becker, E. D.,** *High Resolution NMR: Theory and Chemical Applications,* Academic Press, New York, 1980.
12. **Harris, R. K. and Mann, B. E.,** *NMR and the Periodic Table,* Academic Press, London, 1978.

PROPERTIES OF IMPORTANT NMR NUCLEI

Isotope	Natural abundance	Spin no. I	NMR frequency[a] (MHz) at indicated field strength in kG					Field value[a] (kG) at frequency (MHz) of			Relative sensitivity		Magnetic moment $(eh/4\pi M_r)$	Electric quadrupole moment[b] (b)
			10.000	14.092	21.139	23.487	51.567	4	10	16	Const H	Const ν		
$_1H^1$	99.985	1/2	42.5759	60.0000	90.0000	100.0000	220.0000	0.940	2.349	3.758	1.00	1.00	2.79278	—
$_1H^2$	0.015	1	6.53566	9.21037	13.81555	15.35061	33.77134	6.120	15.30	24.48	9.65×10^{-3}	0.409	0.85742	0.0028
$_1H^{3x}$	—	1/2	45.4129	63.9980	95.9971	106.6634	234.6595	0.881	2.202	3.523	1.21	1.07	2.9789	—
$_6C^{13}$	1.108	1/2	10.7054	15.0866	22.6298	25.1443	55.3174	3.736	9.341	14.946	0.0159	0.252	0.7024	—
$_7N^{14}$	99.635	1	3.0756	4.3343	6.5014	7.2238	15.924	13.01	32.51	52.02	1.01×10^{-3}	0.193	0.4036	0.01
$_7N^{15}$	0.365	1/2	4.3142	6.0798	9.1197	10.1330	22.2925	9.272	23.18	37.09	1.04×10^{-3}	0.101	-0.2831	—
$_8O^{17}$	0.037	5/2	5.772	8.134	12.201	13.557	29.825	6.93	17.3	27.7	0.0291	1.58	-1.8937	-0.026
$_9F^{19}$	100	1/2	40.0541	56.4442	84.6703	94.0769	206.9692	0.999	2.497	3.994	0.834	0.941	2.6288	—
$_{14}Si^{29}$	4.70	1/2	8.4578	11.9191	17.8787	19.8652	43.7035	4.729	11.82	18.92	7.84×10^{-3}	0.199	-0.55477	—
$_{15}P^{31}$	100	1/2	17.235	24.288	36.433	40.481	89.057	2.321	5.802	9.284	0.0665	0.405	1.1317	—
$_{16}S^{33}$	0.76	3/2	3.2654	4.6018	6.9026	7.6696	16.8731	12.25	30.62	49.0	2.26×10^{-3}	0.384	0.6533	-0.055
$_{16}S^{35c}$	—	3/2	5.08	7.16	10.74	11.932	26.250	7.87	19.7	31.5	8.50×10^{-3}	0.597	1.00	0.04
$_{17}Cl^{35}$	75.53	3/2	4.1717	5.8790	8.8184	9.7983	21.5562	9.588	23.97	38.35	4.72×10^{-3}	0.490	0.82183	-0.079
$_{17}Cl^{36c}$	—	2	4.8931	6.8956	10.3434	11.4927	25.2838	8.175	20.44	32.70	0.0122	0.920	1.285	-0.017
$_{35}Br^{76c}$	—	1	4.18	5.89	8.84	9.82	21.60	9.6	24	38	2.52×10^{-3}	0.26	±0.548	±0.25
$_{35}Br^{79}$	50.54	3/2	10.667	15.032	22.549	25.054	55.119	3.750	9.375	15.00	0.0794	1.26	2.106	0.31
$_{35}Br^{81}$	49.46	3/2	11.498	16.204	24.305	27.006	59.413	3.479	8.697	13.92	0.0994	1.35	2.270	0.26
$_{74}W^{183}$	14.40	1/2	1.7716	2.4966	3.7449	4.1610	9.1543	22.58	56.45	90.31	7.3×10^{-5}	0.042	0.117	—

[a] 1 kG = 10^{-10} T, the corresponding SI unit.

[b] 1 b = 10^{-28} m².

[c] Nucleus is radioactive.

GYROMAGNETIC RATIO OF SOME IMPORTANT NUCLEI

The following table lists the gyromagnetic ratio, γ, of some important nuclei which are probed in NMR spectroscopy.[1-3] The gyromagnetic ratio is proportionally constant between the magnetic moment (μ) and the angular momentum: $\mu = \gamma p$.

Nucleus	γ
$_1H^1$	5.5856
$_1H^2$	0.8574
$_1H^3$	5.9575
$_3Li^7$	2.1707
$_5B^{10}$	0.6002
$_5B^{11}$	1.7920
$_6C^{13}$	1.4044
$_7N^{14}$	0.4035
$_7N^{15}$	-0.5660
$_8O^{17}$	-0.7572
$_9F^{19}$	5.2545
$_{14}Si^{29}$	-1.1095
$_{11}Na^{23}$	1.4774
$_{15}P^{31}$	2.2610
$_{16}S^{33}$	0.4284
$_{17}Cl^{35}$	0.5473
$_{17}Cl^{37}$	0.4555
$_{19}K^{39}$	0.2607
$_{35}Br^{79}$	1.3993
$_{35}Br^{81}$	1.5084
$_{74}W^{183}$	0.2324

REFERENCES

1. **Carrington, A. and McLaughlin, A.,** *Introduction to Magnetic Resonance,* Harper & Row, New York, 1967.
2. **Levine, I. M.,** *Molecular Spectroscopy,* John Wiley & Sons, New York, 1975.
3. **Becker, E. D.,** *High Resolution NMR: Theory and Chemical Applications,* Academic Press, New York, 1980.

CHEMICAL SHIFT RANGES OF SOME NUCLEI

The following table gives an approximate chemical shift range in parts per million (ppm) for each of the most popular nuclei. The range is established by the shifts recorded for the most common compounds.[1-4]

Nucleus	Chemical shift range (ppm)	Nucleus	Chemical shift range (ppm)
^1H	15	^{29}Si	400
^7Li	10	^{31}P	700
^{11}B	200	^{33}S	600
^{13}C	250	^{35}Cl	820
^{15}N	930	^{39}K	60
^{17}O	700	^{59}Co	14,000
^{19}F	800	^{119}Sn	2,000
^{23}Na	15	^{133}Cs	150
^{27}Al	270	^{207}Pb	10,000

REFERENCES

1. **Yoder, C. H. and Schaeffer, C. D., Jr.,** *Introduction to Multinuclear NMR,* Benjamin/Cummings, Menlo Park, CA, 1987.
2. **Silverstein, R. M., Bassler, G. C., and Morrill, T. C.,** *Spectrometric Identification of Organic Compounds,* 4th ed., John Wiley & Sons, New York, 1981.
3. **Bruno, T. J. and Svoronos, P. D. N.,** Basic Tables for Chemical Analysis, NBS Technical Note 1096, National Bureau of Standards, U.S. Department of Commerce, Washington, D.C., 1986.
4. **Harris, R. U. and Mann, B. E.,** *NMR and the Periodic Table,* Academic Press, London, 1978.

REFERENCE STANDARDS FOR SELECTED NUCLEI

The following table lists the most popular reference standards used when NMR spectra of various nuclei are being measured. The standards should be inert, soluble in a variety of solvents, and, above all, should produce one singlet peak that appears close to the lowest-frequency end of the chemical shift range. When NMR data are provided, it is always necessary to specify the reference standard employed.[1-4]

Nucleus	Name	Formula
^1H	Tetramethylsilane (TMS)	$(CH_3)_4Si$
	3-(Trimethylsilyl)-1-propanesulfonic acid, sodium salt (DSS)[a]	$(CH_3)_3Si(CH_2)_2CH_2SO_3Na$
^2H	Deuterated chloroform (chloroform-d_3)	$CDCl_3$
^{11}B	Boric acid	H_3BO_3
	Boron trifluoride etherate	$(C_2H_5)_2O \cdot BF_3$
^{13}C	Tetramethylsilane (TMS)	$(CH_3)_4Si$
^{15}N	Ammonium nitrate	NH_4NO_3
	Ammonia	NH_3
	Nitromethane	CH_3NO_2
	Nitric acid	HNO_3
^{19}F	Trichlorofluoromethane (Freon 11, R-11)	CCl_3F
^{31}P	Trimethylphosphite (methyl phosphite)	$(CH_3O)_3P$
	Phosphoric acid (85%)	H_3PO_4
^{35}Cl	Sodium chloride	$NaCl$
^{59}Co	Cobalt(III) hexacyanide anion	$[Co(CN)_6]^{-2}$
^{119}Sn	Tetramethyltin	$(CH_3)_4Sn$
^{195}Pt	Platinum(IV) hexacyanide	$[Pt(CN)_6]^{-2}$
	Dihydrogen platinum(IV) hexachloride	H_2PtCl_6
^{183}W	Sodium tungstate (external)	Na_2WO_4

[a] For aqueous solutions (known also as "water-soluble TMS" or 2,2-dimethyl-2-silapentane-5-sulfonate).

REFERENCES

1. **Yoder, C. H. and Schaeffer, C. D., Jr.,** *Introduction to Multinuclear NMR,* Benjamin/Cummings, Menlo Park, CA, 1987.
2. **Duthaler, R. O. and Roberts, J. D.,** Steric and electronic effects on ^{15}N chemical shifts of piperidine and decahydroquinoline hydrochlorides, *J. Am. Chem. Soc.,* 100, 3889, 1978.
3. **Grim, S. O. and Yankowsky, A. W.,** On the phosphorus-31 chemical shifts of substituted triarylphosphines, *Phosphorus Sulfur,* 3, 191, 1977.
4. **Lambert, J. B., Shurrell, H. F., Verbit, L., Cooks, R. G., and Stout, G. H.,** *Organic Structural Analysis,* Macmillan, New York, 1976.

PROTON NMR ABSORPTION OF MAJOR CHEMICAL FAMILIES

The following tables give the region of the expected nuclear magnetic resonance absorptions of major chemical families. These absorptions are reported in the dimensionless units of parts per million (ppm) vs. the standard compound tetramethylsilane (TMS):

$$
\begin{array}{c}
CH_3 \\
| \\
CH_3-Si-CH_3 \\
| \\
CH_3
\end{array}
$$

The use of this unit of measure makes the chemical shifts independent of the applied magnetic field strength or the radio frequency. For most proton NMR spectra, the protons in TMS are more shielded than almost all other protons. The chemical shift in this dimensionless unit system is then defined by:

$$
\delta \equiv \frac{v_s - v_r}{v_r} \times 10^6
$$

Where v_s and v_r are the absorption frequencies of the sample proton and the reference (TMS) protons (12, magnetically equivalent), respectively. In these tables, the proton(s) whose proton NMR shifts are cited are indicated by italics. For more detail concerning these conventions, the reader is referred to the general references below.[1-3]

REFERENCES

1. **Silverstein, R. M., Bassler, G. C., and Morrill, T. C.,** *Spectrometric Identification of Organic Compounds,* 4th ed., John Wiley & Sons, New York, 1981.
2. **Bruno, T. J. and Svoronos, P. D. N.,** *Basic Tables for Chemical Analysis, Nat. Bur. Stand. Tech. Note 1096,* 1986.
3. **Rahman, A.-U.,** *Nuclear Magnetic Resonance,* Springer Verlag, New York, 1986.

HYDROCARBONS

Family	Proton NMR absorption

Alkanes

CH_3–R ~0.8 ppm

–CH_2–R ~1.1 ppm

–CH–R ~1.4 ppm

(Cyclopropane ~0.2 ppm)

Alkenes

$\underset{>C=C<}{CH_3}$ ~1.6 ppm $\underset{>C=C<}{CH_3-C}$ ~1.0 ppm

$\underset{>C=C<}{-CH_2}$ ~2.1 ppm $\underset{>C=C<}{-CH_2-C}$ ~1.4 ppm

$\underset{>C=C<}{-CH}$ ~2.5 ppm $\underset{>C=C<}{-CH-C}$ ~1.8 ppm

$>C=C<H$ 4.2-6.2 ppm

Alkynes

CH_3–C≡C– ~1.7 ppm CH_3–C–C≡C– ~1.2 ppm

–CH_2–C≡C– ~2.2 ppm   –CH_2–C–C≡C– ~1.5 ppm

–CH–C≡C– ~2.7 ppm   –CH–C–C≡C– ~1.8 ppm

R–C≡C–H ~2.4 ppm

Aromatics

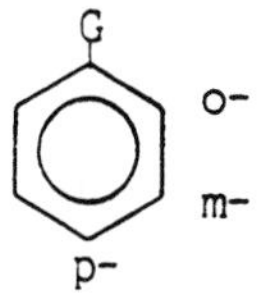

o–
m–
p–

Range: 8.5—6.9 ppm

G = Electron withdrawing
(e.g., –C–, –NO$_2$, –C≡N)
‖
O
o, p — closer to 8.5 ppm
(more downfield)

G = Electron donating
(e.g., –NH$_2$, –OH, –OR, –R)

o, p — closer to 7.0 ppm
(more upfield)

ORGANIC OXYGEN COMPOUNDS

Family   **Proton NMR absorption**

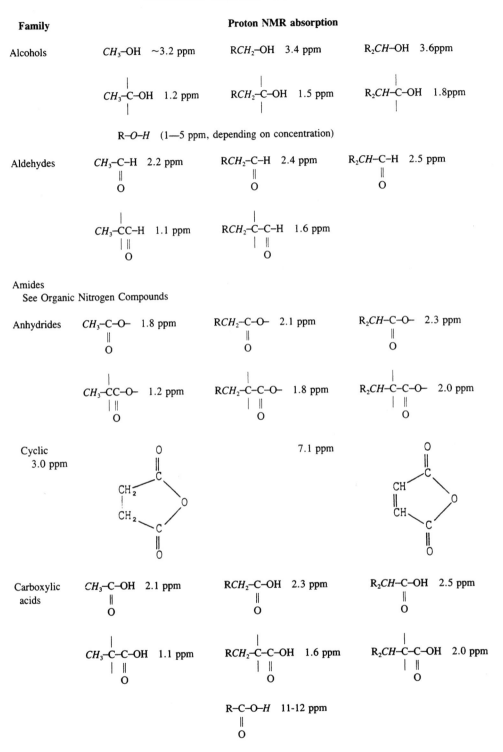

Alcohols CH_3–OH   ~3.2 ppm RCH_2–OH   3.4 ppm R_2CH–OH   3.6ppm

CH_3–C–OH 1.2 ppm RCH_2–C–OH 1.5 ppm R_2CH–C–OH 1.8ppm

R–*O*–*H*   (1—5 ppm, depending on concentration)

Aldehydes CH_3–C–H   2.2 ppm RCH_2–C–H   2.4 ppm R_2CH–C–H   2.5 ppm
                              ‖   ‖   ‖
                              O   O   O

CH_3–CC–H   1.1 ppm RCH_2–C–C–H   1.6 ppm
      ‖   ‖
      O   O

Amides
    See Organic Nitrogen Compounds

Anhydrides CH_3–C–O– 1.8 ppm RCH_2–C–O– 2.1 ppm R_2CH–C–O– 2.3 ppm
                              ‖   ‖   ‖
                              O   O   O

CH_3–CC–O– 1.2 ppm RCH_2–C–C–O– 1.8 ppm R_2CH–C–C–O– 2.0 ppm
      ‖   ‖   ‖
      O   O   O

Cyclic   7.1 ppm
3.0 ppm

Carboxylic CH_3–C–OH 2.1 ppm RCH_2–C–OH 2.3 ppm R_2CH–C–OH 2.5 ppm
acids   ‖   ‖   ‖
                              O   O   O

CH_3–C–C–OH 1.1 ppm RCH_2–C–C–OH 1.6 ppm R_2CH–C–C–OH 2.0 ppm
      ‖   ‖   ‖
      O   O   O

R–C–O–*H* 11-12 ppm
    ‖
    O

ORGANIC OXYGEN COMPOUNDS (continued)

Family   **Proton NMR absorption**

Cyclic ethers

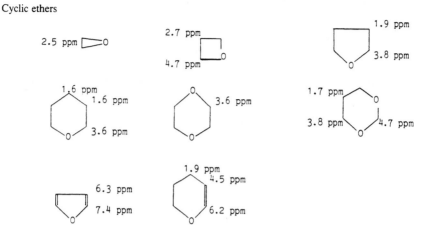

Epoxides
  See Cyclic Ethers

Esters

CH_3–C–OR RCH_2–C–OR R_2CH–COR
  ‖   ‖   ‖
  O   O   O

R = alkyl   1.9 ppm   2.1 ppm   2.3 ppm
R = aryl   2.0 ppm   2.2 ppm   2.4 ppm

CH_3–CC–OR   1.1 ppm RCH_2–C–C–OR 1.7 ppm R_2CH–CC–OR 1.9 ppm

CH_3–O–C–R   3.6 ppm RCH_2–O–C–R 4.1 ppm R_2CH–O–C–R 4.8 ppm
  ‖   ‖   ‖
  O   O   O

CH_3–C–O–C–R 1.3 ppm RCH_2–C–O–C–R 1.6 ppm R_2CH–C–O–C–R 1.8 ppm
        ‖   ‖   ‖
        O   O   O

Esters, cyclic

ORGANIC OXYGEN COMPOUNDS (continued)

Family	**Proton NMR absorption**

Ethers

$$CH_3\text{-O-R} \qquad RCH_2\text{-O-R} \qquad R_2CH\text{-O-R}$$

R = alkyl	3.2 ppm	3.4 ppm	3.6 ppm
R = aryl	3.9 ppm	4.1 ppm	4.5 ppm

$$CH_3\text{-}\underset{|}{\overset{|}{C}}\text{-O-R} \qquad RCH_2\text{-}\underset{|}{\overset{|}{C}}\text{-O-R} \qquad R_2CH\text{-}\underset{|}{\overset{|}{C}}O\text{-R}$$

R = alkyl	1.2 ppm	1.5 ppm	1.8 ppm
R = aryl	1.3 ppm	1.6 ppm	2.0 ppm

Isocyanates
See Nitrogen Compounds

Ketones

$$CH_3\text{-}\underset{O}{\overset{\|}{C}}\text{-} \qquad RCH_2\text{-}\underset{O}{\overset{\|}{C}}\text{-} \qquad R_2CH\text{-}\underset{O}{\overset{\|}{C}}\text{-}$$

1.9 ppm	R = alkyl 2.1 ppm	2.3 ppm
2.4 ppm	R = aryl 2.7 ppm	3.4 ppm

$$CH_3\text{-}\underset{O}{\overset{|}{C}}C\text{-} \qquad RCH_2\text{-}\underset{O}{\overset{|}{C}}C\text{-} \qquad R_2CH\text{-}\underset{O}{\overset{|}{C}}C\text{-}$$

$(CH_2)n \Rightarrow O$

α — hydrogens	2.0—2.3 ppm	(n > 2)
	3.0 ppm	(n = 2)
	1.7 ppm	(n = 1)
β — hydrogens	1.9—1.5 ppm	(n > 2)

Lactones
See Esters, cyclic

Nitrocompounds
See Organic Nitrogen Compounds

Phenols
Ar-O-*H*   9—10 ppm (Ar = aryl)

ORGANIC NITROGEN COMPOUNDS

Amides	Proton NMR absorption		
	Primary $R-C(=O)NH_2$ δ,ppm	Secondary $R-C(=O)NHR_1$ δ,ppm	Tertiary $R-C(=O)NR_1R_2$ δ,ppm
1. N-substitution $\quad\begin{array}{c}\mid\\-C-N-H\\\parallel\\O\end{array}$	5—12	5—12	—
a. Alpha $\quad\begin{array}{c}\mid\\-C-N-CH_3\\\parallel\\O\end{array}$	—	~2.9	~2.9
$\quad\begin{array}{c}\mid\\-C-N-CH_2-\\\parallel\\O\end{array}$	—	~3.4	~3.4
$\quad\begin{array}{c}\mid\;\;\mid\\-C-N-CH-\\\parallel\\O\end{array}$	—	~3.8	~3.8
b. Beta $\quad\begin{array}{c}\mid\;\;\mid\\-C-N-C-CH_3\\\parallel\;\;\;\mid\\O\end{array}$	~1.1	~1.1	~1.1
$\quad\begin{array}{c}\mid\;\mid\\-C-NC-CH_2-\\\parallel\;\;\mid\\O\end{array}$	~1.5	~1.5	~1.5
$\quad\begin{array}{c}\mid\;\;\mid\;\;\mid\\-C-N-C-CH-\\\parallel\;\;\;\mid\\O\end{array}$	~1.9	~1.9	~1.9

ORGANIC NITROGEN COMPOUNDS (continued)

	Proton NMR absorption		
Amides	**Primary** $R-C(=O)NH_2$ δ,ppm	**Secondary** $R-C(=O)NHR_1$ δ,ppm	**Tertiary** $R-C(=O)NR_1R_2$ δ,ppm
2. C-substitution			
a. Alpha			
CH_3-C- $\overset{\|}{\underset{O}{}}$	~1.9	~2.0	~2.1
$-CH_2-C-$ $\overset{\|}{\underset{O}{}}$	~2.1	~2.1	~2.1
$-CH-C-$ $\overset{\|}{\underset{O}{}}$	~2.2	~2.2	~2.2
b. Beta			
CH_3-C-C- $\overset{\|}{\underset{O}{}}$	~1.1	~1.1	~1.1
$-CH_2-CC-$ $\overset{\|}{\underset{O}{}}$	~1.5	~1.5	~1.5
$-CH-C-C$ $\overset{\|}{\underset{O}{}}$	~1.8	~1.8	~1.8

ORGANIC NITROGEN COMPOUNDS (continued)

Amines	Primary R–NH$_2$ δ,ppm	Secondary RN–HR δ,ppm	Tertiary RRRN δ,ppm
1. Alpha protons			
$>$N–CH_3	~2.5	2.3—3.0	~2.2
$>$N–CH_2–	~2.7	2.6—3.4	~2.4
$>$N–$CH<$	~3.1	2.9—3.6	~2.8
2. Beta protons			
$>$N–C–CH_3			~1.1
$>$N–C–CH_2–			~1.4
$>$N–C–$CH<$			~1.7

Cyanocompounds (nitriles)
1. Alpha hydrogens δ, ppm

CH_3–C≡N   ~2.1

–CH_2–C≡N   ~2.5

–CH–C≡N   ~2.9

2. Beta hydrogens δ, ppm

CH_3–C–C≡N   ~1.2

–CH_2–C–C≡N   ~1.6

CH–C–C≡N   ~2.0

Imides
1. Alpha hydrogens δ, ppm

CH_3–C–N–C–   ~2.0
‖ | ‖
O   N   O

CH_2–C–N–C–   ~2.1
‖ | ‖
O   H   O

–CH–C–N–C–   ~2.2
‖ | ‖
O   H   O

ORGANIC NITROGEN COMPOUNDS (continued)

	Proton NMR absorption		
Amides	**Primary** R–C(=O)NH$_2$ δ,ppm	**Secondary** R–C(=O)NHR$_1$ δ,ppm	**Tertiary** R–C(=O)NR$_1$R$_2$ δ,ppm

2. Beta hydrogens δ, ppm

CH_3–C–C–N–C–   ~1.2
        $\overset{|}{\underset{O}{\|}}$ $\overset{|}{H}$ $\overset{|}{\underset{O}{\|}}$

–CH_2–C–C–N–C–   ~1.3
          O H O

–CH–C–N–C–   ~1.4
       O H O

Isocyanates
  Alpha hydrogens δ, ppm
  CH_3–N=C=O   ~3.0
  –CH_2–N=C=O   ~3.3
  –CH–N=C=O   ~3.6
Isocyanides (isonitriles)
  Alpha hydrogens δ, ppm
  CH_3–N≡C: ~2.9
  –CH_2–N≡C: ~3.3
  >CH–N≡C: ~4.9
Isothiocyanates
  Alpha hydrogens δ, ppm
  CH_3–N=C=S   ~3.4
  –CH_2–N=C=S   ~3.7
  >CH–N=C=S   ~4.0
Nitriles δ, ppm
  –CH_2–C≡O   ~4.8
Nitrocompounds δ, ppm

CH_3–NO$_2$ ~4.1	–CH_2–NO$_2$ ~4.2	–CH–NO$_2$ ~4.4
CH_3–C–NO$_2$ ~1.6	–CH_2–C–NO$_2$ ~2.1	–CH–C–NO$_2$ ~2.5

ORGANIC SULFUR COMPOUNDS

Family	Proton NMR absorption			
Benzothiopyrans 2H-1-	sp^{3}CH	~3.3 ~ppm	sp^{2}CH aromatic	5.8—6.4 ppm ~6.8 ppm
4H-1-	sp^{3}CH	~3.2 ppm	sp^{2}CH aromatic	5.9—6.3 ppm ~6.9 ppm
2,3,4H-1-	sp^{3}CH	1.9—2.8 ppm	aromatic	~7.1 ppm
Disulfides	CH_3–S–S–R	~2.4 ppm	CH_3–C–S–S–R	~1.2 ppm
	–CH_2–S–S–R	~2.7 ppm	–CH_2–C–S–S–R	~1.6 ppm
	–CH–S–S–R	~3.0 ppm	–CH–C–S–S–R	~2.0 ppm
Isothiocyanates	CH_3–N=C=S	~2.4 ppm		
	–CH_2–N=C=S	~2.7 ppm		
	–CH–N=C=S	~3.0 ppm		
Mercaptans (thiols)	CH_3–S–H	~2.1 ppm	CH_3–C–S–H	~1.3 ppm
	–CH_2–S–H	~2.6 ppm	–CH_2–C–S–H	~1.6 ppm
	–CH–S–H	~3.1 ppm	CH–C–S–H	~1.7 ppm
S-Methyl salts	$\overset{+}{S}$–CH$_3$	~3.2 ppm		
Sulfates	$(CH_3$–O$)_2$S(=O)$_2$	~3.4 ppm		

ORGANIC SULFUR COMPOUNDS (continued)

Family	Proton NMR absorption			
Sulfides	CH_3–S–	1.8—2.1 ppm	CH_3–CH_2–S–	1.1—1.2 ppm

R–CH_2–S–   1.9—2.4 ppm

$$\begin{array}{c} R \\ | \\ CH_3\text{–CH–S–} \end{array}$$ 0.8—1.2 ppm

$$\begin{array}{c} R \\ | \\ R\text{–}CH\text{–S–} \end{array}$$ 2.8—3.4 ppm

$$\begin{array}{c} Ar \\ | \\ CH_3\text{–CH–S–} \end{array}$$ 1.3—1.4 ppm

Ar–CH_2S–   4.1—4.2 ppm

$$\begin{array}{c} R \\ | \\ CH_3\text{–C–S–} \\ | \\ R \end{array}$$ 1.0 ppm

$$\begin{array}{c} R \\ | \\ Ar\text{–}CH\text{–S–} \end{array}$$ 3.6—4.2 ppm

$$\begin{array}{c} R \\ | \\ Ar\text{–}CH_2\text{–CH–S–} \end{array}$$ 3.0—3.2 ppm

$$\begin{array}{c} Ar \\ | \\ Ar\text{–}CH\text{–S–} \end{array}$$ 5.1—5.2 ppm

$$\begin{array}{c} | \\ >\text{C=C–}CH_2\text{–CH–S–} \\ | \\ Ar \end{array}$$ 2.4—2.6 ppm

$$\begin{array}{c} Ar \\ | \quad | \\ >\text{C=C–}CH_2\text{–C–S–} \\ | \\ Ar \end{array}$$ 2.5 ppm

$$\begin{array}{c} R \\ | \\ R\text{–}CH\text{–}CH_2\text{–S–} \end{array}$$ 2.6—3.0 ppm

$$\begin{array}{c} Ar \\ | \\ Ar\text{–}CH\text{–}CH_2\text{–S–} \end{array}$$ 4.0—4.2 ppm

$$\begin{array}{c} R \\ | \; | \\ >\text{C=C–}CH\text{–CH–S–} \\ | \\ Ar \end{array}$$ 2.3—2.4 ppm

$$\begin{array}{c} R \quad Ar \\ | \; | \quad | \\ >\text{C=C–}CH\text{–C–S–} \\ | \\ Ar \end{array}$$ 2.8—3.2 ppm

ORGANIC SULFUR COMPOUNDS (continued)

Family	Proton NMR absorption	
Sulfilimines	$CH_3-S=N-R^2$ $\|$ R^1	~2.5 ppm
Sulfonamides	$CH_3-SO_2NH_2$	~3.0 ppm
Sulfonates	CH_3-SO_2OR	~3.0 ppm
Sulfones	$CH_3-SO_2R^2$	~2.6 ppm
Sulfonic acids	CH_3-SO_3H	~3.0 ppm
Sulfoxides	$CH_3-S(=O)R$	~2.5 ppm
	$-CH_3-S(=O)R$	~3.1 ppm
Thiocyanates	$CH_3-S-C\equiv N$	~2.7 ppm
	$-CH_3-S-S-C\equiv N$	~3.0 ppm
	$-CH_2-S-C\equiv N$	~3.3 ppm
Thiols	See Mercaptans	

Note: Ar represents aryl.

SOME USEFUL ¹H COUPLING CONSTANTS

This section gives the values of some useful proton NMR coupling constants in hertz (Hz). The data are adapted with permission from the work of Prof. C. F. Hammer, Chemistry Department, Georgetown University, Washington, D.C. The single numbers indicate a typical average, while in some cases, the range is provided.

1. Freely rotating chains.

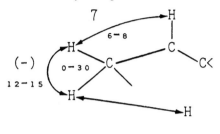

2. Alcohols with no exchange as in DMSO.

$1° =$ triplet
$2° =$ doublet (broad)
$3° =$ singlet

Upon addition of TFA, a sharp singlet results.

3. Alkenes

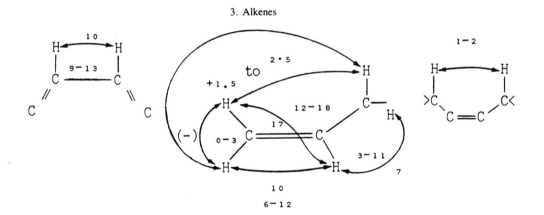

4. Alkynes   5. Aldehydes

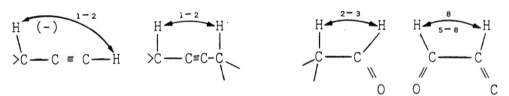

6. Aromatic

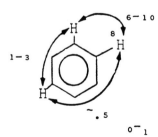

ADDITIVITY RULES IN ^{13}C-NMR CORRELATION TABLES

The wide chemical shift range (~250 ppm) of ^{13}C NMR is responsible for the considerable change of a chemical shift noted when a slight inductive, mesomeric, or hybridization neighboring change occurs. Following the various empirical correlations in ^{1}H NMR.[1-7] D. W. Brown[8] has developed a short set of ^{13}C-NMR correlation tables. This section covers a part of those as adopted by Yoder and Schaeffer[9] and Clerk et al.[10] The reader is advised to refer to Reference 8, should the need for some accurate data on more complicated structures arise.

Alkanes

The chemical shift (in ppm) of C^i can be calculated from the following empirical equation

$$\Delta^i = 2.3 + \sum_i^i A_i$$

where $\sum^i A_i$ is the sum of increments allowed for various substituents depending on their positions $(\alpha, \beta, \gamma, \delta)$ relative to the ^{13}C in question, and (-2.3) is the chemical shift for methane relative to tetramethylsilane (TMS).

^{13}C CHEMICAL SHIFT INCREMENTS FOR A, THE SHIELDING TERM FOR ALKANES AND SUBSTITUTED ALKANES[9,10]

Substituent	Increments			
	α	β	γ	δ
→C (sp^3)	9.1	9.4	−2.5	0.3
>C=C< (sp^2)	19.5	6.9	−2.1	0.4
C≡C− (sp)	4.4	5.6	−3.4	−0.6
C$_6$H$_5$	22.1	9.3	−2.6	0.3
−F	70.1	7.8	−6.8	0.0
−Cl	31.0	10.0	−5.1	−0.5
−Br	18.9	11.0	−3.8	−0.7
−I	−7.2	10.9	−1.5	−0.9
−OH	49.0	10.1	−6.2	0.0
−OR	49.0	10.1	−6.2	0.0
−CHO	29.9	−0.6	−2.7	0.0
−COR	22.5	3.0	−3.0	0.0
−COOH	20.1	2.0	−2.8	0.0
−COO$^-$	24.5	3.5	−2.5	0.0
−COCl	33.1	2.3	−3.6	0.0
−COOR	22.6	2.0	−2.8	0.0
−OOCR	5.5	6.5	−6.0	
	21.4	2.8	−2.5	
−N<	28.3	11.3	−5.1	
−NH$_3^+$	26.0	7.5	−4.6	0.0
−N<−	30.7	5.4	−7.2	−1.4
−ONO	54.3	6.1	−6.5	−0.5
−NO$_2$	61.6	3.1	−4.6	−1.0
−CON<	22.0	2.6	−3.2	−0.4
−NHCO−	31.3	8.3	−5.7	0.0
−C≡N	3.1	2.4	−3.3	−0.5
−NC	31.5	7.6	−3.0	0.0
−S−	10.6	11.4	−3.6	−0.4
−S−CO−	17.0	6.5	−3.1	0.0
−SO−	31.1	9.0	−3.5	0.0

^{13}C CHEMICAL SHIFT INCREMENTS FOR A, THE SHIELDING TERM FOR ALKANES AND SUBSTITUTED ALKANES (continued)[9,10]

	Increments			
Substituent	α	β	γ	δ
–SO$_2$Cl	54.5	3.4	–3.0	0.0
–SCN	23.0	9.7	–3.0	0.0
–C(=S)N–	33.1	7.7	–2.5	0.6
–C=NOH(syn)	11.7	0.6	–1.8	0.0
–C=NOH(anti)	16.1	4.3	–1.5	0.0
	–5.2	4.0	–0.3	0.0

R$_1$R$_2$R$_3$Sn– (R$_1$, R$_2$, R$_3$ = organic substituents).

Thus, the ^{13}C shift for C^i in 2-pentanol is

$$\overset{\beta}{CH_3}-\overset{\alpha}{CH_2}-\overset{i}{CH_2}-\overset{\alpha'}{CH(OH)}-\overset{\beta'}{CH_3}$$

$$\delta^i = (-2.3) + [9.1 + 9.4 + 9.1 + 9.4 + 10.1] = 44.8 \text{ ppm}$$
$$\quad\quad\quad\quad\;\; \alpha \quad\;\; \beta \quad\;\; \alpha' \quad\;\; \beta' \quad\; OH$$

Alkenes

For a simple olefin of the type

$$-C^\gamma-C^\beta-C^\alpha-C^i = C-C^{\alpha'}-C^{\beta'}-C^{\gamma'}-$$

$$\delta^i = 122.8 + \overset{i}{\sum}A_i$$

where

$$A_\alpha = 10.6, \quad A_\beta = 7.2, \quad A_\gamma = -1.5, \quad A_{\alpha'} = -7.9$$

$$A_{\beta'} = -1.8, \quad A_{\gamma'} = 1.5$$

and 122.8 is the chemical shift of the sp^2 carbon in ethene.

If the olefin is in the *cis* configuration, an increment of –1.1 ppm must be added. Thus, the ^{13}C shift for C–3 in *cis*-3-hexene is

$$\overset{\beta}{CH_3}-\overset{\alpha}{CH_2}-\overset{i}{CH_2} = \overset{\alpha'}{CH}-\overset{\beta'}{CH_2}-CH_3$$

$$\delta^i = 122.8 + [10.6 + 7.2 - 1.5 - 7.9] + (-1.1) = 130.1 \text{ ppm}$$
$$\quad\quad\quad\quad\; (\alpha) \quad\; \beta \quad\; (\alpha') \quad (\beta') \quad\quad (cis)$$

Alkynes

For a simple alkyne of the type

$$-C^\beta-C^\alpha-C^i \equiv C-C^{\alpha'}-C^{\beta'}-$$

$$\delta^i = 71.9 + \sum A_i$$

where increments A are given in the next table, and 71.9 is the chemical shift of the sp carbon in acetylene.[9]

¹³CHEMICAL SHIFT INCREMENTS FOR A, THE SHIELDING TERM FOR ALKYNES

Substituents	Increments			
	α	β	α'	β'
–C (sp³)	6.9	4.8	–5.7	2.3
–CH₂OH	11.1		1.9	
–COCH₃	31.4		4.0	
–C₆H₅	12.7		6.4	
–CH=CH₂	10.0		11.0	

Thus, the ¹³C shift for C–1 in 1-phenyl propyne is

$$C_6H_5-C_1^i \equiv C_2-CH_3$$

$$\delta^i = 71.9 + (12.7-5.7) = 78.9 \text{ ppm}$$

while the ¹³C shift for C–2 in the same compound is

$$C_6H_5-C_1 \equiv C_2^i-CH_3$$

$$\delta^i = 71.9 + 6.9 + 6.4 = 85.20 \text{ ppm}$$

Benzenoid Aromatics

For a benzene derivative,

$$C_6H_5-X$$

$$X = \text{substituent}$$

$$\delta^i = 128.5 + \sum_i A_i$$

where $\sum_i A_i$ is the sum of increments given below and 128.5 is the chemical shift of benzene.[9,10]

¹³CHEMICAL SHIFT INCREMENTS FOR A, THE SHIELDING TERM FOR BENZENOID AROMATICS

$$X-C_6H_5$$

$$X = \text{substituent}$$

Substituent X	Increments			
	C^i	*ortho*	*meta*	*para*
–CH₃	9.3	0.8,[9] 0.6[10]	0.0	–2.9,[9] –3.1[10]
–CH₂CH₃	15.8,[9] 15.7[10]	–0.4,[9] –0.6[10]	–0.1	–2.6,[9] –2.8[10]
–CH(CH₃)₂	20.3,[9] 20.1[10]	–1.9,[9] –2.0[10]	0.1,[9] 0.10[10]	–2.4,[9] –2.5[10]
–C(CH₃)₃	22.4,[9] 22.1[10]	–3.1,[9] –3.4[10]	–0.2,[9] 0.4[10]	–2.9,[9] –3.1[10]
–CH=CH₂	7.6	–1.8	–1.8	–3.5
–C≡CH	–6.1	3.8	0.4	–0.2
–C₆H₅	13.0	–1.1	0.5	–1.0
–CHO	8.6,[9] 9.0[10]	1.3,[9] 1.2[10]	0.6,[9] 1.2[10]	5.5,[9] 6.0[10]

^{13}CHEMICAL SHIFT INCREMENTS FOR A, THE SHIELDING TERM FOR BENZENOID AROMATICS (continued)

$$X-C_6H_5$$
$$X = \text{substituent}$$

Substituent X	Increments			
	C^1	*ortho*	*meta*	*para*
–COCH$_3$	9.1,^{9} 9.3^{10}	0.1,^{9} 0.2^{10}	0.0,^{9} 0.2^{10}	4.2
–CO$_2$H	2.1,^{9} 2.4^{10}	1.5,^{9} 1.6^{10}	0.0,^{9} –0.1^{10}	5.1,^{9} 4.8^{10}
–CO$_2^-$	7.6	0.8	0.0	2.8
–CO$_2$R	2.1	1.2	0.0	4.4
–CONH$_2$	5.4	–0.3	–0.9	5.0
–CN	–15.4,^{9} –16.0^{10}	3.6,^{9} 3.5^{10}	0.6,^{9} 0.7^{10}	3.9,^{9} 4.3^{10}
–Cl	6.2,^{9} 6.4^{10}	0.4,^{9} 0.2^{10}	1.3,^{9} 1.0^{10}	–1.9,^{9} –2.0^{10}
–OH	26.9	–12.7	1.4	–7.3
–O–	39.6^{10}	–8.2^{10}	1.9^{10}	–13.6^{10}
–OCH$_3$	31.4,^{9} 30.2^{10}	–14.4,^{9} –14.7^{10}	1.0,^{9} 0.9^{10}	–7.7,^{9} –8.1^{10}
–OC$_6$H$_5$	29.1	–9.5	0.3	–5.3
–OCOCH$_3$	23.0	–6.4	1.3	–2.3
–NH$_2$	18.7,^{9} 19.2^{10}	–12.4	1.3	–9.5
–NHCH$_3$	21.7^{10}	–16.2^{10}	0.7^{10}	–11.8^{10}
–N(CH$_3$)$_2$	22.4	–15.7	0.8	–11.8
–NO$_2$	20.0,^{9} 19.6^{10}	–4.8,^{9} –5.2^{10}	0.9,^{9} 0.8^{10}	5.8,^{9} 6.0^{10}
–SH	2.2	0.7	0.4	–3.1
–SCH$_3$	9.9^{10}	–2.0^{10}	0.1^{10}	–3.7^{10}
–SO$_3$H	15.0	–2.2	1.3	3.8

Note: As an example, the ^{13}C shift for C^i in 3,5-dinitroacetophenone is:

$$C^i = 128.5 + 9.1 + 2(0.9)$$

$$= 132.4 \text{ ppm}$$

on: $CH_3C^iO(C_6H_3)(NO_2)_2$

REFERENCES

1. **Shoolery, J. N.,** *Varian Associates Technical Information Bulletin,* Vol. 2, No. 3, Palo Alto, 1959.
2. **Bell, H. M., Bowles, D. B., and Senese, F.,** Additive NMR chemical shift parameters for deshielded methine protons, *Org. Magn. Res.,* 16, 285, 1981.
3. **Matter, U. E., Pascual, C., Pretsch, E., Pross, A., Simon, W., and Sternhell, S.,** Estimation of the chemical shifts of olefinic protons using additive increments. II. Compilation of additive increments for 43 functional groups, *Tetrahedron,* 25, 691, 1969.
4. **Matter, U. E., Pascual, C., Pretsch, E., Pross, A., Simon, W., and Sternhell, S.,** Estimation of the chemical shifts of olefinic protons using additive increments. III. Examples of utility in N.M.R. studies and the identification of some structural features responsible for deviations from additivity, *Tetrahedron,* 25, 2023, 1969.
5. **Jeffreys, J. A. D.,** A rapid method for estimating NMR shifts for protons attached to carbon, *J. Chem. Educ.,* 56, 806, 1979.
6. **Mikolajczyk, M., Grzeijszczak, S., and Zatorski, A.,** Organosulfur compounds IX: NMR and structural assignments in α,β-unsaturated sulphoxides using additive increments method, *Tetrahedron,* 32, 969, 1976.
7. **Friedrich, E. C. and Runkle, K. G.,** Empirical NMR chemical shift correlations for methyl and methylene protons, *J. Chem. Educ.,* 61, 830, 1984.
8. **Brown, D. W.,** A short set of ^{13}C-NMR correlation tables, *J. Chem. Educ.,* 62, 209, 1985.
9. **Yoder, C. H. and Schaeffer, C. D., Jr.,** *Introduction to Multinuclear NMR,* Benjamin/Cummings Publishing, Menlo Park, 1987.
10. **Clerk, J. T., Pretsch, E., and Seibl, J.,** *Structural Analysis of Organic Compounds by Combined Application of Spectroscopic Methods,* Elsevier, Amsterdam, 1981.

^{13}C-NMR ABSORPTIONS OF MAJOR FUNCTIONAL GROUPS

The table below lists the ^{13}C chemical shifts in parts per million (ppm), in descending order with the corresponding functional groups. A series of typical simple compounds for every family is added to illustrate the correlations above. The shifts for the carbons of interest are given in parentheses — either for each carbon as it appears from left to right in the formula, or by italics.[1-3]

REFERENCES

1. **Yoder, C. H. and Schaeffer, C. D., Jr.,** *Introduction to Multinuclear NMR: Theory and Application,* Benjamin/Cummings, Menlo Park, CA, 1987.
2. **Silverstein, R. M., Bassler, G. C., and Morrill, T. C.,** *Spectrometric Identification of Organic Compounds,* John Wiley & Sons, New York, 1981.
3. **Brown, D. W.,** A short set of ^{13}NMR Correlation Tables, *J. Chem. Educ.,* 62, 209, 1985.

^{13}C-NMR ABSORPTIONS OF MAJOR FUNCTIONAL GROUPS

δ (ppm)	Group	Family	Example (δ of italicized carbon)	
220—165	>C=O	Ketones	$(CH_3)_2CO$	(206.0)
			$(CH_3)_2CHCOCH_3$	(212.1)
		Aldehydes	CH_3CHO	(199.7)
		α,β-Unsaturated	$CH_3CH=CHCHO$	(192.4)
		carbonyls	$CH_2=CHCOCH_3$	(169.9)
		Carboxylic acids	HCO_2H	(166.0)
			CH_3CO_2H	(178.1)
		Amides	$HCONH_2$	(165.0)
			CH_3CONH_2	(172.7)
		Esters	$CH_3CO_2CH_2CH_3$	(170.3)
			$CH_2=CHCO_2CH_3$	(165.5)
140—120	>C=C<	Aromatic	C_6H_6	(128.5)
		Alkenes	$CH_2=CH_2$	(123.2)
			$CH_2=CHCH_3$	(115.9,136.2)
			$CH_2=CHCH_2Cl$	(117.5.133.7)
			$CH_3CH=CHCH_2CH_3$	(132.7)
125—115	—C≡N	Nitriles	$CH_3-C≡N$	(117.7)
80—70	—C≡C—	Alkynes	$HC≡CH$	(71.9)
			$CH_3C≡CH_3$	(73.9)
70—45	—C-O	Esters	$CH_3OOCH_2CH_3$	(57.6,67.9)
		Alcohols	$HOCH_3$	(49.0)
			$HOCH_2CH_3$	(57.0)
40—20	—C-NH$_2$	Amines	CH_3NH_2	(26.9)
			$CH_3CH_2NH_2$	(35.9)
30—15	—S-CH$_3$	Sulfides (thioethers)	$C_6H_5-S-CH_3$	15.6
30—(−2.3)	—C-H	Alkanes, cycloalkanes	CH_4	(−2.3)
			CH_3CH_3	(5.7)
			$CH_3CH_2CH_3$	(15.8,16.3)
			$CH_3CH_2CH_2CH_3$	(13.4,25.2)
			$CH_3CH_2CH_2CH_2CH_3$ (13.9,22.8,34.7)	
			Cyclohexane	(26.9)

^{13}C CHEMICAL SHIFTS OF USEFUL NMR SOLVENTS

The following table gives the expected carbon-13 chemical shift(s) for various useful NMR solvents in parts per million (ppm). In some solvents, slight changes can occur with change of concentration.[1,2]

REFERENCES

1. **Silverstein, R. M., Bassler, G. C., and Morrill, T. C.,** *Spectrometric Identification of Organic Compounds,* John Wiley & Sons, New York, 1981.
2. **Rahman, A-U.,** *Nuclear Magnetic Resonance. Basic Principles,* Springer-Verlag, New York, 1986.

^{13}C CHEMICAL SHIFTS OF USEFUL NMR SOLVENTS

Solvent	Formula	Chemical shift (ppm)
Acetone-d_6	$(CD_3)_2C=O$	29.2 (CD_3) 204.1 ($>C=O$)
Acetonitrile-d_3	$CD_3C\equiv N$	1.3 (CD_3) 117.1 ($C\equiv N$)
Benzene-d_6	C_6D_6	128.4
Carbon disulfide	CS_2	192.3
Carbon tetrachloride	CCl_4	96.0
Chloroform-d_3	$CDCl_3$	77.05
Cyclohexane-d_{12}	C_6D_{12}	27.5
Dichloromethane-d_2	CD_2Cl_2	53.6
Dimethylformamide-d_7	$(CD_3)_2NCOD$	31 (CD_3) 36 (CD_3) 162.4 ($DC=O$)
Dimethylsulfoxide-d_6	$(CD_3)_2S=O$	39.6
Dioxane-d_8	$C_4D_8O_2$	67.4
Methanol-d_4	CD_3OD	49.3
Nitromethane-d_3	CD_3-NO_2	57.3
Pyridine-d_5	C_5D_5N	123.9 (C–3) 135.9 (C–4) 150.2 (C–2)
1,1,2,2-tetrachloroethane-d_2	$CDCl_2-CDCl_2$	75.5
Tetrahydrofuran-d_8	C_4D_8O	25.8 (C–2) 67.9 (C–1)
Trichlorofluoromethane	$CFCl_3$	117.6
Water (heavy)	D_2O	—

^{13}C NMR CHEMICAL SHIFTS OF ORGANIC FAMILIES

The following bibliography should give a good set of references for the various organic families. This collection is by no means complete and should be updated regularly.

REFERENCES

Adamantanes

Maciel, G. E., Dorn, H. C., Greene, R. L., Kleschick, W. A., Peterson, M. R., Jr., and Wahl, G. H., Jr., ^{13}C Chemical shifts of monosubstituted adamantanes, *Org. Magn. Reson.*, 6, 178, 1974.

Amides

Jones, R. G. and Wilkins, J. M., Carbon-13 NMR spectra of a series of parasubstituted *N,N*-dimethylbenzamides, *Org. Magn. Reson.*, 11, 20, 1978.

Benzazoles

Sohr, P., Manyai, G., Hideg, K., Hankovszky, H., and Lex, L., Benzazoles. XIII. Determination of the E and Z configuration of isomeric 2-(2-benzimidazolyl)-di- and tetra-hydrothiophenes by IR, ^{1}H and ^{13}C NMR spectroscopy, *Org. Magn. Reson.*, 14, 125, 1980.

Carbazoles

Giraud, J. and Marzin, C., Comparative ^{13}C NMR study of deuterated and undeuterated dibenzothiophenes, dibenzofurans, carbazoles, fluorenes, and fluorenones, *Org. Magn. Reson.*, 12, 647, 1979.

Chlorinated Compounds

Hawkes, G. E., Smith, R. A., and Roberts, J. D., Nuclear magnetic resonance spectroscopy. Carbon-13 chemical shifts of chlorinated organic compounds, *J. Org. Chem.*, 39, 1276, 1974.

Mark, V. and Weil, E. D., The isomerization and chlorination of decachlorobi-2,4-cyclopentadien-1-yl, *J. Org. Chem.*, 36, 676, 1971.

Diazoles and Diazines

Faure, R., Vincent, E. J., Assef, G., Kister, J., and Metzger, J., Carbon-13 NMR study of substituent effects in the 1,3-diazole and -diazine series, *Org. Magn. Reson.*, 9, 688, 1977.

Disulfides

Takata, T., Iida, K., and Oae, S., ^{13}C-NMR chemical shifts and coupling constants J_{C-H} of six membered ring systems containing sulfur-sulfur linkage, *Heterocytes*, 15, 847, 1981.

Bass, S. W. and Evans, S. A., Jr., Carbon-13 nuclear magnetic resonance spectral properties of alkyl disulfides, thiosulfinates, and thiosulfonates, *J. Org. Chem.*, 45, 710, 1980.

Freeman, F. and Angeletakis, C. N., Carbon-13 nuclear magnetic resonance study of the conformations of disulfides and as their oxide derivatives, *J. Org. Chem.*, 47, 4194, 1982.

Fluorenes and Fluorenones

Giraud, J. and Marzin, C., Comparative ^{13}C NMR study of deuterated and undeuterated dibenzothiophenes, dibenzofurans, carbazoles, fluorenes and fluorenones, *Org. Magn. Reson.*, 12, 647, 1979.

Furans

Giraud, H. and Marzin, C., Comparative ^{13}C NMR study of deuterated and undeuterated dibenzothiophenes, dibenzofurans, carbazoles, fluorenes and fluorenones, *Org. Magn. Reson.*, 12, 647, 1979.

Imines

Allen, M. and Roberts, J. D., Effects of protonation and hydrogen bonding on carbon-13 chemical shifts of compounds containing the >=N– group, *Can. J. Chem.*, 59, 451, 1981.

Oxathianes

Szarek, W. A., Vyas, D. M., Sepulchre, A. M., Gero, S. D., and Lukacs, G., Carbon-13 nuclear magnetic resonance spectra of 1,4-oxathiane derivatives, *Can. J. Chem.,* 52, 2041, 1974.

Murray, W. T., Kelly, J. W., and Evans, S. A., Jr., Synthesis of substituted 1,4-oxathianes. Mechanistic details of diethoxytriphenylphosphorane- and triphenylphosphine/tetra-chloromethane-promoted cyclodehydrations and ^{13}C NMR spectroscopy, *J. Org. Chem.,* 52, 525, 1987.

Oximes

Allen, M. and Roberts, J. D., Effects of protonation and hydrogen bonding on carbon-13 chemical shifts of compounds containing the >C=N– group, *Can. J. Chem.,* 59, 451, 1981.

Polynuclear aromatics (naphthalenes, anthracenes, pyrenes)

Adcock, W., Aurangzeb, M., Kitching, W., Smith, N., and Doddzell, D., Substituent effects of carbon-13 nuclear magnetic resonance: concerning the π-inductive effect, *Aust. J. Chem.,* 27, 1817, 1974.

DuVernet, R. and Boekelheide, V., Nuclear magnetic resonance spectroscopy. Ring-current effects on carbon-13 chemical shifts, *Proc. Natl. Acad. Sci. U.S.A.,* 71, 2961, 1974.

Pyrazoles

Puar, M. S., Rovnyak, G. C., Cohen, A. I., Toeplitz, B., and Gougoutas, J. Z., Orientation of the sulfoxide bond as a stereochemical probe. Synthesis and ^{1}H and ^{13}C NMR of substituted thiopyrano [4,3-c] pyrazoles, *J. Org. Chem.,* 44, 2513, 1979.

Sulfides

Chauhan, M. S. and Still, I. W. J., ^{13}C nuclear magnetic resonance spectra of organic sulfur compounds: cyclic sulfides, sulfoxides, sulfones, and thiones, *Can. J. Chem.,* 53, 2880, 1975.

Gokel, G. W., Gerdes, H. M., and Dishong, D. M., Sulfur heterocycles. III. Heterogenous, phase-transfer, and acid catalyzed potassium permanganate oxidation of sulfides to sulfones and a survey of their carbon-13 nuclear magnetic resonance spectra, *J. Org. Chem.,* 45, 3634, 1980.

Mohraz, M., Jiam-qi, W., Heilbronner, E., Solladie-Cavallo, A., and Matloubi-Moghadam, F., Some comments on the conformation of methyl phenyl sulfides, sulfoxides, and sulfones, *Helv. Chim. Acta,* 64, 97, 1981.

Srinivasan, C., Perumal, S., Arumugam, N., and Murugan, R., Linear free-energy relationship in naphthalene system-substituent effects on carbon-13 chemical shifts of substituted naphthylmethyl sulfides, *Indian J. Chem.,* 25A 227, 1986.

Sulfites

Buchanan, G. W., Cousineau, C. M. E., and Mundell, T. C., Trimethylene sulfite conformations: effects of sterically demanding substituents at C-4, 6 on ring geometry as assessed by ^{1}H and ^{13}C nuclear magnetic resonance, *Can. J. Chem.,* 56, 2019, 1978.

Sulfonamides

Chang, C., Floss, H. G., and Peck, G. E., Carbon-13 magnetic resonance spectroscopy of drugs. Sulfonamides, *J. Med. Chem.,* 18, 505, 1975.

Sulfones (see also other families for the corresponding sulfones)

Fawcett, A. H. Ivin, K. J., and Stewart, C. D., Carbon-13 NMR spectra of monosulphones and disulphones: substitution rules and conformational effects, *Org. Magn. Reson.,* 11, 360, 1978.

Gokel, G. W., Gerdes, H. M., and Dishong, D. M., Sulfur heterocycles. III. Heterogeneous, phase-transfer, and acid catalyzed potassium permanganate oxidation of sulfides to sulfones and a survey of their carbon-13 nuclear magnetic resonance spectra, *J. Org. Chem.,* 45, 3634, 1980.

Balaji, T. and Reddy, D. B., Carbon-13 nuclear magnetic resonance spectra of some new arylcycfloproyl sulphones, *Indian J. Chem.,* 18B, 454, 1979.

Sulfoxides (see also other families for the corresponding sulfoxides)

Gatti, G., Levi, A., Lucchini, V., Modena, G., and Scorrano, G., Site of protonation in sulphoxides: carbon-13 nuclear magnetic resonance evidence, *J. Chem. Soc. Chem. Commun.,* 251, 1973.

Harrison, C. R. and Hodge, P., Determination of the configuration of some penicillin S-oxides by ^{13}C nuclear magnetic resonance spectroscopy, *J. Chem. Soc. Perkin Trans. I,* p. 1772, 1976.

Sulfur Ylides

Matsuyama, H., Minato, H., Kobayashi, M., Electrophilic sulfides(II) as a novel catalyst. V. Structure, nucleophilicity, and steric compression of stabilized sulfur ylides as observed by ^{13}C-NMR spectroscopy, *Bull. Chem. Soc. Jpn.,* 50, 3393, 1977.

Thianes

Willer, R. L. and Eliel, E. L., Conformational analysis. 34. Carbon-13 nuclear magnetic resonance spectra of saturated heterocycles; 6. Methylthianes, *J. Am. Chem. Soc.,* 99, 1925, 1977.

Barbarella, G., Dembech, P., Garbesi, A., and Fara, A., ^{13}C NMR of organosulphur compounds. II. ^{13}C chemical shifts and conformational analysis of methyl substituted thiacyclohexanes, *Org. Magn. Reson.,* 8, 469, 1976.

Murray, W. T., Kelly, J. W., and Evans, S. A., Jr., Synthesis of substituted 1,4-oxathianes. Mechanistic details of diethoxytriphenyl phosphorane and triphenylphosphine/tetrachloromethane-promoted cyclodehydrations and ^{13}C NMR spectroscopy, *J. Org. Chem.,* 52, 525, 1987.

Block, E., Bazzi, A. A., Lambert, J. B., Wharry, S. M., Andersen, K. K., Dittmer, D. C., Patwardhan, B. H., and Smith, J. H., Carbon-13 and oxygen-17 nuclear magnetic resonance studies of organosulfur compounds: the four-membered-ring-sulfone effect, *J. Org. Chem.,* 45, 4807, 1980.

Rooney, R. P. and Evans, S. A., Jr., Carbon-13 nuclear magnetic resonance spectra of *trans*-1-thiadecalin, *trans*-1, 4-dithiadecalin, *trans*-1, 4-oxathiadecalin, and the corresponding sulfoxides and sulfones, *J. Org. Chem.,* 45, 180, 1980.

Thiazines

Fronza, G., Mondelli, R., Scapini, G., Ronsisvalle, G., and Vittorio, F., ^{13}C NMR of *N*-heterocycles: conformation of phenothiazines and 2, 3-diazaphenothiazines, *J. Magn. Reson.,* 23, 437, 1976.

Thiazoles

Harrison, C. R. and Hodge, P., Determination of the configuration of some penicillin S-oxides by ^{13}C nuclear magnetic resonance spectroscopy, *J. Chem. Soc. Perkin Trans. I,* p. 1772, 1976.

Chang, G., Floss, H. G., and Peck, G. E., Carbon-13 magnetic resonance spectroscopy of drugs. Sulfonamides, *J. Med. Chem.,* 18, 505, 1975.

Elguero, J., Faure, R., Lazaro, R., and Vincent, E. J., ^{13}C NMR study of benzothiazole and its nitroderivatives, *Bull Soc. Chim. Belg.,* 86, 95, 1977.

Faure, R., Galy, J. P., Vincent, E. J., and Elguero, J., Study of polyheteroaromatic pentagonal heterocycles by carbon-13 NMR. Thiazoles and thiazolo [2,3-e] tetrazoles, *Can. J. Chem.,* 56, 46, 1978.

Thiochromanones

Chauhan, M. S. and Still, I. W. J., ^{13}C nuclear magnetic resonance spectra of organic sulfur compounds: cyclic sulfides, sulfoxides, sulfones and thiones, *Can. J. Chem.,* 53, 2880, 1975.

Thiones

Chauhan, M. S. and Still, I. W. J., ^{13}C nuclear magnetic resonance spectra of organic sulfur compounds: cyclic sulfides, sulfoxides, sulfones and thiones, *Can. J. Chem.,* 53, 2880, 1975.

Thiophenes

Perjessy, A., Janda, M., and Boykin, D. W., Transmission of substituent effects in thiophenes. Infrared and carbon-13 nuclear magnetic resonance studies, *J. Org. Chem.,* 45, 1366, 1980.

Giraud, J. and Marzin, C., Comparative ^{13}C NMR study of deuterated and undeuterated dibenzothiophenes, dibenzofurans, carbazoles, fluorenes and fluorenones, *Org. Magn. Reson.,* 12, 647, 1979.

Clark, P. D., Ewing, D. F., and Scrowston, R. M., NMR studies of sulfur heterocycles. III. ^{13}C spectra of benzo[b]thiophene and the methylbenzo[b] thiophenes, *Org. Magn. Reson.,* 8, 252, 1976.

Osamura, Y., Sayanagi, O., and Nishimoto, K., C-13 NMR chemical shifts and charge densities of substituted thiophenes — the effect of vacant dπ orbitals, *Bull. Chem. Soc. Jpn.,* 49, 845, 1976.

Balkau, F., Fuller, M. W., and Heffernan, M. L., Deceptive simplicity in ABMX N.M.R. spectra. I. Dibenzothiophen and 9.9'-dicarbazyl, *Aust. J. Chem.,* 24, 2293, 1971.

Geneste, P., Olive, J. L., Ung, S. N., El Faghi, M. E. A., Easton, J. W., Beierbeck, H., and Saunders, J. K., Carbon-13 nuclear magnetic resonance study of benzo[b]thiophenes and benzo[b]thiophene S-oxides and S,S-dioxides, *J. Org. Chem.,* 44, 2887, 1979.

Benassi, R., Folli, U., Iarossi, D., Schenetti, L., and Tadei, F., Conformational analysis of organic carbonyl compounds. III. A ^{1}H and ^{13}C nuclear magnetic resonance study of formyl and acetyl derivatives of benzo [b] thiophen, *J. Chem. Soc. Perkin Trans. II,* p. 911, 1983.

Kiezel, L., Liszka, M., and Rutkowski, M., Carbon-13 magnetic resonance spectra of benzothiophene and dibenzothiophene, *Spectrosc. Lett.,* 12, 45, 1979.

Fujieda, K., Takahashi, K., and Sone, T., The C-13 NMR spectra of thiophenes. II. 2-Substituted thiophenes, *Bull. Chem. Soc. Jpn.,* 58, 1587, 1985.

Satonaka, H. and Watanabe, M., NMR spectra of 2-(2-nitrovinyl) thiophenes, *Bull. Chem. Soc. Jpn.,* 58, 3651, 1985.

Stuart, J. G., Quast, M. J., Martin, G. E., Lynch, V. M., Simmonsen, H., Lee, M. L., Castle, R. N., Dallas, J. L., John B. K., and Johnson, L. R. F., Benzannelated analogs of phenanthro [1,2-b]-[2,1-b] thiophene: synthesis and structural characterization by two-dimensional NMR and x-ray techniques, *J. Heterocycl. Chem.,* 23, 1215, 1986.

Thiopyrans

Senda, Y., Kasahara, A., Izumi, T., and Takeda, T., Carbon-13 NMR spectra of 4-chromanone, 4H-1-benzothiopyran-4-one, 4H-1-benzothiopyran-4-one 1, 1-dioxide, and their substituted homologs, *Bull. Chem. Soc. Jpn.,* 50, 2789, 1977.

Thiosulfinates and Thiosulfonates

Bass, S. W. and Evans, S. A., Jr., Carbon-13 nuclear magnetic resonance spectral properties of alkyl disulfides, thiosulfinates, and thiosulfonates, *J. Org. Chem.,* 45, 710, 1980.

^{15}N CHEMICAL SHIFTS FOR COMMON STANDARDS

The following table lists the ^{15}N chemical shifts in parts per million (ppm) for common standards. The estimated precision is better than 0.1 ppm. Nitromethane, according to Levy and Lichter,[1] is the most suitable primary measurement reference, but has the disadvantage of lying in the low-field end of the spectrum. Thus, ammonia (which lies in the most up-field region) is the most suitable for routine experimental use.[1-6]

REFERENCES

1. **Levy, G. C. and Lichter, R. L.,** *Nitrogen-15 Nuclear Magnetic Resonance Spectroscopy,* John Wiley & Sons, New York, 1979.
2. **Lambert, J. B., Shurvell, H. F., Verbit, L., Cooks, R. G., and Stout, G. H.,** *Organic Structural Analysis,* Macmillan, New York, 1976.
3. **Witanowski, M., Stefaniak, L., Szymanski, S., and Januszewski, H.,** External neat nitromethane scale for nitrogen chemical shifts, *J. Magn. Reson.,* 28, 217, 1977.
4. **Srinivasan, P. R. and Lichter, R. L.,** Nitrogen-15 nuclear magnetic resonance spectroscopy. Evaluation of chemical shift references, *J. Magn. Reson.,* 28, 227, 1977.
5. **Briggs, J. M. and Randall, E. W.,** Nitrogen-15 chemical shifts in concentrated aqueous solutions of ammonium salts, *Mol. Phys.,* 26, 699, 1973.
6. **Becker, E. D.,** Proposed scale for nitrogen chemical shifts, *J. Magn. Reson.,* 4, 142, 1971.

^{15}N CHEMICAL SHIFTS FOR COMMON STANDARDS

Compound	Formula	Conditions	Chemical shift (ppm)
Ammonia	NH_3	Vapor (0.5 MPa)	-15.9
		Liquid (25 °C), anhydrous	0.0
		Liquid (-50 °C)	3.37
Ammonium nitrate	NH_4NO_3	Aqueous HNO_3	21.60
		Aqueous solution (saturated)	20.68
Ammonium chloride	NH_4Cl	2.9 M (in 1 M HCl)	24.93
		1.0 M (in 10 M HCl)	30.31
		Aqueous solution (saturated)	27.34
Tetramethylammonium chloride	$(CH_3)_4N^+Cl^-$	Aqueous (saturated)	43.54
		Chloroform solution (saturated)	45.68
Tetraethylammonium chloride	$(C_2H_5)_4N^+Cl^-$	Aqueous solution (0.3 M)	63.94
		Aqueous solution (saturated)	64.39
		Chloroform solution (0.075 M)	65.69
Tetramethyl urea	$[(CH_3)_2N]_2CO$	Neat	62.50
Dimethylformamide	$(CH_3)_2NCHO$	Neat	103.81
Nitric acid (aqueous solution)	HNO_3	1 M	375.80
		2 M	367.84
		9 M	365.86
		10 M	362.00
		15.7 M	348.92
Sodium nitrate	$NaNO_3$	Aqueous solution (saturated)	376.53
Ammonium nitrate	NH_4NO_3	Aqueous solution (saturated)	376.25
		5 M (in 2 M HNO_3)	375.59
		4 M (in 2 M HNO_3)	374.68
Nitromethane	CH_3NO_2	1:1 (v/v) in $CDCl_3$ 0.03 M Cr(acac)$_3$	379.60
		Neat	380.23

^{15}N CHEMICAL SHIFTS OF MAJOR CHEMICAL FAMILIES

The following table contains ^{15}N chemical shifts of various organic nitrogen compounds. Chemical shifts are expressed relative to different standards (NH_3, NH_4Cl, CH_3NO_2, NH_4NO_3, HNO_3, etc.) and are interconvertible.

Chemical shifts are sensitive to hydrogen bonding and are solvent dependent as seen in case of pyridine (see Note b to table). Consequently, the reference as well as the solvent should always accompany chemical-shift data. No data are given on peptides and other biochemical compounds. All shifts are relative to ammonia unless otherwise specified. A section of "miscellaneous" data gives the chemical shift of special compounds relative to unusual standards.[1-15]

REFERENCES

1. **Levy, G. C. and Lichter, R. L.,** *Nitrogen-15 Nuclear Magnetic Resonance Spectroscopy,* John Wiley & Sons, New York, 1979.
2. **Yoder, C. H. and Schaeffer, C. D., Jr.,** *Introduction to Multinuclear NMR,* Bemjamin/Cummings, Menlo Park, CA, 1987.
3. **Duthaler, R. O. and Roberts, J. D.,** Effects of solvent, protonation, and N-alkylation on the ^{15}N chemical shifts of pyridine and related compounds, *J. Am. Chem. Soc.,* 100, 4969, 1978.
4. **Duthaler, R. O. and Roberts, J. D.,** Steric and electronic effects on ^{15}N chemical shifts of saturated aliphatic amines and their hydrochlorides, *J. Am. Chem. Soc.,* 100, 3889, 1978.
5. **Kozerski, L. and von Philipsborn, W.,** ^{15}N chemical shifts as a conformational probe in enaminones: a variable temperature study at natural isotope abundance, *Org. Magn. Reson.,* 17, 306, 1981.
6. **Duthaler, R. O. and Roberts, J. D.,** Steric and electronic effects on ^{15}N chemical shifts of piperidine and decahydroquinoline hydrochlorides, *J. Am. Chem. Soc.,* 100, 3882, 1978.
7. **Duthaler, R. O., Roberts, J. D.,** Nitrogen-15 nuclear magnetic resonance spectroscopy. Solvent effects on the ^{15}N chemical shifts of saturated amines and their hydrochlorides, *J. Magn. Reson.,* 34, 129, 1979.
8. **Psota, L., Franzen-Sieveking, M., Turnier, J., and Lichter, R. L.,** Nitrogen nuclear magnetic resonance spectroscopy. Nitrogen-15 and proton chemical shifts of methylanilines and methylanilinium ions, *Org. Magn. Reson.,* 11, 401, 1978.
9. **Subramanian, P. K., Chandra Sekara, N., and Ramalingam, K.,** Steric effects on nitrogen-15 chemical shifts of 4-aminooxanes (tetrahydropyrans), 4-aminothianes, and the corresponding *N,N*-dimethyl derivatives. Use of nitrogen-15 shifts as an aid in stereochemical analysis of these heterocyclic systems, *J. Org. Chem.,* 47, 1933, 1982.
10. **Schuster, I. I. and Roberts, J. D.,** Proximity effects on nitrogen-15 chemical shifts of 8-substituted 1-nitronaphthalenes and 1-naphthylamines, *J. Org. Chem.,* 45, 284, 1980.
11. **Kupče, E., Liepinš, E., Pudova, O., and Lukevics, E.,** Indirect nuclear spin-spin coupling constants of nitrogen-15 to silicon-29 in silylamines, *J. Chem. Soc. Chem. Commun.,* 581, 1984.
12. **Allen, M. and Roberts, J. D.,** Effects of protonation and hydrogen bonding on nitrogen-15 chemical shifts of compounds containing the >C=N-group, *J. Org. Chem.,* 45, 130, 1980.
13. **Brownlee, R. T. C. and Sadek, M.,** Natural abundance ^{15}N shifts in substituted benzamides and thiobenzamides, *Magn. Reson. Chem.,* 24, 821, 1986.
14. **Dega-Szafran, Z., Szafran, M., Stefaniak, L., Brevard, C., and Bourdonneau, M.,** Nitrogen-15 nuclear magnetic resonance studies of hydrogen bonding and proton transfer in some pyridine trifluoroacetates in dichloromethane, *Magn. Reson. Chem.,* 24, 424, 1986.
15. **Lambert, J. B., Shurvell, H. F., Verbit, L., Cooks, R. G., and Stout, G. H.,** *Organic Structural Analysis,* Macmillan, New York, 1976.

^{15}N CHEMICAL SHIFTS OF MAJOR CHEMICAL FAMILIES

Chemical-shift range $(\delta)^a$	Family	Example (ppm)
<930	Nitroso compounds	C_6H_5–NO (913, 930)
608	Sodium nitrite	$NaNO_2$
~500	Azo compounds	C_6H_5–N=N–C_6H_5 (510)
380—350	Nitro compounds	$C_6H_5NO_2$ (370.3); CH_3NO_2 (380.2); p-F–C_6H_4–NO_2 (368.5); 1,3(NO_2)$_2C_6H_4$ (365.4)
367	Nitric acid (8.57 M)	HNO_3
360—325	Nitramines (–NO_2)	CH_3NHNO_2 (355.6); $CH_3O_2CNHNO_2$ (334.9)
350—300	Pyridines	C_5H_5N (317)^b (gas); 4–CH_3–C_5H_4N (309.3); 4–NH_2–C_5H_4N (271.5); 4–NC–C_5H_4N (327.9)
~310	Imines	$(C_6H_5)_2C$=NH (308); C_6H_5CH=NCH_3 (318); C_6H_5CH=NC_6H_5 (326)
310.1	Nitrogen (gas)	N_2
250—200	Pyridinium salts	$C_5H_5NH^+$ (215)
250—150	Nitramines (amino N)	CH_3NHNO_2 (157.6)
260—175	Cyanides (nitriles)	CH_3CN (239.5, 245); C_6H_5CN (258.7); KCN (177.8)
~160	Pyrroles	C_4H_4NH (158)
	Isonitriles	CH_3NC (162)
~150	Thioamides	CH_3C(=S)NH_2 (150.2)
120—110	Lactams	HN(CH_2)$_3C$=O (five-membered ring, 114.7); HN(CH_2)$_6C$=O (eight-membered ring, 117.7)
110—100	Amides	$C_6H_5CONH_2$ (100); CH_3CONH_2 (103.4); $CH_3CONHCH_3$ (105.8) $CH_3CON(CH_3)_2$ (103.8); $HCONH_2$ (108.5)
125—90	Sulfonamides	$CH_3SO_2NH_2$ (95); $C_6H_5SO_2NH_2$ (94.3)
~100	Hydrazines	$C_6H_5NHNHC_6H_5$ (96)
110—60	Ureas	$(H_2N)_2CO$ (75, 82); $[(CH_3)_2N]_2CO$ (63.5); $(C_6H_5NH)_2CO$ (107.7)
100—70	Aminophosphines	$C_6H_5NHP(CH_3)_2$ (71.1)
	Aminophosphine Oxides	$C_6H_5NHPO(CH_3)_2$ (86.6)
70—50	Aromatic amines	$C_6H_5NH_2$ (55, 59), (-322.3^c); $C_6H_5NH_3^+$ (48); (-326.4^c) p–$O_2NC_6H_4NH_2$ (70);
40—0	Aliphatic amines	CH_3NH_2 (1.3), (-371^c); $(CH_3)_2NH$ (-363.3^c); (-364.9^d), 6.7; $(CH_3)_3N$ (-356.9^c), (-360.7^d), 13.0
50—10	Isocyanates	CH_3NCO (14.1) C_6H_5NCO (46.5)
65—20	Ammonium salts	NH_4Cl (26.1); CH_3NH_3Cl (24.5); $(CH_3)_2NH_2Cl$ (26.6); $(CH_3)_3NHCl$ (33.8); $(CH_3)_4NCl$ (44.7)

^{15}N CHEMICAL SHIFTS OF MAJOR CHEMICAL FAMILIES (continued)

Chemical-shift range (δ)[a]	Family	Example (ppm)
	Miscellaneous (relative to various standards)	
(−130)—(−110) and ~(212)	Imidazoles	*N*-Methylimidazole (−111.4, pyridine *N* and −215.7, pyrrole *N*)[c]
(−345)—(−310)	Piperidine, hydrochloride salts	Piperidinium hydrochloride (−344.8); 2-methyl piperidinium hydrochloride (−322.1[d])
	Decahydroquinolines, hydrochloride salts	*trans*-Decahydroquinolinium hydrochloride (−322.5); *cis*-decahydroquinolinium hydrochloride (−328.5[d])
(−293)—(−280)	Enaminones	E-CH$_3$COCH=CHNHCH$_3$ [(E)—(−294.2); (Z)—(−285.9)[e]]
35—15	4-Aminotetrahydropyrans 4-Aminotetrahydrothiopyrans	2,6-Diphenyl 4-aminotetrahydropyran (34.5[f]) 2,6-Diphenyl 4-aminotetrahydrothiopyran (33.6)
(−325)—(−310)	1-Naphthylamines	8-Nitro-1-naphthylamine (313.9[g])
(−350)—(−300)	Silylamines	HN[Si(CH$_3$)$_3$]$_2$ (−354.2[h])

[a]   Downfield from anhydrous liquid ammonia, ±0.2 ppm unless otherwise specified.[1]

[b]   Varies with solvent, for instance, cyclohexane (315.5), benzene (312.1), chloroform (304.5), methanol (292.1), water (289), 2,2,2-trifluoroethanol (277.1). All chemical shifts relative to ammonia.[2]

[c]   Upfield from external HNO$_3$ (1 *M*) (CH$_3$OH).[4,6,7]

[d]   Upfield from external HNO$_3$ (1 *M*) (cyclohexane).[6,7]

[e]   Relative to external CH$_3$^{15}NO$_2$.[5]

[f]   With respect to an external standard of 5 *M* ^{15}NH$_4$NO$_3$ in 2 *M* HNO$_3$ (^{15}NH$_4$NO$_3$ = 21.6 ppm relative to anhydrous ammonia).[9]

[g]   In ppm upfield from external 1 *M* D^{15}NO$_3$ in D$_2$O (DMSO).[10]

[h]   Relative to N(SiH$_3$)$_3$ (50% in CDCl$_3$).[11]

SPIN-SPIN COUPLING TO ^{15}N

The following list gives representative spin-spin coupling values (J_{NH} in Hz) to ^{15}N.

REFERENCES

1. **Levy, G. C. and Lichter, R. L.**, *Nitrogen-15 Nuclear Magnetic Resonance Spectroscopy*, John Wiley & Sons, New York, 1979.

^{15}N–H Coupling Constants

Bond type	Family	J_{NH}	Examples
One-bond	Ammonia	(-61.2)	NH_3
	Amines, aliphatic (1°, 2°)	$\sim(-65)$	CH_3NH_2 (-64.5) $(CH_3)_2NH$ (-67.0)
	Ammonium salts	$\sim(-75)$	CH_3NH_3Cl (-75.4); $(CH_3)_2NH_2Cl$ (-76.1); $C_6H_5NH_3^+$ (-76)
	Amines, aromatic (1°, 2°)	$(-78)\!-\!(-95)$	$C_6H_5NH_2$ (-78.5); $p-CH_3O-C_6H_4-NH_2$ (-79.4); $p-O_2N-C_6H_4-NH_2$ (-92.6)
	Sulfonamides	$\sim(-80)$	$C_6H_5SO_2NH_2$ (-80.8)
	Hydrazines	$(-90)\!-\!(-100)$	$C_6H_5NHNH_2$ (-89.6)
	Amides (1°, 2°)	$(-85)\!-\!(-95)$	$HCONH_2$ (-88) (syn); (-92) (anti)
	Pyrroles	$(-95)\!-\!(-100)$	Pyrrole (-96.53)
	Nitriles, salts	$\sim(-135)$	$CH_3C\equiv NH^+$ (-136)
Two-bond	Amines	$\sim(-1)$	CH_3H_2 (-1.0); $(CH_3)_3$ N (-0.85)
	Pyridinium salts	$\sim(-3)$	C_5H_5NH (-3)
	Pyrroles	$\sim(-5)$	C_4H_4NH (-4.52)
	Thiazoles	$\sim(-10)$	C_3H_3NS
	Pyridines	$\sim(-10)$	C_5H_5N (-10.76)
	Oximes, syn	$\sim(-15)$	
	Oximes, anti	$(-2.5)\!-\!(+2.5)$	
Three-bond	Nitriles, salts	$\sim(2-4)$	$CH_3C\equiv NH^+$ (2.8)
	Amides	$\sim(1-2)$	CH_3CONH_2 (1.3)
	Anilines	$\sim(1-2)$	$C_6H_5NH_2$ $(1.5, 1.8)$
	Pyridines	$\sim(0-1)$	C_5H_5N $(0.2$
	Nitriles	$(-1)\!-\!(-2)$	$CH_3C\equiv N$ (-1.7)
	Pyridinium salts	$\sim(-4)$	$C_5H_5NH^+$ (-3.98)
	Pyrroles	$\sim(-5)$	C_4H_4NH (-5.39)

^{15}N–^{13}C Coupling Constants

Bond type	Family	J_{CH}, H_2	Example
One-bond	Amines, aliphatic	$\sim(-4)$	CH_3NH_2 (-4.5); $CH_3(CH_2)_2NH_2$ (-3.9)
	Ammonium salts (aliphatic)	$\sim(-5)$	$CH_3(CH_2)_2NH_3^+$ (-4.4)
	Ammonium salts (aromatic)	$\sim(-9)$	$C_6H_5NH_3^+$ (-8.9)
	Pyrroles	$\sim(-10)$	C_4H_4NH (-10.3)
	Amines, aromatic	$(-11)\!-\!(-15)$	$C_6H_5NH_2$ (-11.43)
	Nitro compounds	$(-10)\!-\!(-15)$	CH_3NO_2 (-10.5); $C_6H_5NO_2$ (-14.5)

SPIN-SPIN COUPLING TO ^{15}N (continued)

^{15}N–^{13}C Coupling Constants

Bond type	Family	J_{CH}, H_2	Examples
	Nitriles	~(−17)	$CH_3C≡N$ (−17.5)
	Amides	~(−14)	$C_6H_5NHCOCH_3$ (−14.3) (CO); (−14.1) (C$_1$)
Two-bond	Amides	7—9	CH_3CONH_2 (9.5)
	Nitriles	~3	$CH_3C≡N$ (3.0)
	Pyridines	~1—3	C_5H_5N (2.53); $C_5H_5NH^+$ (2.01); C_5H_5NO (1.43)
	Amines, aliphatic	~1—2	$CH_3CH_2CH_2NH_2$ (1.2)
	Nitro compounds, aromatic	~(−1)—(−2)	$C_6H_5NO_2$ (−1.67)
	Amines, aromatic	~(−1)—(−2)	$C_6H_5NH_2$ (−2.68); $C_6H_5NH_3^+$ (−1.5)
	Pyrroles	~(−4)	C_4H_4NH (−3.92)
Three-bond	Amides	9	$CH_2=CHCONH_2$ (19)
	Ammonium salts	1—9	$CH_3(CH_2)_2NH_3^+$ (1.3); $C_6H_5NH_3^+$ (2.1)
	Pyridines	~3	C_5H_5N (2.53)
	Amines, aliphatic	1—3	$CH_3(CH_2)_2NH_2$ (1.4)
	Amines, aromatic	~(−1)—(−3)	$C_6H_5NH_2$ (−2.68)
	Nitro compounds	~(−2)	$C_6H_5NO_2$ (−1.67)
	Pyrroles	~(−4)	C_4H_4NH (−3.92)

^{15}N-^{15}N Coupling Constants

Bond type	Family	J_{NN}, H_2	Examples
	Azo compounds	12—25	$C_6H_5N=NC(CH_3)_2C_6H_5$ anti (17) syn (21)
	N-Nitrosamines	~19	$(C_6H_5CH_2)_2N–N=O$(19)
	Hydrazones	~10	p-$O_2NC_6H_4CH=N–NHC_6H_5$ (10.7)
	Hydrazines	~7	$C_6H_5NHNH_2$ (6.7)

^{15}N-^{19}F Coupling Constants

Bond type	Family	J_{NF}, H_2	Example
	Diflurodiazines		
	trans	~190 ($^1J_{NF}$)	F–N=N–F (190)
		~102 ($^2J_{NF}$)	F–N=N–F (102)
	cis	~203 ($^1J_{NF}$)	F–N=N–F (203)
		~52 ($^2J_{NF}$)	F–N=N–F (52)
	Fluoropyridines		
	2-fluoro-		(−52.5)
	3-fluoro-		(+3.6)
	Fluoroanilines		
	2-fluoro-	0	1,2–$C_6H_4F(NH_2)$
	3-fluoro-	0	1,3–$C_6H_4F(NH_2)$
	4-fluoro-	1.5	1,4–$C_6H_4F(NH_2)$
	Fluoroanilinium salts		
	2-fluoro-	1.4	1,2–$C_6H_4F(NH_3^+)$
	3-fluoro-	0.2	1,3–$C_6H_4F(NH_3^+)$
	4-fluoro-	0	1,4–$C_6H_4F(NH_3^+)$

FLUORINE COUPLING CONSTANTS

The following table gives the most important fluorine coupling constants, namely, J_{FH}, J_{FCF}, and J_{CF}, together with some typical examples.[1-7] The coupling-constant values vary with the solvent used.[3] The book by Emsley et al.[1] gives a complete, detailed list of various compounds.

REFERENCES

1. **Emsley, J. W., Phillips, L., and Wray, V.**, *Fluorine Coupling Constants,* Pergamon Press, Oxford, 1977.
2. **Lambert, J. R., Shurvell, H. F., Verbit, L., Cooks, R. G., and Stout, G. H.**, *Organic Structural Analysis,* Macmillan, New York, 1976.
3. **Yoder, C. H. and Schaeffer, C. D., Jr.**, *Introduction to Multinuclear NMR: Theory and Application,* Benjamin/Cummings, Menlo Park, CA, 1987.
4. **Schaeffer, T., Marat, K., Peeling, J., and Veregin, R. P.**, Signs and mechanisms of ^{13}C, ^{19}F spin-spin coupling constants in benzotrifluoride and its derivatives, *Can. J. Chem.,* 61, 2779, 1983.
5. **Adcock, W. and Kok, G. B.**, Polar substituent effects on ^{19}F chemical shifts of aryl and vinyl fluorides: a fluorine-19 nuclear magnetic resonance study of some 1,1-difluoro-2-(4-substituted-bicyclo [2.2.2] oct-1-yl)ethenes, *J. Org. Chem.,* 50, 1079, 1985.
6. **Newmark, R. A. and Hill, J. R.**, Carbon-13-fluorine-19 coupling constants in benzotrifluorides, *Org. Magn. Reson.,* 9, 589, 1977.
7. **Adcock, W. and Abeywickrema, A. N.**, Concerning the origin of substituent-induced fluorine-19 chemical shifts in aliphatic fluorides: carbon-13 and fluorine-19 nuclear magnetic resonance study of 1-fluoro-4-phenylbicyclo [2.2.2] octanes substituted in the arene ring, *J. Org. Chem.,* 47, 2945, 1982.

FLUORINE COUPLING CONSTANTS

^{19}F-H coupling constants

Fluorinated family	J_{FH}	Example
Two-bond		
Alkanes	45—80	CH_3F (45); CH_2F_2 (50); CF_3H (79); C_2H_5F (47); CH_3CHF_2 (57); CH_2FCH_2F (48); CH_2FCHF_2 [J_{CH_2F} = 46; J_{CHF_2} = 54]; CF_3CH_2F (45); $CF_2HCF_2CF_3$ (52)
Alkyl chlorides	49—65	$CFHCl_2$ (53); CF_2HCl (63); $FCHClCHCl_2$ (49); FCH_2CH_2Cl (46)
Alkyl bromides	45—50	$FCHBrCH_3$ (50.5); FCH_2CH_2Br (46); $FCHBrCFHBr$ (49)
Alkenes	45—80	$CHF=CHF$ (*cis*—71.7; *trans*—75.1); CH_2CHF (85); $CF_2=CHF$ (70.5); $FCH_2CH=CH_2$ (47.5); $CF_2=CH-Ar$ (Ar=Aryl; *cis* (5); *trans* (29))
Aromatics	45—75	$ClC_6H_4CFH_2$ (m-47; p-48); $O_2NC_6H_4CFH_2$ (m-47; p-48); $FC_6H_4CFH_2$ (m-48; p-48); p—$BrC_6H_4OCF_2H$ (73)
Ethers	40—75	CF_2HOCH_3 (74); $FCH_2CF_2OCH_3$ (46); $CF_2HOCH(CH_3)_2$ (75); CF_2HCF_2OF (56)
Ketones	45—50	FCH_2COCH_3 (47); CF_2HCOCH_3 (54); $CH_3CH_2CHFCOCH_3$ (50); $CF_2HCOCH(CF_3)_2$ (54)
Aldehydes	~50	$CH_3CH_2CHFCHO$ (51)

FLUORINE COUPLING CONSTANTS (continued)

^{19}F-H coupling constants

Carbon	J_{FH}	Examples
Esters	45—70	$CFH_2CO_2CH_2CH_3$ (47); $CH_3CHFCO_2CH_2CH_3$ (48)
Three-bond		
Alkanes	2—25	CF_2HCH_3 (21); $(CH_3)_3CF$ (20.4); $CH_3CHFCH_2CH_2CH_3$ (23); CF_3CH_3 (13)
Alkyl chlorides	8—20	CF_2HCHCl_2 (8); CF_2ClCH_3 (15)
Alkyl bromides	15—25	CF_2BrCH_2Br (22); CF_2BrCH_3 (16); $FC(CH_3)_2CHBrCH_3$ (21)
Alkenes	(−5)—60 J_{HCF} (cisoid) <20 J_{HCF} (transoid) >20	$CHF=CHF$ (*cis*-19.6; *trans*-2.8); $CH_2=CHF$ (*cis*-19.6; *trans*-51.8) $CHF=CF_2$ (*cis*-(−4.2); *trans*-12.5); $CH_2=CF_2$ (*cis*-0.6; *trans*-33.8)
Alcohols	5—30	CF_3CH_2OH (8); FCH_2CH_2OH (29); CH_3CHFCH_2OH (23.6, 23.6); $CF_3CH(OH)CH_3$ (7.5); $CF_3CH(OH)CF_3$ (6); $FC(CH_3)_2COH(CH_3)_2$ (23)
Ketones	5—25	$CH_3CH_2CHFCOCH_3$ (24); $FC(CH_3)_2COCH_3$ (21); $(CF_3)_2CHCOCH_3$ (8); $CF_2HCOCH(CF_3)_2$ (7)
Aldehydes	10—25	$(CH_3)_2CFCHO$ (22)
Esters		$CH_3CHFCO_2CH_2CH_3$ (23); $(CH_3CH_2)_2CFCO_2CH_3$ (16.5)

^{19}F-^{19}F coupling constants

Carbon	J_{FCF}	Examples
Two-bond		
Saturated (sp^3)	140—250	$CF_3CF_2{}^{a,b}CFHCH_3$ (J_{ab} = 270); $CF_2{}^{a,b}BrCHFSO_2F$ (J_{ab} = 188); $CH_3O-CF_2{}^{a,b}$ $CFHSO_2F$ (J_{ab} = 147); $CH_3O-CF_2{}^{a,b}$ $CFHCl$ (J_{ab} = 142); $CH_3S-CF_2{}^{a,b}$ $CFHCl$ (J_{ab} = 222)
Cycloalkanes	150—240	$F_2C(CH_2)_2$ (150) (3-membered); $F_2C(CH_2)_3$ (200) (4-membered); $F_2C(CH_2)_4$ (240) (5-membered); $F_2C(CH_2)_5$ (228) (6-membered)
Unsaturated (sp^2)	$\leqq 100$	$CF_2=CH_2$ (31,36); $CF_2=CHF$ (87); $CF_2=CBrCl$ (30); $CF_2=CHCl$ (41); $CF_2=CFBr$ (75); $CF_2=NCF_3$ (82); $CF_2=CFCN$ (27); $CF_2=CFCOF$ (7); $CF_2=CFOCH_2CF_3$ (102); $CF_2=CBrCH_2N(CF_3)_2$ (30); $CF_2=CFCOCF_2CF_3$ (12); $CF_2=CHC_6H_5$ (33); $CF_2=CH(CH_2)_5CH_3$ (50); $CF_2=CH-Ar[Ar=aryl]$ (50)

FLUORINE COUPLING CONSTANTS (continued)

^{19}F-^{19}F coupling constants

Carbon	J_{FCF}	Examples
Three-bond		
Saturated (sp³)	0—16	CF_3CH_2F (16); CF_3CF_3 (3.5);
		CF_3CHF_2 (3); CH_2FCH_2F (10—12);
		$CF_2^aHCF^bHCF_2H$ ($J_{ab} = 13$);
		$CF_2^aHCF_2^bCH_2F^b$ ($J_{ab} = 14$);
		$CF_3^aCF_2^bCF^cHCH_3$ ($J_{ab} < 1$; $J_{bc} = 15$)
		$CF_3^aCF^bHCF_2^cH$ ($J_{ab} = 12$; $J_{bc} = 12$);
		$CF_3^aCF_2^bC{\equiv}CCF_3$ ($J_{ab} = 3.3$);
		$(CF_3^a)_2CF^bC{\equiv}CCl$ ($J_{ab} = 10$);
		$CF_3CF_2COCH_2CH_3$ (1);
		$FCH_2CFHCO_2C_2H_5$ (-11.6);
		$CF_3^aCF_2^bCF_2^cCOOH$ ($J_{ab} < 1$; $J_{bc} < 1$)
		$(CF_3^a)_2CF^bS(O)OC_2H_5$ ($J_{ab} = 8$)
Unsaturated (sp²)	>30	$FCH{=}CHF$[cis(-18.7); *trans*
		(-133.5)];
		$CF_2{=}CHBr$ (34.5); $CF_2{=}CHCl$ (41);
		$CF_2{=}CH_2$ (37)

C-^{19}F coupling constants

Fluorinated family	J_{CF}	Examples
One-bond		
Alkanes	150—290	CH_3F (158); CH_2F_2 (237);
		CHF_3 (274); CF_4 (257); CF_3CF_3
		(281); CF_3CH_3 (271); $(CH_3)_3CF$
		(167); $(C^aF_3^b)_2C^cF_2^d$ [$J_{ab} = 285$;
		$J_{cd} = 265$]
Alkenes	250—300	$CF_2{=}CD_2$ (287); $CF_2{=}CCl_2$ (-289);
		$CF_2{=}CBr_2$ (290); $ClFC{=}CHCl$
		[cis (-300); *trans* (-307)];
		$ClFC{=}CClF$ [cis (290);
		trans (290)]
Alkynes	250—260	$C^aF_3^bC{\equiv}CF$ [$J_{ab} = 259$]; $CF_3C{\equiv}CCF_3$
		(256)
Alkyl chlorides	275—350	$CFCl_3$ (337); CF_2Cl_2 (325);
		CF_3Cl (299); $CF_3(CCl_2)_2CF_3$ (286);
		CF_3CH_2Cl (274); $CF_3CCl{=}CCl_2$ (274);
		$CF_2{=}CCl_2$ (-289); CF_3CCl_3 (283)
Alkyl bromides	290—375	$CFBr_3$ (372); CF_2Br_2 (358);
		CF_3Br (324); CF_3CH_2Br (272);
		$CF_2{=}CBr_2$ (290)
Acyl fluorides	350—370	$HCOF$ (369); CH_3COF (353)
Carboxylic acids	245—290	CF_3COOH (283); CF_2HCO_2H (247)
Alcohols	~275	CF_3CH_2OH (278)
Nitriles	~250	CF_2HCN (244)
Esters	~285	$CF_3CO_2CH_2CH_3$ (284)
Ketones	~290	CF_3COCH_3 (289)
Ethers	~265	$(CF_3)_2O$ (265)

VIII. Mass Spectroscopy

NATURAL ABUNDANCE OF IMPORTANT ISOTOPES

The following table lists the atomic masses and relative percent concentrations of naturally occurring isotopes of importance in mass spectroscopy.[1-3]

Element	Total no. of isotopes	More prominent isotopes (mass percent abundance)	
Hydrogen	3	^1H (1.00783, 99.985) ^2H (2.01410, 0.015)	
Boron	6	^{10}B (10.01294, 19.8) ^{11}B (11.00931, 80.2)	
Carbon	7	^{12}C (12.00000, 98.9) ^{13}C (13.00335, 1.1)	
Nitrogen	7	^{14}N (14.00307, 99.6) ^{15}N (15.00011, 0.4)	
Oxygen	8	^{16}O (15.99491, 99.8)	^{18}O (17.9992, 0.2)
Fluorine	6	^{19}F (18.99840, 100.0)	
Silicon	8	^{28}Si (27.97693, 92.2) ^{29}Si (28.97649, 4.7)	^{30}Si (29.97376, 3.1)
Phosphorus	7	^{31}P (30.97376, 100.0)	
Sulfur	10	^{32}S (31.972017, 95.0) ^{33}S (32.97146, 0.7)	^{34}S (33.96786, 4.2)
Chlorine	11	^{35}Cl (34.96885, 75.5)	^{37}Cl (36.96590, 24.5)
Bromine	17	^{79}Br (78.9183, 50.5)	^{81}Br (80.91642, 49.5)
Iodine	23	^{127}I (126.90466, 100.0)	

REFERENCES

1. **Rose, M. E. and Johnstone, R. A. W.,** *Mass Spectrometry for Chemists and Biochemists,* Cambridge University Press, Cambridge, 1982.
2. **Weast, R. C., Ed.,** *Handbook of Chemistry and Physics,* 69th ed., CRC Press, Boca Raton, FL, 1988.
3. **McLafferty, F. W.,** *Interpretation of Mass Spectra,* University Science Books, Mill Valley, CA, 1980.

RULES FOR DETERMINATION OF MOLECULAR FORMULA

The following rules are used in the mass-spectroscopic determination of the molecular formula of an organic compound.[1] These rules should be applied to the molecular ion peak and its isotopic cluster. The molecular ion, in turn, should be the highest mass in the spectrum, must be an odd-electron ion, and must be capable of yielding all other important ions of the spectrum via a logical neutral species loss. The elements that are assumed to (possibly) exist are carbon, hydrogen, nitrogen, the halogens, sulfur, and/or oxygen. The molecular formula that can be derived is not the only possible one and consequently help from nuclear magnetic resonance spectrometry and infrared spectrophotometric data are necessary for the final determination.

Rule 1: An odd molecular ion value suggests the presence of an odd number of nitrogens. An even molecular ion value is due to the presence of zero or an even number of nitrogens. Thus, m/e = 141 suggests 1, 3, 5, 7, etc. nitrogen atoms while m/e = 142 suggests 0, 2, 4, 6, etc. nitrogen atoms (nitrogen rule).

Rule 2: The maximum number of carbon atoms (N_c^{max}) can be calculated from the formula

$$N_c^{max} = \frac{\text{Relative Intensity of M} + 1}{\text{Relative Intensity of M}^+} \times \frac{100}{1.1}$$

where M + 1 is the peak next to that of the molecular ion (M$^+$). This rule gives the *maximum* number of carbons, but not necessarily the *actual* number. If, for example, the relative intensities of M$^+$ and M + 1 are 100 and 9%, respectively, then the *maximum* number of carbons is

$$\frac{9}{100} \times \frac{100}{1.1} = 8$$

There is a possibility for seven, six, etc. carbons, but not for nine or more.

Rule 3: The maximum numbers of sulfur atoms (N_s^{max}) can be calculated from the formula

$$N_s^{max} = \frac{\text{Relative Intensity of M} + 2}{\text{Relative Intensity of M}^+} \times \frac{100}{4.4}$$

where M + 2 is the peak two units higher than that of the molecular ion M$^+$.

Rule 4: The number of chlorine atoms and/or bromine atoms can be derived from the table (which follows) showing the relative ratio of the isotope cluster of the molecular ion.

Rule 5: The difference should be only oxygen and hydrogen. These rules assume the absence of phosphorus, silicon, or any other elements.

REFERENCE

1. **McLafferty, F. W.**, *Interpretation of Mass Spectra*, University Science Books, Mill Valley, CA, 1980.

CHLORINE-BROMINE COMBINATION ISOTOPE INTENSITIES

Due to the distinctive mass-spectral patterns caused by the presence of chlorine and bromine in a molecule, interpretation can be much easier if the results of the relative isotopic concentrations are known. The following table provides peak intensities (relative to the molecular ion (M^+) at an intensity normalized to 100%) for various combinations of chlorine and bromine atoms, assuming the absence of all other elements except carbon and hydrogen.[1-3] The mass abundance calculations were based upon the atomic mass data of Weast.[4]

REFERENCES

1. **McLafferty, F. W.,** *Interpretation of Mass Spectra,* University Science Books, Mill Valley, CA, 1980.
2. **Silverstein, R. H., Bassler, G. C., and Morrill, T. C.,** *Spectrometric Identification of Organic Compounds,* John Wiley & Sons, New York, 1981.
3. **Williams, D. H. and Fleming, I.,** *Spectroscopic Methods in Organic Chemistry,* McGraw-Hill, London, 1973.
4. **Weast, R. C., Ed.,** *Handbook of Chemistry and Physics,* 63rd ed., CRC Press, Boca Raton, FL, 1984.

RELATIVE INTENSITIES OF ISOTOPE PEAKS FOR COMBINATIONS OF BROMINE AND CHLORINE ($M^+ = 100\%$)

		Br_0	Br_1	Br_2	Br_3	Br_4
Cl_0	P + 2		98.0	196.0	294.0	390.8
	P + 4			96.1	288.2	547.7
	P + 6				94.1	375.3
	P + 8					92.0
Cl_1	P + 2	32.5	130.6	228.0	326.1	424.6
	P + 4		31.9	159.0	383.1	704.2
	P + 6			31.2	187.4	564.1
	P + 8				30.7	214.8
	P + 10					30.3
Cl_2	P + 2	65.0	163.0	261.1	359.3	456.3
	P + 4	10.6	74.4	234.2	490.2	840.3
	P + 6		10.4	83.3	312.8	791.6
	P + 8			10.2	91.7	397.5
	P + 10				9.8	99.2
	P + 12					10.1
Cl_3	P + 2	97.5	195.3	294.0	393.3	
	P + 4	31.7	127.0	99.7	609.8	
	P + 6	3.4	34.4	159.4	473.8	
	P + 8		3.3	37.1	193.9	
	P + 10			3.2	39.6	
	P + 12				3.0	
	P + 14					
Cl_4	P + 2	130.0	228.3	326.6	4.2	
	P + 4	63.3	190.9	414.9	735.3	
	P + 6	13.7	75.8	263.1	670.0	
	P + 8	1.2	14.4	88.8	347.1	
	P + 10		1.1	15.4	102.2	
	P + 12			1.3	16.2	
	P + 14				0.7	
Cl_5	P + 2	162.6	260.7	358.9		
	P + 4	105.7	265.3	520.8		
	P + 6	34.3	137.9	397.9		
	P + 8	5.5	39.3	174.5		
	P + 10	0.3	5.8	44.3		
	P + 12		0.3	5.7		
	P + 14			0.5		
Cl_6	P + 2	195.3				
	P + 4	158.6				
	P + 6	68.8				
	P + 8	16.6				
	P + 10	2.1				
	P + 12	0.1				
Cl_7	P + 2	227.8				
	P + 4	222.1				
	P + 6	120.3				
	P + 8	39.0				
	P + 10	7.5				
	P + 12	0.8				
	P + 14	0.05				

REFERENCE COMPOUNDS UNDER ELECTRON IMPACT CONDITIONS IN MASS SPECTROMETRY

The following table lists the most popular reference compounds for use under electron impact conditions in mass spectrometry. For accurate mass measurements, the reference compound is introduced and ionized concurrently with the sample and the reference peaks are resolved from sample peaks. Reference compounds should contain as few heteroatoms and isotopes as possible "to facilitate the assignment of reference masses and minimize the occurrence of unresolved multiplets within the reference spectrum itself".[1] An approximate upper mass limit should assist in the selection of the appropriate reference.[1,2]

Reference compound	Formula	Upper mass limit
Perfluoro-2-butyltetrahydrofuran	$C_8F_{16}O$	416
Decafluorotriphenyl phosphine (Ultramark 433; DFTPP)	$(C_6F_5)_3P$	443
Heptacosafluorotributylamine (perfluoro tributylamine; heptacosa; PFTBA)	$(C_4F_9)_3N$	671
Perfluoro kerosene, low-boiling (perfluoro kerosene-L)	$CF_3(CF_2)_nCF_3$	600
Perfluoro kerosene, high-boiling (perfluoro kerosene-H)	$CF_3(CF_2)_nCF_3$	800—900
tris-(Trifluoromethyl)-*s*-triazine	$C_3N_3(CF_3)_3$	285
tris-(Pentafluoroethyl)-*s*-triazine	$C_3N_3(CF_2CF_3)_3$	435
tris-(Heptafluoropropyl)-*s*-triazine	$C_3N_3(CF_2CF_2CF_3)_3$	585
tris-(Perfluoroheptyl)-*s*-triazine	$C_3N_3[(CF_2)_6CF_3]_3$	1185
tris-(Perfluorononyl)-*s*-triazine	$C_3N_3[(CF_2)_8CF_3]_3$	1485
Ultramark 1621 (fluoralkoxy cyclotriphosphazine mixture)	$P_3N_3[OCH_2(CF_2)_nH]_6$	~2,000
Fomblin diffusion pump fluid (Ultramark F series; perfluoro polyether)	$CF_3O[CF(CF_3)CF_2O]_m(CF_2O)_nCF_3$	≥3,000

REFERENCES

1. **Chapman, J. R.**, *Computers in Mass Spectrometry*, Academic Press, London, 1978.
2. **Chapman, J. R.**, *Practical Organic Mass Spectrometry*, John Wiley & Sons, Chichester, U.K., 1985.

MAJOR REFERENCE MASSES IN THE SPECTRUM OF
HEPTACOSAFLUOROTRIBUTYLAMINE (PERFLUOROTRIBUTYLAMINE)

The following list tabulates the major reference masses (with their relative intensities and formulas) of the mass spectrum of heptacosafluorotributylamine (perfluorotributylamine).[1]

Accurate mass	Relative intensity	Formula	Accurate mass	Relative intensity	Formula
613.9647	2.6	$C_{12}F_{24}N$	180.9888	1.9	C_4F_7
575.9679	1.7	$C_{12}F_{22}N$	175.9935	1.0	C_4F_6N
537.9711	0.4	$C_{12}F_{20}N$	168.9888	3.6	C_3F_7
501.9711	8.6	$C_9F_{20}N$	163.9935	0.7	C_3F_6N
463.9743	3.8	$C_9F_{18}N$	161.9904	0.3	C_4F_6
425.9775	2.5	$C_9F_{16}N$	149.9904	2.1	C_3F_6
413.9775	5.1	$C_8F_{16}N$	130.9920	31	C_3F_5
375.9807	0.9	$C_8F_{14}N$	118.9920	8.3	C_2F_5
325.9839	0.4	$C_7F_{12}N$	113.9967	3.7	C_2F_4N
313.9839	0.4	$C_6F_{12}N$	111.9936	0.7	C_3F_4
263.9871	10	$C_5F_{10}N$	99.9936	12	C_2F_4
230.9856	0.9	C_5F_9	92.9952	1.1	C_3F_3
225.9903	0.6	C_5F_8N	68.9952	100	CF_3
218.9856	62	C_4F_9	49.9968	1.0	CF_2
213.9903	0.6	C_4F_8N	30.9984	2.3	CF

REFERENCE

1. **Chapman, J. R.**, *Practical Organic Mass Spectrometry*, John Wiley & Sons, Chichester, U.K., 1985.

COMMON FRAGMENTATION PATTERNS OF FAMILIES OF ORGANIC COMPOUNDS

The following table provides a guide to the identification and intrepretation of commonly observed mass-spectral fragmentation patterns for common organic functional groups.[1-6] It is of course highly desirable to augment mass-spectroscopic data with as much other structural information as possible. Especially useful in this regard will be the confirmatory information of infrared and ultraviolet spectrophotometry as well as nuclear magnetic resonance spectrometry.

REFERENCES

1. **Pasto, D. J. and Johnson, C. R.**, *Organic Structure Determination*, Prentice-Hall, Englewood Cliffs, NJ, 1969.
2. **McLafferty, F. W.**, *Interpretation of Mass Spectra*, University Science Books, Mill Valley, CA, 1980.
3. **Rose, M. E. and Johnstone, R. A. W.**, *Mass Spectrometry for Chemical and Biochemists*, Cambridge University Press, Cambridge, 1982.
4. **Silverstein, R. M., Bassler, G. C., and Morrill, T. C.**, *Spectrometric Identification of Organic Compounds*, John Wiley & Sons, New York, 1981.
5. **Smakman, R. and deBoer, T. J.**, The mass spectra of some aliphatic and alicyclic sulphoxides and sulphones, *Org. Mass Spec.*, 3, 1561, 1970.
6. **Bowie, J. H., Williams, D. H., Lawesson, S. O., Madsen, J. O., Nolde, C., and Schroll, G.**, Studies in mass spectrometry. XV. Mass spectra of sulphoxides and sulphones. The formation of C–C and C–O bonds upon electron impact, *Tetrahedron*, 22, 3515, 1966.

COMMON FRAGMENTATION PATTERNS OF FAMILIES OF ORGANIC COMPOUNDS

Family	Molecular ion peak	Common fragments; characteristic peaks
Acetals		Cleavage of all C–O, C–H, and C–C bonds around the original aldehydic carbon
Alcohols	Weak for 1 and 2°; not detectable for 3°; strong for benzyl alcohols	Loss of 18 (H_2O — usually by cyclic mechanism); loss of H_2O and olefin simultaneously with four (or more) carbon-chain alcohols; prominent peak at m/e = 31 $(CH_2\overset{+}{O}H)^+$ for 1° alcohols; prominent peak at m/e = $(RCH\overset{+}{O}H)^+$ for 2° and m/e = $(R_2C\overset{+}{O}H)^+$ for 3° alcohols
Aldehydes	Low intensity	Loss of aldehydic hydrogen (strong M-1 peak, especially with aromatic aldehydes); strong peak at m/e = 29 ($HC \equiv O^+$); loss of chain attached to alpha carbon (beta cleavage); McLafferty rearrangement via beta cleavage if gamma hydrogen is present
Alkanes		
Chain	Low intensity	Loss of 14 units (CH_2)
Branched	Low intensity	Cleavage at the point of branch; low-intensity ions from random rearrangements
Alicyclic	Rather intense	Loss of 28 units ($CH{=}CH_2$) and side chains
Alkenes (olefins)	Rather high intensity (loss of π-electron), especially in the case of cyclic olefins	Loss of units of general formula C_nH_{2n-1}; formation of fragments of the composition C_nH_{2n} (via McLafferty rearrangement); retro Diels-Alder fragmentation
Alkyl halides	Abundance of molecular ion F < Cl < Br < I; intensity decreases with increase in size and branching	Loss of fragments equal to the mass of the halogen until all halogens are cleaved off
Fluorides	Very low intensity	Loss of 20 (HF); loss of C_2H_2 in case of fluorobenzenes
Chlorides	Low intensity; characteristic isotope cluster	Loss of 35 (Cl) or 36 (HCl); loss of chain attached to the gamma carbon to the carbon carrying the Cl
Bromides	Low intensity; characteristic isotope cluster	Loss of 79 (Br); loss of chain attached to the gamma carbon to the carbon carrying the Br
Iodides	Higher than other corresponding halides	Loss of 127 (I)
Alkynes	Rather high intensity (loss of π-electron)	Fragmentation similar to that of alkenes
Amides	Rather high intensity	Strong peak at m/e indicate of a 1° amide ($O = C = \overset{+}{N}H_2$); base peak at m/e = 59 ($CH_2 = C(OH)\overset{+}{N}H_2$); possibility of McLafferty rearrangement; loss of C_2H_2O for amides of the form $RNHCOCH_3$ when R is an aromatic ring
Amines	Hardly detectable in case of acyclic aliphatic amines; high intensity for aromatic and cyclic amines	Beta cleavage yielding $>C = \overset{+}{N}<$; base peak for all 1° amines at m/e = 30 ($CH_2{=}\overset{+}{N}H_2$); moderate M-1 peak for aromatic amines; loss of 27 (HCN) in aromatic amines; fragmentation at alpha carbons in cyclic amines
Aromatic hydrocarbons (arenes)	Rather intense	Loss of side chain; formation of RCH = CHR′ (via McLafferty rearrangement); cleavage at the bonds beta to the aromatic ring; peaks at m/e = 77 (benzene ring; especially mono-substituted), 91 (tropyllium); the ring position of alkyl substitution has very little effect on the spectrum
Carboxylic acids	Weak for straight-chain monocarboxylic acids; large if aromatic acids	Base peak at m/e = 60 ($CH_2{=}C(OH)_2$) if δ-hydrogen is present; peak at m/e = 45 (COOH); loss of 17 (–OH) in case of aromatic or short-chain acids

COMMON FRAGMENTATION PATTERNS OF FAMILIES (continued)
OF ORGANIC COMPOUNDS

Family	Molecular ion peak	Common fragments; characteristic peaks
Disulfides	Rather low intensity	Loss of olefins (m/e equal to R–S–S–H‡); strong peak at m/e = 66 (HSSH‡)
Esters R–C–OR′ ‖ O	Rather weak intensity	Base peak at m/e equal to the mass of R–C ≡ O^{+}; peaks at m/e equal to the mass of $^{+}$O ≡ C–OR′, the mass of OR′, and R′; McLafferty rearrangement possible in case of (1) presence of a beta hydrogen in R′ (peak at m/e equal to the mass of R–C(=$\overset{+}{O}$H)OH, and (2) presence of a gamma hydrogen in R (peak at m/e equal to the mass of (CH$_2$=C($^{+}$OH)OR); loss of 42 (CH$_2$=C=O) in case of benzyl esters; loss of ROH via the ortho effect in case of *o*-substituted benzoates
Ethers	Rather low intensity	Cleavage of a bond beta to the oxygen and formation of CH$_2$=$\overset{+}{O}$–R species; cleavage of C–O bond and loss of R-groups; loss of (CH$_2$O), (CHO), and (CH$_3$) in case of anisoles; loss of m/e = 1 (H), 28 (CO), and 29 (CHO) in diphenyl ethers
Ketones	Rather high intensity	Loss of R-groups attached to the >C = O (alpha cleavage); peak at m/e = 43 for all methyl ketones (CH$_3$CO^{+}); McLafferty rearrangement via beta cleavage if gamma hydrogen is present; loss of m/e = 28 (C = O) for cyclic ketones after initial alpha cleavage and McLafferty rearrangement
Mercaptans (thiols)	Rather low intensity but higher than that of corresponding alcohol	Similar to those of alcohols (–OH substituted by –SH); loss of m/e = 45 (CHS) and m/e = 44 (CS) for aromatic thiols
Nitriles	Unlikely to be detected except in case of acetonitrile (CH$_3$CN) and propionitrile (C$_2$H$_5$CN)	M + 1 ion may appear (especially at higher pressures); M − 1 peak is weak but detectable (R–CH=C=N^{+}); base peak at m/e = 41 (CH$_2$=C=Ṅ–H); McLafferty rearrangement possible; loss of HCN in case of cyanobenzenes
Nitrites	Absent (or very weak at best)	Base peak at m/e = 30 (NO^{+}); large peak at m/e = 60 (CH$_2$=$\overset{+}{O}$NO) in all unbranched nitrites at the alpha carbon; absence of m/e = 46 permits differentiation from nitrocompounds
Nitro compounds	Seldom observed	Loss of 30 (NO); subsequent loss of CO (in case of aromatic nitrocompounds); loss of NO$_2$ from molecular ion peak
Phenols	High intensity (base peak generally)	Loss of 28 (C=O) and 29 (CHO); strong peak at m/e = 65 (C$_5$H$_5$$^{+}$)
Sulfides (thioethers)	Rather low intensity but higher than that of corresponding ether	Similar to those of ethers (–O– substituted by –S–); aromatic sulfides show strong peaks at m/e = 109 (C$_6$H$_5$S‡); 65 (C$_5$H$_5$‡); 91 (tropyllium ion)
Sulfonamides	Rather intense	Loss of m/e = 64 (SONH$_2$) and m/e = 27 (HCN) in case of benzenesulfonamide
Sulfones	High intensity	Similar to sulfoxides; loss of mass equal to RSO$_2$; aromatic heterocycles show peaks at M-32 (sulfur), M-48(SO), and M-64(SO$_2$)
Sulfoxides	High intensity	Loss of 17 (OH); loss of alkene (m/e equal to RSO‡); peak at m/e = 63 (CH$_2$=$^{+}$SOH); aromatic sulfoxides show peak at m/e = 125 ($^{+}$S–CH=CHCH=CHC=O), 97 (C$_5$H$_5$S^{+}), 93(C$_6$H$_5$$\overset{\cdot}{O}$H); aromatic heterocycles show peaks at M-16 (oxygen), M-29 (COH), and M-48 (SO)

COMMON FRAGMENTS LOST

The following table gives a list of neutral species most commonly lost in the mass spectra of organic compounds.

The list is suggestive rather than comprehensive and should be used with caution.[1-3] The listed fragments include only combinations of carbon, hydrogen, oxygen, nitrogen, sulfur, and the halogens.

REFERENCES

1. **Silverstein, R. M., Bassler, G. C., and Morrill, T. C.,** *Spectrometric Identification of Organic Compounds*, John Wiley & Sons, New York, 1981.
2. **Hamming, M. and Foster, N.,** *Interpretation of Mass Spectra of Organic Compounds*, Academic Press, New York, 1972.
3. **McLafferty, F. W.,** *Interpretation of Mass Spectra*, University Science Books, Mill Valley, CA, 1980.

COMMON FRAGMENTS LOST

Mass lost	Fragment lost	Mass lost	Fragment lost
1	H·	49	·CH$_2$Cl
15	CH$_3$·	51	·CHF$_2$
17	OH·	52	C$_4$H$_4$; C$_2$N$_2$
18	H$_2$O	53	C$_4$H$_5$
19	F·	54	CH$_2$=CHCH=CH$_2$
20	HF	55	CH$_2$=CH–CHCH$_3$
26	HC≡CH; ·C≡N	56	CH$_2$=CH–CH$_2$CH$_3$;
27	CH$_2$=CH·; HC≡N		CH$_3$CH=CHCH$_3$; CO (2 mol)
28	CH$_2$=CH$_2$; :C=O; (HCN and H·)	57	C$_4$H$_9$·
29	CH$_3$CH$_2$·; H–·C=O	58	·NCS; (CH$_3$)$_2$C=O; (NO and CO)
30	·CH$_2$NH$_2$; HCHO; NO	59	CH$_3$OC=O; CH$_3$CONH$_2$; C$_2$H$_3$S·
31	CH$_3$O·; ·CH$_2$OH; CH$_3$NH$_2$	60	C$_3$H$_7$OH
32	CH$_3$OH; S	61	CH$_3$CH$_2$S·; (CH$_2$)$_2$S·H
33	HS·	62	[H$_2$S and CH$_2$=CH$_2$]
34	H$_2$S	63	·CH$_2$CH$_2$Cl
35	Cl·	64	S$_2$; SO$_2$; C$_5$H$_4$
36	HC; 2H$_2$O	68	CH$_2$=CHC(CH$_3$)=CH$_2$
37	H$_2$Cl	69	CF$_3$·; C$_5$H$_9$·
38	C$_3$H$_2$·; C$_2$N; F$_2$	71	C$_5$H$_{11}$·
39	C$_3$H$_3$; HC$_2$N	73	CH$_3$CH$_2$OC·=O
40	CH$_3$C≡CH	74	C$_4$H$_9$OH
41	CH$_2$=CHCH$_2$·	75	C$_6$H$_3$
42	CH$_2$=CHCH$_3$; CH$_2$=C=O; (CH$_2$)$_3$; NCO; NCNH$_2$	76	C$_6$H$_4$; CS$_2$
		77	C$_6$H$_5$; HCS$_2$
43	C$_3$H$_7$·; CH$_3$C=O; CH$_2$=CH–Ö; HCNO	78	C$_6$H$_6$; H$_2$CS$_2$; C$_5$H$_4$N
44	CH$_2$=CHOH; CO$_2$; N$_2$O; CONH$_2$; NHCH$_2$CH$_3$	79	Br·; C$_5$H$_5$N
		80	HBr
45	CH$_3$CHOH; CH$_3$CH$_2$Ö; CO$_2$H; CH$_3$CH$_2$NH$_2$	85	·CClF$_2$
		100	CF$_2$=CF$_2$
46	CH$_3$CH$_2$OH; ·NO$_2$	119	CF$_3$CF$_2$·
47	CH$_3$S·	122	C$_6$H$_5$CO$_2$H
48	CH$_3$SH; SO; O$_3$	127	I·
		128	HI

IMPORTANT PEAKS IN THE MASS SPECTRA OF COMMON SOLVENTS

The following table gives the most important peaks that appear in the mass spectra of the most common solvents which might occur as an impurity in organic samples. The solvents are classified in ascending order of their M^+ peaks. The highest intensity peaks are indicated with (100%).[1-3]

REFERENCES

1. **Clerc, J. T., Pretsch, E., and Seibl, J.**, *Studies in Analytical Chemistry*, Vol. I. *Structural Analysis of Organic Compounds by Combined Application of Spectroscopic Methods*, Elsevier, Amsterdam, 1981.
2. **McLafferty, F. W.**, *Interpretation of Mass Spectra*, University Science Books, Mill Valley, CA, 1980.
3. **Pasto, D. J. and Johnson, C. R.**, *Organic Structure Determination*, Prentice-Hall, Englewood Cliffs, NJ, 1969.

IMPORTANT PEAKS IN THE MASS SPECTRA OF COMMON SOLVENTS

Solvents	Formula	M^+	Important peaks (m/e)
Water	H_2O	18 (100%)	17
Methanol	CH_3OH	32	31 (100%), 29, 15
Acetonitrile	CH_3CN	41 (100%)	40, 39, 38, 28, 15
Ethanol	CH_3CH_2OH	46	45, 31 (100%), 27, 15
Dimethyl ether	CH_3OCH_3	46 (100%)	45, 29, 15
Acetone	CH_3COCH_3	58	43 (100%), 42, 39, 27, 15
Acetic acid	CH_3CO_2H	60	45, 43, 18, 15
Ethylene glycol	$HOCH_2CH_2OH$	62	43, 33, 31 (100%), 29, 18, 15
Furan	C_4H_4O	68 (100%)	42, 39, 38, 37, 29, 18
Tetrahydrofuran	C_4H_8O	72	71, 43, 42 (100%), 41, 40, 39, 27, 18, 15
n-Pentane	C_5H_{12}	72	57, 43 (100%), 42, 41, 39, 29, 28, 27, 15
Dimethyformamide (DMF)	$HCON(CH_3)_2$	73 (100%)	58, 44, 42, 30, 29, 28, 18, 15
Diethylether	$(C_2H_5)_2O$	74	59, 45, 41, 31 (100%), 29, 27, 15
Methylacetate	$CH_3CO_2CH_3$	74	59, 43 (100%), 42, 32, 29, 28, 15
Carbon disulfide	CS_2	76 (100%)	64, 44, 38, 32
Benzene	C_6H_6	78 (100%)	77, 52, 51, 50, 39, 28
Pyridine	C_5H_5N	79 (100%)	80, 78, 53, 52, 51, 50, 39, 26
Dichloromethane	CH_2Cl_2	84	86, 51, 49 (100%), 48, 47, 35, 28
Cyclohexane	C_6H_{12}	84	69, 56, 55, 43, 42, 41, 39, 27
n-Hexane	C_6H_{14}	86	85, 71, 69, 57 (100%), 43, 42, 41, 39, 29, 28, 27
p-Dioxane	$C_4H_8O_2$	88 (100%)	87, 58, 57, 45, 43, 31, 30, 29, 28
Tetramethylsilane (TMS)	$(CH_3)_4Si$	88	74, 73, 55, 45, 43, 29
1,2-Dimethoxy ethane	$(CH_3OCH_2)_2$	90	60, 58, 45 (100%), 31, 29
Toluene	$C_6H_5CH_3$	92	91 (100%), 65, 51, 39, 28
Chloroform	$CHCl_3$	118	120, 83, 81 (100%), 47, 35, 28
Chloroform-d$_1$	$CDCl_3$	119	121, 84, 82 (100%), 48, 47, 35, 28
Carbon tetrachloride	CCl_4	152 (not seen)	121, 119, 117 (100%), 84, 82, 58.5, 47, 35, 28
Tetrachloroethene	$CCl_2{=}CCl_2$	164 (not seen)	168, 166 (100%), 165, 164, 131, 128, 129, 95, 94, 82, 69, 59, 47, 31, 24

PROTON AFFINITY OF REACTANT GASES IN CHEMICAL IONIZATION MASS SPECTROMETRY

The following table gives a series of popular reactant gases in chemical ionization mass spectrometry with their corresponding proton affinities (PA) and references.[1-4] Proton affinity is defined as the heat of reaction, ΔH, for the "reverse" ionization of the protonated reactant gas. Thus, if the reactant gas has the hypothetical formula BH, ΔH is the heat of reaction for the equation:

$$BH_2^+ \rightarrow BH + H^+$$

The value for PA is given in kilocalories per mole (kcal/mol) for convenience; to convert to the appropriate SI unit (kilojoules per mole [kJ/mol]), multiply by 4.1845.

Reactant gas	PA (kcal/mol)	Reactant gas	PA (kcal/mol)
H_2	100[1], 100.7[2]	N_2O	137.0[2]
O_2	100.4[2]	CO	141.4[2]
Kr	101.4[2]	C_2H_4	160[1], 173.5[2]
N_2	717.4[2]	H_2O	165[1], 173[2]
Xe	118.0[2]	H_2S	170[1], 176.6[2]
CH_4	127[1], 130.5[2]	CH_3OH	182[1], 184.9[2]
CO_2	128.6[2]	$(CH_3)_3CH$	195[1]
		NH_3	207[1]

REFERENCES

1. **Field, F. H.,** Chemical ionization mass spectrometry, *Acc. Chem. Res.,* 1, 42, 1968.
2. **Harrison, A. G.,** *Chemical Ionization Mass Spectrometry,* CRC Press, Boca Raton, FL, 1983.
3. **Michnowicz, J. A.,** Reactant Gas Selection in Chemical Ionization Mass Spectrometry, Application Note, Hewlett-Packard, Palo Alto, CA, October, 1977.
4. **Chapman, J. R.,** *Practical Organic Mass Spectrometry,* John Wiley & Sons, Chichester, 1985.

PROTON AFFINITIES OF SOME SIMPLE MOLECULES

The following table gives the proton affinities (PA) of some simple molecules. For the occurrence of proton transfer (or reaction) between a reactant ion and a sample molecule, the reaction must be exothermic. Thus,

$$\Delta H \text{ reaction } = \text{PA (reactant gas)} - \text{PA (sample)} < 0$$

The more exothermic the reaction, the greater the degree of fragmentation. Endothermic reactions do not yield a protonated form of a sample; therefore the sample compound cannot be recorded. One can choose the proper reactant gas that will give the correct fragmentation pattern of a desired compound out of a mixture of compounds.[1-3] Chapman[4] lists positive-ion chemical ionization applications by reagent gas and by compounds analyzed until 1982. The values are provided in kilocalories per mole (kcal/mol) for convenience; to convert to the appropriate SI unit (kilojoules per mole [kJ/mol]), multiply by 4.184.

REFERENCES

1. **Field, F. H.**, Chemical ionization mass spectrometry, *Acc. Chem. Res.*, 1, 42, 1968.
2. **Harrison, A. G.**, *Chemical Ionization Mass Spectrometry*, CRC Press, Boca Raton, FL, 1983.
3. **Michnowicz, J. A.**, Reactant Gas Selection in Chemical Ionization Mass Spectrometry, Application Note, Hewlett-Packard, Palo Alto, CA, October, 1977.
4. **Chapman, J. R.**, *Practical Organic Mass Spectrometry*, John Wiley & Sons, Chichester, 1985.

PROTON AFFINITIES OF SOME SIMPLE MOLECULES

Family	Typical examples (PA in kcal/mol)
Alcohols	CH_3OH (184.9); CH_3CH_2OH (190.3); $CH_3CH_2CH_2OH$ (191.4); $(CH_3)_3COH$ (195.0); CF_3CH_2OH (174.9)
Aldehydes	HCHO (177.2); CH_3CHO (188.9); CH_3CH_2CHO (191.4); $CH_3CH_2CH_2CHO$ (193.3)
Alkanes	CH_4 (130.5); $(CH_3)_3CH$ (195)
Alkenes	$H_2C=CH_2$ (163.5); $CH_3CH=CH_2$ (184.9); $(CH_3)_2C=CH_2$ (196.9); *trans*-$CH_3CH=CHCH_3$ (182.0)
Aromatics, substituted C_6H_5–G	G=–H (182.8); –Cl (181.7); –F (181.5); –CH_3 (191.2); –C_2H_5 (192.2); –$CH_2CH_2CH_3$ (191.0); –$CH(CH_3)_2$ (191.4); –$C(CH_3)_3$ (191.6); –NO_2 (193.8); –OH (196.2); –CN (196.3); –CHO (200.3); –OCH_3 (200.6); –NH_2 (211.5)
Amines	1°: NH_3 (205.0); CH_3NH_2 (214.1); $C_2H_5NH_2$ (217.1); $CH_3CH_2CH_2NH_2$ (218.5); $CH_3CH_2CH_2CH_2NH_2$ (219.0) 2°: $(CH_3)_2NH$ (220.5); $(C_2H_5)_2NH$ (225.1); $(CH_3CH_2CH_2)_2NH$ (227.4) 3°: $(CH_3)_3N$ (224.3); $(C_2H_5)_3N$ (231.2); $(CH_3CH_2CH_2)_3N$ (233.4)
Carboxylic acids	HCO_2H (182.8); CH_3CO_2H (190.7); $CH_3CH_2CO_2H$ (193.4); CF_3CO_2H (176.0)
Dienes	$CH_2=CHCH=CH_2$ (193); E–$CH_2=CHCH=CHCH_3$ (201.8); E–$CH_2=CHC(CH_3)=CHCH_3$ (205.7); cyclopentadiene (200.0)
Esters	HCO_2CH_3 (190.4); $HCO_2C_2H_5$ (194.2); $HCO_2CH_2CH_2CH_3$ (195.2); $CH_3CO_2C_2H_5$ (201.3); $CH_3CO_2CH_2CH_2CH_3$ (202.0)
Ethers	$(CH_3)_2O$ (193.1); $(C_2H_5)_2O$ (200.4); $(CH_3CH_2CH_2)_2O$ (202.9); $(CH_3CH_2CH_2CH_2)_2O$ (203.9); tetrahydrofuran (199.6); tetrahydropyran (200.7)
Ketones	CH_3COCH_3 (197.2); $CH_3COC_2H_5$ (199.4)
Nitriles (cyano compounds)	HCN (178.9); CH_3CN (190.9); C_2H_5CN (192.8); $CH_3CH_2CH_2CN$ (193.8)
Sulfides	$(CH_3)_2S$ (200.7); $(C_2H_5)_2S$ (205.6); $[(CH_3)_2CH]_2S$ (209.3)
Thiols	H_2S (176.6); CH_3SH (188.6); C_2H_5SH (192.0); $[(CH_3)_2CH]_2SH$ (194.7)

PROTON AFFINITIES OF SOME ANIONS

The following table lists the proton affinities of common anions (X^-). The reaction of an anion (X^-) with a proton (H^+)

$$X^- + H^+ \rightarrow H\text{–}X$$

is exothermic and can be used to generate other anions with smaller proton affinity by the addition of the corresponding neutral species.[1]

Anion	Proton affinity (kJ/mol)
NH_2^-	1689
H^-	1676
OH^-	1636
$O^{\cdot -}$	1595
CH_3O^-	1583
$(CH_3)_2CHO^-$	1565
$^-CH_2CN$	1556
F^-	1554
$C_5H_5^-$	1480
$O_2^{\cdot -}$	1465
CN^-	1462
Cl^-	1395

REFERENCE

1. **Chapman, J. R.**, *Practical Organic Mass Spectrometry*, John Wiley & Sons, Chichester, U.K., 1985.

IX. Atomic Absorption and Emission Spectrometry

INTRODUCTION
FOR ATOMIC SPECTROMETRIC TABLES

The tables presented in this section are designed to aid in the area of atomic spectrometric methods of analysis. The following conventions for abbreviation are recommended by the International Union of Pure and Applied Chemistry:[1] Atomic Emission Spectrometry (AES); Atomic Absorption Spectrometry (AAS); Flame Atomic Emission Spectrometry (FAES); Flame Atomic Absorption Spectrometry (FAAS); Electrothermal Atomic Absorption Spectrometry (EAAS); and Inductively Coupled Plasma Atomic Emission Spectrometry (ICP-AES). Other variations such as cold vapor and hydride generation are not abbreviated but spelled out, e.g., cold vapor AAS, hydride-generation FAAS, etc. These abbreviations are used whenever appropriate throughout the section.

Several of these tables have appeared in Parsons et al.[2] in one form or another. They have been updated insofar as possible, and the wavelength values have been made to conform to those in the National Standard Reference Data System-National Bureau of Standards (NSRDS-NBS) 68[3] wherever possible.

Because several of the tables cite the same references, all cited references will be listed at the end of this introduction instead of being repeated at the end of each table.

REFERENCES

1. Commission on Spectrochemical and other Optical Procedures for Analysis, Nomenclature, symbols, units and their usage in spectrochemical analysis I. General atomic emission spectroscopy; II. Data interpretation; III. Analytical flame spectroscopy and associated procedures, *Spectrochim. Acta, Part B*, 33, 219, 1978.
2. **Parsons, M. L., Smith, B. W., and Bentley, G. E.,** *Handbook of Flame Spectroscopy,* Plenum Press, New York, 1975.
3. **Reader, J., Corliss, C. H., Wiese, W. L., and Martin, G. A.,** Wavelengths and Transition Probabilities for Atoms and Atomic Ions, NSRDS-NBS 68, U.S. Government Printing Office, Washington, D.C., 1980.
4. **Smith, B. W. and Parsons, M. L.,** Preparation of standard solutions: critically selected compounds, *J. Chem. Ed.,* 50, 679, 1973.
5. **Dean, J. A. and Rains, T. C.,** in *Flame Emission and Atomic Absorption Spectrometry,* Vol. 2, Marcel Dekker, New York, 1971.
6. Guide to Analytical Values for TJA Spectrometers, Thermo Jarrell Ash Corp., Waltham, MA, 1987.
7. **Anderson, T. A. and Parsons, M. L.,** ICP emission spectra III. The spectra of the Group IIIA elements and spectral interferences due to Group IIA and IIIA elements, *Appl. Spectrosc.,* 38, 625, 1984; **Parsons, M. L., Forster, A., and Anderson, D.,** *An Atlas of Spectral Interferences in ICP Spectroscopy,* Plenum Press, New York, 1980.
8. **Park, D. A.,** Further Investigations of Spectra and Spectral Interferences Due to Group A Elements in ICP Spectroscopy: Groups IVA and VA, Ph.D. thesis, Arizona State University, Tempe; **Parsons, M. L.,** unpublished data, Los Alamos National Laboratory, Los Alamos, NM, 1987.
9. Mercury/Hydride System, Rep. No. 1876/6.79, Perkin-Elmer Corporation, Norwalk, CT, 1987.
10. **Lovett, R. J., Welch, D. L., and Parsons, M. L.,** On the importance of spectral interferences in atomic absorption Spectrosc., *Appl. Spectrosc.,* 29, 470, 1975.
11. **Layman, L., Palmer, B., and Parsons, M. L.,** Unpublished data taken with the Los Alamos National Laboratory FTS Facility, Los Alamos, NM, 1987.
12. **Sneddon, J.,** Background correction techniques in atomic spectroscopy, *Spectroscopy,* 2(5), 38, 1987.
13. **Wittenberg, G. K., Haun, D. V., and Parsons, M. L.,** The use of free-energy minimization for calculating beta factors and equilibrium compositions in flame spectroscopy, *Appl. Spectrosc.,* 33, 626, 1979.
14. **Parsons, M. L., Smith, B. W., and McElfresh, P. M.,** On the selection of analysis lines in atomic absorption spectrometry, *Appl. Spectrosc.,* 27, 471, 1973.
15. **Parker, L. R., Jr., Morgan, S. L., and Deming, S. N.,** Simplex optimization of experimental factors in atomic absorption spectrometry, *Appl. Spectrosc.,* 29, 429, 1975.
16. **Parsons, M. L. and Winefordner, J. D.,** Optimization of the critical instrumental parameters for achieving maximum sensitivity and precision in flame-spectrometric methods of analysis, *Appl. Spectrosc.,* 21, 368, 1967.
17. **Wiese, W. L., Smith, M. W., and Glennon, B. M.,** Atomic Transition Probabilities, Vol. I, Hydrogen through Neon, NSRDS-NBS 4, U.S. Government Printing Office, Washington, D.C., 1966.

STANDARD SOLUTIONS — SELECTED COMPOUNDS AND PROCEDURES

The compounds selected for Table 1 were chosen using a rather stringent set of criteria, including stability, purity, ease of preparation, availability, high molecular mass, and toxicity. It is very important to have a compound that is pure, and can be dried, weighed, and dissolved with comparative ease. The list of compounds provided here meets those goals insofar as possible. No attempt was made to include all compounds that met these criteria. Nor are all the compounds in this list trivial to dissolve; some require a rather long time and/or vigorous conditions.

In this table the significant figures in all columns represent the accuracy with which the atomic masses of the elements are known.

This table was compiled from References 4 and 5.

Table 1
STANDARD SOLUTIONS — SELECTED COMPOUNDS AND PROCEDURES

Element	Compound	Relative formula mass	Weight for 1000 µg/L (ppm)-g/L	Solvent	Note
Aluminum	Al-metal	26.982	1.0000	Hot dil. HCl-2M	APS
Antimony	$KSbOC_4H_4O_6$	324.92	2.6687	Water	—[d]
	·1/2 H_2O		(antimony potassium tartarate)		
	Sb-metal	121.75	1.0000	Hot aq.reg.	
Arsenic	As_2O_3	197.84	1.3203	1:1 NH_3	PS, NBS[a]
Barium	$BaCO_3$	197.35	1.4369	Dil. HCl	—[f]
	$BaCl_2$	208.25	1.5163	Water	—[e]
Beryllium	Be-metal	9.0122	1.0000	HCl	—[a]
	$BeSO_4·4H_2O$	177.135	19.6550	Water + acid	—[g]
Bismuth	Bi_2O_3	465.96	1.1148	HNO_3	
	Bi-metal	208.980	1.00000	HNO_3	
Boron	H_3BO_3	61.84	5.720	Water	PS, NBS[k]
Bromine	KBr	119.01	1.4894	Water	APS
Cadmium	CdO	128.40	1.1423	HNO_3	
	Cd-metal	112.40	1.0000	Dil. HCl	
Calcium	$CaCO_3$	100.09	2.4972	Dil. HCl	—[f]
Cerium	$(NH_4)_2Ce(NO_3)_6$	548.23	3.9126	Water	
Cesium	Cs_2SO_4	361.87	1.3614	Water	
Chlorine	NaCl	58.442	1.6485	Water	PS
Chromium	$K_2Cr_2O_7$	294.19	2.8290	Water	PS, NBS
	Cr-metal	51.996	1.0000	HCl	
Cobalt	Co-metal	58.933	1.0000	HNO_3	APS
Copper	Cu-metal	63.546	1.0000	Dil. HNO_3	APS
	CuO	79.545	1.2517	Hot HCl	APS
	$CuSO_4·5H_2O$	249.678	3.92909	Water	
Dysprosium	Dy_2O_3	373.00	1.477	Hot HCl	—[c]
Erbium	Er_2O_3	382.56	1.1435	Hot HCl	—[c]
Europium	Eu_2O_3	351.92	1.1579	Hot HCl	—[c]
Fluorine	NaF	41.988	2.2101	Water	—[h]
Gadolinium	Gd_2O_3	362.50	1.1526	Hot HCl	—[c]
Gallium	Ga-metal	69.72	1.000	Hot HNO_3	—[i]
Germanium	GeO_2	104.60	1.4410	Hot 1 *M* NaOH or 50 g oxalic acid + water	
Gold	Au-metal	196.97	1.0000	Hot aq. reg.	APS, NBS
Hafnium	Hf-metal	178.49	1.0000	HF, Fusion	—[j]
Holmium	Ho_2O_3	377.86	1.1455	Hot HCl	—[c]
Indium	In_2O_3	277.64	1.2090	Hot HCl	
	In-metal	114.82	1.0000	Dil. HCl	

Table 1 (continued)
STANDARD SOLUTIONS — SELECTED COMPOUNDS AND PROCEDURES

Element	Compound	Relative formula mass	Weight for 1000 µg/L (ppm)-g/L	Solvent	Note
Iodine	KIO_3	214.00	1.6863	Water	PS
Iridium	Na_3IrCl_6	473.8	2.466	Water	
Iron	Fe-metal	55.847	1.0000	Hot HCl	APS
Lanthanum	La_2O_3	325.82	1.1728	Hot HCl	—[c]
Lead	$Pb(NO_3)_2$	331.20	1.5985	HCl	APS, NBS
Lithium	Li_2CO_3	73.890	5.3243	Dil. HCl	APS[f]
Lutetium	Lu_2O_3	397.94	1.1372	Hot HCl	—[c]
Magnesium	MgO	40.311	1.6581	HCl	
	Mg-metal	24.312	1.0000	Dil. HCl	
Manganese	$MnSO_4 \cdot H_2O$	169.01	3.0764	Water	—[m]
Mercury	$HgCl_2$	271.50	1.3535	Water	—[a]
	Hg-metal	200.59	1.0000	$5\ M\ HNO_3$	
Molybdenum	MoO_3	143.94	1.5003	$1\ M$ NaOH or $2\ M\ NH_3$	
Neodymiun	Nd_2O_3	336.48	1.1664	HCl	—[c]
Nickel	Ni-metal	58.71	1.000	Hot HNO_3	APS
Niobium	Nb_2O_5	265.81	1.4305	HF, fusion	—[n,o]
	Nb-metal	92.906	1.0000	$HF + H_2SO_4$	—[o]
Osmium	Os-metal	190.20	1.0000	Hot H_2SO_4	—[b]
Palladium	Pd-metal	106.40	1.0000	Hot HNO_3	
Phosphorus	KH_2PO_4	136.09	4.3937	Water	
	$(NH_3)_2HPO_4$	209.997	6.77983	Water	
Platinum	K_2PtCl_4	415.12	2.1278	Water	
	Pt-metal	195.05	1.0000	Hot Aq. Reg.	APS, NBS
Potassium	KCl	74.555	1.9067	Water	PS, NBS
	$KHC_6H_4O_4$ (potassium hydrogen phthalate)	204.22	5.2228	Water	PS, NBS
	$K_2Cr_2O_7$	294.19	3.7618	Water	PS, NBS
Praseodymium	Pr_6O_{11}	1021.43	1.20816	HCl	—[c]
Rhenium	Re-metal	186.2	1.000	HNO_3	
	$KReO_4$	289.3	1.554	Water	
Rhodium	Rh-metal	102.91	1.0000	Hot H_2SO_4	
Rubidium	Rb_2SO_4	267.00	1.5628	Water	
Ruthenium	RuO_4	165.07	1.6332	Water	
Samarium	Sm_2O_3	348.70	2.3193	Hot HCl	—[c]
Scandium	Sc_2O_3	137.91	1.5339	Hot HCl	
Selenium	Se-metal	78.96	1.000	Hot HNO_3	
	SeO_2	110.9	1.405	Water	
Silicon	Si-metal	28.086	1.0000	NaOH, conc	
	SiO_2	60.085	2.1393	HF	
Silver	$AgNO_3$	169.875	1.57481	Water	APS[p]
	Ag-metal	107.870	1.00000	HNO_3	
Sodium	NaCl	58.442	2.5428	Water	PS
	$Na_2C_2O_4$ (sodium oxalate)	134.000	2.91432	Water	PS, NBS
Strontium	$SrCO_3$	147.63	1.6849	Dil. HCl	APS[f]
Sulfur	K_2SO_4	174.27	5.4351	Water	
	$(NH_4)_2SO_4$	114.10	3.5585	Water	
Tantalum	Ta_2O_5	441.893	1.22130	HF, fusion	—[n,o]
	Ta-metal	180.948	1.00000	$HF + H_2SO_4$	—[o]
Tellurium	TeO_2	159.60	1.2507	HCl	
Terbium	Tb_2O_3	365.85	1.1512	Hot HCl	—[c]

Table 1 (continued)
STANDARD SOLUTIONS — SELECTED COMPOUNDS AND PROCEDURES

Element	Compound	Relative formula mass	Weight for 1000 µg/L (ppm)-g/L	Solvent	Note
Thallium	Tl_2CO_3	468.75	1.1468	Water	APS[a]
	$TlNO_3$	266.37	1.3034	Water	
Thorium	$Th(NO_3)_4 \cdot 4H_2O$	552.118	2.37943	HNO_3	
Thulium	Tm_2O_3	385.87	1.1421	Hot HCl	—[c]
Tin	Sn-metal	118.69	1.0000	HCl	
	SnO	134.69	1.1348	HCl	
Titanium	Ti-metal	47.90	1.000	1:1 H_2SO_4	APS
Tungsten	$Na_2WO_4 \cdot 2H_2O$	329.86	1.7942	Water	—[q]
	Na_2WO_4	293.83	1.5982	Water	—[d]
Uranium	UO_2	270.03	1.1344	HNO_3	PS, NBS
	U_3O_5	842.09	1.1792	HNO_3	
	$UO_2(NO_3)_2 \cdot 6H_2O$	502.13	2.1095	Water	
Vanadium	V_2O_5	181.88	1.7852	Hot HCl	
	NH_4VO_3	116.98	2.2963	Dil. HNO_3	
Ytterbium	Yb_2O_3	394.08	1.1386	Hot HCl	—[c]
Yttrium	Y_2O_3	225.81	1.2700	Hot HCl	—[c]
Zinc	ZnO	81.37	1.245	HCl	APS
	Zn-metal	65.37	1.000	HCl	APS, NBS
Zirconium	Zr-metal	91.22	1.000	HF, Fusion	—[n]
	$ZrOCl_2 \cdot 8H_2O$	322.2	3.533	HCl	

Note: PS, primary standard; APS, compounds which approach primary standard quality; NBS, compounds sold as primary standards by the National Bureau of Standards, Office of Standard Reference Materials, Washington, D.C.

[a]   Highly toxic
[b]   Very highly toxic
[c]   The rare earth oxides, because they absorb CO_2 and water vapor from the atmosphere, should be freshly ignited prior to weighing.
[d]   Loses water at 100 °C. Water is only slowly regained, but rapid weighing and dessicator storage are required.
[e]   Drying at 250 °C, rapid weighing, and dessicator storage are required.
[f]   Add a quantity of water, then add dilute acid and swirl until the CO_2 has ceased to bubble out, then dilute.
[g]   Dissolve in water, then add 5 mL of concentrated HCl and dilute.
[h]   Sodium fluoride solutions will etch glass and should be freshly prepared.
[i]   Because the melting point is 29.6 °C, the metal may be warmed and weighed as a liquid.
[j]   Zr and Hf compounds were not investigated in the laboratory of Reference 4.
[k]   Boric acid may be weighed directly from the bottle. It loses 1 H_2O and 100 °C, but it is difficult to dry to a constant mass.
[l]   Several references suggest that the addition of acid will help stabilize the solution.
[m]   This compound may be dried at 100 °C without losing the water of hydration.
[n]   Nb and Ta are slowly soluble in 40% HF. The addition of H_2SO_4, accelerates the dissolution process.
[o]   Dissolve in 20 mL hot HF in a platinum dish, add 40 mL H_2SO_4, and evaporate to fumes; dilute with 8 M H_2SO_4.
[p]   When kept dry, silver nitrate crystals are not affected by light. Solutions should be stored in brown bottles.
[q]   Sodium tungstate loses both water molecules at 100 °C. The water is not rapidly regained, but the compound should be kept in a dessicator after drying and should be weighed quickly once it is removed.

LIMITS OF DETECTION FOR COMMON ANALYTICAL TRANSITIONS IN AES AND AAS

The next five tables (Tables 2 through 7) present the common transitions for analysis and the detection limits for AES and AAS where appropriate for the specific atom cell indicated. The detection limits are from the literature cited and are given in parts per billion (ppb), or nanograms per milliliter (ng/mL) of aqueous solution. The limits of detection (LOD) are generally defined as a signal-to-noise ratio of two or three. This generally relates to a concentration that produces a signal of two or three times the standard deviation of the measurement. These are measured in dilute aqueous solution and represent the best that the system was capable of measuring. In most cases, the detection limit in real samples will be one or two orders of magnitude higher, or worse, than those stated here. The Type designation is I for free atom and II for single ion. In all cases, *NO* means that no observation was made for the situation indicated, and *NA* means that either AES or AAS was observed but no detection limit was reported.

In all cases where it was possible, the wavelengths of the transitions were made to conform with Reference 3; any wavelength below 200 nm is the wavelength given in vacuum; all others are in air.

Table 2
LIMITS OF DETECTION FOR THE AIR-HYDROCARBON FLAME[a]

Element	Symbol	Wavelength (nm)	Type	LOD-AAS (ppb)
Antimony	Sb	217.581	I	100
		231.147	I	100
Bismuth	Bi	223.061	I	50
Calcium	Ca	422.673	I	2
Cesium	Cs	455.5276	I	600
		852.1122	I	50
Chromium	Cr	357.869	I	5
Cobalt	Co	240.725	I	5
Copper	Cu	324.754	I	50
		327.396	I	50
Gallium	Ga	287.424	I	70
Gold	Au	242.795	I	20
Indium	In	303.936	I	50
Iridium	Ir	208.882	I	15,000
		263.971	I	2,000
Iron	Fe	248.3271	I	5
Lead	Pb	283.3053	I	10
Lithium	Li	670.776	I	5
Magnesium	Mg	285.213	I	0.3
Manganese	Mn	279.482	I	2
		403.076	I	2
Mercury	Hg	253.652	I	500
Molybdenum	Mo	313.259	I	30
Nickel	Ni	232.003	I	5
Osmium	Os	290.906	I	17.000
Palladium	Pd	244.791	I	2,000
		247.642	I	30
Platinum	Pt	265.945	I	100
Potassium	K	766.490	I	5
Rhodium	Rh	343.489	I	30
Rubidium	Rb	420.180	I	NA
		780.027	I	5
Ruthenium	Ru	349.894	I	300
		372.803	I	3,000
Selenium	Se	196.09	I	100
		203.98	I	2,000
Silver	Ag	328.068	I	5
		338.289	I	200
Sodium	Na	330.237	I	NA
		588.9950	I	2
		589.5924	I	2
Strontium	Sr	407.771	II	NA
		460.733	I	10
Tellurium	Te	214.281	I	100
Thallium	Tl	276.787	I	30
		377.572	I	2,400
Tin	Sn	224.605	I	30
Zinc	Zn	213.856	I	2

Note: These date were taken from Reference 2.

[a] Flames formed from air combined with the lighter hydrocarbons such as methane, propane, butane, or natural gas, behave in a very similar fashion with similar temperatures, similar chemical, properties, etc.

Table 3
LIMITS OF DETECTION FOR THE AIR-ACETYLENE FLAME

Element	Symbol	Wavelength (nm)	Type	LOD-AES (ppb)	LOD-AAS (ppb)
Aluminum	Al	308.2153	I	NO	700
		309.2710	I	NO	500
		396.1520	I	NA	600
Antimony	Sb	206.833	I	NA	50
		217.581	I	NA	40
		231.147	I	3,000	40
		259.805	I	NA	NO
Arsenic	As	193.759	I	10,000	140
Barium	Ba	455.403	II	NA	NO
		553.548	I	NA	NO
Bismuth	Bi	223.061	I	3,000	25
Boron	B	249.677	I	NA	NO
Cadmium	Cd	228.8022	I	500	1
		326.1055	I	NA	NA
Calcium	Ca	393.366	II	NO	5,000
		396.847	II	NO	5,000
		422.673	I	0.5	0.5
Cesium	Cs	455.5276	I	NA	NO
		852.1122	I	NA	8
Chromium	Cr	357.869	I	NA	3
		425.435	I	NA	200
Cobalt	Co	240.725	I	NO	4
		352.685	I	NA	125
Copper	Cu	324.754	I	NA	1
		327.396	I	NA	120
Gallium	Ga	287.424	I	NO	50
		294.364	I	NA	50
		417.204	I	NA	1,500
Germanium	Ge	265.1172	I	7,000	
Gold	Au	242.795	I	NA	6
		267.595	I	NA	90
Indium	In	303.936	I	NA	30
		325.609	I	NA	20
		451.131	I	NA	200
Iodine	I	183.038	I	NO	8,000
		206.163	I	2,500,000	NO
Iridium	Ir	208.882	I	NO	600
		263.971	I	NO	2,500
Iron	Fe	248.3271	I	NO	5
		371.9935	I	NA	700
Lead	Pb	217.000	I	NO	9
		283.3053	I	NA	240
		368.3462	I	NA	NO
Lithium	Li	670.776	I	NA	0.3
Lutetium	Lu	451.857	I	NO	NA
Magnesium	Mg	279.553	II	NO	NA
		280.270	II	NO	NA
		285.213	I	NA	0.1
Manganese	Mn	279.482	I	NA	2
		403.076	I	NA	600
Mercury	Hg	253.652	I	NA	140
Molybdenum	Mo	313.259	I	NO	20
		379.825	I	80,000	900
		390.296	I	100	1,600
Nickel	Ni	232.003	I	NO	2
		352.454	I	NA	350

Table 3 (continued)
LIMITS OF DETECTION FOR THE AIR-ACETYLENE FLAME

Element	Symbol	Wavelength (nm)	Type	LOD-AES (ppb)	LOD-AAS (ppb)
Niobium	Nb	309.418	II	NO	NA
Osmium	Os	290.906	I	NA	1,200
Palladium	Pd	244.791	I	NO	20
		247.642	I	NO	20
		340.458	I	NA	660
		363.470	I	NA	300
Phosphorus	P	213.547	I	NO	30,000
Platinum	Pt	214.423	I	NO	350
		265.945	I	NA	50
Potassium	K	766.490	I	NA	1
Rhenium	Re	346.046	I	NO	800
Rhodium	Rh	343.489	I	NA	2
		369.236	I	NA	70
Rubidium	Rb	420.180	I	NA	NO
		780.027	I	NA	0.3
Ruthenium	Ru	349.894	I	NA	400
		372.803	I	NA	250
Selenium	Se	196.09	I	NA	50
		203.98	I	50,000	10,000
Silver	Ag	328.068	I	NA	1
		338.289	I	NA	70
Sodium	Na	330.237	I	NO	NA
		588.9950	I	NA	1
		589.5924	I	NA	0.2
Strontium	Sr	407.771	II	NA	400
		421.552	II	NO	NA
		460.733	I	NA	2
Sulfur	S	180.7311	I	NO	30,000
Tellurium	Te	214.281	I	500	30
		238.578	I	NO	NA
Thallium	Tl	276.787	I	NA	30
		377.572	I	NA	1,200
		535.046	I	NA	12,000
Tin	Sn	224.605	I	NO	10
		235.484	I	2,000	600
		283.999	I	NA	1,000
		326.234	I	NA	NO
Tungsten	W	255.135	I		3,000
		400.875	I	90,000	
Uranium	U	591.539	I	NA	NO
Vanadium	V	318.540	I	NA	NO
		437.924	I	300	NO
Ytterbium	Yb	398.799	I	NO	80
Zinc	Zn	213.856	I	7,000	1
Zirconium	Zr	351.960	I	NO	NA

Note: These data were taken from References 2 and 6.

Table 4
LIMITS OF DETECTION FOR THE NITROUS OXIDE-ACETYLENE FLAME

Element	Symbol	Wavelength (nm)	Type	LOD-AES (ppb)	LOD-AAS (ppb)
Aluminum	Al	308.2153	I	NA	NO
		309.2710	I	NA	20
		396.1520	I	3	900
Barium	Ba	553.548	I	1	8
Beryllium	Be	234.861	I	100	1
Boron	B	208.891	I	NO	NA
		208.957	I	NO	24,000
		249.677	I	NO	700
		249.773	I	NO	1,500
Cadmium	Cd	326.1055	I	800	NO
Calcium	Ca	422.673	I	0.1	1
Cesium	Cs	455.5276	I	600	NO
		852.1122	I	0.02	NO
Chromium	Cr	425.435	I	1	NO
Cobalt	Co	352.685	I	200	NO
Copper	Cu	324.754	I	30	NO
		327.396	I	3	NO
Dysprosium	Dy	353.170	II	NO	800
		404.597	I	20	500
		421.172	I	NO	50
Erbium	Er	337.271	II	NO	100
		400.796	I	20	40
Europium	Eu	459.403	I	0.2	30
Gadolinium	Gd	368.413	I	NO	2,000
		440.186	I	1,000	NO
Gallium	Ga	417.204	I	5	NO
Germanium	Ge	265.1172	I	400	50
Gold	Au	267.595	I	500	NO
Hafnium	Hf	307.288	I	NO	2,000
Holmium	Ho	345.600	II	NO	3,000
		405.393	I	10	400
		410.384	I	NO	40
Indium	In	303.936	I	NO	1,000
		325.609	I	NO	700
		451.131	I	1	3,500
Iridium	Ir	208.882	I	NO	500
Iron	Fe	371.9935	I	10	NO
Lanthanum	La	408.672	II	NO	7,500
		550.134	I	4,000	2,000
Lead	Pb	368.3462	I	0.2	NO
Lithium	Li	670.776	I	0.001	NO
Lutetium	Lu	261.542	II	NO	3,000
		451.857	I	400	NO
Magnesium	Mg	285.213	I	1	NO
Manganese	Mn	403.076	I	1	NO
Mercury	Hg	253.652	I	10,000	NO
Molybdenum	Mo	313.259	I	10	25
		379.825	I	300	NO
		390.296	I	10	NO
Neodymium	Nd	463.424	I		600
		492.435	I	200	700
Nickel	Ni	352.454	I	20	NO
Niobium	Nb	334.906	I	NO	1,000
		405.894	I	60	5,000
Osmium	Os	290.906	I	NO	80

Table 4 (continued)
LIMITS OF DETECTION FOR THE NITROUS OXIDE-ACETYLENE FLAME

Element	Symbol	Wavelength (nm)	Type	LOD-AES (ppb)	LOD-AAS (ppb)
Palladium	Pd	363.470	I	40	NO
Phosphorus	P	177.499	I	NO	30,000
		213.547	I	NO	29,000
Platinum	Pt	265.945	I	2,000	2,000
Potassium	K	766.490	I	0.01	NO
Praseodymium	Pr	495.137	I	500	2,000
Rhenium	Re	346.046	I	200	200
Rhodium	Rh	343.489	I	NO	700
		369.236	I	10	1,400
Rubidium	Rb	780.027	I	8	NO
Ruthenium	Ru	372.803	I	300	NO
Samarium	Sm	429.674	I	NO	500
		476.027	I	50	14,000
Scandium	Sc	391.181	I	10	20
Selenium	Se	196.09	I	100,000	NO
Silicon	Si	251.6113	I	3,000	20
		288.1579	I	NO	NA
Silver	Ag	328.068	I	2	NO
Sodium	Na	588.9950	I	0.01	NO
		589.5924	I	0.01	NO
Strontium	Sr	460.733	I	0.1	50
Tantalum	Ta	271.467	I	NO	800
		474.016	I	4,000	NO
Terbium	Tb	432.643	I	NA	600
Thallium	Tl	377.572	I	50	NO
		535.046	I	2	NO
Thorium	Th	324.4448	I	NO	181,000
		491.9816	II	10,000	NO
Thulium	Tm	371.791	I	4	10
Tin	Sn	224.605	I	NO	3,000
		235.484	I	NO	90
		283.999	I	100	NO
Titanium	Ti	334.941	II	NO	NA
		364.268	I	NA	10
		365.350	I	30	500
Tungsten	W	255.135	I	NO	500
		400.875	I	200	7,500
Uranium	U	358.488	I	NO	7,000
Vanadium	V	318.540	I	200	20
		437.924	I	7	100
Ytterbium	Yb	398.799	I	0.2	5
Yttrium	Y	410.238	I	NO	50
Zinc	Zn	213.856	I	10,000	NO
Zirconium	Zr	351.960	I	1,2000	NO
		360.119	I	3,000	1,000

Note: These data were taken from References 2 and 6.

Table 5
LIMITS OF DETECTION FOR GRAPHITE-FURNACE AAS[a]

Element	Symbol	Wavelength (nm)	Type	LOD (ppb)
Aluminum	Al	308.2153	I	NA
		309.2710	I	0.01
		396.1520	I	600
Antimony	Sb	206.833	I	NA
		217.581	I	0.08
		231.147	I	NA
Arsenic	As	189.042	I	NA
		193.759	I	0.12
Barium	Ba	553.548	I	0.04
Beryllium	Be	234.861	I	0.003
Bismuth	Bi	223.061	I	0.01
Cadmium	Cd	228.8022	I	0.0002
Calcium	Ca	422.673	I	0.01
Chromium	Cr	357.869	I	0.004
Cobalt	Co	240.725	I	8
Copper	Cu	324.754	I	0.005
		327.396	I	NA
Erbium	Er	400.796	I	0.3
Gadolinium	Gd	440.186	I	0.3
Gallium	Ga	287.424	I	0.01
Germanium	Ge	265.1172	I	0.1
Gold	Au	242.795	I	0.01
Holmium	Ho	345.600	II	NA
		405.393	I	NA
Indium	In	303.936	I	0.02
Iodine	I	183.038	I	40,000
Iridium	Ir	208.882	I	0.5
Iron	Fe	248.3271	I	0.01
		371.9935	I	NA
Lanthanum	La	550.134	I	0.5
Lead	Pb	217.000	I	0.007
		283.3053	I	NA
Lithium	Li	670.776	I	0.01
Magnesium	Mg	285.213	I	0.0002
Manganese	Mn	279.482	I	0.0005
		403.076	I	NA
Mercury	Hg	253.652	I	0.2
Molybdenum	Mo	313.259	I	0.03
Nickel	Ni	232.003	I	0.05
Osmium	Os	290.906	I	2
Palladium	Pd	247.642	I	0.05
Phosphorus	P	177.499	I	NA
		213.547	I	20
		253.561	I	NA
Platinum	Pt	265.945	I	0.2
Potassium	K	766.490	I	0.004
Rhenium	Re	346.046	I	10
Rhodium	Rh	343.489	I	0.1
Rubidium	Rb	780.027	I	NA
Selenium	Se	196.09	I	0.05
Silicon	Si	251.6113	I	0.6
Silver	Ag	328.068	I	0.001
Sodium	Na	588.9950	I	0.004
Strontium	Sr	460.733	I	0.01

Table 5 (continued)
LIMITS OF DETECTION FOR GRAPHITE-FURNACE AAS[a]

Element	Symbol	Wavelength (nm)	Type	LOD (ppb)
Sulfur	S	180.7311	I	NA
		182.0343	I	NA
		216.89		NA
Tellurium	Te	214.281	I	0.03
Thallium	Tl	276.787	I	0.01
Tin	Sn	235.484	I	0.03
		283.999	I	NA
Titanium	Ti	364.268	I	0.3
		365.350	I	NA
Uranium	U	358.488	I	30
Vanadium	V	318.540	I	0.4
Ytterbium	Yb	398.799	I	0.01
Yttrium	Y	410.238	I	10
Zinc	Zn	213.856	I	0.001

[a]   The detection limits for the graphite-furnace AAS are calculated using 100μL of sample. In graphite-furnace AAS, additional chemicals are often added to aid in determining certain elements. Walter Slavin has published an excellent guide to these and has provided an excellent bibliography: Slavin, W., Graphite Furnace Source Book, Perkin-Elmer Corporation, Ridgefield, CT, 1984; Slavin, W. and Manning, D. C., Furnace interferences, a guide to the literature, *Prog. Anal. At. Spectrosc.*, 5, 243, 1982.

Table 6
LIMITS OF DETECTION FOR ICP-AES

Element	Symbol	Wavelength (nm)	Type	LOD (ppb)	Ref.
Aluminum	Al	167.0787	II	1	6
		308.2153	I	0.4	7
		309.2710	I	0.02	8
		396.1520	I	0.2	7
Antimony	Sb	206.833	I	10	7
		217.581	I	15	7
		231.147	I	61	7
		259.805	I	107	7
Arsenic	As	189.042	I	136	8
		193.759	I	2	7
		197.262	I	76	7
		234.984	I	90	7
Barium	Ba	455.403	II	0.001	8
		493.409	II	0.3	7
		553.548	I	2	7
Beryllium	Be	234.861	I	0.003	7
		313.042	II	0.1	6
		313.107	II	0.01	8
Bismuth	Bi	223.061	I	0.03	8
		289.798	I	10	7
Boron	B	208.891	I	5	8
		208.957	I	3	8
		249.677	I	0.1	8
		249.773	I	2	8
Bromine	Br	470.486	II	NA	8
		827.244	I	NA	8

Table 6 (continued)
LIMITS OF DETECTION FOR ICP-AES

Element	Symbol	Wavelength (nm)	Type	LOD (ppb)	Ref.
Cadmium	Cd	214.441	II	0.1	8
		226.502	II	0.05	8
		228.8022	I	0.08	8
		326.1055	I	3	8
Calcium	Ca	364.441	I	0.5	8
		393.366	II	0.0001	8
		396.847	II	0.002	8
		422.673	I	0.2	8
Carbon	C	193.0905	I	40	6
		247.856	I	100	8
Cerium	Ce	394.275	II	2	8
		413.765	II	40	6
		418.660	II	0.4	7
Chlorine	Cl	413.250	II	NA	7
		837.594	I	NA	8
Chromium	Cr	205.552	II	0.009	8
		267.716	II	0.08	8
		357.869	I	0.1	8
		425.435	I	5	8
Cobalt	Co	228.615	II	0.3	8
		238.892	II	0.1	7
Copper	Cu	213.5981	II	7	8
		324.754	I	0.01	8
		327.396	I	0.06	8
Dysprosium	Dy	353.170	II	1	8
Erbium	Er	337.271	II	1	8
		400.796	I	1	7
Europium	Eu	381.967	II	0.06	7
Fluorine	F	685.603	I	NA	8
Gadolinium	Gd	342.247	II	0.4	7
Gallium	Ga	287.424	I	78	7
		294.364	I	3	8
		417.204	I	0.6	8
Germanium	Ge	199.8887	I	0.6	8
		209.4258	I	11	8
		265.1172	I	4	7
Gold	Au	242.795	I	2	8
		267.595	I	0.9	7
Hafnium	Hf	277.336	II	2	8
		339.980	II	5	6
Holmium	Ho	345.600	II	1	6
		389.102	II	0.9	8
Hydrogen	H	486.133	I	NA	8
		656.2852	I	NA	7
Indium	In	230.605	II	30	8
		303.936	I	15	8
		325.609	I	15	6
		451.131	I	30	7
Iodine	I	183.038	I	NA	7
		206.163	I	10	8
Iridium	Ir	224.268	II	0.6	8
		263.971	I	0.6	8
Iron	Fe	238.204	II	0.004	8
		259.9396	II	0.09	7
		371.9935	I	0.3	7

Table 6 (continued)
LIMITS OF DETECTION FOR ICP-AES

Element	Symbol	Wavelength (nm)	Type	LOD (ppb)	Ref.
Lanthanum	La	333.749	II	2	6
		408.672	II	0.1	8
Lead	Pb	217.000	I	30	8
		220.3534	II	0.6	8
		283.3053	I	2	7
		368.3462	I	20	8
Lithium	Li	670.776	I	0.02	7
Lutetium	Lu	261.542	II	0.1	7
		261.542	I	8	7
Magnesium	Mg	279.553	II	0.003	7
		280.270	II	0.01	7
		285.213	I	0.2	7
Manganese	Mn	257.610	II	0.01	7
		403.076	I	0.6	7
Mercury	Hg	184.905	II	1	7
		194.227	II	10	6
		253.652	I	1	7
Molybdenum	Mo	202.030	II	0.3	8
		313.259	I	NA	8
		379.825	I	0.2	7
		390.296	I	80	8
Neodymium	Nd	401.225	II	0.3	7
Nickel	Ni	221.648	II	2	8
		232.003	I	6	8
		352.454	I	0.2	7
Niobium	Nb	309.418	II	0.2	7
Nitrogen	N	174.2729	I	1,000	8
		821.634	I	27,000	8
Osmium	Os	225.585	II	4	8
		290.906	I	6	8
Oxygen	O	436.825	I	NA	8
		777.194	I	NA	8
Palladium	Pd	340.458	I	2	8
		363.470	I	1	8
Phosphorus	P	177.499	I	NA	8
		213.547	I	16	6
		253.561	I	15	7
Platinum	Pt	214.423	I	16	6
		265.945	I	0.9	7
Potassium	K	766.490	I	5	8
Praseodymium	Pr	390.805	II	0.3	8
		422.535	II	10	7
Rhenium	Re	197.3	?	6	7
		221.426	II	4	6
Rhodium	Rh	233.477	II	30	7
		343.489	I	8	6
		369.236	I	7	8
Rubidium	Rb	420.180	I	38,000	8
		780.027	I	100	6
Ruthenium	Ru	240.272	II	8	6
		349.894	I	NA	8
		372.803	I	60	7
Samarium	Sm	359.260	II	0.5	8
		373.912	II	2	7
Scandium	Sc	361.384	II	0.1	8

Table 6 (continued)
LIMITS OF DETECTION FOR ICP-AES

Element	Symbol	Wavelength (nm)	Type	LOD (ppb)	Ref.
Selenium	Se	196.09	I	0.1	8
		203.98	I	0.03	8
Silicon	Si	251.6113	I	2	7
		288.1579	I	10	7
Silver	Ag	328.068	I	0.8	8
		338.289	I	7	8
Sodium	Na	330.237	I	100	8
		588.9950	I	0.1	7
		589.5924	I	0.5	8
Strontium	Sr	407.771	II	0.2	6
		421.552	II	0.1	8
		460.733	I	0.4	8
Sulfur	S	180.7311	I	15	6
		182.0343	I	30	7
		216.89		NA	7
Tantalum	Ta	226.230	II	15	8
		240.063	II	13	6
		296.513	II	5	7
Tellurium	Te	214.281	I	0.7	8
		238.578	I	2	8
Terbium	Tb	350.917	II	0.1	7
		367.635	II	1.5	8
Thallium	Tl	190.864	II	4	8
		276.787	I	27	6
		377.572	I	17	8
Thorium	Th	283.7295	II	8	6
		401.9129	II	1.3	8
Thulium	Tm	313.126	II	0.9	6
		346.220	II	0.2	7
Tin	Sn	189.991	II	0.05	8
		235.484	I	9	8
		283.999	I	10	8
		326.234	I	0.5	8
Titanium	Ti	334.941	II	0.1	8
		365.350	I	230	8
		368.520	II	0.2	8
Tungsten	W	207.911	II	7	8
		276.427	II	0.8	7
		400.875	I	3	7
Uranium	U	263.553	II	70	6
		385.957	II	2	7
Vanadium	V	309.311	II	0.06	7
		311.062	II	0.06	7
		437.924	I	0.2	7
Ytterbium	Yb	328.937	II	0.01	8
		369.419	II	0.02	7
Yttrium	Y	371.030	II	0.04	7
		377.433	II	0.1	8
Zinc	Zn	202.548	II	0.6	8
		213.856	I	0.07	8
Zirconium	Zr	343.823	II	0.06	7

Note: These data were taken from References 7 and 8.

ALL WAVELENGTHS FOR ALL AES/AAS METHODS

This table presents all of the transitions listed in the previous five tables and indicates which transition has been used for which technique. The table is organized in order of increasing wavelength. *Y* means that a signal is observed for the particular technique and an *N* means it is not. An asterisk beside the Y in any of the technique columns indicates that the transition is the most sensitive for the indicated element by that technique.

Table 7
ALL WAVELENGTHS FOR ALL AES/AAS METHODS

Wavelength (nm)	Element	Symbol	AES	AAS	EAAS	ICP-AES
167.0787	Aluminum	Al	N	N	N	Y
174.2729	Nitrogen	N	N	N	N	Y*
177.499	Phosphorus	P	N	Y	Y	Y
180.7311	Sulfur	S	N	Y*	Y	Y*
182.0343	Sulfur	S	N	N	Y	Y
183.038	Iodine	I	N	Y*	Y*	Y
184.905	Mercury	Hg	N	N	N	Y*
189.042	Arsenic	As	N	Y	Y	Y
189.991	Tin	Sn	N	N	N	Y*
190.864	Thallium	Tl	N	N	N	Y*
193.0905	Carbon	C	N	N	N	Y*
193.759	Arsenic	As	N	Y*	Y*	Y*
194.227	Mercury	Hg	N	N	N	Y
196.09	Selenium	Se	Y	Y*	Y*	Y
197.262	Arsenic	As	N	Y	N	Y
197.3	Rhenium	Re	N	N	N	Y
199.8887	Germanium	Ge	N	N	N	Y*
202.030	Molybdenum	Mo	N	N	N	Y
202.548	Zinc	Zn	N	N	N	Y
203.98	Selenium	Se	Y*	Y	N	Y*
205.552	Chromium	Cr	N	N	N	Y*
206.163	Iodine	I	Y*	N	N	Y*
206.833	Antimony	Sb	Y	Y	Y	Y*
207.911	Tungsten	W	N	N	N	Y
208.882	Iridium	Ir	N	Y*	Y*	N
208.891	Boron	B	N	Y	N	Y
208.957	Boron	B	N	Y	N	Y
209.4258	Germanium	Ge	N	Y	N	Y
213.547	Phosphorus	P	N	Y	Y*	Y
213.5981	Copper	Cu	N	N	N	Y
213.856	Zinc	Zn	Y*	Y*	Y*	Y*
214.281	Tellurium	Te	Y*	Y*	Y*	Y*
214.423	Platinum	Pt	N	N	N	Y
214.441	Cadmium	Cd	N	N	N	Y
216.89	Sulfur	S	N	N	Y	Y
217.000	Lead	Pb	N	Y*	Y*	Y
217.581	Antimony	Sb	Y	Y*	Y*	Y

Table 7 (continued)
ALL WAVELENGTHS FOR ALL AES/AAS METHODS

Wavelength (nm)	Element	Symbol	AES	AAS	EAAS	ICP-AES
220.3534	Lead	Pb	N	N	N	Y*
221.426	Rhenium	Re	N	N	N	Y*
221.648	Nickel	Ni	N	N	N	Y
223.061	Bismuth	Bi	Y*	Y*	Y*	Y*
224.268	Iridium	Ir	N	N	N	Y*
224.605	Tin	Sn	N	Y*	N	N
225.585	Osmium	Os	N	N	N	Y*
226.230	Tantalum	Ta	N	N	N	Y
226.502	Cadmium	Cd	N	N	N	Y*
228.615	Cobalt	Co	N	N	N	Y
228.8022	Cadmium	Cd	Y*	Y*	Y*	Y
230.605	Indium	In	N	N	N	Y
231.147	Antimony	Sb	Y*	Y	Y	Y
232.003	Nickel	Ni	N	Y*	Y*	Y
233.477	Rhodium	Rh	N	N	N	Y
234.861	Beryllium	Be	Y*	Y*	Y*	Y*
234.984	Arsenic	As	Y	N	N	Y
235.484	Tin	Sn	Y	Y	Y*	Y
238.204	Iron	Fe	N	N	N	Y*
238.578	Tellurium	Te	Y	Y	N	Y
238.892	Cobalt	Co	N	N	N	Y*
240.063	Tantalum	Ta	N	N	N	Y
240.272	Ruthenium	Ru	N	N	N	Y*
240.725	Cobalt	Co	Y	Y	Y*	N
242.795	Gold	Au	Y	Y*	Y*	Y
244.791	Palladium	Pd	Y	Y*	N	N
247.642	Palladium	Pd	N	Y*	Y*	N
247.856	Carbon	C	N	N	N	Y
248.3271	Iron	Fe	N	Y*	Y*	N
249.677	Boron	B	N	Y	N	Y
249.773	Boron	B	N	Y	N	Y
251.6113	Silicon	Si	Y*	Y*	Y*	Y*
253.561	Phosphorus	P	Y	Y*	Y	Y*
253.652	Mercury	Hg	Y*	Y*	Y*	Y
255.135	Tungsten	W	N	Y*	N	N
257.610	Manganese	Mn	N	N	N	Y*
259.805	Antimony	Sb	Y	N	N	Y
259.9396	Iron	Fe	N	N	N	Y
261.542	Lutetium	Lu	N	Y	N	Y
263.553	Uranium	U	N	N	N	Y
263.971	Iridium	Ir	N	Y	N	Y*
265.1172	Germanium	Ge	Y*	Y*	Y*	Y
265.945	Platinum	Pt	Y*	Y*	Y*	Y*
267.595	Gold	Au	Y*	Y	N	Y*
267.716	Chromium	Cr	N	N	N	Y

Table 7 (continued)
ALL WAVELENGTHS FOR ALL AES/AAS METHODS

Wavelength (nm)	Element	Symbol	AES	AAS	EAAS	ICP-AES
271.467	Tantalum	Ta	N	Y*	N	N
276.427	Tungsten	W	N	N	N	Y*
276.787	Thallium	Tl	Y	Y*	Y*	Y
277.336	Hafnium	Hf	N	N	N	Y*
279.482	Manganese	Mn	Y	Y*	Y*	N
279.553	Magnesium	Mg	N	Y	N	Y*
280.270	Magnesium	Mg	N	Y	N	Y
283.3053	Lead	Pb	Y	Y	Y	Y
283.7295	Thorium	Th	N	N	N	Y
283.999	Tin	Sn	Y*	Y	Y	Y
285.213	Magnesium	Mg	Y*	Y*	Y*	Y
286.044	Arsenic	As	Y	N	N	N
287.424	Gallium	Ga	N	Y*	Y*	Y
288.1579	Silicon	Si	N	N	N	Y
289.798	Bismuth	Bi	N	N	N	Y
290.906	Osmium	Os	N	Y*	Y*	Y
294.364	Gallium	Ga	N	Y	N	Y
296.513	Tantalum	Ta	N	N	N	Y*
303.936	Indium	In	N	Y	Y*	Y*
307.288	Hafnium	Hf	N	Y*	N	N
308.2153	Aluminum	Al	Y	Y	Y	Y
309.2710	Aluminum	Al	Y	Y*	Y*	Y*
309.311	Vanadium	V	N	N	N	Y*
309.418	Niobium	Nb	N	Y	N	Y*
311.062	Vanadium	V	N	N	N	Y*
313.042	Beryllium	Be	N	N	N	Y
313.107	Beryllium	Be	N	N	N	Y
313.126	Thulium	Tm	N	N	N	Y
313.259	Molybdenum	Mo	Y*	Y*	Y*	Y
318.540	Vanadium	V	Y	Y*	Y*	N
324.4448	Thorium	Th	N	Y*	N	N
324.754	Copper	Cu	Y	Y*	Y*	Y*
325.609	Indium	In	Y	Y*	N	Y*
326.1055	Cadmium	Cd	Y	N	N	Y
326.234	Tin	Sn	Y	N	N	Y
327.396	Copper	Cu	Y*	Y	Y	Y
328.068	Silver	Ag	Y*	Y*	Y*	Y*
328.937	Ytterbium	Yb	N	N	N	Y*
330.237	Sodium	Na	Y	Y	N	Y
333.749	Lanthanum	La	N	N	N	Y
334.906	Niobium	Nb	N	Y	N	N
334.941	Titanium	Ti	N	N	N	Y*
337.271	Erbium	Er	N	N	N	Y*
338.289	Silver	Ag	Y	Y	N	Y

Table 7 (continued)
ALL WAVELENGTHS FOR ALL AES/AAS METHODS

Wavelength (nm)	Element	Symbol	AES	AAS	EAAS	ICP-AES
339.980	Hafnium	Hf	N	N	N	Y
340.458	Palladium	Pd	Y	Y	N	Y
342.247	Gadolinium	Gd	N	N	N	Y*
343.489	Rhodium	Rh	Y	Y*	Y*	Y
343.823	Zirconium	Zr	N	N	N	Y*
345.600	Holmium	Ho	N	N	Y	Y
346.046	Rhenium	Re	Y*	Y*	Y*	N
346.220	Thulium	Tm	N	N	N	Y*
349.894	Ruthenium	Ru	Y	Y	N	Y
350.917	Terbium	Tb	N	N	N	Y*
351.960	Zirconium	Zr	Y*	Y	N	N
352.454	Nickel	Ni	Y*	Y	N	Y*
352.685	Cobalt	Co	Y*	Y*	N	N
353.170	Dysprosium	Dy	Y	Y	N	Y*
357.869	Chromium	Cr	Y	Y	Y*	Y
358.488	Uranium	U	N	Y*	Y*	N
359.260	Samarium	Sm	N	N	N	Y*
360.119	Zirconium	Zr	Y	Y*	N	N
361.384	Scandium	Sc	N	N	N	Y*
362.094	Yttrium	Y	Y	N	N	N
363.470	Palladium	Pd	Y*	Y	N	Y*
364.268	Titanium	Ti	Y	Y*	Y*	N
364.441	Calcium	Ca	N	N	N	Y
365.350	Titanium	Ti	Y*	Y	Y	Y
367.635	Terbium	Tb	N	N	N	Y
368.3462	Lead	Pb	Y*	N	N	Y
368.413	Gadolinium	Gd	Y	Y*	N	N
368.520	Titanium	Ti	N	N	N	Y
369.236	Rhodium	Rh	Y*	Y	N	Y*
369.419	Ytterbium	Yb	Y	N	N	Y
371.030	Yttrium	Y	Y	N	N	Y*
371.791	Thulium	Tm	Y*	Y*	N	N
371.9935	Iron	Fe	Y*	Y	Y	Y
372.803	Ruthenium	Ru	Y*	Y*	N	Y
373.912	Samarium	Sm	N	N	N	Y
377.433	Yttrium	Y	Y	N	N	Y
377.572	Thallium	Tl	Y	Y	N	Y
379.825	Molybdenum	Mo	Y	Y	N	Y*
381.967	Europium	Eu	Y	N	N	Y*
385.957	Uranium	U	N	N	N	Y*
389.102	Holmium	Ho	N	N	N	Y*
390.296	Molybdenum	Mo	Y*	Y	N	Y
390.805	Praseodymium	Pr	N	N	N	Y*
391.181	Scandium	Sc	Y*	Y*	N	N
393.366	Calcium	Ca	N	Y	N	Y*

Table 7 (continued)
ALL WAVELENGTHS FOR ALL AES/AAS METHODS

Wavelength (nm)	Element	Symbol	AES	AAS	EAAS	ICP-AES
394.275	Cerium	Ce	N	N	N	Y
396.1520	Aluminum	Al	Y*	Y	Y	Y
396.847	Calcium	Ca	N	Y	N	Y
398.799	Ytterbium	Yb	Y*	Y*	Y*	N
400.796	Erbium	Er	Y*	Y*	Y*	Y*
400.875	Tungsten	W	Y*	Y	N	Y
401.225	Neodymium	Nd	N	N	N	Y*
401.9129	Thorium	Th	N	N	N	Y*
403.076	Manganese	Mn	Y*	Y	Y	Y
404.597	Dysprosium	Dy	Y*	Y	N	N
405.393	Holmium	Ho	Y*	Y	Y	N
405.894	Niobium	Nb	Y	Y	N	N
407.771	Strontium	Sr	Y	Y	N	Y
408.672	Lanthanum	La	N	Y	N	Y*
410.238	Yttrium	Y	Y	Y*	Y*	N
410.384	Holmium	Ho	Y	Y*	N	N
413.250	Chlorine	Cl	N	N	N	Y
413.765	Cerium	Ce	N	N	N	Y
417.204	Gallium	Ga	Y*	Y	N	Y*
418.660	Cerium	Ce	N	N	N	Y*
420.180	Rubidium	Rb	Y	Y	N	Y
421.172	Dysprosium	Dy	Y	Y*	N	N
421.552	Strontium	Sr	Y	Y	N	Y*
422.535	Praseodymium	Pr	N	N	N	Y
422.673	Calcium	Ca	Y*	Y*	Y*	Y
425.435	Chromium	Cr	Y*	Y*	N	Y
429.674	Samarium	Sm	Y	Y*	N	N
432.643	Terbium	Tb	Y	Y*	N	N
436.825	Oxygen	O	N	N	N	Y
437.924	Vanadium	V	Y*	Y	N	Y
440.186	Gadolinium	Gd	Y	Y*	Y*	N
442.047	Osmium	Os	Y*	Y	N	N
451.131	Indium	In	Y*	Y	N	Y
451.857	Lutetium	Lu	Y	Y	N	Y
455.403	Barium	Ba	Y	N	N	Y*
455.5276	Cesium	Cs	Y	Y	N	N
459.403	Europium	Eu	Y*	Y*	N	N
460.733	Strontium	Sr	Y*	Y*	Y*	Y
463.424	Neodymium	Nd	Y	Y	N	N
470.486	Bromine	Br	N	N	N	Y
474.016	Tantalum	Ta	Y*	N	N	N
476.027	Samarium	Sm	Y*	Y	N	N
486.133	Hydrogen	H	N	N	N	Y
491.9816	Thorium	Th	Y*	N	N	N
492.453	Neodymium	Nd	Y	Y	N	N

Table 7 (continued)
ALL WAVELENGTHS FOR ALL AES/AAS METHODS

Wavelength (nm)	Element	Symbol	AES	AAS	EAAS	ICP-AES
493.409	Barium	Ba	N	N	N	Y
495.137	Praseodymium	Pr	Y	Y	N	N
535.046	Thallium	Tl	Y*	N	N	N
550.134	Lanthanum	La	Y*	Y*	Y*	N
553.548	Barium	Ba	Y*	Y*	Y*	Y
588.9950	Sodium	Na	Y	Y	Y*	Y
589.5924	Sodium	Na	Y*	Y*	N	Y*
591.539	Uranium	U	Y*	N	N	N
656.2852	Hydrogen	H	N	N	N	Y
670.776	Lithium	Li	Y*	Y*	Y*	Y*
685.603	Fluorine	F	N	N	N	Y
766.490	Potassium	K	Y*	Y*	Y*	Y*
777.194	Oxygen	O	N	N	N	Y
780.027	Rubidium	Rb	Y*	Y*	Y*	Y*
821.634	Nitrogen	N	N	N	N	Y
827.244	Bromine	Br	N	N	N	Y
837.594	Chlorine	Cl	N	N	N	Y
852.1122	Cesium	Cs	Y*	Y*	N	N

DETECTION LIMITS BY HYDRIDE GENERATION AND COLD-VAPOR AAS

In addition to the AAS methods in flames or graphite furnaces, the elements listed below (Table 8) are detected and determined at extreme sensitivity by introduction into a flame or a hot quartz cell by AAS.

Table 8
DETECTION LIMITS BY HYDRIDE GENERATION AND COLD-VAPOR AAS

Element	Wavelength[a] (nm)	LOD[b] (ppb)
Antimony, Sb	217.581	0.1
Arsenic, As	193.759	0.02
Bismuth, Bi	223.061	0.02
Mercury, Hg	313.652	0.02
Selenium, Se	196.09	0.02
Tellurium, Te	214.281	0.02
Tin, Sn	235.484	0.5

Note: These data were taken from Reference 9.

[a]   It has been assumed that the transitions used for these detection limits were the most sensitive cited for AAS.
[b]   The detection limits are based on 50-mL sample solution volumes.

SPECTRAL OVERLAPS

In FAES and FAAS, the analytical results will be totally degraded if there is a spectral overlap of an analyte transition. This can result from an interfering matrix element with a transition close to that of the analyte. Table 9 presents a list of those overlaps that have been observed and those which are predicted to happen. In many cases the interferent element has been present in great excess when compared to the analyte species. Therefore, if the predicted interferent element is a major component of the matrix, a careful investigation for spectral overlap should be made. Excitation sources other than flames were not covered in this study.

Table 9
SPECTRAL OVERLAPS

Analyte element	Wavelength (nm)	Interfering element	Wavelength (nm)
		Observed Overlaps	
Aluminum	308.2153	Vanadium	308.211
Antimony	217.023	Lead	217.000
Antimony	231.147	Nickel	231.096
Cadmium	228.8022	Arsenic	228.812
Calcium	422.673	Germanium	422.6562
Cobalt	252.136	Indium	252.137
Copper	324.754	Europium	324.755
Gallium	403.299	Manganese	403.307
Iron	271.9027	Platinum	271.904
Manganese	403.307	Gallium	403.299
Mercury	253.652	Cobalt	253.649
Silicon	250.690	Vanadium	250.690
Zinc	213.856	Iron	213.859
		Predicted Overlaps	
Boron	249.773	Germanium	249.7962
Bismuth	202.121	Gold	202.138
Cobalt	227.449	Rhenium	227.462
Cobalt	242.493	Osmium	242.497
Cobalt	252.136	Tungsten	252.132
Cobalt	346.580	Iron	346.5860
Cobalt	350.228	Rhodium	350.252
Cobalt	351.348	Iridium	351.364
Copper	216.509	Platinum	216.517
Gallium	294.417	Tungsten	294.440
Gold	242.795	Strontium	242.810
Hafnium	295.068	Niobium	295.088
Hafnium	302.053	Iron	302.0639
Indium	303.936	Germanium	303.9067
Iridium	208.882	Boron	208.891
Iridium	248.118	Tungsten	248.144
Iron	248.3271	Tin	248.339
Lanthanum	370.454	Vanadium	370.470
Lead	261.3655	Tungsten	261.382
Molybdenum	379.825	Niobium	379.812

Table 9 (continued)
SPECTRAL OVERLAPS

Analyte element	Wavelength (nm)	Interfering element	Wavelength (nm)
Osmium	247.684	Nickel	247.687
Osmium	264.411	Titanium	264.426
Osmium	271.464	Tantalum	271.467
Osmium	285.076	Tantalum	285.098
Osmium	301.804	Hafnium	301.831
Palladium	363.470	Ruthenium	363.493
Platinum	227.438	Cobalt	227.449
Rhodium	350.252	Cobalt	350.262
Scandium	298.075	Hafnium	298.081
Scandium	298.895	Ruthenium	298.895
Scandium	393.338	Calcium	393.366
Silicon	252.4108	Iron	252.4293
Silver	328.068	Rhodium	328.055
Strontium	421.552	Rubidium	421.553
Tantalum	263.690	Osmium	263.713
Tantalum	266.189	Iridium	266.198
Tantalum	269.131	Germanium	269.1341
Thallium	291.832	Hafnium	291.858
Thallium	377.572	Nickel	377.557
Tin	226.891	Aluminum	226.910
Tin	266.124	Tantalum	266.134
Tin	270.651	Scandium	270.677
Titanium	264.664	Platinum	264.689
Tungsten	265.654	Tantalum	265.661
Tungsten	271.891	Iron	271.9027
Vanadium	252.622	Tantalum	252.635
Zirconium	301.175	Nickel	301.200
Zirconium	386.387	Molybdenum	386.411
Zirconium	396.826	Calcium	396.847

Note: These data were taken from Reference 10.

RELATIVE INTENSITIES OF ELEMENTAL TRANSITIONS FROM HOLLOW CATHODE LAMPS

In AAS, the hollow cathode lamp (HCL) is the most important excitation source for most of the elements determined. However, sufficient light must reach the detector for the measurement to be made with good precision and detection limits. For elements in this table (Table 10) with intensities of less than 100, HCLs are probably inadequate, and other sources such as electrodeless discharge lamps should be investigated.

Table 10
RELATIVE INTENSITIES OF ELEMENTAL TRANSITIONS FROM HOLLOW CATHODE LAMPS

Element	Fill gas	Wavelength (nm)	Relative emission intensity[a]
Aluminum	Ne	309.2710}	1,200
		309.2839}	
		396.1520	800
Antimony	Ne	217.581	250
		231.147	250
Arsenic	Ar	193.759	125
		197.262	125
Barium	Ne	553.548	400
		350.111	200
Beryllium	Ne	234.861	2,500
Bismuth	Ne	223.061	120
		306.772	400
Boron	Ar	249.773	400
Cadmium	Ne	228.8022	2,500
		326.1055	5,000
Calcium	Ne	422.673	1,400
Cerium	Ne	520.012}	8
		520.042}	
		569.699	8
Chromium	Ne	357.869	6,000
		425.435	5,000
Cobalt	Ne	240.725	1,000
		345.350	1,500
		352.685	1,300
Copper	Ne	324.754	7,000
		327.396	6,000
Dysprosium	Ne	404.597	2,000
		418.682	2,000
		421.172	2,500

Table 10 (continued)
RELATIVE INTENSITIES OF ELEMENTAL TRANSITIONS FROM
HOLLOW CATHODE LAMPS

Element	Fill gas	Wavelength (nm)	Relative emission intensity[a]
Erbium	Ne	400.796	1,600
		386.285	1,600
Europium	Ne	459.403	1,000
		462.722	950
Gadolinium	Ne	368.413	350
		407.870	700
Gallium	Ne	287.424	400
		417.204	1,100
Germanium	Ne	265.1172 ⎫	500
		265.1568 ⎭	
		259.2534	250
Gold	Ne	242.795	750
		267.595	1,200
Hafnium	Ne	307.288	300
		286.637	200
Holmium	Ne	405.393	2,000
		410.384	2,200
Indium	Ne	303.936	500
		410.176	500
Iridium	Ne	263.971	400
Iron	Ne	248.3271	400
		371.9935	2,400
Lanthanum	Ne	550.134	120
		392.756	45
Lead	Ne	217.000	200
		283.3053	1,000
Lithium	Ne	670.776	700
Lutetium	Ar	335.956	30
		337.650	25
		356.784	15
Magnesium	Ne	285.213	6,000
		202.582	130
Manganese	Ne	279.482	3,000
		280.106	2,200
		403.076	14,000
Mercury	Ar	253.652	1,000

Table 10 (continued)
RELATIVE INTENSITIES OF ELEMENTAL TRANSITIONS FROM HOLLOW CATHODE LAMPS

Element	Fill gas	Wavelength (nm)	Relative emission intensity[a]
Molybdenum	Ne	313.259	1,500
		317.035	800
Neodymium	Ne	463.424	300
		492.453	600
Nickel	Ne	232.003	1,000
		341.476	2,000
Niobium	Ne	405.894	400
		407.973	360
Osmium	Ar	290.906	400
		301.804	200
Palladium	Ne	244.791	400
		247.642	300
		340.458	3,000
Phosphorus	Ne	213.547 ⎱	30
		213.618 ⎰	
		214.914	20
Platinum	Ne	265.945	1,500
		299.797	1,000
Potassium	Ne	766.490	6
		404.414	300
Praseodymium	Ne	495.137	100
		513.342	70
Rhenium	Ne	346.046	1,200
		346.473	900
Rhodium	Ne	343.489	2,500
		369.236	2,000
		350.732	200
Rubidinum	Ne	780.027	1.5
		420.180	80
Ruthenium	Ar	349.894	600
		392.592	300
Samarium	Ne	429.674	600
		476.027	800
Scandium	Ne	391.181	3,000
		390.749	2,500
		402.040	1,800
		402.369	2,100

Table 10 (continued)
RELATIVE INTENSITIES OF ELEMENTAL TRANSITIONS FROM
HOLLOW CATHODE LAMPS

Element	Fill gas	Wavelength (nm)	Relative emission intensity[a]
Selenium	Ne	196.09	50
		203.98	50
Silicon	Ne	251.6113	500
		288.1579	500
Silver	Ar	328.068	3,000
		338.289	3,000
Sodium	Ne	588.9950	2,000
		330.237 ⎫ 330.298 ⎭	40
Strontium	Ne	460.733	1,000
Tantalum	Ar	271.467	150
		277.588	100
Tellurium	Ne	214.281	60
		238.578	50
Terbium	Ne	432.643 ⎫ 432.690 ⎭	110
		431.883	90
		433.841	60
Thallium	Ne	276.787	600
		258.014	50
Thulium	Ne	371.791	40
		409.419	50
		410.584	70
Tin	Ne	224.605	100
		286.332	250
Titanium	Ne	364.268	600
		399.864	600
Tungsten	Ne	255.100 ⎫ 255.135 ⎭	200
		400.875	1,400
Uranium	Ne	358.488	300
		356.659	200
		351.461	200
		348.937	150
Vanadium	Ne	318.341 ⎫ 318.398 ⎭	600
		385.537 ⎫ 385.584 ⎭	200
Ytterbium	Ar	398.799	2,000
		346.437	800

Table 10 (continued)
**RELATIVE INTENSITIES OF ELEMENTAL TRANSITIONS FROM
HOLLOW CATHODE LAMPS**

Element	Fill gas	Wavelength (nm)	Relative emission intensity[a]
Yttrium	Ne	407.738	500
		410.238	600
		414.285	300
Zinc	Ne	213.856	2,500
		307.590	2,000

Note: These data were taken from Reference 2. They were obtained using Westinghouse HCLs and a single experimental setup. No correction has been made for the spectral response of the monochromator/photomultiplier tube system.

[a] The most intense line is the Mn 403.076 transition with a relative intensity of 14,000.

INERT GASES

In AAS, the excitation source inert gas emission offers a potential background spectral interference. The most common inert gases used in HCLs are Ne and Ar. The data taken for Tables 11 and 13 are from HCLs; however, electrodeless discharge lamps emit very similar spectra. The emission spectra for Ne and Ar HCLs (Tables 11 and 13) and close lines that must be resolved for accurate analytical results (Tables 12 and 14) are provided in the following four tables. This information was obtained for HCLs and flame atom cells and should not be considered with respect to plasma sources. In the Type column, *I* indicates that the transition originates from an atomic species, and *II* indicates a singly ionized species.

Table 11
NEON HCL SPECTRUM

Wavelength (nm)	Type	Relative intensity[a]	Wavelength (nm)	Type	Relative intensity[a]
323.237	II	5.4	357.461	II	5.9
330.974	II	2.8	359.3526	I	19.
331.972	II	8.7	360.0169	I	3.5
332.374	II	28.	363.3665	I	3.6
332.916	II	1.7	366.407	II	1.9
333.484	II	5.2	369.421	II	3.5
334.440	II	17.	370.962	II	4.9
335.502	II	3.5	372.186	II	3.1
336.060	II	1.7	404.264	I	1.4
336.9908}	I	7.8	533.0778	I	1.6
336.9908}			534.920	I	1.6
337.822	II	17.	540.0562	I	3.3
339.280	II	8.3	576.4419	I	2.3
341.7904	I	16.	585.2488	I	100.
344.7703	I	12.	588.1895	I	8.7
345.4195	I	15.	594.4834	I	14.
346.0524	I	6.6	597.4627}	I	2.6
346.6579	I	12.	597.5534}		
347.2571	I	12.	602.9997	I	2.8
349.8064	I	2.9	607.4338	I	11.
350.1216	I	3.8	609.6163	I	15.
351.5191	I	3.6	614.3063	I	20.
352.0472	I	61.	616.3594	I	5.2
356.850	II	7.8			

Note: These data are from Reference 2 and were taken with a Varian Copper HCL operated at 10 mA. The Cu 324.7-nm transition was a factor of 2.9 more intense than the 585.249-nm Ne transition. The spectrum was taken with a IP28 photomultiplier tube (PMT). The relative intensities were not corrected for the instrumental/PMT response.

[a]   These data are referenced to the Ne transition at 585.2488 nm which has been assigned the value of 100.

Table 12
NEON LINES WHICH MUST BE RESOLVED FOR ACCURATE AAS MEASUREMENTS

Analyte element	Wavelength (nm)	Neon Line (nm)	Required resolution[a] (nm)
Chromium	357.869	357.461	0.20
	359.349	359.3526	0.002
	360.533	360.0169	0.26
Copper	324.754	323.237	0.75
Dysprosium	404.597	404.264	0.17
Gadolinium	371.357	370.962	0.20
	371.748	372.186	0.22
Lithium	670.776	335.502 in 2nd order is 671.004	0.11
Lutetium	335.956	336.060	0.05
Niobium	405.894	404.264	0.82
Rhenium	346.046	346.0524	0.003
	346.473	346.6579	0.11
	345.188	345.4195	0.12
Rhodium	343.489	344.7703	0.64
	369.236	369.421	0.09
Ruthenium	372.803	372.186	0.31
Scandium	402.369	404.264	0.94
Silver	338.289	337.822	0.23
Sodium	588.995	588.1895	0.40
	589.592	588.1895	0.70
Thulium	371.792	372.186	0.19
Titanium	337.145	336.9808 and 336.9908	0.08
	364.268	363.3665	0.45
	365.350	366.407	0.53
Uranium	356.660	356.850	0.09
	358.488	359.3526	0.43
Ytterbium	346.436	346.6579	0.11
Zirconium	351.960	352.0472	0.04
	360.119	360.0169	0.05

Note: These data were taken from Reference 10.

[a] The monochromator settings must be at least one half of the separation of the analyte and interferent transitions.

Table 13
ARGON HCL SPECTRUM

Wavelength (nm)	Type	Relative intensity[a]	Wavelength (nm)	Type	Relative intensity[a]
294.2893	II	3.5	401.3857	II	4.3
297.9050	II	1.9	403.3809	II	2.5
329.3640	II	1.5	403.5460	II	2.3
330.7228	II	1.5	404.2894	II	1.6
335.0924	II	2.2	404.4418	I	9.0
337.6436	II	2.2	405.2921	II	21
338.8531	II	1.8	407.2005	II	34
347.6747	II	3.7	407.2385	II	5.5
349.1244	II	2.0	407.6628	II	2.0
349.1536	II	7.2	407.9574	II	4.4
350.9778	II	7.0	408.2387	II	3.2
351.4388	II	4.0	410.3912	II	10
354.5596	II	11	413.1724	II	61
354.5845	II	12	415.6086	II	2.4
355.9508	II	16	415.8590	I	1.4
356.1030	II	1.8	416.4180	I	4.4
357.6616	II	11	418.1884	I	6.9
358.1608	II	3.9	419.0713	I	9.1
358.2355	II	8.5	419.1029	I	8.9
358.8441	II	1.2	419.8317	I	38
360.6522	I	2.0	420.0674	I	38
362.2138	II	1.3	421.8665	II	2.2
363.9833	II	3.3	422.2637	II	4.3
371.8206	II	5.5	422.6988	II	5.6
372.9309	II	1.3	422.8158	II	12
373.7889	II	9.8	423.7220	II	15
376.5270	II	5.1	425.1185	I	2.3
376.6119	II	6.8	425.9326	I	42
377.0520	II	1.7	426.6286	I	11
378.0840	II	4.0	426.6527	II	7.4
380.3172	II	5.5	427.2169	I	18
380.9456	II	1.8	427.7528	II	100
383.4679	I	2.6	428.2898	II	2.5
385.0581	II	1.2	430.0101	I	13
386.8528	II	7.0	430.0650	II	3.3
392.5719	II	9.9	430.9239	II	6.6
392.8623	II	6.9	433.1200	II	17
393.2547	II	3.2	433.2030	II	4.2
394.6097	II	14	433.3561	I	12
394.8979	I	5.2	433.5338	I	5.8
397.9356	II	5.6	434.5168	I	3.7
399.4792	II	6.2	434.8064	II	1.5

Table 13 (continued)
ARGON HCL SPECTRUM

Wavelength (nm)	Type	Relative intensity[a]	Wavelength (nm)	Type	Relative intensity[a]
435.2205	II	5.4	476.4865	II	1.5
436.2066	II	3.3	480.6020	II	36
436.7832	II	9.6	484.7812	II	1.6
437.0753	II	32	486.5910	II	1.3
437.1329	II	6.5	487.9864	II	58
437.5954	II	10	488.9042	II	11
437.9667	II	20	490.4752	II	3.5
438.5057	II	6.7	493.3209	II	4.1
440.0097	II	5.7	496.5080	II	28
440.0986	II	15	500.9334	II	5.5
442.6001	II	1.6	501.7163	II	12
443.0189	II	1.2	506.2037	II	5.9
443.0996	II	5.4	509.0495	II	2.9
443.3838	II	5.3	514.1783	II	5.7
443.9461	II	5.3	514.5308	II	3.7
444.8879	II	7.9	516.2285	I	3.8
447.4759	II	19	516.5773	II	1.8
448.1811	II	33	518.7746	I	3.8
451.0733	I	20	522.1271	I	1.2
452.2323	I	2.0	545.1652	I	1.7
453.0552	II	3.2	549.5874	I	3.1
454.5052	II	1.3	555.8702	I	4.0
457.9350	II	1.4	557.2541	I	1.9
458.9898	II	47	560.6733	I	4.9
459.6097	I	1.8	565.0704	I	1.7
460.9567	II	1.3	588.8584	I	1.9
462.8441	I	1.4	591.2085	I	4.1
			592.8813	I	1.4
463.7233	II	5.5	603.2127	I	4.1
465.7901	II	1.9	604.3223	I	1.6
470.2316	I	2.5			
			611.4923	II	2.2
472.6868	II	43	617.2278	II	1.1
473.2053	II	9.7	696.5431	I	3.2
473.5906	II	1.3			
			706.7218	I	1.7
			738.3980	I	1.2
			750.3869	I	2.7

Note: These data were taken from an Ar-filled Ga HCL at the Los Alamos Fourier Transform Spectrometer facility.[11]

[a] These data are referenced to the Ar transition at 427.7528 nm which has been assigned the value of 100.

Table 14
ARGON LINES THAT MUST BE RESOLVED FOR ACCURATE AAS MEASUREMENTS

Analyte element	Wavelength (nm)	Argon line (nm)	Required resolution[a] (nm)	Analyte element	Wavelength (nm)	Argon line (nm)	Required resolution[a] (nm)
Barium	455.403	454.5052	0.05	Palladium	363.470	363.9833	0.26
Calcium	393.366	393.2547	0.06	Rubidium	420.180	420.0674	0.06
	422.673	422.6988	0.01	Ruthenium	349.894	349.1536	0.37
Cesium	455.5276	454.5052	0.51		372.803	372.9309	0.06
Chromium	357.869	357.6616	0.10	Samarium	429.674	430.0101	0.17
	425.435	425.1185	0.16		476.027	476.4865	0.23
Dysprosium	404.597	404.4418	0.08	Silver	328.068	329.3640	0.65
	421.172	421.8665	0.35		338.2068	338.8531	0.32
Erbium	337.271	337.6436	0.19	Sodium	330.237	330.7228	0.24
	400.796	401.3857	0.30		588.9950	588.8584	0.07
Europium	459.403	459.6097	0.10		589.5924	588.8584	0.37
Gadolinium	440.186	440.0986	0.04	Strontium	407.771	407.6628	0.05
Gallium	294.364	294.2893	0.04		421.552	421.8665	0.16
	417.204	416.4180	0.39		460.733	460.9567	0.11
Holmium	405.393	405.2921	0.05	Tantalum	474.016	473.5906	0.21
	410.384	410.3912	0.004	Terbium	432.643	433.1200	0.24
Indium	451.131	451.0733	0.03	Thallium	377.572	378.0840	0.26
Iron	371.9935	371.8206	0.09	Thulium	371.791	371.8206	0.015
Lanthanum	408.672	408.2387	0.22	Titanium	334.941	335.0924	0.08
	550.134	549.5874	0.27		364.268	363.9833	0.14
Lutetium	451.857	452.2323	0.19	Tungsten	400.875	401.3857	0.26
Manganese	403.076	403.3809	0.15	Uranium	358.488	358.2355	0.13
Molybdenum	379.825	380.3172	0.25		591.539	591.2085	0.16
Neodymium	463.424	463.7233	0.15	Vanadium	437.924	437.9667	0.02
	492.453	493.3209	0.43	Ytterbium	398.799	397.9356	0.43
Nickel	352.454	351.4388	0.51	Yttrium	410.238	410.3912	0.08
Niobium	334.906	335.0924	0.093	Zirconium	351.960	351.4388	0.26
	405.894	405.2921	0.301		360.119	360.6522	0.27
Osmium	442.047	442.6001	0.28				

Note: These data were derived from Tables 2 through 7, the information in Table 13, and Reference 11.

[a] The monochromator settings must be at least one half of the separation of the analyte and interferent transitions.

CLOSE LINES FOR BACKGROUND CORRECTION

In AAS, it is possible to make background corrections in many cases by measuring a normally nonabsorbing transition near the analytical transition. This table presents a list of suitable transitions for such a background measurement. It is often desirable to check the background absorbance by more than one method even if there is a built-in background measurement by some other means such as the continuum or Zeeman methods. In the table below, the first two columns give the analyte element and wavelength of the analytical transition, and the last two columns give the transition useful for the background measurement and its source. If the source is Ne and the HCL is Ne filled, the same HCL can be used for the background measurement; if not, a different HCL must be placed in the spectometer to make the measurement.

Table 15
CLOSE LINES FOR BACKGROUND CORRECTION

Element	Analysis line (nm)		Background line (nm)		Source
Aluminum	309.2710	I	306.614	I	Al
Antimony	217.581	I	217.919	I	Sb
	231.147	I	231.398	I	Ni
Arsenic	193.759	I	191.294	II	As
Barium	553.548	I	540.0562	I	Ne
			553.305	I	Mo
			557.742	I	Y
Beryllium	234.861	I	235.484	I	Sn
Bismuth	223.061	I	226.502	II	Cd
	306.772	I	306.614	I	Al
Bromine	148.845	I	149.4675	I	N
Cadmium	228.8022	I	226.502	II	Cd
Calcium	422.673	I	421.9360	I	Fe
			423.5936	I	Fe
Cesium	852.1122	I	854.4696	I	Ne
Chromium	357.869	I	352.0472	I	Ne
			358.119	I	Fe
Cobalt	240.206	I	238.892	II	Co
			242.170	I	Sn
Copper	324.754	I	324.316	I	Cu
Dysprosium	421.172	I	421.645	II	Fe
			421.096	I	Ag
Erbium	400.796	I	394.442	I	Er
Europium	459.403	I	460.102	I	Cr
Gallium	287.424	I	283.999	I	Sn
			283.690	I	Cd
Gold	242.795	I	242.170	I	Sn
Indium	303.936	I	306.614	I	Al
Iodine	183.038	I	184.445	I	I
Iron	248.3271	I	249.215	I	Cu
Lanthanum	550.134	I	550.549	I	Mo
			548.334	I	Co
Lead	283.3053	I	280.1995	I	Pb
			283.6900	I	Cd
	217.000	I	220.3534	II	Pb
Lithium	670.791	I	671.7043	I	Ne
Magnesium	285.213	I	283.690	I	Cd
			283.999	I	Sn
Manganese	279.482	I	282.437	I	Cu
			280.1995	I	Pb

Table 15 (continued)
CLOSE LINES FOR BACKGROUND CORRECTION

Element	Analysis line (nm)		Background line (nm)		Source
Mercury	253.652	I	249.215	I	Cu
Molybdenum	313.259	I	312.200	II	Mo
Nickel	232.003	I	232.138	I	Ni
Palladium	247.642	I	249.215	I	Cu
Phosphorus	213.618	I	213.856	I	Zn
Potassium	766.490	I	769.896	I	K
			767.209	I	Ca
Rhodium	343.489	I	350.732	I	Rh
			352.0472	I	Ne
Rubidium	780.027	I	778.048	I	Ba
Ruthenium	249.894	I	352.0472	I	Ne
Selenium	196.09	I	199.51	I	Se
Silicon	251.6113	I	249.215	I	Cu
Silver	328.068	I	332.374	II	Ne
			326.234	I	Sn
Sodium	588.9950	I	588.833	I	Mo
Strontium	460.733	I	460.500	I	Ni
Tellurium	214.281	I	213.856	I	Zn
			217.581	I	Sb
Thallium	276.787	I	280.1995	I	Pb
Tin	224.605	I	226.502	II	Cd
	286.332	I	283.999	I	Sn
Titanium	364.268	I	361.939	I	Ni
	365.350	I	361.939	I	Ni
Uranium	358.488	I	358.119	I	Fe
Vanadium	318.398	I	324.754	I	Cu
	318.540	I	324.754	I	Cu
Zinc	213.856	I	212.274	II	Zn

Note: These data were taken from Reference 12.

BETA VALUES FOR THE AIR-ACETYLENE AND NITROUS OXIDE-ACETYLENE FLAMES

Beta values represent the fraction of free atoms present in the hot flame gases of the flame indicated. These values have been taken from various sources and were either experimentally measured or calculated from thermodynamic data using the assumption of local thermodynamic equilibrium in the flame. These values do not have very good agreement within each element; however, the values do provide an indication of the probable sensitivity of the particular flame.

Table 16
BETA VALUES FOR THE AIR-ACETYLENE AND NITROUS OXIDE-ACETYLENE FLAMES

Element	Symbol	Beta A/AC Flame	Beta N/AC Flame	Element	Symbol	Beta A/AC Flame	Beta N/AC Flame
Aluminum	Al	<0.0001	0.13	Indium	In	0.10	0.37
		<0.00005	0.29			0.67	0.93
		0.0005	0.97[a]			0.67	
			0.5	Iridium	Ir	0.1	
Antimony	Sb	0.03		Iron	Fe	0.38	0.83
Arsenic	As	0.0002				0.66	0.91
Barium	Ba	0.0009	0.074			0.84	1.00
		0.002	0.074			0.66	
		0.003	0.98	Lead	Pb	0.44	0.84
		0.0018				0.77	
Beryllium	Be	0.0004	0.095	Lithium	Li	0.21	0.34[a]
		0.00006	0.98			0.26[a]	0.96[a]
			0.98			0.20[a]	0.041
Bismuth	Bi	0.17	0.35			0.08	0.91[a]
Boron	B	<0.0006	0.0035	Magnesium	Mg	0.59	0.88
		<0.000001	0.2			1.05	0.99
Cadmium	Cd	0.38	0.56			0.62	0.92
		0.50	0.60				0.99[a]
		0.80		Manganese	Mn	0.45	0.37
Calcium	Ca	0.066	0.34			0.93	0.77
		0.14	0.52[a]			1.0	
		0.05[a]	0.98	Mercury	Hg	0.04	
		0.018		Molybdenum	Mo	0.03	
Cesium	Cs	0.02	0.0004	Nickel	Ni	1	
		0.0057		Palladium	Pd	1	
Chromium	Cr	0.071	0.63	Platinium	Pt	0.4	
		0.13	1.02	Potassium	K	0.7[a]	0.12[a]
		0.53	1.00			0.25	0.0004
		0.042				0.45	0.17[a]
Cobalt	Co	0.023	0.11			0.59[a]	
		0.28	0.25			0.45	0.17[a]
		0.41				0.59[a]	
Copper	Cu	0.4	0.49	Rhodium	Rh	1	
		0.82	0.66	Rubidium	Rb	0.16	
		0.98	1.00[a]	Ruthenium	Ru	0.3	
Gallium	Ga	0.16	0.73	Selenium	Se	0.0001	
		0.16		Silicon	Si	<0.001	0.55
Germanium	Ge	0.001				<0.0000001	0.12
Gold	Au	0.21	0.16				0.36
		0.40	0.27	Silver	Ag	0.66	0.57
		0.63				0.70	

Table 16 (continued)
BETA VALUES FOR THE AIR-ACETYLENE AND NITROUS OXIDE-ACETYLENE FLAMES

Element	Symbol	Beta A/AC Flame	Beta N/AC Flame	Element	Symbol	Beta A/AC Flame	Beta N/AC Flame
Sodium	Na	0.63	0.32	Tin	Sn	<0.0001	0.35
		1.00	0.97[a]			0.043	0.82
		1.00[a]	0.012			0.078	
		0.56	0.80			0.061	
Strontium	Sr	0.068	0.26	Titanium	Ti	<0.001	0.11
		0.10	0.57				0.33
		0.13	0.99				0.49
		0.021		Tungsten	W	0.004	0.71
Tantalum	Ta		0.045	Vanadium	V	0.0004	0.32
Thallium	Tl	0.36	0.55			0.015	0.99
		0.52				0.000001	
				Zinc	Zn	0.66	0.49
						0.45	

Note: These data were taken from References 2 and 13.

[a] Ionization has been suppressed for these measurements/calculations.

LOWER-ENERGY-LEVEL POPULATIONS (IN PERCENT) AS A FUNCTION OF TEMPERATURE

It is possible to calculate the relative number of atoms in the ground energy level(s) using the following equation:

$$\% \text{ Atoms (ith level)} = n_1/n_t \times 100 = g_1/Z \times \exp(-E_1/kT)$$

where n_1 is the number of atoms in the ith level per unit volume of atom cell, n_t is the total number of atoms per unit volume of atom cell, g_1 is the statistical weight for energy level i, Z is the electronic partition function, E_1 is the energy of the ith level, k is the Boltzmann constant, and T is the absolute temperature. Of course, all of the data must be in consistent units.

In utilizing these data, it should be remembered that, other things being equal, the larger the percentage of atoms in the ground or lower level of a transition, the larger the absorption signal from that transition should be. For example, a transition with 100% of the atoms in the ground state should be ten times more sensitive than one with 10%. Also, these data refer to the percent of atoms in the atomic state only; therefore, this information should be used in conjunction with that in Table 16.

Table 17
LOWER ENERGY LEVEL POPULATIONS (IN PERCENT) AS A FUNCTION OF TEMPERATURE

Element	Energy level (cm^{-1})	Percent population at temperature (°C)		
		2,000	2,500	3,000
Aluminum	0.0	35.1	34.8	34.5
	112.040	64.9	65.2	65.5
Antimony	0.0	99.7	98.7	97.0
	8,512.100	0.2	0.7	1.6
	9,854.100	0.1	0.5	1.7
Arsenic	0.0	99.9	99.5	98.5
	10,592.500	0.0	0.2	0.6
	10,914.600	0.1	0.3	0.8
Barium	0.0	98.0	92.6	82.6
	9,033.985	0.4	1.5	3.3
	9,215.518	0.6	2.3	5.0
	9,596.551	0.7	2.6	5.8
	11,395.382	0.1	0.7	1.8
Beryllium	0.0	100.0	100.0	100.0
Bismuth	0.0	100.0	99.8	99.5
Boron	0.0	33.5	33.5	33.5
	16.0	66.5	66.5	66.5
Cadmium	0.0	100.0	100.0	100.0
Calcium	0.0	100.0	99.8	99.3
Cesium	0.0	99.9	99.5	98.3
	11,178.240	0.0	0.2	0.5
	11,732.350	0.0	0.2	0.7
Chromium	0.0	98.6	95.9	91.5
	7,593.160	0.3	0.9	1.7
	7,750.780	0.1	0.2	0.3
	7,810.820	0.1	0.2	0.3
	7,927.470	0.1	0.4	0.9
	8,095.210	0.2	0.7	1.3
	8,307.570	0.3	0.8	1.7

Table 17 (continued)
LOWER ENERGY LEVEL POPULATIONS (IN
PERCENT) AS A FUNCTION OF
TEMPERATURE

Element	Energy level (cm⁻¹)	Percent population at temperature (°C)		
		2,000	**2,500**	**3,000**
Cobalt	0.0	51.8	45.6	40.9
	816.000	23.1	22.8	22.1
	1,406.840	11.3	12.2	12.5
	1,809.330	5.6	6.4	6.9
	3,482.820	4.2	6.1	7.7
	4,142.660	2.1	3.4	4.5
	4,690.180	1.1	1.8	2.6
	5,075.830	0.5	1.0	1.4
	7,442.410	0.2	0.5	0.9
Copper	0.0	99.9	99.4	98.3
	11,202.565	0.1	0.5	1.9
Gallium	0.0	47.5	44.6	42.6
	826.240	52.5	55.4	57.4
Germanium	0.0	20.6	18.2	16.6
	557.100	41.4	39.7	38.7
	1,409.900	37.4	40.5	42.4
	7,125.260	0.6	1.5	2.7
Gold	0.0	99.5	98.5	96.4
	9,161.300	0.4	1.5	3.5
Hafnium	0.0	73.9	64.0	55.9
	2,356.680	19.0	23.1	25.3
	4,567.640	5.0	8.3	11.3
	5,521.780	0.3	0.5	0.8
	5,638.620	1.3	2.5	3.7
	6,572.550	0.4	0.9	1.4
Indium	0.0	71.0	64.1	59.1
	2,212.560	29.0	35.9	40.9
Iridium	0.0	85.1	77.2	69.9
	2,834.980	11.1	15.1	18.0
	4,078.940	1.8	3.0	4.0
	5,784.620	0.8	1.7	2.6
	6,323.910	0.7	1.6	2.7
	7,106.610	0.4	1.0	1.9
Iron	0.0	46.2	43.4	41.0
	415.933	26.7	26.6	26.1
	704.003	15.5	16.1	16.3
	888.123	8.1	8.7	8.9
	978.074	2.5	2.7	2.9
	6.928.280	0.4	1.0	1.8
	7,376.775	0.2	0.6	1.2
	7,728.071	0.1	0.4	0.8
	7,985.795	0.1	0.2	0.5
Lanthanum	0.0	42.5	34.2	28.3
	1,053.200	29.9	28.0	25.6
	2,668.200	6.2	7.4	7.9
	3,010.010	7.3	9.1	10.0
	3,494.580	6.9	9.2	10.6
	4,121.610	5.5	8.0	9.8
	7,011.900	0.4	0.9	1.5
	7,231.360	0.1	0.3	0.4
	7,490.460	0.2	0.5	0.8
	7,679.940	0.3	0.6	1.1

Table 17 (continued)
LOWER ENERGY LEVEL POPULATIONS (IN
PERCENT) AS A FUNCTION OF
TEMPERATURE

Element	Energy level (cm^{-1})	Percent population at temperature (°C)		
		2,000	2,500	3,000
Lead	0.0	98.7	95.7	90.8
	7,819.350	1.1	3.2	6.4
	10,650.470	0.2	1.1	2.8
Lithium	0.0	100.0	99.9	99.8
Magnesium	0.0	100.0	100.0	100.0
Manganese	0.0	100.0	100.0	99.8
Mercury	0.0	100.0	100.0	100.0
Molybdenum	0.0	99.9	99.4	98.1
Nickel	0.0	39.5	36.4	34.2
	204.786	26.6	25.2	24.1
	879.813	11.7	12.2	12.4
	1,332.153	11.8	13.2	14.0
	1,713.080	3.9	4.5	5.0
	2,216.519	4.5	5.7	6.6
	3,409.925	1.9	2.8	3.7
Niobium	0.0	7.5	6.5	5.7
	154.190	13.4	11.8	10.6
	391.990	17.0	15.4	14.2
	695.250	18.2	17.3	16.4
	1,050.260	17.6	17.7	17.3
	1,142.790	6.6	6.7	6.6
	1,586.900	7.2	7.8	8.0
	2,154.110	6.4	7.5	8.1
	2,805.360	5.0	6.4	7.5
	4,998.170	0.2	0.4	0.5
	5,297.920	0.3	0.6	0.9
	5,965.450	0.3	0.6	1.0
Osmium	0.0	86.5	78.3	70.4
	2,740.490	6.7	9.0	10.5
	4,159.320	3.4	5.6	7.5
	5,143.920	2.6	5.0	7.3
	5,766.140	0.5	0.9	1.5
	6,092.790	0.1	0.3	0.4
	8,742.830	0.2	0.5	1.1
Palladium	0.0	91.7	80.3	67.2
	6,464.110	6.2	13.8	21.6
	7,754.990	1.8	4.7	8.3
	10,093.940	0.2	0.7	1.6
	11,721.770	0.1	0.5	1.2
Platinum	0.0	47.0	43.8	41.5
	775.900	19.2	20.0	20.4
	823.700	33.4	35.1	36.0
	6,140.00	0.1	0.2	0.3
	6,567.5000	0.3	0.7	1.3
Potassium	0.0	100.0	99.8	99.4
Rhenium	0.0	99.9	99.5	98.3
Rhodium	0.0	69.0	60.0	53.6
	1,529.970	18.1	19.9	20.6
	2,598.030	6.3	8.1	9.3
	3,309.860	3.8	5.4	6.6
	3,472.680	2.2	3.3	4.1

Table 17 (continued)
LOWER ENERGY LEVEL POPULATIONS (IN PERCENT) AS A FUNCTION OF TEMPERATURE

Element	Energy level (cm^{-1})	Percent population at temperature (°C)		
		2,000	2,500	3,000
Rhodium (cont.)	5,657.970	0.5	0.9	1.4
	5,690.970	0.9	1.8	2.8
Rubidium	0.0	100.0	99.8	99.3
Ruthenium	0.0	62.4	55.2	49.3
	1,190.640	21.7	22.8	22.8
	2,091.540	8.8	10.6	11.5
	2,713.240	4.0	5.3	6.1
	3,105.490	1.8	2.5	3.0
	6,545.030	0.5	1.0	1.8
	7,483.070	0.2	0.6	1.1
	8,084.120	0.1	0.3	0.7
	9,183.660	0.0	0.1	0.3
Scandium	0.0	42.9	42.2	41.5
	168.340	57.0	57.4	57.4
Selenium	0.0	85.0	80.5	76.8
	1,989.490	12.2	15.4	17.8
	2,534.350	2.7	3.7	4.6
	9,576.080	0.1	0.3	0.8
Silicon	0.0	12.3	11.9	11.6
	77.150	34.8	34.2	33.5
	223.310	52.2	52.3	52.1
	6,298.810	0.7	1.6	2.8
Silver	0.0	100.0	100.0	100.0
Sodium	0.0	100.0	100.0	99.9
Strontium	0.0	100.0	99.8	98.9
Tantalum	0.0	64.9	53.8	45.0
	2,010.00	23.0	25.4	25.8
	3,963.920	7.5	11.0	13.5
	5,621.040	2.9	5.3	7.6
	6,049.420	0.4	0.8	1.2
	9,253.430	0.1	0.4	0.8
	9,705.380	0.1	0.4	0.9
Technetium	0.0	27.2	25.7	24.0
	170.132	33.8	32.6	31.0
	386.873	37.1	37.0	35.9
	6,556.860	0.1	0.4	0.6
	6,598.830	0.2	0.6	1.0
	6,661.000	0.3	0.8	1.4
	6,742.790	0.4	1.0	1.7
	6,843.000	0.4	1.1	2.0
	7,255.290	0.1	0.4	0.7
Tungsten	0.0	28.4	20.5	15.9
	1,670.300	25.6	23.6	21.5
	2,951.290	23.8	26.3	27.1
	3,325.530	13.0	15.2	16.2
	4,830.000	6.2	8.9	11.0
	6,219.330	2.9	5.2	7.3
Vanadium	0.0	14.1	12.6	11.5
	137.380	19.2	17.5	16.2
	323.420	22.4	20.9	19.7
	553.020	23.7	22.9	22.1

Table 17 (continued)
LOWER ENERGY LEVEL POPULATIONS (IN
PERCENT) AS A FUNCTION OF
TEMPERATURE

Element	Energy level (cm^{-1})	Percent population at temperature (°C)		
		2,000	2,500	3,000
Vanadium	2,112.320	1.5	1.9	2.1
	2,153.200	3.0	3.7	4.1
	2,220.130	4.3	5.3	6.0
	2,311.370	5.4	6.7	7.6
	2,424.890	6.2	7.8	9.0
Yttrium	0.0	49.3	47.2	45.4
	530.360	50.5	52.2	52.8
Zinc	0.0	100.0	100.0	100.0
Zirconium	0.0	34.2	29.1	25.1
	570.410	31.7	29.3	26.7
	1,240.840	25.2	25.6	24.9
	4,186.110	1.7	2.6	3.4
	4,196.850	0.3	0.5	0.7
	4,376.280	0.9	1.4	1.8
	4,870.530	0.6	1.1	1.5
	5,023.410	0.9	1.6	2.3
	5,101.680	0.9	1.5	2.2
	5.249.070	1.1	2.0	2.8
	5,540.540	1.1	2.2	3.2
	5,888.930	1.1	2.2	3.3
	8,057.300	0.2	0.5	1.0

Note: These data were taken from Reference 2.

CRITICAL OPTIMIZATION PARAMETERS FOR AES/AAS METHODS

In most multiparameter instrumental techniques, the parameters can be classified into two types: independent and dependent. Independent parameters can be optimized independently from all other parameters and can therefore be subjected to a univariate approach, i.e., the variable can be adjusted until the largest signal-to-noise ratio (SNR) is obtained and set at that value for the best instrumental performance. This is the simplest situation and can be handled in a very straightforward manner.

Dependent parameters are an entirely different matter. Most dependent parameters have optimum values that depend on the value of the other parameters. If the value of any variable is changed, then the optimum for the parameter under question will be different.

Table 18 lists the parameters for FAAS, EAAS, and FAES which are both dependent and independent. "Yes" in any column indicates that the listed parameter is appropriate for that technique. If an optimization is necessary when independent parameters are involved, it is important to use a systematic approach which permits one to vary all parameter values to develop the optimum for each. If the variables are simply varied one at a time, false optimum values will be obtained and poor results will be obtained. Experimental design techniques are required for good results; one of the best approaches is the SIMPLEX technique which has been fully discussed in the literature.[15]

Table 18
CRITICAL OPTIMIZATION PARAMETERS FOR AES/AAS METHODS

Parameter	FAAS	EAAS	FAES
Independent Parameters			
Excitation source power	Yes	Yes	NA[a]
Photomultiplier voltage[b]	Yes	Yes	Yes
Readout gain[c]	Yes	Yes	Yes
Noise-suppression setting[d]	Yes	Yes	Yes
Dependent (Interdependent) Parameters			
Oxidant gas-flow rate	Yes	NA	Yes
Fuel-to-oxidant ratio	Yes	NA	Yes
Sheath gas-flow rate[e]	Yes	Yes	Yes
Solution flow rate[f]	Yes	NA	Yes
Sample size	NA	Yes	NA
Height of optical measurement	Yes	Yes	Yes
Monochromator slit setting	Yes	Yes	Yes
Burner variables[g]	Yes	NA	Yes
Furnace variables[h]	NA	Yes	NA

Note: This information was taken from Reference 16.

a   NA stands for "not applicable".
b   The photomultiplier tube voltage does not affect the SNR unless extreme voltages are used. It will specify the level of signal which is observed.
c   The gain does not affect the SNR until electronic noise becomes important. It also specifies the level of signal which is observed.
d   This specifies the frequency response of the system and is accompanied by a time requirement. More noise filtering requires a longer measurement.
e   Most commercial burners do not use a sheath gas; however, there is always the possibility of a sheath gas in EAAS.
f   This is important if the sample solution flow rate is controlled by a pump rather than by the oxidant gas flow rate.
g   Some burners have additional variables such as bead position and nebulizer position.
h   The timing cycle and temperature are always critical variables for the graphite furnaces.

FLAME TEMPERATURES AND REFERENCES ON TEMPERATURE MEASUREMENTS

Table 19
FLAME TEMPERATURES

Flame type	Experimental measurement range (K)	Calculated stoichiometric temperature (K)	Typical[a] (K)
Hydrocarbon/air	1900—2150	2228	2000
Acetylene/air	2360—2600	2523	2450
Acetylene/nitrous oxide	2830—3070	3148	2950
Hydrogen/air	2100—2300	2373	2300
Hydrogen/oxygen	2500—2900	3100	2800
Acetylene/oxygen	2900—3300	3320	3100

Note: These data were taken from Reference 2.

[a] This value represents the value most often cited for flames used in analytical spectroscopy.

REFERENCES

Techniques of Temperature Measurement

1. **Gaydon, A. G. and Wolfhard, H. G.,** *Flames, Their Structure, Radiation, and Temperature,* Chapman and Hall, London, 1970.
2. **Fristrom, R. M. and Westenberg, A. A.,** *Flame Structure,* McGraw-Hill, New York, 1965.
3. **Tourin, R. H.,** *Spectroscopic Gas Temperature Measurement,* Elsevier, Amsterdam, 1966.
4. **Gaydon, A. G. and Wolfhard, H. G.,** The spectrum-line reversal method of measuring flame temperature, *Proc. Phys. Soc. London,* 65A, 19, 1954.

Flame Temperatures and Temperature Measurements

5. **Browner, R. F. and Winefordner, J. D.,** Measurement of flame temperatures by a two-line atomic absorption method, *Anal. Chem.,* 44, 247, 1972.
6. **Omenetto, N., Benetti, P., and Rossi, G.,** Flame temperature measurements by means of atomic fluorescence spectrometry, *Spectrochim. Acta,* 27B, 253, 1972.
7. **Herzfield, C. M.,** Temperature, Its Measurement and Control in Science and Industry, Vol. III (Part 2), Dahl, I., Ed., Van Nostrand Reinhold, New York, 1962.
8. **Alkemade, C. Th. J., Hollander, Tj., Snelleman, W., and Zeegers, P. J. Th.,** *Metal Vapours in Flames,* Pergamon Press, New York, 1982.

FUNDAMENTAL DATA FOR THE COMMON TRANSITIONS

Insofar as possible, the fundamental data for the transitions commonly used with the methods discussed in this section are given in Table 20. The transition in nanometers (nm), the type of transition (*I* indicates atomic and *II* indicates ionic), the lower and upper energy levels, E-low and E-high, in cm^{-1}, the statistical weight, $g(i)$, of the lower level (i), the transition probability, $A(ji)$, in s^{-1}, and the merit and reference for the transition probability are listed. In some cases the $g(i)$ and the $A(ji)$ were available only in the multiplied form, and in these cases the "$gA = xx$" format was used. If a blank appears, no information was available for that specific column.

Table 20
FUNDAMENTAL DATA FOR THE COMMON TRANSITIONS

Element	Symbol	Wavelength (nm)	Type	E-low (cm^{-1})	E-high (cm^{-1})	g(i)	A(ji) $10^8 s^{-1}$	Merit[a]	Ref.
Aluminum	Al	167.0787	II						
		308.2153	I	0	32,435	4	0.63	C	3
		309.2710	I	112	32,437	6	0.73	C	3
		396.1520	I	112	25,348	2	0.98	C	3
Antimony	Sb	206.833	I	0	48,332	6	42	E	2
		217.581	I	0	45,945	4	13.8	E	2
		231.147	I	0	43,249	2	3.75	E	2
		259.805	I	8,512	46,991	2	32	E	2
Arsenic	As	189.042	I		52,898	6	2.0	D	3
		193.759	I	0	51,610	4	2.0	D	3
		197.262	I	0	50,694	2	2.0	D	3
		234.984	I	10,592	53,136	4	3.1	D	3
		286.044	I	18,186	53,136	2	0.55	D	3
Barium	Ba	455.403	II	0	21,952	4	1.17	A	3
		493.409	II	0	20,262	2	0.955	B	3
		553.548	I	0	18,080	3	1.15	B	3
Beryllium	Be	234.861	I	0	42,565	3	5.56	B	3
		313.042	II	0	31,935	4	1.14	B	3
		313.107	II	0	31,929	2	1.15	B	3
Bismuth	Bi	223.061	I	0	44,817	4	0.25	D	3
		289.798	I	11,418	45,916	2	1.53	C	3
Boron	B	208.891	I	0	47,857	4	0.28	D	3
		208.957	I	16	47,857	6	0.33	D	3
		249.677	I	0	40,040	2	0.84	C	3
		249.773	I	16	40,040	2	1.69	C	3
Bromine	Br	470.486	II		115,176	7	1.1	D	3
		827.244	I						
Cadmium	Cd	214.441	II	0	46,619	4	2.8	C	3
		226.502	II	0	44,136	2	3.0	C	3
		228.8022	I	0	43,692	3	0.24	D	3
		326.1055	I	0	30,656	3	0.004	C	3
Calcium	Ca	364.441	I	15,316	42,747	7	0.355	C	3
		393.366	II	0	25,414	4	1.47	C	3
		396.847	II	0	25,192	2	1.4	C	3
		422.673	I	0	23,652	3	2.18	B	3
Carbon	C	193.0905	I		61,982	3	3.7	D	3
		247.856	I	21,648	61,982	3	0.18	D	3
Cerium	Ce	394.275	II	6,913	32,269	gA = 19		E	2
		413.765	II	4,166	28,327	gA = 4.8		E	2
		418.660	II	6,968	30,847	gA = 18		E	2

Table 20 (continued)
FUNDAMENTAL DATA FOR THE COMMON TRANSITIONS

Element	Symbol	Wavelength (nm)	Type	E-low (cm⁻¹)	E-high (cm⁻¹)	g(i)	A(ji) 10⁸ s⁻¹	Merit[a]	Ref.
Cesium	Cs	455.5276	I	0	21,946	4	0.019	C	3
		852.1122	I	0	11,732	4	0.32	E	2
Chlorine	Cl	413.250	II		153,259	5	1.6	D	3
		837.594	I						
Chromium	Cr	205.552	II	0	48,632	gA =	9.1	E	2
		267.716	II	12,304	49,646	gA =	132	E	2
		357.869	I	0	27,935	gA =	8.3	E	2
		425.435	I	0	23,499	9	0.315	B	3
Cobalt	Co	228.615	II	3,350	47,078	gA =	169	E	2
		238.892	II	3,350	45,198	gA =	278	E	2
		240.725	I	0	41,529	12	3.08	E	2
		352.685	I	0	28,346	10	0.12	C	3
Copper	Cu	213.5981	II	0	30,784	4	1.39	B	3
		324.754	I	0	30,784	4	1.39	B	3
		327.396	I	0	30,535	2	1.37	B	3
Dysprosium	Dy	353.170	II	0	28,307	gA =	19	E	2
		404.597	I	0	24,709	15	1.5	D	3
		421.172	I	0	23,737	19	2.08	C	3
Erbium	Er	337.271	II	0	29,641	gA =	13	E	2
		400.796	I	0	24,943	15	26	D	3
Europium	Eu	381.967	II	0	26,173	gA =	4.8	E	2
		459.403	I	0	21,761	10	1.4	D	3
Fluorine	F	685.603	I		116,987	8	0.42	D	3
Gadolinium	Gd	342.247	II	1,935	31,146	gA =	19	E	2
		368.413	I	0	27,136	gA =	12	E	2
		440.186	I	1,719	24,430	gA =	4.2	E	2
Gallium	Ga	287.424	I	0	34,782	4	1.2	C	3
		294.364	I	826	34,788	6	1.4	C	3
		417.204	I	826	24,789	2	0.92	C	3
Germanium	Ge	199.8887	I	1,410	51,438	5	0.55	C	3
		209.4258	I	1,410	49,144	7	0.97	C	3
		265.1172	I	1,410	39,118	5	2.0	C	3
Gold	Au	242.795	I	0	41,174	4	1.5	D	3
		267.595	I	0	37,359	2	1.1	D	3
Hafnium	Hf	277.336	II	6,344	42,391	gA =	14	E	2
		307.288	I	0	32,533	gA =	3.2	E	2
		339.980	II	0	29,405	gA =	1.1	E	2
Holmium	Ho	345.600	II						
		389.102	II	637	26,331				
		405.393	I	0	24,660				
		410.384	I	0	24,361				
Hydrogen	H	486.133	I	82,259	102,824	32	0.084	A	17
		656.2852	I	82,259	97,492	18	0.441	A	17
Indium	In	230.605	II	0	43,349	gA =	0.032	E	2
		303.936	I	0	32,892	gA =	7.1	E	2
		325.609	I	2,213	32,915	6	1.3	D	3
		451.131	I	2,213	24,373	2	1.02	C	3
Iodine	I	183.038	I		56,093	4	2.71	C	3
		206.163	I						
Iridium	Ir	208.882	I	0	47,858	12	28	E	2
		224.268	II	0	44,576				
		263.971	I	0	37,872	10	0.56	E	2

Table 20 (continued)
FUNDAMENTAL DATA FOR THE COMMON TRANSITIONS

Element	Symbol	Wavelength (nm)	Type	E-low (cm⁻¹)	E-high (cm⁻¹)	g(i)	A(ji) 10⁸ s⁻¹	Meritᵃ	Ref.
Iron	Fe	238.204	II	0	41,968	gA	= 92	E	2
		248.3271	I	0	40,257	11	4.9	C	3
		259.9396	II	0	38,459	10	2.22	C	3
		371.9935	I	0	26,875	11	0.163	B	3
Lanthanum	La	333.749	II	3,250	33,204	gA	= 3.5	E	2
		408.672	II	0	24,463	5	0.20	E	2
		550.134	I	0	18,172	4	0.08	E	2
Lead	Pb	217.000	I	0	46,068	3	1.5	D	3
		220.3534	II	14,081	59,448	gA	= 5.7	E	2
		283.3053	I	0	35,287	3	0.58	D	3
		368.3462	I	7,819	34,960	1	1.5	D	3
Lithium	Li	670.776	I	0	14,904	4	0.372	B	3
Lutetium	Lu	261.542	II	0	38,223	gA	= 5.8	E	2
		451.857	I	0	22,125	4	0.21	B	3
Magnesium	Mg	279.553	II	0	35,761	4	4.0	C	3
		280.270	II	0	35,669	2	2.6	C	3
		285.213	I	0	35,051	3	5.3	D	3
Manganese	Mn	257.610	II	0	38,807	9	8.89	E	2
		279.482	I	0	35,770	8	3.7	C	3
		403.076	I	0	24,802	8	0.19	C	3
Mercury	Hg	184.905	II						
		194.227	II						
		253.652	I	0	39,412	3	0.13	D	3
Molybdenum	Mo	202.030	II	0	49,481	gA	= 24	E	2
		313.259	I	0	31,913	9	1.09	E	2
		379.825	I	0	26,321	9	0.49	E	2
		390.296	I	0	25,614	5	0.42	E	2
Neodymium	Nd	401.225	II	5,086	30,002	20	0.55	D	3
		463.424	I	0	21,572	gA	= 2.0	E	2
		492.453	I	0	20,301	gA	= 2.0	E	2
Nickel	Ni	221,648	II		53,496	12	5.5	D	3
		232.003	I	0	43,090	11	6.9	C	3
		352.454	I	205	28,569	5	1.0	C	3
Niobium	Nb	309.418	II	4,146	36,455	13	1.1	E	2
		334.906	I	2,154	32,005	10	0.45	E	2
		405.894	I	1,050	25,680	12	0.65	E	2
Nitrogen	N	174.2729	I						
		821.634	I						
Osmium	Os	225.585	II	0	44,315				
		290.906	I	0	34,365	11	1.0	E	2
		442.047	I	0	22,616	9	0.034	E	2
Oxygen	O	436.825	I						
		777.194	I		86,631	7	0.34	B	3
Palladium	Pd	244.791	I	0	40,839	3	0.28	E	2
		247.642	I	0	40,369	3	0.37	E	2
		340.458	I	6,564	35,928	9	1.33	E	2
		363.470	I	6,564	34,069	5	1.24	E	2
Phosphorus	P	177.499	I	0	56,340	6	2.17	C	3
		213.547	I	11,362	58,174	4	0.211	C	3
		253.561	I	18,748	58,174	4	0.20	C	3
Platinum	Pt	214.423	I	0	46,622	7	5.14	E	2
		265.945	I	0	37,591	9	0.91	E	2
Potassium	K	766.490	I	0	13,043	4	0.387	B	3

Table 20 (continued)
FUNDAMENTAL DATA FOR THE COMMON TRANSITIONS

Element	Symbol	Wavelength (nm)	Type	E-low (cm^{-1})	E-high (cm^{-1})	g(i)	A(ji) 10^8 s^{-1}	Merit[a]	Ref.
Praseodymium	Pr	390.805	II						
		422.535	II	0	23,660		gA = 1.4	E	2
		495.137	I	0	20,190				
Rhenium	Re	197.3							
		221.426	II	0	45,148		gA = 15	E	2
		346.046	I	0	28,890				
Rhodium	Rh	233.477	II	16,885	59,702		gA = 44	E	2
		343.489	I	0	29,105	12	0.34	E	2
		369.236	I	0	27,075	8	0.35	E	2
Rubidium	Rb	420.180	I	0	23,793	4	0.018	C	3
		780.027	I	0	12,817	4	0.370	B	3
Ruthenium	Ru	240.272	II	9,152	50,758		gA = 247	E	2
		349.894	I	0	28,572	13	0.46	E	2
		372.803	I	0	26,816	11	0.42	E	2
Samarium	Sm	359.260	II	3,053	30,880		gA = 6.3	E	2
		373.912	II	326	27,063				
		429.674	I	4,021	27,288		gA = 21	E	2
		476.027	I	812	21,813		gA = 3.3	E	2
Scandium	Sc	361.384	II	178	27,841	9	0.14	D	3
		391.181	I	168	25,725	8	1.37	C	3
Selenium	Se	196.09	I	0	50,997	2	100	E	2
		203.98	I	1,989	50,997	2	65	E	2
Silicon	Si	251.6113	I	223	39,955	5	1.21	C	3
		288.1579	I	6,299	40,992	3	1.89	C	3
Silver	Ag	328.068	I	0	30,473	4	1.4	B	3
		338.2068	I	0	29,552	2	1.3	B	3
Sodium	Na	330.237	I	0	30,273	4	0.028	C	3
		588.9950	I	0	16,973	4	0.622	A	3
		589.5924	I	0	16,956	2	0.618	A	3
Strontium	Sr	407.771	II	0	24,517	4	1.42	C	3
		421.552	II	0	23,715	2	1.27	C	3
		460.733	I	0	21,698	3	2.01	B	3
Sulfur	S	180.7311	I		55,331	3	3.8	C	3
		182.0343	I		55,331	3	2.2	C	3
		216.89							
Tantalum	Ta	226.230	II	2,642	46,831		gA = 35	E	2
		240.063	II	6,187	47,830		gA = 516	E	2
		271.467	I	0	36,826	6	1.17	E	2
		296.513	II	0	33,715		gA = 7.8	E	2
		474.016	I	9,976	31,066	4	0.028	E	2
Tellurium	Te	214.281	I	0	46,653	3	38	E	2
		238.578	I	4,751	46,653	3	5.47	E	2
Terbium	Tb	350.917	II	0	28,488				
		367.635	II	1,016	28,209				
		432.643	I	0	23,107		gA = 7.2	E	2
Thallium	Tl	190.864	II						
		276.787	I	0	36,118	4	1.26	C	3
		377.572	I	0	26,478	2	0.625	B	3
		535.046	I	7,793	26,478	2	0.705	B	3
Thorium	Th	283.7295	II	6,214	41,448				
		324.4448	I	0	30,813		gA = 0.12	E	2
		401.9129	II	0	24,874		gA = 0.66	E	2
		491.9816	II	6,168	26,489		gA = 0.08	E	2

Table 20 (continued)
FUNDAMENTAL DATA FOR THE COMMON TRANSITIONS

Element	Symbol	Wavelength (nm)	Type	E-low (cm^{-1})	E-high (cm^{-1})	g(i)	A(ji) 10^8 s^{-1}	Merit[a]	Ref.
Thulium	Tm	313.126	II	0	31,927	gA = 4.6		E	2
		346.220	II	0	28,875	gA = 2.5		E	2
		371.791	I	0	26,889	gA = 8.3		E	2
Tin	Sn	189.991	II				1.6	D	3
		224.605	I	0	44,509	3	1.7	D	3
		235.484	I	1,692	44,145	5	1.7	D	3
		283.999	I	3,428	38,629	5	2.7	D	3
		326.234	I	8,613	39,257	3			
Titanium	Ti	334.941	II	393	30,241	12	1.3	D	3
		364.268	I	170	27,615	9	0.67	C	3
		365.350	I	387	27,750	11	0.66	C	3
		368.520	II	4,898	32,026				
Tungsten	W	207.911	II	6,147	54,229	gA = 93		E	2
		255.135	I	0	39,183	7	1.17	E	2
		276.427	II	0	36,165	gA = 6.9		E	2
		400.875	I	2,951	27,890	9	0.20	E	2
Uranium	U	263.553	II						
		358.488	I	0	27,887	15	0.10	B	3
		385.957	II	289	26,191	gA = 2.6		E	2
		591.539	I	0	16,900	gA = 0.12		E	2
Vanadium	V	309.311	II	3,163	35,483	13	1,8	D	3
		311.062	II	2,809	34,947	9	1.5	D	3
		318.540	I	553	31,937	12	1.4	D	3
		437.924	I	2,425	25,254	12	1.2	D	3
Ytterbium	Yb	328.937	II	0	30,392	4	1.8	C	3
		369.419	II	0	27.062	2	1.4	C	3
		398.799	I	0	25,068	3	1.76	C	3
Yttrium	Y	362.094	I	530	28,140	4	1.55	E	2
		371.030	II	1,450	28,394				
		377.433	II	1,045	27,532				
		410.238	I	530	24,900	8	0.64	E	2
Zinc	Zn	202.548	II	0	49,355	4	3.3	C	3
		213.856	I	0	46,745	3	7.09	B	3
Zirconium	Zr	343.823	II	763	29,840	gA = 13		E	2
		351.960	I	0	28,404	7	0.71	E	2
		360.119	I	1,241	29,002	11	0.91	E	2

[a] The key for the merit of the A(ji) values follows that given in Reference 3 as follows: A, within 3%; B, within 10%; C, within 25%; D, within 50%; and E, greater than 50%.

X. Qualitative Tests

ORGANIC GROUP QUALITATIVE TESTS

The following flow charts and notes provide a step-by-step process for the identification of functional groups which may be present in an unknown sample.[1-11] These are meant to augment and confirm information obtainable using instrumental methods of analysis. It will usually be necessary to use gas or liquid chromatography before these "wet" chemical tests in order to determine the number of components present in a given sample. Since many of these tests require the use of dangerous compounds, the strictest rules of laboratory safety must be observed at all times. The use of a fume hood is recommended. The book by Feigl et al. is an excellent guide for spot rests.[12]

REFERENCES

1. **Pasto, D. J. and Johnson, C. R.**, *Laboratory Text for Organic Chemistry*, Prentice-Hall, Englewood Cliffs, NJ, 1979.
2. **Sarlo, E. and Svoronos, P. D. N.**, *Experiments for Organic Chemistry*, Burgess Publishing, Minneapolis, 1983.
3. **Roberts, R. M., Gilbert, J. C., Rodewald, L. B., and Wingrove, A. S.**, *Modern Experimental Organic Chemistry*, Sanders, New York, 1985.
4. Hopkins and Williams Research Laboratory, *Organic Reagents for Organic Analysis*, Chemical Publishing, Brooklyn, 1946.
5. **Hodgman, C. D., Weast, R. C., and Selby, S. M.**, *Tables for Identification of Organic Compounds*, Chemical Rubber Publishing, Cleveland, 1950.
6. **Kamm, O.**, *Qualitative Organic Analysis*, John Wiley & Sons, New York, 1932.
7. **McElvain, S. M.**, *The Characterization of Organic Compounds*, rev. ed., Macmillan, New York, 1953.
8. **Shriner, R. L. and Fuson, R. C.**, *The Systematic Identification of Organic Compounds, a Laboratory Manual*, 3rd ed., John Wiley & Sons, New York, 1978.
9. **Vogel, A. I.**, *A Textbook of Practical Organic Chemistry*, Longmans, Green, London, 1951.
10. **Behforouz, M.**, Getting the acid out of your 2,4-DNPH, *J. Chem. Ed.*, 63, 723, 1986.
11. **Durst, H. D. and Gokel, G. W.**, *Experimental Organic Chemistry*, McGraw-Hill, New York, 1987.
12. **Feigl, F. Anger, V., and Oesper, R. E.**, *Spot Tests in Organic Analysis*, Elsevier, Amsterdam, 1966.

Protocol for Chemical Tests for Organic Compounds

The following section gives a suggested protocol for some of the most popular chemical tests used in the identification of organic compounds. Variations of the procedures are possible, but these protocols have been used successfully for most organic identifications.[1-8]

Basic hydrolysis — Reflux 0.5 g of the compound in 5 mL of a 10% sodium hydroxide solution.

Bromine test — The compound to be tested is treated with a few drops of 5% Br_2/CCl_4 solution. A positive test is indicated by decolorization of the bromine color.

Ceric ammonium nitrate — To 1 to 2 ml of 5% ceric ammonium nitrate add ten drops of the compound to be tested. A change to an orange/red color is indicative of an alcohol (detection limit 100 mg; compounds tested C_1 to C_{10}).

Dichromate test — Add ten drops of the alcohol to be tested to a mixture of 1 ml 1% $Na_2Cr_2O_7$ and 5 drops concentrated H_2SO_4. A blue-green solution is positive test for a 1 or 2° alcohol; 3° alcohols do not react and, therefore, the solution stays orange (detection limit 20 μg; compounds tested C_1 to C_8).

2,4-Dinitrophenylhydrazine test (2,4-DNP) — Add ten drops of the compound to be tested to 1 mL of the 2,4-DNP reagent. A yellow to orange-red precipitate is considered a positive test. The crystals can be purified by washing them with 5% $NaHCO_3$ and then with water, and finally recrystallized from ethanol. The 2,4-DNP reagent can be prepared by dissolving 1 g 2,4-dinitrophenylhydrazine in 5 mL concentrated H_2SO_4 and then mixing it with 8 mL of water and 20 mL of 95% ethanol. The solution should be filtered before reacting it with the unknown compound (detection limit 20 μg; compounds tested C_1 to C_8).

Ferric chloride test — Add ten drops of 3% aqueous $FeCl_3$ solution to 1 mL of a 5% aqueous (ethanol) solution of the compound in question. Phenols give red, blue, purple, or green colorations. The same test can be done by using chloroform as a solvent (detection limit 50 μg).

Hinsberg test — To 0.5 mL of the amine (0.5 g, if solid) in a test tube, add 1 g of *p*-toluenesulfonyl chloride and 8 mL 10% NaOH. Stopper the tube and shake for 3 to 5 min. Remove the stopper and warm the tube with shaking in a hot-water bath (70°C) for about 1 min. No reaction is indicative of a 3° amine, which is usually soluble upon acidification (pH = 2 to 4) with 10% HCl. If a precipitate is present in the alkaline solution, dilute with 5 to 8 mL of H_2O and shake. If the precipitate does not dissolve, the original amine is probably, a 2° one. If the solution is clear, acidify (pH = 4) with 10% HCl. The formation of a precipitate is an indication of a 1° amine (detection limit 100 mg; compounds tested C_1 to C_{10}.)

Iodoform test — The reagent calls for the mixture of 10 g I_2 and 20 g KI in 100 mL of water. The reagent is then added dropwise to a mixture of ten drops of the compound in question in 2 mL of water (or dioxane, to facilitate the solubility) and 1 mL 10% aqueous NaOH solution until a *persistent* brown color remains (even when heating in a hot-water bath at 60 °C.) A yellow precipitate is indicative of iodoform (CHI_3) formation and is characteristic of a methyl ketone, acetaldehyde, or an alcohol of the general formula

$$CH_3CHOH$$
$$|$$
$$R$$

(R = alkyl, hydrogen). Aldols [$RCOCH_2CH(OH)R'$] also give a positive iodoform test by a retro aldo condensation first yielding $RCOCH_3$ and RCHO (detection limit 100 mg).

Lucas test — The reagent is made by dissolving 16 g of anhydrous $ZnCl_2$ in 10 mL concentrated HCl with cooling to avoid HCl loss. Add 10 to 15 drops of the alcohol to 2 mL of the reagent; 3° alcohols form an emulsion (the alkyl halide) almost immediately; 2° alcohols form it after 2 to 5 min; while 1° alcohols react after a very long time.

Permanganate test — The compound to be tested is treated with 10 to 15 drops of 5% $KMnO_4$ solution. A positive test is indicated by the formation of a black (MnO_2) precipitate.

Silver nitrate test — The compound to be tested is treated with a few drops of 1% alcoholic silver nitrate. A white precipitate indicates a positive reaction.

Sodium fusion test — Fifty milligrams of the compound to be analyzed are mixed with a fresh piece of sodium metal of the size of a small pea in 4-in. test tube. The test tube is warmed gently until melting of the sodium metal and decomposition (indicated by charring) of the compound occurs. When it appears that all the volatile material has been decomposed, the test tube is strongly heated until the residue becomes red. After 3 min of constant heating, the mixture is left to cool to room temperature, then a few drops of methanol are added. If no smoke appears, and excess of sodium metal is not present and incomplete conversion of the elements (N, S, halogens) to their anions (CN^-, S^{-2}, halides) is very likely. Addition of another tiny piece of sodium metal and repetition of the heating process is necessary. If smoke appears, then the test tube is plunged in a small beaker containing 10 to 15 ml *distilled water* and covered with a watch glass or a wire gauze. The test tube might shatter and, therefore, having the small beaker inside a larger one is recommended. The contents of the test tube together with the broken glass are ground in a mortar using a pestle, then transferred in the small beaker and heated for a few minutes. The solution is then filtered and divided into *two* larger parts and *one* 1-ml part. Further tests follow:

1.   **Detection of nitrogen** — To one of the larger parts add 0.5 mL of 6 *M* NaOH (pH

adjusted to 13), five drops of saturated $Fe(NH_4)_2(SO_4)_2$ and five drops of 30% KF solution. The mixture is boiled for 30 s and immediately acidified with 6 M H_2SO_4 with stirring until the colloidal iron hydroxides are dissolved. The formation of a blue color is indicative of the presence of nitrogen.

2. **Detection of sulfur** — To the 1-mL part, add 10 drops of 6 M acetic acid and 2 to 3 drops of 5% lead acetate solution. A black precipitate is indicative of sulfur presence.

3. **Detection of halogens** — To the other larger part add 10% H_2SO_4 until the solution is acidic. Boil off the solution to one third of its volume to secure evaporation of H_2S and HCN gases. Formation of a precipitate upon addition of 10% $AgNO_3$ solution is indicative of the presence of a halogen — white for a chloride (which is soluble in 6 M NH_4OH), pale yellow for a bromide (which is only slightly soluble in 6 M NH_4OH), and canary yellow for an iodide (which is insoluble in 6 M NH_4OH). Should the color of the precipitate be such as to make satisfactory identification of the halogen difficult, proceed as follows: The working solution, which has been acidified with 10% H_2SO_4 and boiled down, is treated with four to give drops of 0.1 N $KMnO_4$ solution, with enough oxalic acid added to discharge the color of excess permanganate and 0.5 mL of carbon disulfide. The presence of a color in the carbon-disulfide layer indicates the presence of bromine or iodine — purple if iodine is present, red-brown if bromine is present, colorless if chlorine is present. Should the compound to be tested carry *both* bromine and iodine, the identification will be difficult (red-brown to purple carbon-disulfide layer). Addition of a few drops of allyl alcohol decolorizes bromine but does not decolorize iodine.

Tollen's test — The reagent is actually made by mixing two solutions (A and B). Solution A is a 10% aqueous $AgNO_3$ solution and solution B is a 10% aqueous NaOH solution. When the test is required, 1 mL of Solution A and 1 mL of Solution B are mixed, and the silver oxide thus formed is dissolved by dropwise addition of 10% aqueous NH_4OH. To the clear solution, ten drops of the compound to be tested are added. A silver mirror is indicative of the presence of an aldehyde. The reagent mixture (A + B) is to be prepared *immediately prior to use*, otherwise explosive silver fulminate might be formed. The silver mirror is usually deposited on the walls of the test tube immediately or after a short warming period in a hot-water bath. It is to be disposed of immediately with water and/or diluted HNO_3 (detection limit 50 mg; compounds tested C_1 to C_6).

REFERENCES

1. **Shriner, R. L. and Fuson, R. C.,** *Identification of Organic Compounds,* John Wiley & Sons, New York, 1948.
2. **Vogel, A. I.,** *A Textbook of Practical Organic Chemistry,* Longmans, Green, London, 1951.
3. **Pasto, D. J. and Johnson, C. R.,** *Laboratory Text for Organic Chemistry,* Prentice-Hall, Englewood Cliffs, NJ, 1979.
4. **Roberts R. M., Gilbert, J. C., Rodewald, L. B., and Wingrove, A. S.,** *Modern Experimental Organic Chemistry,* W. B. Saunders, New York, 1985.
5. **Uamm, O.,** *Qualitative Organic Analysis,* John Wiley & Sons, New York, 1932.
6. **Behforouz, M.,** Getting the acid out of your 2,4-DNPH, *J. Chem. Ed.,* 63, 723, 1986.
7. **Durst, H. D. and Gokel, G. W.,** *Experimental Organic Chemistry,* McGraw-Hill, New York, 1987.
8. **Fieser, L. F. and Freser, M.,** *Reagents for Organic Synthesis,* John Wiley & Sons, New York, 1968.

ORGANIC GROUP QUALITATIVE TESTS

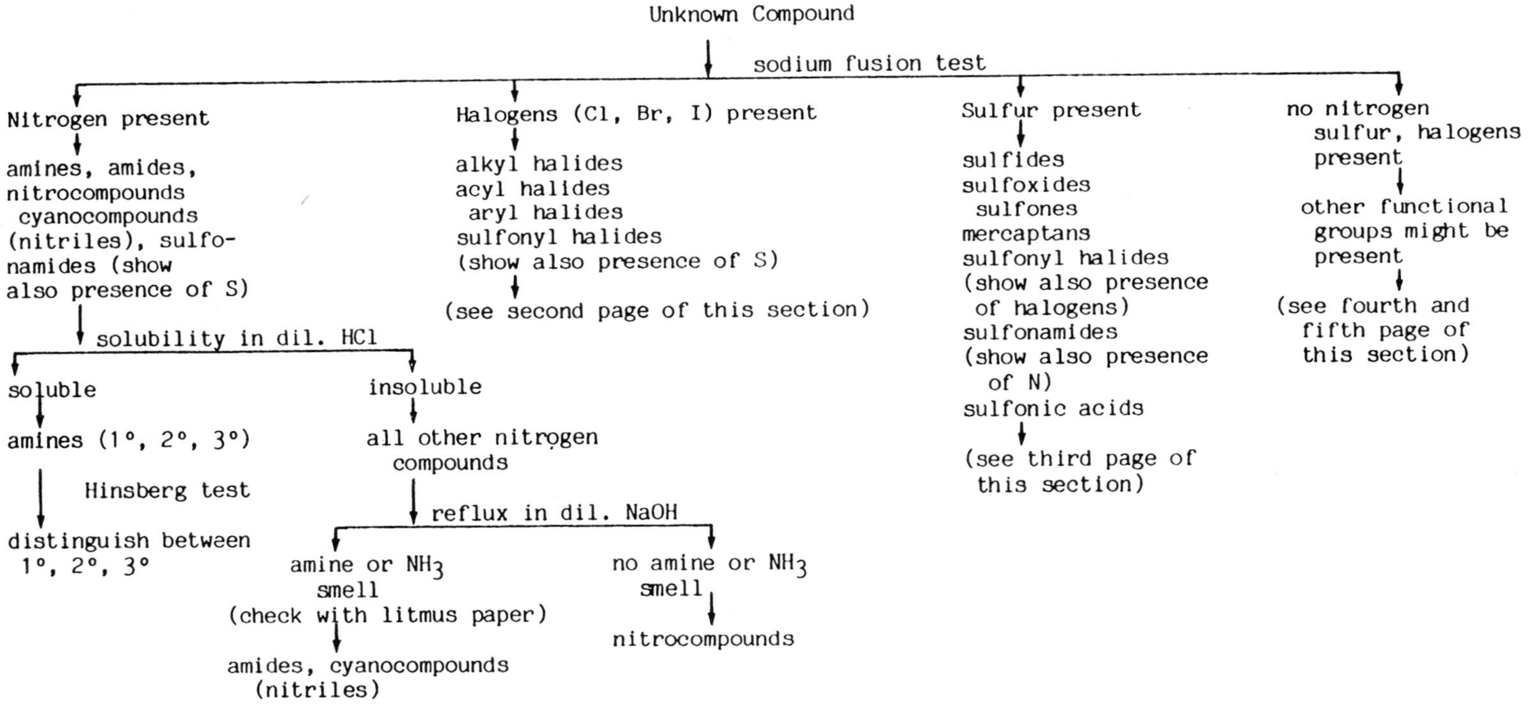

Unknown Compound

sodium fusion test

Nitrogen present

amines, amides,
nitrocompounds
cyanocompounds
(nitriles), sulfo-
namides (show
also presence of S)

solubility in dil. HCl

soluble — amines (1°, 2°, 3°)

Hinsberg test

distinguish between
1°, 2°, 3°

insoluble — all other nitrogen
compounds

reflux in dil. NaOH

amine or NH₃
smell
(check with litmus paper)

amides, cyanocompounds
(nitriles)

no amine or NH₃
smell

nitrocompounds

Halogens (Cl, Br, I) present

alkyl halides
acyl halides
aryl halides
sulfonyl halides
(show also presence of S)

(see second page of this section)

Sulfur present

sulfides
sulfoxides
sulfones
mercaptans
sulfonyl halides
(show also presence
of halogens)
sulfonamides
(show also presence
of N)
sulfonic acids

(see third page of
this section)

**no nitrogen
sulfur, halogens
present**

other functional
groups might be
present

(see fourth and
fifth page of
this section)

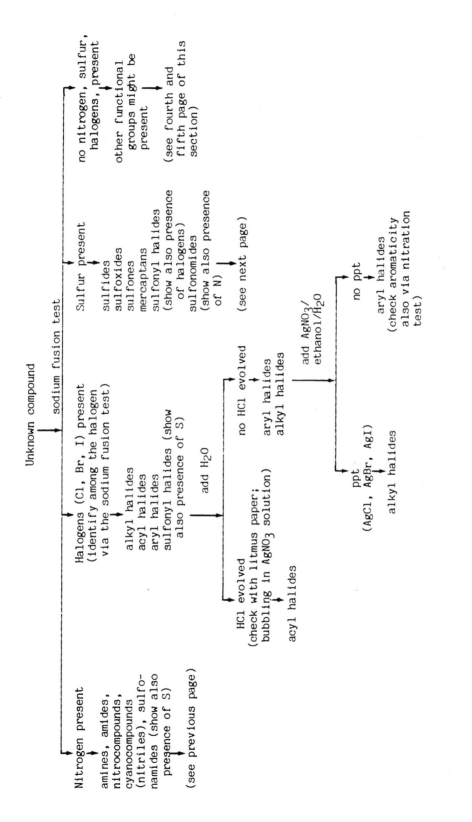

ORGANIC GROUP QUALITATIVE TESTS (continued)

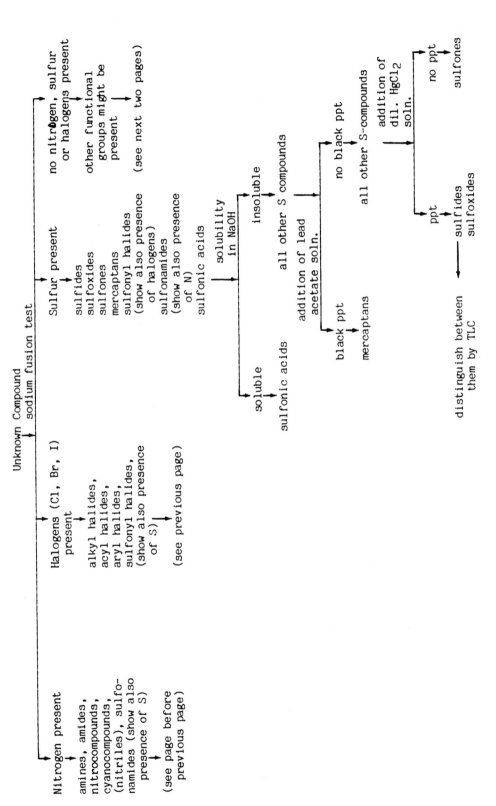

Unknown Compound
sodium fusion test

Nitrogen present

amines, amides, nitrocompounds, cyanocompounds, (nitriles), sulfonamides (show also presence of S)

(see page before previous page)

Halogens (Cl, Br, I) present

alkyl halides, acyl halides, aryl halides, sulfonyl halides, (show also presence of S)

(see previous page)

Sulfur present

sulfides
sulfoxides
sulfones
mercaptans
sulfonyl halides (show also presence of halogens)
sulfonamides (show also presence of N)
sulfonic acids

no nitrogen, sulfur or halogens present

other functional groups might be present

(see next two pages)

solubility in NaOH

soluble

sulfonic acids

insoluble

all other S compounds

addition of lead acetate soln.

black ppt

mercaptans

no black ppt

all other S-compounds

addition of dil. HgCl2 soln.

ppt

sulfides
sulfoxides

no ppt

sulfones

distinguish between them by TLC

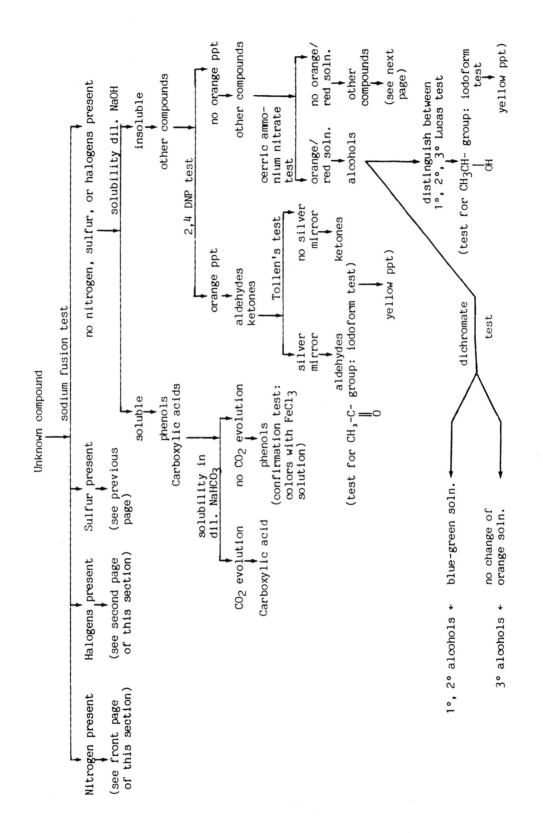

ORGANIC GROUP QUALITATIVE TESTS (continued)

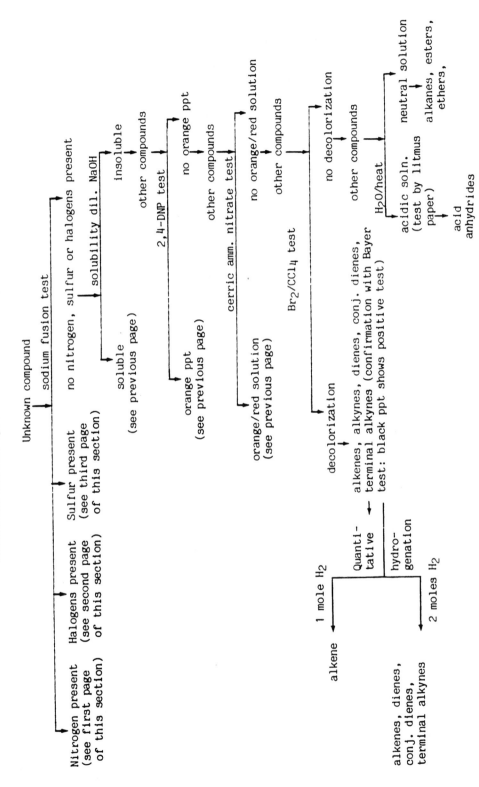

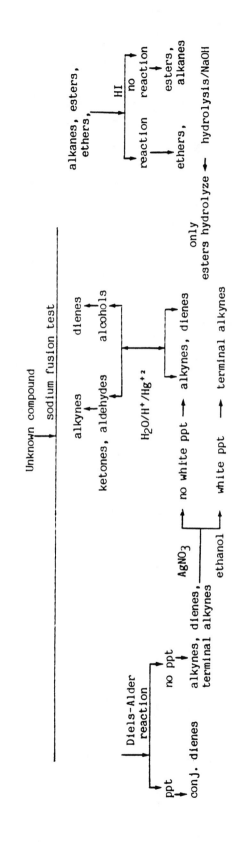

ORGANIC FAMILIES AND CHEMICAL TESTS

The following section gives the major organic families and their most important confirmatory chemical tests. This part serves as a complement to the previous section (Organic Group Qualitative Tests).[1-7]

REFERENCES

1. **Pasto, D. J. and Johnson, C. R.,** *Laboratory Text for Organic Chemistry,* Prentice-Hall, Englewood Cliffs, NJ, 1979.
2. **Roberts, R. M., Gilbert, J. C., Rodewald, L. B., and Wingrove A. S.,** *Modern Experimental Organic Chemistry,* Saunders, New York, 1985.
3. **Kamm, O.,** *Qualitative Organic Analysis,* John Wiley & Sons, New York, 1932.
4. **Sarlo, E. and Svoronos, P. D. N.,** *Experiments for Organic Chemistry,* Burgess Publishing, Minneapolis, 1983.
5. **Vogel, A. I.,** *A Textbook of Practical Organic Chemistry,* Longmans, Green, London, 1951.
6. **Durst, H. D. and Gokel, G. W.,** *Experimental Organic Chemistry,* McGraw-Hill, New York, 1980.
7. **Feigl, F., Anger, and Oesper, R. E.,** *Spot Tests in Organic Analysis,* Elsevier, Amsterdam, 1966.

ORGANIC FAMILIES AND CHEMICAL TESTS

Family	Test	Notes
Alcohols	Ceric ammonium nitrate	Positive for all alcohols
	Dichromate test	Positive for 1 and 2° alcohols; negative for 3° alcohols
	Iodoform test	Positive for all alcohols of the general formula $CH_3CH(OH)R$
	Lucas test	Immediate reaction for 3° or benzylic alcohols; slower reaction (5 min) for 2°; no reaction for 1° alcohols
Aldehydes	Benedict's test	Positive for all aldehydes
	Dichromate test	Positive for all aldehydes
	2,4-Dinitrophenylhydrazine (2,4-DNP)	Positive for all aldehydes (and ketones)
	Fehling test	Positive for all aldehydes
	Iodoform test	Positive only for acetaldehyde
	Oxime test	Positive for all aldehydes (and ketones)
	Permanganate test	Positive for all aldehydes
	Semicarbazone test	Positive for all aldehydes (and ketones)
	Tollen's test	Positive for all aldehydes
Alkanes	No test	—
Alkenes	Bromine test	Positive for all alkenes
	Permanganate test	Positive for all alkenes
	Sulfuric acid	All alkanes dissolve
Alkynes	Bromine test	Positive for all alkynes
	Permanganate test	Positive for all alkynes
	Silver nitrate test	Positive for all terminal alkynes only
	Sodium metal test	Positive for all terminal alkynes only
	Sulfuric acid	Positive for all alkynes
Amides	Basic hydrolysis	All amides yield ammonia or the corresponding amine
Amines	Diazotization test	All 1° amines give red azeo dyes with β-naphthol
	Hinsberg test	Distinguishes between 1, 2, or 3°
	PTC-Hoffman test	Positive only for 1° amines
	Carbylamine test	
	Solubility in dilute hydrochloric acid	All amines soluble

ORGANIC FAMILIES AND CHEMICAL TESTS (continued)

Family	Test	Notes
Arenes	Aluminum chloride-chloroform	Positive for all arenes
Aryl halides	Aluminum chloride-chloroform	Positive for all arylhalides
Carboxylic acids	Solubility in dilute sodium bicarbonate	All carboxylic acids are soluble
	Solubility in dilute sodium hydroxide	All carboxylic acids are soluble
Ketones	2,4-Dinitrophenyl-hydrazine (2,4 DNP) test	Positive for all ketones (and aldehydes)
	Hydrazine test	Positive for all ketones (and aldehydes)
	Iodoform test	Positive for methyl ketones
	Oxime test	Positive for all ketones (and aldehydes)
	Semicarbazone test	Positive for all ketones (and aldehydes)
Nitriles	Basic hydrolysis	Positive for all nitriles
Phenols	Acetylation	Ppt., of a characteristic melting point[a]
	Benzoylation	Ppt., of a characteristic melting point[a]
	Sulfonation	Ppt., of a characteristic melting point[a]
	Ferric chloride test	Variety of colors
	Solubility in aqueous base	Most phenols are soluble in dilute sodium hydroxide but insoluble in dilute sodium bicarbonate; Phenols with strong electron withdrawing groups (e.g., picric acid) are soluble in sodium bicarbonate
Sulfonamides	Basic hydrolysis	Positive for all sulfonamides
Sulfonic acids	Sodium fusion test	Presence of sulfur
	Solubility in aqueous base	Most sulfonic acids are soluble in dilute sodium hydroxide and generate carbon dioxide with sodium bicarbonate

[a]   Ppt., precipitate.

INORGANIC GROUP QUALITATIVE TESTS

The following tables list some simple chemical tests which will indicate the presence or absence of a given inorganic cation or anion.[1-4] For most of these tests, the anion or cation must be present at a relatively high concentration; the approximate lower bound is 0.05% unless otherwise specified. It may therefore be necessary to concentrate more dilute samples before successful results can be obtained. These tests should be used in conjunction with other methods such as the chromatographic methods or spectrometry and spectrophotometry. Since many of these tests require the use of dangerous compounds, the strictest rules of laboratory safety must be observed at all times.* The use of a fume hood is strongly recommended. All of the test reagents specified in this section are assumed to be in aqueous solution, unless otherwise designated. The reader is referred to the work of Svehla[3] for details on reagent preparation.

REFERENCES

1. **Barber, H. H . and Taylor, T. I.**, *Semimicro Qualitative Analysis*, Harper Brothers, New York, 1953.
2. **Bruno, T. J. and Svoronos, P. D. N.**, Basic Tables for Chemical Analysis, NBS Tech. Note 1096, National Bureau of Standards, U.S. Department of Commerce, Washington, D.C., 1986.
3. **Svehla, G.**, *Vogels Qualitative Inorganic Analysis*, John Wiley & Sons, New York, 1987.
4. **De, A. K.**, *Separation of Heavy Metals*, Pergamon Press, New York, 1961.

---

* It is illegal to dispose of cyanide solutions in municipal sewer systems (POTW-publicly owned treatment works.) The cyanide ion must be destroyed prior to disposal. This is easily done using an aqueous solution of chlorine.

$$CN^- + Cl_2 \rightarrow CNCl + Cl^-$$
$$CNCl + 2OH^- \rightarrow CNO^- + Cl^- + H_2O$$
$$2CNO^- + 4OH^- + 3Cl_2 + 2CO_2 + 2H_2O + N_2 + 6Cl^-$$

The first step will occur at all pH levels; the second requires high pH; the third requires a pH near 7. Thus, careful acidification using HCl will complete the reaction, producing harmless nitrogen and carbon dioxide. Actually, it may be permissible (by local ordinance) to dispose of CNO^- after the second step, since cyanate is far less toxic than cyanide.

TESTS FOR ANIONS

Acetates, CH_3COO^-

Sulfuric acid, dilute	Evolution of acetic acid (vinegar-like ordor); concentrated sulfuric acid also evolves sulfur dioxide upon mild heating
Ethanol and concentrated sulfuric acid	Evolves ethyl acetate (fruity odor); isoamyl alcohol may be substituted
Silver nitrate	White precipitate of silver acetate, in cold, concentrated solutions, soluble in dilute ammonia solution
Iron(III) chloride	Deep-red coloration (coagulates on boiling, forming a brownish-red precipitate)

Benzoates, $C_6H_5COO^-$ (or $C_7H_5O_2^-$)

Dilute sulfuric acid	White precipitate of benzoic acid forms in cold solutions
Silver nitrate	White precipitate of silver benzoate from cold solutions, soluble in hot water and also in dilute ammonia solution
Iron(III) chloride	Buff-colored precipitate of iron(III) benzoate from neutral solution, soluble in hydrochloric acid

Borates, BO_3^{3-}, $B_4O_7^{2-}$, BO_2^-

Concentrated sulfuric acid	Upon heating solution, white fumes of boric acid are evolved.
Silver nitrate	White precipitate of silver metaborate, soluble in dilute ammonia solution and in acetic acid
Barium chloride	White precipitate of barium metaborate, soluble in excess reagent, dilute acids, and in ammonium salt solutions

Bromates, BrO_3^-

Concentrated sulfuric acid	Evolution of bromine and oxygen in cold solution
Silver nitrate	White precipitate of silver bromate, soluble in dilute ammonia

Note: Bromates are reduced to bromides by sulfur dioxide, hydrogen sulfide, or sodium nitrite solution.

Bromides, Br^-

Concentrated sulfuric acid	Reddish-brown coloration; later, reddish-brown vapors (hydrogen bromide + bromine) are produced
Manganese dioxide + sulfuric acid	Reddish-brown bromine vapors evolve upon mild heating
Silver nitrate	Pale-yellow, curdy precipitate of silver bromide, slightly soluble in ammonia solution; insoluble in nitric acid
Lead acetate	White crystalline precipitate of lead bromide, soluble in hot water

Special tests: The addition of an aqueous solution of chlorine (or sodium hypochlorite) will liberate free bromine, which can be dissolved in a layer of carbon tetrachloride or carbon disulfide.

Carbonates, CO_3^{2-}

Hydrochloric acid	Decomposition with effervescence, evolution of carbon dioxide (odorless)
Barium chloride	White precipitate of barium carbonate, soluble in HCl (calcium chloride may be substituted)
Silver nitrate	Gray precipitate of silver carbonate

Special tests: Effervescence with all acids, giving carbon dioxide which renders limewater cloudy.

Chlorates, ClO_3^-

Concentrated sulfuric acid	Liberates chlorine dioxide gas (green); solids decrepitate (crackle explosively) when warmed; DANGER — may develop into a violent explosion
Concentrated hydrochloric acid	Chlorine dioxide gas evolved, imparts yellow color to acid
Manganese(II) sulfate + phosphoric acid	Violet coloration due to diphosphatomanganate formation; peroxodisulfate nitrates, bromates, iodates, and periodates, react similarly

Special tests: 1. Solids give off oxygen when heated.
               2. Acidified solutions bleach litmus paper.

TESTS FOR ANIONS (continued)

Chlorides, Cl^-
Concentrated sulfuric acid	Evolution of hydrogen chloride gas (pungent odor)
Silver nitrate	White precipitate of silver chloride, soluble in ammonia solution (reprecipitate in HNO_3)
Lead acetate	White precipitate of lead bromide, soluble in boiling water

Special tests: 1. Maganese dioxide + sulfuric acid evolve Cl_2 gas.
2. An aqueous solution of chlorine + carbon disulfide produces no coloration.

Chromates, CrO_4^{2-}; dichromates, $Cr_2O_7^{2-}$
Barium chloride	Pale-yellow precipitate of barium chromate, soluble in dilute mineral acids, insoluble in water and in acetic acid
Silver nitrate	Brownish-red precipitate of silver chromate, soluble in dilute nitric acid and in ammonia solution; insoluble in acetic acid
Lead acetate	Yellow precipitate of lead chromate, soluble in dilute nitric acid, insoluble in acetic acid
Hydrogen peroxide	Deep-blue coloration in acidic solution, which quickly turns green, with the liberation of oxygen
Hydrogen sulfide	Dirty yellow deposit of sulfur is produced in acidic solutions

Citrates, $C_6H_5O_7^{3-}$
Concentrated sulfuric	Evolution of carbon dioxide and carbon monoxide (HIGHLY POISONOUS)
Silver nitrate	White precipitate of silver citrate, soluble in dilute ammonia solution
Cadmium acetate	White gelatinous precipitate of cadmium citrate, practically insoluble in boiling water, soluble in warm acetic acid

Cyanates, OCN^-
Sulfuric acid, concentrated and dilute	Vigorous effervescence, due largely to evolution of carbon dioxide, with concentrated acid producing a more dramatic effect
Silver nitrate	Curdy white precipitate of silver cyanate
Copper sulfate-pyridine test	Lilac-blue precipitate (interference by thiocyanates); reagent is prepared by adding 2—3 drops of pyridene to 0.25 M $CuSO_4$ solution

Cyanides, CN^-
Dilute hydrochloric acid, cold	Liberation of hydrogen cyanide (odor of bitter almond); CAUTION — highly toxic
Silver nitrate	White precipitate of silver cyanide
Concentrated sulfuric acid, hot	Liberation of carbon monoxide (CAUTION)
Mercury(I) nitrate	Gray precipitate of mercury
Copper sulfide	Formation of colorless tetracyanocuprate (I) ions; (this test can be done on a section of filter paper)

Dithionites, $S_2O_4^{2-}$
Dilute sulfuric acid	Orange coloration which disappears quickly, accompanied by evolution of sulfur dioxide gas and precipitation of pale-yellow sulfur
Concentrated sulfuric acid	Fast evolution of sulfur dioxide and precipitation of pale-yellow sulfur
Silver nitrate	Black precipitate of silver
Copper sulfate	Red precipitate of copper
Mercury(II) chloride	Gray precipitate of mercury
Methylene blue	Decolorization in cold solution
Potassium hexacyanoferrate(II) and iron(II) sulfate	White precipitate of dipotassium iron(II) hexacyanoferrate(II); turns from white to Prussian blue

TESTS FOR ANIONS (continued)

Fluorides, F⁻

Concentrated sulfuric acid	Evolution of hydrogen fluoride dimer
Calcium chloride	White, slimy precipitate of calcium fluoride, slightly soluble in dilute hydrochloric acid
Iron(III) chloride	White precipitate

Special tests: HF etches glass (only visible after drying).

Formates, HCOO⁻

Dilute sulfuric acid	Formic acid evolved (pungent odor)
Concentrated sulfuric acid	Carbon monoxide (HIGHLY POISONOUS) is evolved on warming
Ethanol and concentrated H_2SO_4	Ethyl formate evolved (pleasant odor)
Silver nitrate	White precipitate of silver formate in neutral solutions, forming a black deposit of elemental silver upon mild heating
Iron(III) chloride	Red coloration due to complex formation
Mercury(II) chloride	White precipitate of calomel produced on warming; upon boiling, a black deposit of elemental mercury is produced

Hexacyanoferrate(II) ions, $[Fe(CN)_6]^{4-}$

Silver nitrate	White precipitate of silver hexacyanoferrate(II)
Iron(III) chloride	In neutral or acid conditions, Prussian blue precipitate which is decomposed by alkali bases
Iron(II) sulfate (aq)	White precipitate of potassium iron(II) hexacyanoferrate which turns blue by oxidation
Copper sulfate	Brown precipitate of copper hexacyanoferrate(II)
Thorium nitrate	White precipitate of thorium hexacyanoferrate(III)

Hexacyanoferrate(III) ions, $[Fe(CN_6)]^{3-}$

Silver nitrate	Orange-red precipitate of silver hexacyanoferrate(III), which is soluble in ammonia solution, but not in nitric acid
Iron(II) sulfate	Dark-blue precipitate in neutral or acid solution (Prussian or Turnbull's blue)
Iron(III) chloride	Brown coloration
Copper sulfate	Green precipitate of copper(II) hexacyanoferrate(III)
Concentrated hydrochloric acid	In cold solution, brown precipitate of hexacyanoferric acid

Hexafluorosilicates (silicofluorides), $[SiF_6]^{2-}$

Barium chloride	White, crystalline precipitate of barium hexafluorosilicate, insoluble in dilute HCl, slightly soluble in water
Potassium chloride	White gelatinous precipitate of potassium hexafluorosilicate, slightly soluble in water
Ammonia solution	Gelatinous precipitate of silica acid

Hydrogen peroxide, H_2O_2

Potassium iodide and starch	If sample is previously acidified by dilute sulfuric acid, a deep-blue coloration due to production of iodine
Potassium permanganate	Decolorization, evolution of oxygen
Titanium(IV) chloride	Orange-red coloration; very sensitive test

Special test: $4H_2O_2 + PbS \rightarrow PbSO_4 \downarrow + 4H_2O$

Black lead sulfide reacts to produce white lead sulfate.

Hypochlorites, OCl⁻

Dilute hydrochloric acid	Yellow coloration; later, chlorine gas is liberated
Lead acetate or lead nitrate	Brown lead dioxide forms upon heating
Cobalt nitrate	Black precipitate of cobalt(II) hydroxide
Mercury	On shaking slightly acidified solution of a hypochlorite with Hg, a brown precipitate of mercury(II) chloride is formed

TESTS FOR ANIONS (continued)

Hypophosphites, $H_2PO_2^-$

Silver nitrate	White precipitate of silver hypophosphite
Mercury(II) chloride	White precipitate of calomel in cold solution, which darkens upon warming
Copper sulfate	On warming the solution, red precipitate of copper(I) hydride forms
Potassium permanganate	Immediate decolorization under cold conditions

Iodates, IO_3^-

Silver nitrate	White, curdy precipitate of silver iodate, soluble in dilute ammonia solution
Barium chloride	White precipitate of barium iodate, sparingly soluble in hot water or dilute nitric acid; insoluble in ethanol and methanol
Mercury(II) nitrate	White precipitate of mercury(II) iodate

Iodides, I^-

Concentrated sulfuric acid	Evolves hydrogen iodide and iodine
Silver nitrate	Yellow precipitate of silver iodide, slightly soluble in ammonia solution; insoluble in dilute nitric acid
Lead acetate	Yellow precipitate of lead iodide, soluble in excess hot water
Potassium dichromate and concentrated sulfuric acid	Liberation of iodine
Sodium nitrite	Liberation of iodine
Copper sulfate	Brown mixed precipitate
Mercury(II) chloride	Scarlet precipitate of mercury(II) iodide

Special tests: 1. $MnO_2 + H_2SO_4$ produces I_2
2. Cl_2 (aq) + CS_2 produces I_2
3. Starch paste + Cl_2 (aq) yields deep-blue coloration

Metaphosphates, PO_3^-

Silver nitrate	White precipitate, soluble in dilute nitric acid, in dilute ammonia solution, and in dilute acetic acid
Albumin and dilute acetic acid	Coagulation
Zinc sulfate solution	White precipitate on warming; soluble in dilute acetic acid

Nitrates, NO_3^-

Concentrated sulfuric acid	Solid nitrate with concentrated sulfuric acid evolves reddish-brown vapors of nitrogen dioxide + nitric acid vapors when heated

Special tests: 1. Add iron(II) sulfate, shake; then add concentrated sulfuric acid; produces brown ring.
2. White precipitate is formed upon addition of nitron reagent ($C_{20}H_{16}N_4$); test is not specific to only nitrates; however, see table of precipitation reagents.

Nitrites, NO_2^-

Dilute hydrochloric acid	Cautious addition of acid to a solid nitrite in cold gives a transient pale (of nitrous acid or the anhydride) blue liquid and evolves brown fumes of nitrogen dioxide
Silver nitrate	White precipitate of silver nitrite
Iron(II) sulfate solution (25%, acidified with either acetic or sulfuric acid)	A brown ring forms at the junction of the two liquids due to the formation of a complex
Acidified potassium permanganate	Decolorization with no gas evolution
Ammonium chloride (solid)	Boiling with excess of solid reagent causes nitrogen to be evolved
Concentrated sulfuric acid	Liberates nitrogen dioxide (brown)

Special tests: Acidifed solutions of nitrites liberate iodine from potassium iodide (brown color produced).

TESTS FOR ANIONS (continued)

Orthophosphates, PO_4^{3-}

Silver nitrate	Yellow precipitate of silver orthophosphate, soluble in dilute ammonia and in dilute nitric acid
Barium chloride	White precipitate of barium hydrogen phosphate, soluble in dilute mineral acids and acetic acid
Magnesium nitrate reagent or magnesia mixture	White crystalline precipitate of magnesium ammonium phospahte, soluble in acetic acid and mineral acids, practically insoluble in 2.5% ammonia solution
Ammonium molybdate	Addition or 2—3 ml excess reagent to approximately 0.5 mL of sample gives yellow precipitate of ammonium phosphomolybdate; precipitate soluble in ammonia solution and in solutions of caustic alkalis; large quantities of hydrochloric acid interfere.
Iron(III) chloride	Yellowish-white precipitate of iron(III) phosphate, soluble in mineral acids, insoluble in dilute acidic acid
Ammonium molybdate-quinine	Yellow precipitate of unknown composition, reducing agents interfere

Note: The orthophosphates are salts of orthophosphoric acid, H_3PO_4, and are simply referred to as phosphates.

Oxalates, $(COO)_2^{2-}$

Silver nitrate	White precipitate of silver oxalate, soluble in ammonia solution and dilute nitric acid
Calcium chloride	White precipitate of calcium oxalate, insoluble in dilute acetic acid, oxalic acid, and in ammonium oxalate solution; soluble in dilute hydrochloric acid and in dilute nitric acid
Potassium permanganate	Decolorization upon warming to 60—70 °C, in acidified solution

Perchlorates, ClO_4^-

Potassium chloride	White precipitate of potassium perchlorate, insoluble in alcohol

Special tests: 1. Neutral ClO_4^- + cadmium sulfate in concentrated ammonia produces $[Cd(NH_3)_4](ClO_4)_2$ (white precipitate).
2. Cautious heating of solids evolves oxygen.

Peroxidisulfates, $S_2O_8^{2-}$

Water	On boiling, decomposes into the sulfate, free sulfuric acid, and oxygen
Silver nitrate	Black precipitate of silver peroxide
Barium chloride	On boiling or standing for some time, forms precipitate of barium sulfate
Manganese(II) sulfate	Brown precipitate of hydrate complex in neutral or alkaline test solution

Phosphites, HPO_3^{2-}

Silver nitrate	White precipitate of silver phosphite, which yields black metallic silver on standing
Barium chloride	White precipitate of barium phosphite; soluble in dilute acids
Mercury(II) choloride	White precipitate in cold solutions, which yields gray metallic mercury on warming
Copper sulfate	Light-blue precipitate; dissolves when boiled with acetic acid
Lead acetate	White precipitate of lead hydrogen phosphite

Pyrophosphates, $P_2O_7^{4-}$

Silver nitrate	White precipitate, soluble in dilute nitric acid and in dilute acetic acid
Copper sulfate	Pale-blue precipitate
Magnesia mixture or magnesium reagent	White precipitate, soluble in excess reagent but reprecipitated on boiling
Cadmium acetate and dilute acetic acid	White precipitate
Zinc sulfate	White precipitate, insoluble in dilute acetic acid; soluble in dilute ammonia solution, yielding a white precipitate on boiling

TESTS FOR ANIONS (continued)

Salicylates, $C_6H_4(OH)COO^-$ (or $C_7H_5O_3^-$)

Concentrated sulfuric acid	Evolution of carbon monoxide and sulfur dioxide (poisonous)
Concentrated sulfuric acid and methanol	0.5 g of sample + 3 ml of reagent + heat evolves methyl salicylate (odor of wintergreen)
Dilute hydrochloric acid	Crystalline precipitate of salicylic acid
Silver nitrate	Heavy crystalline precipitate of silver salicylate, soluble in boiling water; recrystallization upon cooling
Iron(III) chloride	Violet-red coloration which clears upon the addition of dilute mineral acids

Silicates, SiO_3^{2-}

Dilute hydrochloric acid	Gelatinous precipitate of metasilicic acid; insoluble in concentrated acids, soluble in water and dilute acids
Ammonium chloride or ammonium carbonate	Gelatinous precipitate of silicic acid
Silver nitrate	Yellow precipitate of silver silicate, soluble in dilute acids and ammonia solution
Barium chloride	White precipitate of barium silicate, soluble in dilute nitric acid

Succinates, $C_4H_4O_4^{2-}$

Silver nitrate	White precipitate of silver succinate, soluble in dilute ammonia solution
Iron(III) chloride	Light-brown precipitate of iron(III) succinate
Barium chloride	White precipitate of barium succinate
Calcium chloride	Precipitate of calcium succinate appears slowly

Sulfates, SO_4^{2-}

Barium chloride	White precipitate of barium sulfate, insoluble in warm dilute hydrochloric acid and in dilute nitric acid, slightly soluble in boiling hydrochloric acid
Lead acetate	White precipitate of lead sulfate; soluble in hot concentrated sulfuric acid, ammonium acetate, ammonium tartrate, and sodium hydroxide
Silver nitrate	White precipitate of silver sulfate
Mercury(II) nitrate	Yellow precipitate of mercury(II) sulfate; very sensitive test

Sulfides, S^{2-}

Dilute hydrochloric acid or sulfuric acid	Hydrogen sulfide gas is evolved and detected using lead acetate paper
Silver nitrate	Black precipitate of silver sulfide, soluble in hot, dilute nitric acid
Lead acetate	Black precipitate of lead sulfide
Sodium nitroprusside solution ($Na_2[Fe(CN)_5NO]$)	Transient purple color in the presence of solutions of alkalis

Special tests: Catalysis of iodine-azide reaction: solution of sodium azide, (NaN_3) and iodine reacts with a trace of a sulfide to evolve nitrogen. Thiosulfates and thiocyanates act similarly and therefore must be absent.

Sulfites, SO_3^{2-}

Dilute hydrochloric acid	Decomposition, which becomes more rapid on warming; evolution of sulfuric dioxide (odor of burning sulfur)
Barium chloride, strontium chloride	White precipitate of the respective sulfite, with the precipitate being soluble in dilute hydrochloric acid
Silver nitrate	At first, no change; upon addition of more reagent, white crystalline precipitate of silver sulfite forms, which darkens to metallic silver upon heating
Potassium permanganate solution acidified with dilute sulfuric acid	Decolorization (Fuchsin test)
Potassium dichromate acidified with dilute sulfuric acid	Green coloration
Lead acetate or lead nitrate solution	White precipitate of lead sulfite

TESTS FOR ANIONS (continued)

Zinc and sulfuric acid	Hydrogen sulfide gas evolved, detected by holding lead acetate paper to mouth of test tube
Concentrated sulfuric acid	Evolution of sulfur-dioxide gas
Sodium nitroprusside–zinc sulfate	Red compound of unknown composition

Tartrates, $C_4H_4O_6^{2-}$

Concentrated sulfuric acid	When sample is heated, the evolution of carbon monoxide, carbon dioxide, and sulfur dioxide (burnt-sugar odor) results
Silver nitrate	White precipate of silver tartrate
Calcium chloride	White precipitate of calcium tartrate, soluble in dilute acetic acid, dilute mineral acids, and in cold alkali solutions
Potassium chloride	White precipitate; the reaction is $C_4H_4O_6^{2-} + K^+ + CH_3COOH \rightarrow KH \cdot C_4H_4O_6 \downarrow + CH_3COO^-$

Special test: One drop of 25% iron(II) sulfate and two to three drops of hydrogen peroxide produces deep violet-blue color (Fenton's Test).

Thiocyanates, SCN^-

Sulfuric acid	In cold solution, yellow coloration is produced; upon warming, violent reaction occurs and carbonyl sulfide is formed
Silver nitrate	White precipitate of silver thiocyanate
Copper sulfate	First a green coloration, then black precipitate of copper(II) thiocyanate is formed
Mercury(II) nitrate	White precipitate of mercury(II) thiocyanate
Iron(III) chloride	Blood-red coloration due to complex formation
Dilute nitric acid	Upon warming, red coloration is observed, with nitrogen oxide and hydrogen cyanide (POISONOUS) being evolved
Cobalt nitrate	Blue coloration due to complex ion formation

Thiosulfates, $S_2O_3^{2-}$

Iodine solution	Decolorized; a colorless solution of tetrathionate ions is formed
Barium chloride	White precipitate of barium thiosulfate
Silver nitrate	White precipitate of silver thiosulfate
Lead acetate or lead nitrate solution	First no change; on further addition of reagent, a white precipitate of lead thiosulfate forms
Iron(III) chloride solution	Dark-violet coloration due to complex formation
Nickel ethylenediamine nitrate $[Ni(NH_2 \cdot CH_2 \cdot CH_2 \cdot NH_2)_3](NO_3)_2$	Violet complex precipitate forms; hydrogen sulfide and ammonium sulfide interfere

Special tests: Blue-ring test: When solution of thiosulfate mixed with ammonium molybdate solution, is poured slowly down the side of a test tube which contains concentrated sulfuric acid, a blue ring is formed briefly at the contact zone.

TESTS FOR CATIONS

This table provides a summary of the common tests for cations, primarily in aqueous solution. The cations are grouped according to the usual convention of reactivity to a set of common reagents. Abbreviations: conc, concentrated; dil, dilute; g, gaseous: ETOH, ethanol.

Group I: Pb^{2+}, Ag^{1+}, Hg^{1+} [a]

Lead, Pb_2^+
Potassium chromate	Yellow precipitate of lead chromate
Potassium iodide	Yellow precipitate of lead iodide
Sulfuric acid, dilute	White precipitate of lead sulfate
Hydrogen sulfide gas	Black precipitate of lead sulfide
Potassium cyanide	White precipitate of lead cyanide
Tetramethyldiaminodiphenyl-methane	Blue oxidation product (Bi, Ce, Mn, Th, Co, Ni, Fe, Cu interfere)
Gallocyanine	Deep-violet precipitate, unknown composition (Bi, Cd, Cu, Ag interfere)
Diphenylthiocarbazone	Brick-red complex in neutral or ammonical solution

Silver Ag^+
Potassium chromate	Reddish-brown precipitate of silver chromate
Potassium iodide	Yellow precipitate of silver iodide
Hydrogen sulfide gas	Black precipitate of silver sulfide
Disodium hydrogen phosphate	Yellow precipitate of silver phosphate
Sodium carbonate	Yellow-white precipitate of silver carbonate, forming the brown oxide upon heating
p-Dimethylaminobenzylidenerhodanine	Reddish-violet precipitate in acidic solution
Ammonia solution	Brown precipitate of silver oxide, dissolving in excess to form Ag_3N, which is explosive

Mercury (I), Hg_2^{+2}
Potassium carbonate	Red precipitate of mercury(I) chromate
Potassium iodide	Green precipitate of mercury(I) iodide
Dilute sulfuric acid	White precipitate of mercury(I) sulfate
Elemental copper, aluminum, or zinc	Amalgamation occurs
Hydrogen sulfide	Black precipitate (in neutral or acid medium) of mercury(I) sulfide + mercury
Ammonia solution	Black precipitate of $HgO \cdot Hg(NH_2)(NO_3)$
Diphenylcarbazide (1% in ethanol, with 0.2 M nitric acid)	Violet-colored complex results (high sensitivity and selectivity)
Potassium cyanide	Mercury(I) cyanide solution, with a precipitation of elemental mercury, (mercury(II) interferes)

Group II: HG^{2+}, Cu^{2+}, Bi^{3+}, Cd^{2+}, As^{3+}, As^{5+}, Sb^{3+}, Sb^{5+}, Sn^{2+}, Sn^{4+} [b]

Bismuth(III), Bi^{3+}
Potassium iodide	Black precipitate of bismuth(III) iodide
Potassium chromate	Yellow precipitate of bismuth(III) chromate
Ammonia solution	White precipitate of variable composition, approximate formula: $Bi(OH)_2NO_3$
Pyrogallol (10%)	Yellow precipitate of the pyrogallate
8-Hydroxyquinoline (5%) + potassium iodide (6M)	Red precipitate of the tetraiodobismuthate (characteristic in the absence of Cl^-, F^-, Br^-)
Sodium hydroxide	White precipitate of bismuth(III) hydroxide

Copper, (II)Cu^2
Potassium iodide	Brown precipitate of copper(I) iodide, colored brown due to I_3^-
Potassium cyanide	Yellow precipitate of copper(II) cyanide, which then decomposes
Potassium thiocyanate	Black precipitate of copper(II) thiocyanate, which then decomposes
α-Benzionoxime (or cupfon), 5% in ETCH	Green precipitate of the benzonoxime

TESTS FOR CATIONS (continued)

Salicylaldoxime (1%)	Greenish-yellow precipitate of the copper complex
Rubeanic acid (0.5%) (dithiooxamide)	Black precipitate of the rubeanate, from weakly acid solution
Cadmium(II), Cd^{2+}	
Ammonia solution	White precipitate of cadmium hydroxide dissolving in excess
Potassium cyanide	White precipitate of cadmium cyanide dissolving in excess
Sodium hydroxide	White precipitate of cadmium hydroxide, insoluble in excess
Dinitro-*p*-diphenyl carbizide	Forms brown product with cadmium hydroxide
Arsenic(III), As^{3+}	
Silver nitrate	Yellow precipitate of silver arsenite in neutral solution
Copper(II) sulfate	Green precipitate of copper(II) arsenite $Cu_3(AsO_3)_2 \cdot xH_2O$
Potassium triiodide ($KI + I_2$)	Becomes decolorized upon oxidation
Tin(II) chloride + concentrated hydrochloric acid	Black precipitate forms in the presence of excess reagent
Arsenic(V), As^{5+}	
Silver nitrate	Brownish-red precipitate of silver arsenate from neutral solutions
Ammonium molybdate	Yellow precipitate (in presence of excess reagent0 of ammonium arsenomolybdate $(NH_4)_3AsMo_{12}O_4$
Potassium iodide + concentrated hydrochloric acid	Precipitation of iodine

Note: Small amounts of As(III) or As(V) can be identified by the response to the Marsh, Gutzeit, or Fleitmann tests (see references at the beginning of this section).

Antimony(III), Sb^{3+}	
Sodium hydroxide	White precipitate of the hydrated oxide $Sb_2O_3 \cdot xH_2O$
Elemental zinc or tin	Black precipitate of antimony
Potassium iodide	Yellow color of $[SbI_6]^{3-}$ ion
Phosphomolybdic acid, $H_3[PMo_{12}O_{40}]$	Blue color produced; Sn(II) interferes; 0.2-μg sensitivity
Antimony(V), Sb^{5+}	
Water	White precipitate of basic salts, and ultimately antimonic acid, H_3SbO_4
Potassium iodide	Removal of iodine as a floating precipitate
Elemental zinc or tin	Black precipitate of antimony (in the presence of hydrochloric acid)

Note: Small amounts of antimony can be identified using Marsh's and/or Gutzeit's test (see references at the beginning of this section for details).

Tin(II), Sn^{2+}	
Mercury(II) chloride	White precipitate of mercury(I) chloride (in an excess of tin ions, precipitate turns grey)
Bismuth nitrate	Black precipitate of bismuth metal
Coacotheline (nitro-derivative of brucine, $C_{21}H_{21}O_7N_3$)	Violet coloration with stannous salts; the following interfere: strong reducing aents (hydrogen sulfide, dithionites, sulfites, and selenites), also U, V, Te, Hg, Bi, Au, Pd, Se, Sb
Diazine green (dyestuff formed by coupling diazotized safranine with dimethylaniline)	Color change blue → violet → red.
Tin(IV), Sn^{4+}	
Metallic iron	Reduces Sn(IV) to Sn(II)

Group III: Fe^{2+}, Fe^{3+}, Al^{3+}, Cr^{3+}, Cr^{6+}, Ni^{2+}, Co^{2+}, Mn^{2+}, Mn^{7+}, and Zn^{2+} c

Iron, Fe(II)	
Ammonia solution	Precipitation of iron(II) hydroxide; if large amounts of ammonium ion are present, precipitation does not occur
Ammonium sulfide	Black precipitate of iron(II) sulfide

TESTS FOR CATIONS (continued)

Potassium cyanide(POISON)	Yellowish-brown precipitate of iron (II) cyanide, soluble in excess reagent, forming the hexacyanoferrate (II) ion
Potassium hexacyanoferrate(II) solution	In complete absence of air, white precipitate of potassium iron(II) hexacyanoferrate(II); if air is present, a pale-blue precipitate is formed
Potassium hexacyanoferrate(III) solution	Dark-blue precipitate, called Turnball's blue
α,α′ - Dipyridyl	Deep-red bivalent cation $[Fe(C_5H_4N)_2]^{2+}$ formed with iron(II) salts in mineral acid solution; sensitivity: 0.3 μg
Dimethylglyoxime (DMG)	Red, iron(II) dimethylglyoxime; nickel, cobalt, and large quantities of copper salts interfere; sensitivity: 0.04 μg
o-Phenanthroline (0.1 wt% in water)	Red coloration due to the complex cation $[Fe(C_{12}H_8N_2)_3]^{2+}$, in slightly acidic conditions

Iron, Fe(III)

Ammonia solution	Reddish-brown gelatinous precipitate of iron(III) hydroxide
Ammonium sulfide	Black precipitate of iron(II) sulfide + sulfur
Potassium cyanide	When added slowly, gives reddish-brown precipitate of iron(III) cyanide, dissolving in excess to give yellow solution
Potassium hexacyanoferrate(III)	A brown coloration is produced due to formation of iron(III) hexacyanoferate
Disodium hydrogen phosphate	Yellowish-white precipitate of iron(III) phosphate
Sodium acetate solution	Reddish-brown coloration caused by a complex
Cupferron ($C_6H_5N(NO)ONH_4$) aqueous solution, freshly prepared	Reddish-brown precipitate formed in presence of hydrochloric acid
Ammonium thiocyanate + dilute acid	Deep-red coloration of iron(III) thiocyanate complex
7-Iodo-8-hydroxyquinoline-5- sulfonic acid (ferron)	Green or greenish-blue coloration in slightly acidic solutions; sensitivity: 0.5 μg Fe^{3+}

Cobalt, Co

Ammonia solution	In the absence of ammonium salts, small amounts of $Co(OH)NO_3$ precipitate which are solube in excess
Ammonium sulfide	Black precipitate of cobalt(II) sulfide, from neutral or alkaline solutions
Potassium cyanide (POISON)	Reddish-brown precipitate of cobalt(II) cyanide, which dissolves in excess
Potassium nitrite	Yellow precipitate of potassium hexacyanocobaltate(III), $K_3[Co(NO_2)_6]$
Ammonium thiocyanate (crystals)	Gives blue coloration when added to neutral or acid solution of cobalt, due to a complex formation (Vogel's reaction); sensitivity: 0.5 μg Co
α-Nitroso-β-naphthol (1% in 50% acetic acid)	Red-brown (chelate) precipitate, extractable using carbon tetrachloride; sensitivity: 0.05 μg Co

Nickel, Ni

Ammonia solution	Green precipitate of nickel(II) hydroxide which dissolves in excess
Potassium cyanide (POISON)	Green precipitate of nickel(II) cyanide which dissolves in excess
Dimethylglyoxime ($C_4H_8O_2N_2$)	Red precipitate of nickel-DMG chelate complex in ammoniacal solution; sensitivity 0.16 μg Ni

Manganese, Mn(II)

Ammonia solution	Partial precipitation of white manganese(II) hydroxide
Ammonium sulfide	Pink precipitate of manganese(II) sulfide, which is soluble in mineral acids
Sodium phosphate (in the presence of ammonia or ammonium ions)	Pink precipitate of manganese ammonium phosphate, $Mn(NH_4)PO_4 \cdot 7H_2O$, which is soluble in acids

Aluminum, Al^{3+}

Ammonia	White gelatinous precipitate of aluminum hydroxide
Sodium hydroxide	White precipitate of aluminum hydroxide, soluble in excess reagent
Ammonium sulfide	White precipitate of aluminum hydroxide
Sodium acetate	Upon boiling with excess reagent, a precipitate of basic aluminum acetate, $Al(OH)_2CH_3COO$, is formed

TESTS FOR CATIONS (continued)

Sodium phosphate	White gelatinous precipitate of aluminum phosphate
"Aluminon" (a solution of the ammonium salt of aurine tricarboxylic acid)	Bright-red coloration
Quinalizarin, alizarin-S, alizarin	Red precipitate or "lake"

Chromium, Cr(III)

Ammonia solution	Gray-green to -blue gelatinous precipitate of chromium(III) hydroxide
Sodium carbonate	Precipitate of chromium(III) hydroxide

Zinc, Zn^{2+}

Ammonia solution	White precipitate of zinc hydroxide, soluble in excess
Disodium hydrogen phosphate	White precipitate of zinc phosphate, soluble in dilute acids
Potassium hexacyanoferrate(II)	White precipitate of variable composition, which is soluble in sodium hydroxide
Ammonium tetrathiocyanatomercurate(II) — copper sulfate, slightly acidic	Solution is treated with 0.1 mL of 0.25 M copper sulfate solution followed by 2 mL ammonium tetrathiocyanatomercurate to give a violet precipitate

Group IV: Ba^{2+}, Sr^{2+}, Ca^{2+} [d]

Barium, Ba^{2+}

Ammonium carbonate	White precipitate of barium carbonate, soluble in dilute acids
Ammonium oxalate	White precipitate of barium oxalate, soluble in dilute acids
Dilute sulfuric acid	Heavy, white, finely divided precipitate of barium sulfate
Saturated calcium sulfate (or strontium sulfate)	White precipitate of barium sulfate or strontium sulfate)
Sodium rhodizonate	Red-brown precipitate; sensitivity: 0.25 μg Ba

Strontium, Sr

Ammonium carbonate	White precipitate of strontium carbonate
Dilute sulfuric acid	White precipitate of strontium sulfate
Saturated calcium sulfate	White precipitate of strontium sulfate
Potassium chromate	Yellow precipitate of strontium chromate
Ammonium oxalate	White precipitate of strontium oxalate, soluble in mineral acids

Calcium, Ca

Ammonium carbonate	White precipitate of calcium carbonate
Dilute sulfuric acid	White precipitate of calcium sulfate
Ammonium oxalate	White precipitate of calcium oxalate, soluble in mineral acids
Potassium chromate	Yellow precipitate of calcium chromate, soluble in mineral acids
Sodium rhodizonate	Red-brown precipitate; sensitivity: 4 μg

Group V: Mg^{2+}, Na^+, K^+, NH_4^+ [e]

Magnesium, Mg^{2+}

Ammonia solution, sodium hydroxide	Partial precipitation of white magnesium hydroxide
Ammonium carbonate	White precipitate of magnesium carbonate, only in the absence of ammonia salts
Oxine + ammoniacal ammonium chloride solution	Yellow precipitate of the complex salt: Mg $(C_9H_6NO)_2 \cdot 4H_2O$
Quinalizarin	Blue precipitate, or blue-colored solution, which can be cleared by a few drops of bromine-water

Sodium, Na^+

Uranyl magnesium acetate solution (in 30 v/v ETOH)	Yellow precipitate of sodium magnesium uranylacetate
Uranyl zinc acetate solution	Yellow precipitate of sodium zinc uranyl acetate; sensitivity: 12.5 μg Na

Potassium, K^+

Sodium hexanitrocobaltate (III) ($Na_3[Co(NO_2)_6]$)	Yellow precipitate of potassium hexanitrocobaltate(III); insoluble in acetic acid
Tartaric acid solution (sodium acetate buffered)	White precipitate of potassium hydrogen tartrate

TESTS FOR CATIONS

Perchloric acid	White precipitate of potassium perchlorate
Dipicrylamine	Orange-red complex precipitate (NH_4^+ interferes); sensitivity: 3 μg K
Sodium tetraphenylboron + acetic acid	White precipitate of the potassium substitution salt
Ammonium, NH_4^+	
Sodium hydroxide	Evolution of ammonia gas
Potassium tetraiodomercurate (Nessler's reagent)	Brown-yellow color, or brown precipitate of mercury(II) amidoiodide; high sensitivity; all other metals (except Na and K) interfere
Tannic acid-silver nitrate	Precipitate of black elemental silver, from neutral solution; very sensitive
p-Nitrobenzene-diazonium chloride	Red-colored solution results in the presence of sodium hydroxide; sensitivity: 0.7 μg NH_4^+

Note: Ammonium ions will cause a similar reaction to that of potassium in the presence of sodium hexanitrocobaltate(III) sodium hydrogen tartrate.

[a]   All members are precipitated by dilute HCl, to give lead chloride ($PbCl_2$), silver chloride (AgCl), or mercury chloride (Hg_2Cl_2).

[b]   All members show no reaction with HCl; all form a precipitate with H_2S.

[c]   All members are precipitated by H_2S in the presence of ammonia and ammonium chloride, or ammonium sulfide solutions.

[d]   All members of this group react with ammonium carbonate.

[e]   No common reaction or reagent.

ORGANIC PRECIPITATION REAGENTS FOR INORGANIC IONS

The following table lists the most important organic reagents used for precipitating various inorganic species from solution.[1,2] Many of these reagents are subject to the serious disadvantage caused by lack of selectivity. Thus, many of the listed reagents will precipitate more than one species. The selectivity of some of the reagents can be controlled to a certain extent by adjustment of pH, reagent concentrations, and the use of masking reagents. The first two factors, pH and concentration, are the most critical. A number of these reagents form rather large, bulky complexes. While this can serve to enhance sensitivity (especially for gravimetric procedures), it can also impose rather stringent concentration limits. The reader is referred to several excellent "recipe" texts for further guidance.[3-7]

REFERENCES

1. **Kennedy, J. H.,** *Analytical Chemistry,* Harcourt, Brace, Jovanovich, San Diego, CA, 1984.
2. **Christian, G. D.,** *Analytical Chemistry,* 4th ed., John Wiley & Sons, New York, 1986.
3. **Barber, H. H. and Taylor, T. I.,** *Semimicro Qualitative Analysis,* Harper Brothers, New York, 1953.
4. **Greenfield, S. and Clift, M.,** *Analytical Chemistry of the Condensed Phosphates,* Pergamon Press, Oxford, 1975.
5. **Ryabchikov, D. I. and Gol'Braikh, E. K.,** *The Analytical Chemistry of Thorium,* Macmillan, New York, 1963.
6. **Jungreis, E.,** *Spot Test Analysis,* Interscience, New York, 1985.
7. **Svehla, G.,** *Vogel's Qualitative Inorganic Analysis,* John Wiley & Sons, New York, 1987.

ORGANIC PRECIPITATION REAGENTS

	Structure/formula	Applications and notes
Alizarin-S (sodium alizarin sulfonate)	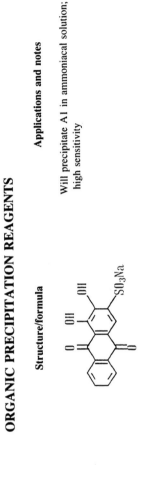	Will precipitate A1 in ammoniacal solution; high sensitivity
Ammonium nitroso-phenylhydroxyl-amine (cupferron)		Will precipitate Fe(III), V(V), Ti(IV), Zr(IV), Sn(IV), and U(IV) in the presence of moderate acidity; will also precipitate rare earths
Anthranilic acid		Used to precipitate Cu, Cd, Ni, Co, Pb, and Zn in acetic acid or nearly neutral solution
α-Benzoinoxime (cupron)		Will precipitate Cu(II) in the presence of NH₃, or tartrate; Mo(VI), W(VI) in acidic environment
Dimethylglyoxime (DMG)		Will precipitate Ni(II) in the presence of NH₃, or buffered acetic acid; Pd(II) in HCl solution; the addition of tartaric acid to the reagent will mask Fe(III) and Cr(III) interferences; Pd(II) and Bi(III) are also precipitated

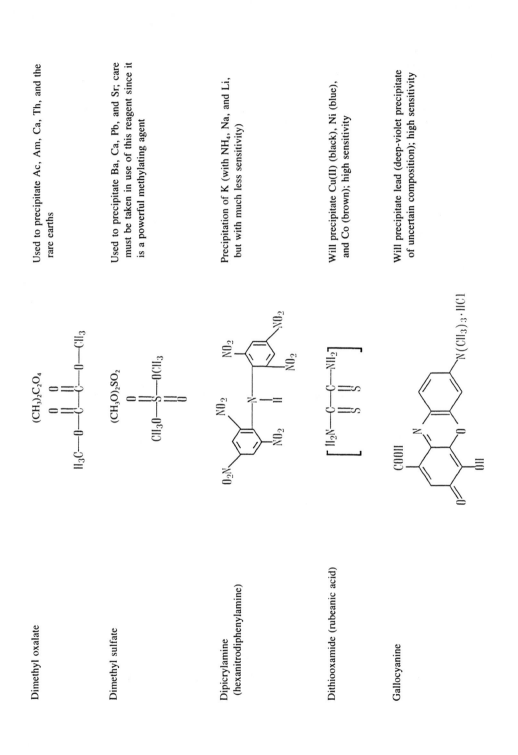

Dimethyl oxalate

$(CH_3)_2C_2O_4$

Used to precipitate Ac, Am, Ca, Th, and the rare earths

Dimethyl sulfate

$(CH_3O)_2SO_2$

Used to precipitate Ba, Ca, Pb, and Sr; care must be taken in use of this reagent since it is a powerful methylating agent

Dipicrylamine (hexanitrodiphenylamine)

Precipitation of K (with NH$_4$, Na, and Li, but with much less sensitivity)

Dithiooxamide (rubeanic acid)

Will precipitate Cu(II) (black), Ni (blue), and Co (brown); high sensitivity

Gallocyanine

Will precipitate lead (deep-violet precipitate of uncertain composition); high sensitivity

ORGANIC PRECIPITATION REAGENTS

Reagent	Structure/formula	Applications and notes
8-Hydroxyquinoline (oxine)		Useful for precipitation of Al (III) at pH 4—5, and Mg(II) in the presence of NH$_3$; will precipitate Be, Bi, Cd, Cu, Ga(I), Hf, Fe, In, Mg, Hg, Nb, Pd, Sc, Ta, Ti, Th, U, Zn, Zr, and W at pH 4—5 or in the presence of NH$_3$, with pH control providing a measure of selectivity
Nitron reagent (in acetic acid)	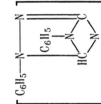	Used to precipitate nitrate, bromide, iodide, nitrite, chromate chloride, perchlorate, thiocyanate, oxalate, and picrate
α-Nitroso-β-naphthol	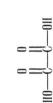	Will precipitate Co(II), Ni(II), Fe(III), and Pd(II) in a weakly acidic solution
Oxalic acid		Used to precipitate Ca; high concentrations of Mg will interfere

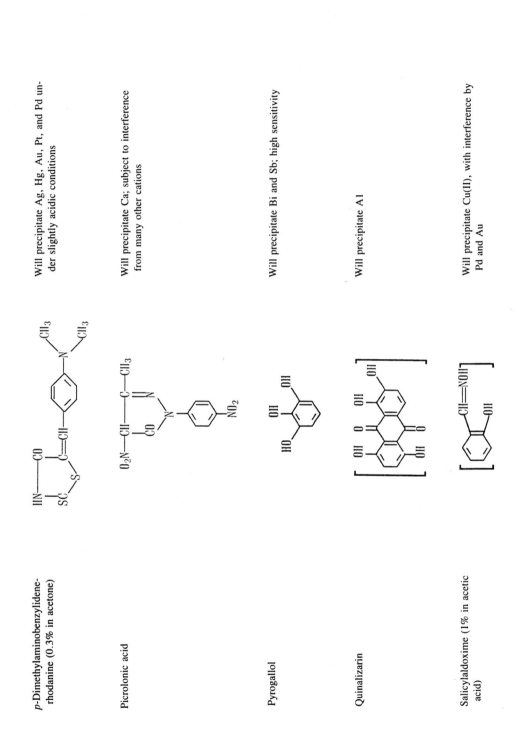

p-Dimethylaminobenzylidene-
rhodanine (0.3% in acetone)

Will precipitate Ag, Hg, Au, Pt, and Pd under slightly acidic conditions

Picrolonic acid

Will precipitate Ca; subject to interference
from many other cations

Pyrogallol

Will precipitate Bi and Sb; high sensitivity

Quinalizarin

Will precipitate Al

Salicylaldoxime (1% in acetic
acid)

Will precipitate Cu(II), with interference by
Pd and Au

ORGANIC PRECIPITATION REAGENTS

Reagent	Structure/formula	Applications and notes
Sodium diethylthiocarbamate		Useful for the precipitation of many metals
Sodium dihydroxytartrate osazone		Will precipitate Ca; subject to interference from many other cations
Sodium rhodizonate		Will precipitate Ba and Sr; subject to interference by all H_2S reactive cations
Sodium tetraphenylboron		Used to precipitate K, Rb, Cs, Tl, Ag, Hg(I), Cu(I), NH_4^+, RNH_3^+, $R_2NH_2^+$, R_3NH^+, and R_4N^+ in cold acidic solution; selectivity is high for K^+ and NH_4^+
Tetraphenyl arsonium chloride		Used to precipitate $Cr_2O_7^{2-}$, MnO_4^-, ReO_4^-, MoO_4^{2-}, WO_4^{2-}, ClO_4^-, and I_3^- in acidic solution

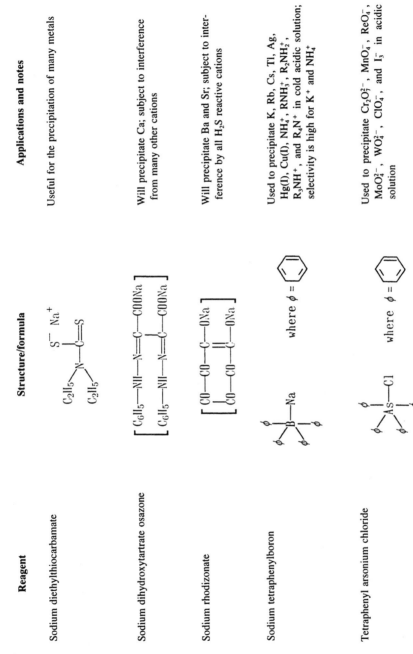

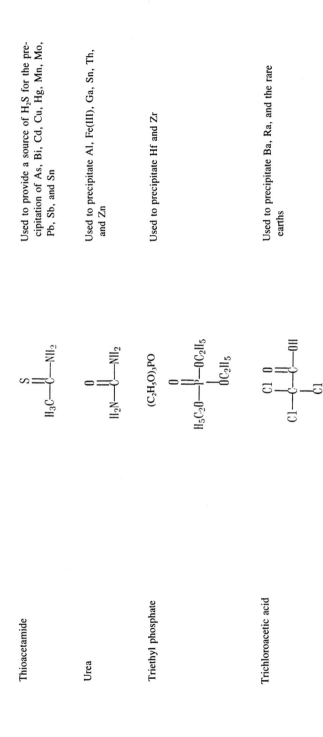

Thioacetamide

$$H_3C-\overset{\displaystyle S}{\underset{\displaystyle \|}{C}}-NH_2$$

Used to provide a source of H_2S for the precipitation of As, Bi, Cd, Cu, Hg, Mn, Mo, Pb, Sb, and Sn

Urea

$$H_2N-\overset{\displaystyle O}{\underset{\displaystyle \|}{C}}-NH_2$$

Used to precipitate Al, Fe(III), Ga, Sn, Th, and Zn

Triethyl phosphate

$(C_2H_5O)_3PO$

$$H_5C_2O-\overset{\displaystyle O}{\underset{\displaystyle \|}{P}}-OC_2H_5$$
$$\overset{\displaystyle |}{OC_2H_5}$$

Used to precipitate Hf and Zr

Trichloroacetic acid

$$\underset{\displaystyle Cl}{\overset{\displaystyle Cl}{\underset{\displaystyle |}{\overset{\displaystyle |}{C}}}}-\overset{\displaystyle O}{\underset{\displaystyle \|}{C}}-OH$$

Used to precipitate Ba, Ra, and the rare earths

XI. Miscellaneous Tables

UNIT CONVERSIONS

The international system of units is described in detail in National Bureau of Standards Special Publication 330,[1] and lists of physical constants and conversion factors are tabulated by Mechtly.[2] The Americal Society for Testing and Materials (ASTM), the American National Standards Institute (ANSI), and the Institute of Electrical and Electronics Engineers (IEEE) have published a joint standard for metric practice.[3] The most recent guidelines for use of the metric system are given in References 1, 2, and 3. Selected unit conversions[1,4] are given in the following tables. The conversions are presented in matrix format when all of the units are of a convenient order of magnitude. When some of the unit conversions are of little value (such as the conversion between metric tons and grains), tabular form is followed, with the less useful units omitted.

REFERENCES

1. The International System of Units, NBS Special Publ. No. SP-330, National Bureau of Standards, U.S. Department of Commerce, Washington, D.C., 1977.
2. **Mechtly, E. A.,** The International System of Units, Physical Constants and Conversion Factors, NASA Special Publ. SP-7012, National Air and Space Administration, Washingtron, D.C., 1964.
3. Standard for Metric Practice, ANSI/ASTM E-380-76, IEEE Std. 268-1976, July, 1976, American Society for Testing and Materials, Philadelphia, 1976.
4. **Chiu, Y.,** A Dictionary for Unit Conversion, School of Engineering and Applied Science, George Washington University, Washington, D.C., 1975.

AREA

Multiply	By	To obtain
Square millimeters	0.00155	Square inches (U.S.)
	1×10^{-6}	Square meters
	0.01	Square centimeters
	1.2732	Circular millimeters
Square centimeters	1.196×10^{-4}	Square yards
	0.00108	Square feet
	0.15500	Square inches
	1×10^{-4}	Square meters
	100	Square millimeters
Square kilometers	0.38610	Square miles (U.S.)
	1.1960×10^{6}	Square yards
	1.0764×10^{7}	Square feet
	1×10^{6}	Square meters
	247.10	Acres (U.S.)
Square inches (U.S.)	0.00694	Square feet
	0.00077	Square yards
	6.4516×10^{-4}	Square meters
	6.4516	Square centimeters
	645.15	Square millimeters
Square feet (U.S.)	3.5870×10^{-8}	Square miles
	0.11111	Square yards
	144	Square inches
	0.09290	Square meters
	2.2957×10^{-5}	Acres
Square miles	640	Acres
	3.0967×10^{6}	Square yards
	2.7878×10^{-7}	Square feet
	2.5900	Square kilometers

DENSITY

kg/m³	g/cm³	lb/ft³
16.018	0.016018	1
1	0.001	0.062428
1,000	1	62.428
2,015.9	2.0159	125.85

ENTHALPY, HEAT OF VAPORIZATION, HEAT OF CONVERSION, SPECIFIC ENERGIES

Kj/kg (J/g)	cal/g	Btu/lb
2.3244	0.55556	1
1	0.23901	0.43022
4.1840	1	1.8

LENGTH

Multiply	By	To obtain
Angstroms	1×10^{-10}	Meters
	3.9370×10^{-9}	Inches (U.S.)
	1×10^{-4}	Micrometers
	1×10^{-8}	Centimeters
	0.1	Manometers
Nanometers	1×10^{-9}	Meters
	1×10^{-7}	Centimeters
	10	Angstroms
Micrometers (μm)	3.9370×10^{-5}	Inches (U.S.)
	1×10^{-6}	Meters
	1×10^{-4}	Centimeters
	1×10^{4}	Angstroms
Millimeters	0.03937	Inches (U.S.)
	1000	Micrometers
Centimeters	0.39370	Inches (U.S.)
	1×10^{4}	Micrometers (μm)
	1×10^{7}	Manometers
	1×10^{8}	Angstroms
Meters	6.2137×10^{-4}	Miles (statute)
	1.0936	Yards (U.S.)
	39.370	Inches (U.S.)
	1×10^{9}	Millimicrons
	1×10^{10}	Angstroms
Kilometers	0.53961	Miles (nautical)
	0.62137	Miles (statute)
	1093.6	Yards
	3280.8	Feet
Inches (U.S.)	0.02778	Yards
	2.5400	Centimeters
	2.5400×10^{8}	Angstroms
Feet (U.S.)	0.30480	Meters
	30.480	Centimeters
Yards (U.S.)	5.6818×10^{-4}	Miles
	0.91440	Meters
	91.440	Centimeters
Miles (nautical)	1.1516	Statute miles
	2026.8	Yards
	1.8533	Kilometers
Miles (U.S. statute)	320	Rods
	0.86836	Nautical miles
	1.6094	Kilometers
	1609.4	Meters

PARTS PER MILLION

PPM	vs.	Percent
1	=	0.0001
10	=	0.001
100	=	0.01
1 000	=	0.1
10 000	=	1.0
100 000	=	10.0
1 000 000	=	100.0

PRESSURE

MPa	atm	torr (mmHg)	bar	lb/in.^{2} (psi)
6.8948×10^{-3}	0.068046	51.715	6.8948×10^{-2}	1
1	9.8692	7,500.6	10.0	145.04
0.101325	1	760.0	1.01325	14.696
1.3332×10^{-4}	1.3158×10^{-3}	1	1.332×10^{-3}	0.019337
0.1	0.98692	750.06	1	14.504

SPECIFIC HEAT, ENTROPY

kJ/(kg-K) J/(g-k)	Btu/(°R·lb)
4.184	1
1	0.23901

SPECIFIC VOLUME

m^3/kg (L/g)	cm^3/g	ft^3/lb
0.062428	62.428	1
1	1,000	16.018
0.001	1	0.016018

SURFACE TENSION

N/m	dyn/cm	lb/in.
175.13	175.13×10^3	1
1	1 000	5.7102×10^{-6}
0.001	1	5.7102×10^{-3}

TEMPERATURE

T (Rankine) = 1.8T (Kelvin)
T (Celsius) = T (Kelvin) − 273.15
T (Fahrenheit) = T (Rankine) − 459.67
T (Fahrenheit) = 1.8T (Celsius) + 32

THERMAL CONDUCTIVITY

mW/(cm − K)	J/(s − cm − K)	cal/(s − cm − K)	Btu/(ft − h − °R)
17.296	0.017296	0.0041338	1
1	0.001	2.3901×10^{-4}	0.057816
1,000	1	0.23901	57.816
4,184	4.184	1	241.90

VELOCITY

Multiply	By	To obtain
Feet per minute	0.01136	Miles per hour
	0.01829	Kilometers per hour
	0.5080	Centimeter per second
	0.01667	Feet per second
Feet per second	0.6818	Miles per hour
	1.097	Kilometers per hour
	30.48	Centimeter per second
	0.3048	Meters per second
	0.5921	Knots
Knots (Br.)	1.0	Nautical miles per hour
	1.6889	Feet per second
	1.1515	Miles per hour
	1.8532	Kilometers per hour
	0.5148	Meters per second
Meters per second	3.281	Feet per second
	2.237	Miles per hour
	3.600	Kilometers per hour
Miles per hour	1.467	Feet per second
	0.4470	Meters per second
	1.609	Kilometers per hour
	0.8684	Knots

VELOCITY OF SOUND

m/s	ft/s
0.3048	1
1	3.2808

VISCOSITY

kg/(m − s) (N − s/m², Pa·s)	cP (10⁻²g/(cm − s))	lb − s/ft² (slug/(ft − s))	lb/(ft − s)
1.48816	1 488.16	0.31081	1
1	1 000	0.020885	0.67197
0.001	1	2.0885×10^{-5}	6.7197×10^{-4}
47.881	4.7881×10^{-4}	1	32.175

VOLUME

Multiply	By	To obtain
Barrels (pet)	42	Gallons (U.S.)
	34.97	Gallons (Br.)
Cubic centimeters	10^{-3}	Liters
	0.0610	Cubic inches
Cubic feet	28317	Cubic centimeters
	1728	Cubic inches
	0.03704	Cubic yards
	7.481	Gallons (U.S., liq.)
	28.317	Liters

VOLUME

Multiply	By	To obtain
Cubic inches	16.387	Cubic centimeters
	0.016387	Liters
	4.329×10^{-3}	Gallons (U.S., liq.)
	0.01732	Quarts (U.S., liq.)
Gallons, imperial	277.4	Cubic inches
	1.201	Gallons (U.S.)
	4.546	Liters
Gallons (U.S.)		
(Liquid)	231	Cubic inches
	0.1337	Cubic feet
	3.785	Liters
	0.8327	Imperial gallons
	128	Fluid ounces (U.S.)
Ounces, fluid	29.57	Cubic centimeters
	1.805	Cubic inches
Liters	0.2642	Gallons
	0.0353	Cubic feet
	1.0567	Quarts (U.S. liq.)
	61.025	Cubic inches
Quarts (U.S.)		
(Liquid)	0.0334	Cubic feet
	57.749	Cubic inches
	0.9463	Liters

MASS (WEIGHT)

Multiply	By	To obtain
Milligrams	2.2046×10^{-6}	Pounds (avoirdupois)
	3.5274×10^{-5}	Ounces (avoirdupois)
	0.01543	Grains
	1×10^{-6}	Kilograms
Micrograms	1×10^{-6}	Grams
Grams	0.00220	Pounds (avoirdupois)
	0.03527	Ounces (avoirdupois)
	15.432	Grains
	1×10^{6}	Micrograms
Kilograms	0.00110	Tons (short)
	2.2046	Pounds (avoirdupois)
	35.274	Ounces (avoirdupois)
	1.5432×10^{4}	Grains
Grains	1.4286×10^{-4}	Pounds (avoirdupois)
	0.00229	Ounces (avoirdupois)
	0.06480	Grams
	64.799	Milligrams
Ounces (avoirdupois)	3.1250×10^{-5}	Tons (short)
	0.06250	Pounds (avoirdupois)
	437.50	Grains
	28.350	Grams
Pounds (avoirdupois)	5×10^{-4}	Tons (short)
	16	Ounces (avoirdupois)
	7000	Grains
	0.45359	Kilograms
	453.59	Grams
Tons (short, U.S.)	2000	Pounds (avoirdupois)
	3.200×10^{4}	Ounces (avoirdupois)
	907.19	Kilograms
Tons (long)	2240	Pounds (avoirdupois)
	1016	Kilograms
Tons (metric)	1000	Kilograms
	2205	Pounds (avoirdupois)
	1.102	Tons (short)

PREFIXES FOR SI UNITS

	Prefix	Symbol
Fraction		
10^{-1}	deci	d
10^{-2}	centi	c
10^{-3}	milli	m
10^{-6}	micro	μ
10^{-9}	nano	n
10^{-12}	pico	p
10^{-15}	femto	f
10^{-18}	atto	a
Multiple		
10	deka	da
10^{2}	hecto	h
10^{3}	kilo	k
10^{6}	mega	M
10^{9}	giga	G
10^{12}	tera	T
10^{15}	peta	P
10^{18}	exa	E

RECOMMENDED VALUES OF PHYSICAL CONSTANTS

Physical constant	Symbol	Value
Avogadro consant	N_A	$6.022\ 05 \times 10^{23}\ \text{mol}^{-1}$
Boltzmann constant	k	$1.380\ 66 \times 10^{-23}\ \text{J·K}^{-1}$
Charge-to-mass ratio	e/m	$1.758\ 796 \times 10^{11}\ \text{C·kg}^{-1}$
Electronic charge	e	$1.602\ 19 \times 10^{-19}\ \text{C}$
Faraday constant	F	$9.648\ 46 \times 10^4\ \text{C·mol}^{-1}$
Gas constant	R	$8.314\ \text{J K}^{-1}\ \text{mol}^{-1}$
"Ice point" temperature	T_{ice}	$273{\cdot}150\ \text{K (exactly)}$
Molar volume of ideal gas (stp)	V_m	$2.241\ 38 \times 10^{-2}\ \text{m}^3{\cdot}\text{mol}^{-1}$
Permittivity of vacuum	ϵ_o	$8.854\ 188 \times 10^{-12}\ \text{kg}^{-1}\ \text{m}^{-3}{\cdot}\text{s}^4{\cdot}\text{A}^2$ (F·m^{-1})
Planck constant	h	$6.626\ 2 \times 10^{-34}\ \text{J·s}$
Standard atmosphere pressure	p	$101\ 325\ \text{N·m}^{-2}\ \text{(exactly)}$
Atomic mass unit	m_u	$1.660\ 566 \times 10^{-27}\ \text{kg}$
Speed of light in vacuum	c	$2.997\ 925 \times 10^8\ \text{m·s}^{-1}$

STANDARDS FOR LABORATORY WEIGHTS

The following table provides a summary of the requirements for (metric) weights and mass standards commonly used in chemical analysis.[1] The actual specifications are under the jurisdiction of ASTM Committee E-41 on General Laboratory Apparatus and are the direct responsibility of Subcommittee E-41.06 on weighing devices. These standards do not generally refer to instruments used in commerce.

Weights are classified according to type (either Type I or Type II), grade (S,O,P, or Q) and class (1,2,3,4,5, or 6). Information on these weight standards is presented to allow the user to make appropriate choices when using analytical weights for the calibration of electronic balances, for making large-scale weight measurements (such as those involving gas cylinders), and in the use of dead-weight pressure balances.

REFERENCES

1. Annual Book of ASTM Standards, ANSI/ASTM E617-78, American Society of Testing and Materials, Philadephia, 1986.
2. **Battino, R. and Williamson, A. G.,** Single-pan balances, buoyancy, and gravity or "a mass of confusion", *J. Chem. Educ.,* 61(1), 51, 1984.

TYPE—CLASSIFICATION BY DESIGN

Type I: One-piece construction; contains no added adjusting material; used for highest accuracy work.

Type II: Can be of any appropriate and convenient design, incorporating plugs, knobs, rings, etc.; adjusting material can be added if it is contained so that it cannot become separated from the weight

GRADE—CLASSIFICATION BY PHYSICAL PROPERTY

Grade S
    Density   7.7 to 8.1 g/cm^3 (for 50 mg and larger)
    Surface area   Not to exceed that of a cylinder of equal height and diameter
    Surface finish   Highly polished
    Surface protection None permitted
    Magnetic properties No more magnetic than 300 series
                     Stainless steels
    Corrosion resistance Same as 303 stainless steel
    Hardness   At least as hard as brass

Grade O
    Density   7.7 to 9.1 g/cm^3 (for 1 g and larger)
    Surface area   Same as grade S
    Surface finish   Same as grade S
    Surface protection May be plated with suitable material such as platinum or rhodium
    Magnetic properties Same as grade S
    Corrosion resistance Same as grade S
    Hardness   At least as hard as brass when coated; smaller weights at least as hard as
                     aluminum

Grade P
    Density   7.2 to 10 g/cm^3 (for 1 g or larger)
    Surface area   No restriction
    Surface finish   Smooth, no irregularities
    Surface protection May be plated or lacquered
    Magnetic properties Same as grades S and O
    Corrosion resistance Surface must resist corrosion and oxidation
    Hardness   Same as grade O

Grade Q
    Density   7.22 to 10 g/cm^3 (for 1 g or larger)
    Surface area   Same as grade P
    Surface finish   Same as grade P
    Surface protection May be plated, lacquered, or painted
    Magnetic properties No more magnetic than unhardened unmagnetized steel
    Corrosion resistance Same as grade P
    Hardness   Same as grades O and P

TOLERANCE — CLASSIFICATION BY DEVIATION[a]

Class 1

(g)	Individual tolerance (mg)	Group tolerance (mg)
500	1.2	
300	0.75	
		1.35
200	0.50	
100	0.25	
50	0.12	
30	0.074	
		0.16
20	0.074	
10	0.050	
5	0.034	
3	0.034	
		0.065
2	0.034	
1	0.034	

Class 2

(g)	Individual tolerance (mg)	Group tolerance (mg)
500	2.5	
300	1.5	
		2.7
200	1.0	
100	0.50	
50	0.25	
30	0.15	
		0.29
20	0.10	
10	0.074	
5	0.054	
3	0.054	
		0.105
2	0.054	
1	0.054	

Class 3

(g)	Tolerance (mg)
500	5.0
300	3.0
200	2.0
100	1.0
50	0.60
30	0.45
20	0.35
10	0.25

Class 4

(g)	Tolerance (mg)
500	10
300	6.0
200	4.0
100	2.0
50	1.2
30	0.90
20	0.70
10	0.50

Class 5

(g)	Tolerance (mg)
500	30
300	20
200	15
100	9
50	5.6
30	4.0
20	3.0
10	2.0
5	1.3
3	0.95
2	0.75
1	0.50

Class 6

(g)	Tolerance (mg)
500	50
300	30
200	20
100	10
50	7
30	5
20	3
10	2
5	2
3	2
2	2
1	2

[a] In simple terms, the permitted deviation between the assigned nominal mass value of the weight and the actual mass of the weight. Verification of tolerance should be possible on reasonably precise equipment, without using a buoyancy correction, within the political jurisdication or organizational bounds of a given weight specification.

APPLICATIONS FOR WEIGHTS AND MASS STANDARDS[a]

Application	Type	Grade	Class
Reference standards for calibrating other weights	I	S	1,2,3, or 4[a]
High-precision standards for calibration of weights and precision balances	I or II[b]	S or O[b]	1 or 2[c]
Working standards for calibration and precision analytical work, dead-weight pressure balances	I or II[b]	S or O	2
Laboratory weights for routine analytical work	II	O	2 or 3
Built-in weights, high-quality analytical balances	I or II	S	2
Moderate-precision laboratory balances	II	P	3 or 4
Dial scales and trip balances	II	Q	4 or 5
Platform scales	II	Q	5 or 6

[a] Primary standards are for reference use only and should be calibrated. Since the actual values for each weight are stated, close tolerances are neither required nor desirable.
[b] Type I and Grade S will have a higher constancy but will probably be higher priced.
[c] Since working standards are used for the calibration of measuring instruments, the choice of tolerance depends upon the requirements of the instrument. The weights are usually used at the assumed nominal values and appropriate tolerances should be chosen.

Reprinted (with modification) with permission of the American Society for Testing and Materials, 1916 Race Street, Philadelphia, PA 19103.

STANDARD CGA FITTINGS

Gas	Fitting
Acetylene	510
Air	346
Carbon dioxide	320
Carbon monoxide	350
Chlorine	660
Ethane	350
Ethylene	350
Ethylene oxide	510
Helium	580
Hydrogen	350
Hydrogen chloride	330
Methane	350
Neon	580
Nitrogen	580
Nitrous oxide	326
Oxygen	540
Sulfur dioxide	660
Sulfur hexafluoride	590
Xenon	580

Reproduced from the CGA Pamphet V-1-87, American National, Canadian, and Compressed Gas Association Standard for Compressed Gas Cylinder Valve Outlet and Inlet Connections, ANSI B57.1; CSA B96, by permission of the Compressed Gas Association, Inc., Arlington, Virginia.

STANDARD CGA FITTINGS

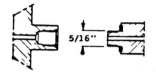

CONNECTION 110 – LECTURE BOTTLE OUTLET FOR CORROSIVE GASES – 5/16"- 32 RH INT.

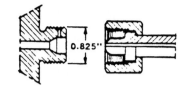

CONNECTION 326 – 0.825"- 14 RH EXT.

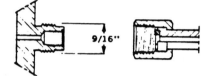

CONNECTION 170 – LECTURE BOTTLE OUTLET FOR NON-CORROSIVE GASES – 9/16"- 18 RH EXT. and 5/16"- 32 RH INT.

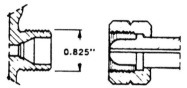

CONNECTION 350 – 0.825"- 14 LH EXT.

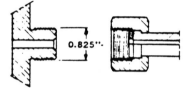

CONNECTION 320 – 0.825"- 14 RH EXT.

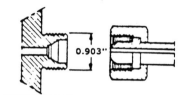

CONNECTION 540 – 0.903"- 14 RH EXT.

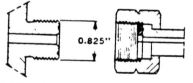

CONNECTION 330 – 0.825"- 14 LH EXT.

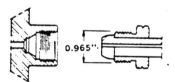

CONNECTION 590 – 0.965"- 14 LH INT.

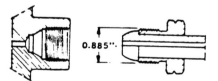

CONNECTION 510 – 0.885"- 14 LH INT.

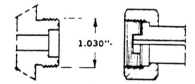

CONNECTION 660 – 1.030"- 14 RH EXT.

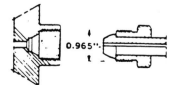

CONNECTION 580 – 0.965"- 14 RH INT.

BUFFERS

The following table gives the necessary information for preparing various buffers at different pHs. These buffers are suitable for use either in enzymatic or histochemical studies.[1,2] The accuracy of the tables is within ± 0.05 pH at 23°C and the pH values do not change considerably even at 37 °C. The recommended mixture of the various solutions is given under the corresponding pH with the "final solution volume" indicated. This assumes addition of water to the necessary dilution. A list of stock solutions follows the buffer/pH table. The approximate composition of buffers can be calculated from the equation:

$$pH = pK + \log ([\text{conc of salt}]/[\text{conc of acid}])$$

REFERENCES

1. **Colowick, S. P. and Kaplan, N. O., Eds.,** *Methods in Enzymology,* Vol. 1, Academic Press, New York, 1955.
2. **Perrin, D. D. and Dempsey, B.,** *Buffers for pH and Metal Ion Control,* Chapman and Hall, London, 1974.

BUFFERS

Buffer	pH	1.0	1.1	1.2	1.3	1.4	Final soln vol
Hydrochloric acid/potassium chloride		50.0 A + 97.0 B	50.0 A + 78.0 B	50.0 A + 64.5 B	50.0 A + 51.0 B	50.0 A + 41.5 B	200

Buffer	pH	1.5	1.6	1.7	1.8	1.9	Final soln vol
Hydrochloric acid/potassium chloride		50.0 A + 33.3 B	50.0 A + 26.3 B	50.0 A + 20.6 B	50.0 A + 16.6 B	50.0 A + 13.2 B	200

Buffer	pH	2.0	2.1	2.2	2.3	2.4	Final soln vol
Hydrochloric acid/potassium chloride		50.0 A + 10.6	50.0 A + 8.4 B	50.0 A + 6.7 B			200
Glycine/hydrochloric acid				50.0 C + 44.0 B		50.0 C + 32.4 B	200
Phthalate/hydrochloric acid				50.0 D + 46.7 B		50.0 D + 39.6 B	200

Buffer	pH	2.5	2.6	2.7	2.8	2.9	Final soln vol
Glycine/hydrochloric acid			50.0 C + 24.2 B		50.0 C + 16.8 B		200
Phthalate/hydrochloric acid				50.0 D + 33.0 B		50.0 D + 26.4 B	200
Aconitate		20.0 E + 15.0 F	20.0 E + 18.0 F	20.0 E + 21.0 F	20.0 E + 24.6 F	20.0 E + 28.0 F	200
Citrate/phosphate			44.6 G + 5.4 K		42.2 G + 7.8 K		100
Succinate							
Phthalate/sodium hydroxide							
Maleate							

BUFFERS (continued)

Buffer	pH 3.0	3.1	3.2	3.3	3.4	Final soln vol
Glycine/hydrochloric acid	50.0 C + 11.4		50.0 C + 8.2 B		50.0 C + 6.4 B	200
Phthalate/hydrochloric acid	50.0 D + 20.3 B		50.0 D + 14.7 B		50.0 D + 9.9 B	200
Aconitate	20.0 E + 32.0 F	20.0 E + 36.0 F	20.0 E + 40.0 F	20.0 E + 44.0 F	20.0 E + 48.0 F	200
Citrate	46.5 G + 3.5 H		43.7 G + 6.3 H		40.0 G + 10.0 H	100
Citrate/Phosphate	39.8 G + 10.2 K		37.7 G + 12.3 K		35.9 G + 14.1 K	100

Buffer	pH 3.5	3.6	3.7	3.8	3.9	Final Soln vol
Glycine/hydrochloric acid		50.0 C + 5.0 B		50.0 D + 2.63 B		200
Phthalate/hydrochloric acid		50.0 D + 6.0 B				200
Aconitate	20.0 E + 52.0 F	20.0 E + 56.0 F	20.0 E + 60.0 F	20.0 E + 64.0 F	20.0 E + 68.0 F	200
Citrate		37.0 G + 13.0 H		35.0 G + 15.0 H		100
Acetate		46.3 I + 3.7 J		44.0 I + 6.0 J		100
Citrate/phosphate		33.9 G + 16.1 K		32.3 G + 17.7 K		100
Succinate				25.0 L + 7.5 F		100

Buffer	pH 4.0	4.1	4.2	4.3	4.4	Final soln vol
Aconitate	20.0 E + 72.0 F	20.0 E + 76.0 F	20.0 E + 79.6 F	20.0 E + 83.0 F	20.0 E + 86.6 F	200
Citrate	33.0 G + 17.0 H		31.5 G + 18.5 H		28.0 G + 22.0 H	100
Acetate	41.0 I + 9.0 J		36.8 I + 13.2 J		30.5 I + 19.5 J	100
Citrate/phosphate	30.7 G + 19.3 K		29.4 G + 20.6 K		27.8 G + 22.2 K	100
Succinate	25.0 L + 10.0		25.0 L + 13.3 F		25.0 L + 16.7 F	100
Phthalate/sodium hydroxide			50.0 D + 3.7 F		50.0 D + 7.5 F	200

Buffer	pH 4.5	4.6	4.7	4.8	4.9	Final soln vol
Aconitate	20.0 E + 90.0 F	20.0 E + 93.6 F	20.0 E + 97.0 F	20.0 E + 100.0 F	20.0 E + 103.0 F	200
Citrate		25.5 G + 24.5 H		23.0 G + 27.0 H		200
Acetate		25.5 I + 24.5 J		20.0 I + 30.0 J		100
Citrate/phosphate		26.7 G + 23.3 K		25.2 F + 24.8 K		100
Succinate		25.0 L + 20.0 F		25.0 L + 23.5 F		100
Phthalate/sodium hydroxide		50.0 D + 12.2 F		50.0 D + 17.7 F		200

Buffer	pH 5.0	5.1	5.2	5.3	5.4	Final soln vol
Aconitate	20.0 W + 105.6 F	20.0 E + 108.0 F	20.0 E + 110.6 F	20.0 E + 113.0 F	20.0 E + 116.0 F	200
Citrate	20.5 G + 29.5 H		18.0 G + 32.0 H		16.0 G + 34.0 H	100
Acetate	14.8 I + 35.2 J		10.5 I + 39.5 J		8.8 I + 41.2 J	100
Citrate/phosphate	24.3 G + 25.7 K		23.3 G + 26.7 K		22.2 G + 27.8 K	100
Succinate	25.0 L + 26.7		25.0 L + 30.3 F		25.0 L + 34.2 F	100
Phthalate/sodium hydroxide	50.0 D + 23.9 F		50.0 D + 30.0 F		50.0 D + 35.5 F	200
Maleate			50.0 M + 7.2 F		50.0 M + 10.5 F	200
Cacodylate	50.0 N + 47.0 B		50.0 N + 45.0 B		50.0 N + 43.0	200
Tris-Maleate			50.0 Q + 7.0 F		50.0 Q + 10.8 F	200

Buffer	pH 5.5	5.6	5.7	5.8	5.9	Final soln vol
Aconitate	20.0 E + 119.0 F	20.0 E + 122.6 F	20.0 E + 126.0 F			200
Citrate		13.7 G + 36.3 H		11.8 G + 38.2 H		100
Acetate		4.8 I + 45.2 J				100
Citrate/phosphate		21.0 G + 29.0 K		19.7 G + 30.3 K		100
Succinate		25.0 L + 37.5 F		25.0 L + 40.7 F		100
Phthalate/sodium hydroxide		50.0 D + 39.8 F		50.0 D + 43.0 F		200
Maleate		50.0 M + 15.3 F		50.0 M + 20.8 F		200
Cacodylate		50.0 N + 39.2 B		50.0 N + 34.8 B		200
Phosphate			93.5 O + 6.5 P	92.0 O + 8.0 P	90.0 O + 10.0 P	200
Tris-Maleate		50.0 Q + 15.5 F		50.0 Q + 20.5 F		200

Buffer	pH 6.0	6.1	6.2	6.3	6.4	Final soln vol
Citrate	9.5 G + 41.5 H		7.2 G + 42.8 H			100
Acetate						100
Citrate/phosphate	17.9 G + 32.1 K		16.9 G + 33.1 K		15.4 G + 34.6 K	100
Succinate	25.0 L + 43.5 F					100
Phthalate/sodium hydroxide	50.0 D + 45.5 F					200
Maleate	50.0 M + 26.9 F		50.0 M + 33.0 F		50.0 M + 38.0 F	200
Cacodylate	50.0 N + 29.6 B		50.0 N + 23.8 B		50.0 N + 18.3 B	200
Phosphate	87.7 O + 12.3 P	85.0 O + 15.0 P	81.5 O + 18.5 P	77.5 O + 22.5 P	73.5 O + 26.5 P	200
Tris-Maleate	50.0 Q + 26.0 F		50.0 Q + 31.5 F		50.0 Q + 37.0 F	200

BUFFERS (continued)

Buffer	pH 6.5	6.6	6.7	6.8	6.9	Final soln vol
Barbital				50.0 R + 45.0 B		200
Citrate/phosphate	13.6 G + 36.4 K			9.1 G + 40.9 K		100
Maleate		50.0 M + 41.6 F		50.0 M + 44.4 F		200
Cacodylate		50.0 N + 13.3 B		50.0 N + 9.3 B		200
Phosphate	68.5 O + 31.5 P	62.5 O + 37.5 P	56.5 O + 43.5 P	51.0 O + 49.0 P	45.0 O + 55.0 P	200
Tris-Maleate		50.0 Q + 42.5 F		50.0 Q + 45.0 F		200

Buffer	pH 7.0	7.1	7.2	7.3	7.4	Final soln vol
Barbital	50.0 R + 43.0		50.0 R + 39.0 B		50.0 R + 32.5	200
Tris			50.0 S + 44.2 B		50.0 S + 41.4 B	200
Citrate/phosphate	6.5 G + 43.6 K					100
Cacodylate	50.0 N + 6.3		50.0 N + 4.2 B		50.0 N + 2.7 B	200
Phosphate	39.0 O + 61.0 P	33.0 O + 67.0 P	28.0 O + 72.0 P	23.0 O + 77.0 P	19.0 O + 81.0 P	200
Tris-Maleate	50.0 Q + 48.0 F		50.0 Q + 51.0 F		50.0 Q + 54.0 F	200

Buffer	pH 7.5	7.6	7.7	7.8	7.9	Final soln vol
Barbital		50.0 R + 27.5 B		50.0 R + 22.5 B		200
Tris		50.0 S + 38.4 B		50.09 S + 32.5 B		200
Boric acid/borax		50.0 T + 2.0 U		50.0 T + 3.1 U		200
Ammediol				50.0 V + 43.5 P		200
Phosphate	16.0 O + 84.0 P	13.0 O + 87.0 P	10.5 O + 90.5 P	8.5 O + 91.5 P	7.0 O + 93.0 P	200
Tris-Maleate		50.0 Q + 58.0		50.0 Q + 63.5 F		200

Buffer	pH 8.0	8.1	8.2	8.3	8.4	Final soln vol
Barbital	50.0 R + 17.5 B		50.0 R + 12.7 B		50.0 R + 9.0 B	200
Tris	50.0 S + 26.8 B		50.0 S + 21.9 B		50.0 S + 16.5 B	200
Boric acid/borax	50.0 T + 4.9 U		50.0 T + 7.3 U		50.0 T + 11.5 U	200
Ammediol	50.0 V + 41.0 B		50.0 V + 37.7 B		50.0 V + 34.0 B	200
Phosphate	5.3 O + 94.7 P					200
Tris-Maleate	50.0 Q + 69.0 F		50.0 Q + 75.0 F		50.0 Q + 81.0 F	200

Buffer	pH	8.5	8.6	8.7	8.8	8.9	Final soln vol
Barbital		50.0 R + 2.5 B	50.0 R + 6.0 B		50.0 R + 4.0 B		200
Tris		50.0 S + 5.0 B	50.0 S + 12.2 B		50.0 S + 8.1 B		200
Boric acid/borax		50.0 T + 59.0 U	50.0 T + 17.5 U	50.0 T + 22.5 U	50.0 T + 30.0 U	50.0 T + 42.5 U	200
Ammediol		50.0 V + 16.7 B	50.0 V + 29.5 B		50.0 V + 22.0 B		200
Glycine/sodium hydroxide		50.0 C + 8.8 F	50.0 C + 4.0 F		50.0 C + 6.0 F		200
Tris-Maleate			50.0 Q + 86.5 F				200

Buffer	pH	9.0	9.1	9.2	9.3	9.4	Final soln vol
Barbital				50.0 R + 1.5 B			200
Tris							200
Boric acid/borax			50.0 T + 83.0 U	50.0 T + 115.0 U			200
Ammediol				50.0 V + 12.5 B		50.0 V + 8.5 B	200
Glycine/sodium hydroxide				50.0 C + 12.0 F		50.0 C + 16.8 F	200
Borax/sodium hydroxide				50.0 U (pH = 9.28)		50.0 U + 11.0 F	200
Carbonate/bicarbonate				4.0 W + 46.0 X	7.5 W + 42.5 X	9.5 W + 40.5 X	200

Buffer	pH	9.5	9.6	9.7	9.8	9.9	Final soln vol
Ammediol			50.0 V + 5.7 B		50.0 V + 3.7 B		200
Glycine/sodium hydroxide			50.0 C + 22.4 F		50.0 C + 27.2 F		200
Borax/sodium hydroxide		50.0 U + 17.6 F	50.0 U + 23.0 F	50.0 U + 29.0 F	50.0 U + 34.0 F	50.0 U + 38.0 F	200
Carbonate/bicarbonate		13.0 W + 37.0 X	16.0 W + 34.0 X	19.5 W + 30.5 X	22.0 W + 28.0 X	25.0 W + 25.0 X	200

Buffer	pH	10.0	10.1	10.2	10.3	10.4	Final soln vol
Ammediol		50.0 V + 2.0 B					200
Glycine/sodium hydroxide		50.0 C + 32.0 F				50.0 C + 45.5 F	200
Borax/sodium hydroxide		50.0 U + 43.0 F	50.0 U + 46.0 F				200
Carbonate/bicarbonate		27.5 W + 22.5 X	30.0 W + 20.0 X	33.0 W + 17.0 X	35.5 W + 14.5 X	38.5 W + 11.5 X	200

BUFFERS (continued)

Buffer	pH	10.5	10.6	10.7	10.8	10.9	Final soln vol
Glycine/sodium hydroxide		40.5 W + 9.5 X	42.5 W + 7.5 X	45.0 W + 5.0 X			200
Carbonate/bicarbonate			50.0 C + 45.5 F				200

Note: All quantities expressed in milliliters (mL).

STOCK SOLUTIONS

A = 0.2 M potassium chloride (14.91 g in 1000 mL)

B = 0.2 M hydrochloric acid

C = 0.2 M glycine (15.01 g in 1000 mL)

D = 0.2 M potassium acid phthalate (40.84 g in 1000 mL)

E = 0.5 M aconitic acid (87.05 g in 1000 mL)

F = 0.2 M sodium hydroxide

G = 0.1 M citric acid (21.01 g in 1000 mL)

H = 0.1 M sodium citrate dihydrate (29.41 in 1000 mL); avoid using any other hydrated salt

I = 0.2 M acetic acid

J = 0.2 M anhydrous sodium acetate (16.4 g in 1000 mL) or 0.2 M sodium acetate trihydrate (27.2 g in 1000 mL)

K = 0.2 M dibasic sodium phosphate heptahydrate (53.65 g in 1000 mL) or 0.2 M dibasic sodium phosphate dodecahydrate (71.7 g in 1000 mL)

L = 0.2 M succinic acid (23.6 g in 1000 mL)

M = 0.2 M maleate (8.0 g NaOH + 23.2 g maleic acid or 19.6 g maleic anhydride in 1000 mL)

N = 0.2 M sodium cacodylate (42.8 g sodium cacodylate trihydrate in 1000 mL)

O = 0.2 M monobasic sodium phosphate (27.8 g in 1000 mL)

P = 0.2 M dibasic sodium phosphate (53.65 g dibasic sodium phosphate heptahydrate or 71.7 g dibasic sodium phosphate dedecahydrate in 1000 mL)

Q = 0.2 M tris acid maleate (24.2 g tris (hydroxymethyl) amino methane + 23.2 g maleic acid or 19.6 g maleic anhydride in 1000 mL)

R = 0.2 M sodium barbital (veronal) (41.2 g in 1000 mL)

S = 0.2 M tris (hydroxymethyl) aminomethane (24.2 g in 1000 mL)

T = 0.2 M boric acid (12.4 g in 1000 mL)

U = 0.05 M borax (19.05 g in 1000 mL)

V = 0.2 M 2-amino-2-methyl-1,3-propanediol (21.03 g in 1000 mL)

W = 0.2 M anhydrous sodium carbonate (21.2 g in 1000 mL)

X = 0.2 M anhydrous sodium bicarbonate (16.8 g in 1000 mL)

APPROXIMATE pK VALUES OF COMPOUNDS USEFUL IN BUFFER SYSTEMS

The following table provides the pK values of acids and bases needed to make the most popular buffers. The approximate composition of buffers can be calculated from the equation

$$pH = pK + \log \frac{[salt]}{[acid]}$$

pKa	Compound	Formula
2.12	(K_1) Phosphoric acid	H_3PO_4
2.35	(K_1) Glycine	$H_2NCH_2CO_2H$
2.95	(K_1) Phthalic acid	$C_6H_4-1,2-(CO_2H)_2$
3.22	(K_1) Citric acid	$HOC(COOH)(CH_2COOH)_2$
3.66	(K_1) β,β′-Dimethyl glutamic acid	$[HO_2CCH_2^-]_2C(CH_3)_2$
4.21	(K_1) Succinic acid	$HO_2CCH_2CH_2CO_2H$
4.76	Acetic acid	CH_3CO_2H
4.84	(K_2) Citric acid	$HOC(COOH)(CH_2COOH)_2$
5.41	(K_2) Phthalic acid	$C_6H_4-1,2-(CO_2H)_2$
5.64	(K_2) Succinic acid	$HO_2CCH_2CH_2CO_2H$
6.4	(K_1) Carbonic acid	H_2CO_3
6.15	(K_1) Cacodylic acid	$(CH_3)_2As(O)OH$
6.2	(K_2) β,β′-Dimethyl glutaric acid	$[HO_2CCH_2]_2C(CH_3)_2$
6.33	(K_2) Maleic acid	$HO_2CCH=CHCO_2H$
6.39	(K_3) Citric acid	$HOC(COOH)(CH_2COOH)_2$
7.21	(K_2) Phosphoric acid	H_3PO_4
8.07	(K_1) *tris*-(Hydroxymethyl)-aminomethane	$(HOCH_2)_3CNH_2$
8.67	(K_1) 2-Amino-2-methyl-1,3-propanediol	$(HOCH_2)_2C(CH_3)NH_2$
9.23	(K_1) Boric acid	H_3BO_3
9.78	(K_2) Glycine	$H_2NCHG_2CO_2H$
10.33	(K_2) Carbonic acid	H_2CO_3
12.32	(K_3) Phosphoric acid	H_3PO_4

REFERENCES

1. **Ramette, R. W.**, *Chemical Equilibrium and Analysis*, Addison-Wesley, Reading, MA, 1981.
2. **Skoog, D. A. and West, D. W.**, *Fundamentals of Analytical Chemistry*, 2nd ed., Holt, Reinhart & Winston, New York, 1969.

COMMON DRYING AGENTS FOR ORGANIC LIQUIDS

The following table gives the suggested common agents for drying various organic liquids. Those squares marked "X" are the best combination of organic family/drying agent. Those marked "never" are the worst combinations, primarily due to possible chemical reactions. For instance, alcohols and sodium metal react vigorously. Consequently one should look for other drying agents. Those that are blank might be efficient, but are not recommended for use, unless the suggested drying agents are not available. Some combinations do not give efficient results due to complexation (footnoted as *d* in table).[1-4]

REFERENCES

1. **Vogel, A. I.,** *A Textbook of Practical Organic Chemistry,* Longmans, Green, London, 1951.
2. **Brewster, R. Q., Vanderwerf, C. A., and McEwen, W. E.,** *Unitized Experiments in Organic Chemistry,* D Van Nostrand, New York, 1977.
3. **Gordon, A. J. and Ford, R. A.,** *The Chemists Companion; A Handbook of Practical Data, Techniques, and References,* John Wiley & Sons, New York, 1972.
4. **Bruno, T. J. and Svoronos, P. D. N.,** Basic Tables for Chemical Analysis, NBS Tech. Note 1096, National Bureau of Standards, U.S. Department of Commerce, Washington, D.C., 1986.

COMMON DRYING AGENTS FOR ORGANIC LIQUIDS

Family	Na_2CO_3[a]	K_2CO_3[a]	$MgSO_4$[b]	$CaSO_4$[c]	Na_2SO_4[a]	$CaCl_2$	Na	P_2O_5	NaOH (solid)	KOH (solid)	Quicklime	CaH_2	$LiAlH_4$
Alcohols		X	X	X		—[d]	Never	Never	Never	Never	X	—[e]	Never
Aldehydes			X	X	X	—[d]	Never	Never	Never	Never		Never	Never
Alkyl halides						X	Never	X	Never	Never		X	Never
Amines						—[d]	Never	Never	X	X	X	—[f]	Never
Anhydrides								X					
Aryl halides	Never					X	Never	X				X	X
Carboxylic acids		Never	X	X	X	—[g]	Never	X	Never	Never	Never	Never	Never
Esters			X		X	—[d]			Never	Never		X	Never
Ethers				X		X	X	X			X	X	X
Hydrocarbons, aromatic	X	X		X	Poor	X	X	X			X	X	X
Hydrocarbons, saturated	X	X		X	Poor	X	X	X				X	X
Hydrocarbons, unsaturated	X	X			Poor	X	X	X				X	X
Ketones	X	X	X	X	X	—[d]	Never	Never	Never	Never	Never	Never	Never
Nitriles	X	X	X	X	X			X	Never	Never	Never	Never[h]	Never

[a] Excellent in salting out.

[b] Best all-purpose drying agent.

[c] High-capacity, but slow reacting

[d] Forms complexes.

[e] Lime (common impurity) reacts with acidic hydrogen.

[f] Only for 3° amines (R₃N).

[g] Only for C₄ and higher alcohols.

[h] Never if α-hydrogen is present.

COMMON RECRYSTALLIZATION SOLVENTS

The following table gives a list of solvents (and their useful properties) in order of decreasing polarity, and the organic compounds they are capable of recrystallizing. In choosing a solvent one should consider the following criteria: (1) low toxicity, (2) low cost, (3) ease of separation of the solvent from the crystals (relatively high degree of volatility), (4) the ability to dissolve the crystals while hot, but not while cold, with impurities being either soluble or insoluble both in hot *and* cold, and (5) a boiling point that is lower than the melting point of the compound. While not all of these factors may be optimized with each application, an attempt should be made to achieve optimization of as many as possible. For the same compound, a variety of recrystallizing solvents can be employed, based on the type of impurities that are present.[1-4]

REFERENCES

1. **Gordon, A. J. and Ford, R. A.,** *The Chemist's Companion: a Handbook of Practical Data, Techniques, and References,* John Wiley & Sons, New York, 1972.
2. **Roberts, R. M., Gilbert, J. C., Rodewald, L. B., and Wingrove, A. S.,** *An Introduction to Modern Experimental Organic Chemistry,* Holt, Rinehart, & Winston, New York, 1969.
3. **Sarlo, E. and Svoronos, P. D. N.,** *Experiments for Organic Chemistry,* Burgess Publishing, Minneapolis, 1983.
4. **Weast, R. C., Ed.,** *Handbook of Chemistry and Physics,* 63rd ed., CRC Press, Boca Raton, FL, 1983.

COMMON RECRYSTALLIZATION SOLVENTS

Solvent	bp[a]	ε[b,c]	Flammability[d]	Toxicity[d]	Good for	Second solvent in mixture[e]	Comments
Water	100	78.5[c]	0	0	Amides, salts, some carboxylic acids	Methanol, ethanol, acetone, dioxane, acetonitrile	Difficult to remove from crystals
Acetic acid	118	6.15[b]	1	2	Amides, some carboxylic acids, some sulfoxides	Water	Difficult to remove from crystals
Acetonitrile	81.6	37.5[b]	3	3	Some carboxylic acids, hydroquinones	Water, ether, benzene	
Methanol	64.5	32.63[c]	3	1	Nitro-compounds, esters, bromo-compounds, some sulfoxides, sulfones and sulfilimines, anilines	Water, ether, benzene	
Ethanol	78.3	24.30[c]	3	0	Same as methanol	Water, ethyl acetate, hydrocarbons, methylene chloride	
Acetone	56	20.7[c]	3	1	Nitrocompounds, osazones	Water, ether, hydrocarbons	
Methyl cellosolve	124		2	2	Carbohydrates	Water, ether, benzene	
Pyridine	116	123[c]	3	3	Quinones, thiazoles, oxazoles	Water, methanol	Difficult to remove from crystals
Methyl acetate	57	6.68[b]	4	2	Esters, carbonyl compounds, sulfide derivatives, carbinols	Water, ether	
Ethyl acetate	77.1	6.02[c]	3	1	Same as methyl acetate	Water, ether, chloroform, methylene chloride	
Methylene chloride (dichloromethane)	40	9.08[b]	0	2	Low-melting compounds	Ethanol, hydrocarbons	Easily removed
Ether (diethylether)	34.5	4.34[b]	4	2	Low-melting compounds	Acetone, acetonitrile, methanol, ethanol, acetate esters	Easily removed; can create peroxides
Chloroform	61.7	4.81[b]	0	4	Polar compounds	Ethanol, acetate esters, hydrocarbons	Easily removed; suspected carcinogen[f]
Dioxane	102	2.21[c]	3	2	Amides	Water, hydrocarbons, benzene	Can form complexes with ethers

COMMON RECRYSTALLIZATION SOLVENTS (continued)

Solvent	bp[a]	ε[b,c]	Flammability[d]	Toxicity[d]	Good for	Second solvent in mixture[e]	Comments
Carbon tetrachloride	76.5	2.24[b]	0	4	Acid chlorides, anhydrides	Ether, benzene, hydrocarbons	Can react with strong organic bases; suspected carcinogen[f]
Toluene	110.6	2.38[c]	3	2	Aromatics, hydrocarbons	Ether, ethyl acetate, hydrocarbons	A little difficult to remove from crystals
Benzene	80.1	2.28[b]	3	4	Aromatics, hydrocarbons, molecular complexes, sulfides, ethers	Ether, ethyl acetate, hydrocarbons	Suspected carcinogen[f]
Ligroin (naphtha solvent)	90—110	—	3	1	Hydrocarbons, aromatic heterocycles	Ethyl acetate, benzene, methylene chloride	
Petroleum ether (ACS)	35—60	—	4	1	Hydrocarbons	Any solvent less polar than ethanol	Easy to separate
n-Pentane	36.1	1.84[b]	4	1	Hydrocarbons	Any solvent less polar than ethanol	Easy to separate
n-Hexane	69	1.89[b]	4	1	Hydrocarbons	Any solvent less polar than ethanol	
Cyclohexane	80.7	2.02[b]	4	1	Hydrocarbons	Any solvent less polar than ethanol	
n-Heptane	98.4		4	1	Hydrocarbons	Any solvent less polar than ethanol	

[a]   Normal boiling point (°C).
[b]   Dielectric constant (20 °C).
[c]   Dielectric constant (25 °C).
[d]   Scale varies from 4 (highly flammable, highly toxic) to 0 (not flammable, not toxic).
[e]   Second solvent used to facilitate dissolving the crystals in a solvent mixture.
[f]   See Carcinogen Table.

PROPERTIES OF HAZARDOUS SOLIDS

The following table lists some of the more important properties of hazardous room-temperature solids commonly used in the analytical laboratory.[1,2]

REFERENCES

1. **Turner, C. F. and McCreery, J. W.**, *The Chemistry of Fire and Hazardous Materials*, Allyn & Bacon, Boston, 1981.

2. Data Safety Sheet, E. I. du Pont de Nemours & Co., Inc., Wilmington, DL, 1985.

PROPERTIES OF HAZARDOUS SOLIDS

Name	Formula	Specific gravity (at 20 °C)	Melting point (°C)	Boiling point (°C)	Flash point (°C)	Autoignition point (°C)	Ignition/explosion mechanism	Emergency procedures[a]
Acetyl peroxide	$(CH_3CO)_2O_2$	1.2	30	63	—	—	Heat, shock	a,c
Adipic acid	$(CH_2)_4(COOH)_2$	1.3	152	330	196	420	Heat	a,b,c
Aluminum	Al	2.7	660	2270	—	—	Mixing with iron oxides	a
Aluminum chlorate	$Al(ClO_3)_3$	—	—	—	—	—	Heat, impact, reducing agents	a
Aluminum chloride	$AlCl_3$	2.4	192	180	—	—	Heat, moisture	a
Ammonium nitrate	NH_4NO_3	1.7	169	210	—	—	Heat	b
Ammonium nitrite	NH_4NO_2	1.7	Decomposes	—	70	—	Heat, shock, impact	a
Ammonium perchlorate	NH_4ClO_4	—	Decomposes	—	—	—	Shock, impact	—
Antimony	Sb	6.7	630	1375	—	—	Heat, water	a
Antimony trisulfide	SbS_3	4.6	—	—	—	—	Heat, strong organic acids, oxidizers	b
Antimony pentasulfide	SbS_5	4.1	—	—	—	—	Heat, strong oxidizers, acids	a
Barium	Ba	3.6	850	1530	—	—	Heat	a
Beryllium	Be	1.87	1280	1500	—	—	Heat, friction	a

PROPERTIES OF HAZARDOUS SOLIDS (continued)

Name	Formula	Specific gravity (at 20 °C)	Melting point (°C)	Boiling point (°C)	Flash point (°C)	Autoignition point (°C)	Ignition/explosion mechanism	Emergency procedures[a]
Cadmium	Cd	8.6	321	765	—	—	Heat	a
Calcium hypochlorite	CaCl(ClO)·4H$_2$O	—	dec	—	—	—	Heat, contact with combustible material, acid	—
Camphor	C$_{10}$H$_{16}$O	1.0	177	Sublimes	66	466	High conc in air	a,c
Cesium	Cs	1.9	29	670	—	—	Water	f
Iodine	I$_2$	4.9	113	183	—	—	Heat	c
Lithium	Li	0.53	179	1335	—	—	Water, inorganic acids	a
Magnesium	Mg	1.75	651	1107	—	—	Water	a
Phosphorus, red	P$_4$	2.2	600	Sublimes	—	260	Heat, oxidizers	b,e
Phosphorus, white	P$_4$	1.82	44	279	Ambient	30	Heat, oxidizers, dry atmosphere	b
Phosphorus pentachloride	PCl$_5$	4.7	167	Sub	—	—	Moist air, heat	a,c
Phosphorus pentasulfide	P$_2$S$_5$	2.03	276	514	142	287	Water, acids	—
Potassium chlorate	KClO$_3$	2.3	368	Decomposes	—	—	Charcoal, sulfur and phosphorous	b
Potassium nitrate	KNO$_3$	2.1	334	400	—	999	Friction, contact with organics	b
Potassium nitrite	KNO$_2$	1.9	388	Decomposes	—	999	Friction, impact	a
Sodium	Na	0.97	98	890	—	—	Moisture	a
Sodium hydride	NaH	0.9	800	—	—	—	Water, oxidizers	a
Sodium nitrate	NaNO$_3$	2.3	307	379	—	—	Contact with organics	b
Sodium nitrite	NaNO$_2$	2.17	271	318	—	—	Contact with organics	b
Sodium styrene sulfonate	C$_8$H$_7$SO$_3$Na	—	225 (decomposes)	—	—	462	Hot surfaces, flames, sparks	a,b,c
Sulfur	S/S$_8$	2.07	115	445	207	232	Heat	a,d
Triphenylboron	(C$_6$H$_5$)$_3$B	—	136	347	—	220	Water (produces benzene), heat	a,b,c,d

[a]   a, dry chemicals; b, H$_2$O; c, CO$_2$; d, H$_2$O as a fog (not a stream); e, wet sand; f, chlorinated hydrocarbons.

FLAMMABILITY HAZARDS OF COMMON LIQUIDS

The following table lists relevant data regarding the flammability of common organic liquids.[1,2]

REFERENCES

1. **Turner, C. F. and McCreery, J. W.,** *The Chemistry of Fire and Hazardous Material,* Allyn & Bacon, Boston, 1981.
2. Data Safety Sheet, E. I. du Pont de Nemours & Co., Inc., Wilmington, DE, 1985.

Solvent	Formula	Specific gravity	Boiling point (°C)	Flash point (°C)	Auto-ignition point (°C)	How to extinguish flames[a]
Acetaldehyde	CH_3CHO	0.8	21	−38	185	a,b,c
Acetone	$(CH_3)_2CO$	0.8	57	−18	538	a,b
Acetonitrile	$CH_3C\equiv N$	0.79	82	6	—	a,c,d
Acetylacetone	$CH_3COCH_2COCH_3$	1.0	139	41	—	a,b,c
Acrolein	$CH_2=CHCHO$	0.8	53	−26	277	a,b,c
Acrylonitrile	$CH_2=CH-CHCN$	0.81	77	0	481	a,c,d
Allylamine	$CH_2=CHCH_2NH_2$	0.8	53	−29	374	a,b
Amylmercapatan	$CH_3(CH_2)_4SH$	0.8	127	18	—	a,b
Aniline	$C_6H_5NH_2$	1.0	184	70	768	a,b,c (use masks)
Anisole	$C_6H_5OCH_3$	1.0	154	52	—	a,b,c
Benzaldehyde	C_6H_5CHO	1.1	179	65	192	a,b,c
Benzene	C_6H_6	0.88	79	−11	563	a,b,c
Bromine	Br_2	3.0	59	—	—	CO_2 (never H_2O)
Butyl alcohol	C_4H_9OH	0.8	117	29	366	a,b,c
t-Butylperacetate	$CH_3CO(O_2)C(CH_3)_3$	—	—	<27	—	b,c
t-Butylperbenzoate	$C_6H_5CO(O_2)C(CH_3)_3$	>1.0	112	88	8	a,b,c
Butyraldehyde	$CH_3(CH_2)_2CHO$	0.8	76	7	230	a,b,c
Carbon disulfide	CS_2	1.3	47	−30	100	b,d (use masks)
Crotonaldehyde	$CH_3CH=CHCHO$	0.9	104	13	232	a,b,c
Cumene hydroperoxide	—	1.0	153	175	—	a,b,c
Cyclohexanone		0.9	156	43	420	a,b,c
Diacetyl	$(CH_3CO)_2$	1.0	88	27	—	a,b,c
Diethanolamine	$(HOCH_2CH_2)_2NH$	1.0	269	152	662	b,c
Diethylene glycol diethylether	$CH_3(CH_2OCH_2)_3CH_3$	0.9	189	83	—	a, halons
Diethylether	$(C_2H_5)_2O$	0.7	34	−45	180	a,b, halons
Diethylketone	$(C_2H_5)_2CO$	0.8	101	13	452	a,b,c
Dimethyl sulfate	$(CH_3)_2SO_4$	1.3	188	83	188	a,b,c,d
Dimethyl sulfide	$(CH_3)_2S$	0.8	37	−18	206	b,c
1,4-Dioxane		1.0	101	2	180	a,b,c

FLAMMABILITY HAZARDS OF COMMON LIQUIDS (continued)

Solvent	Formula	Specific gravity	Boiling point (°C)	Flash point (°C)	Auto-ignition point (°C)	How to extinguish flames[a]
Ethanol	C_2H_5OH	0.8	78	13	423	a,b,c
Ethylacetone (2-pentanone)	$CH_3COCH_2CH_2CH_3$	0.8	102	7	504	a,b,c
Ethylamine	$C_2H_5NH_2$	0.7	31	−18	384	a,b,c
Ethylenediamine	$H_2NCH_2CH_2NH_2$	0.9	117	34	385	a,b,c
Ethylene glycol	$HOCH_2CH_2OH$	1.1	198	111	413	a,b,c,d
Formaldehyde	HCHO	1.0	99	88	427	a,b,c
Furfural	(furan ring)—CHO	1.2	162	60	316	a,b,c,d
Furfuryl alcohol	(furan ring)—CH_2OH	1.1	171	75	491	a,b,c
Gasoline	C_7H_{16} (isomers)	<1.0	38—218	−43	257	a,b,c
Hexylamine	$C_6H_{13}NH_2$	0.8	132	29	—	a,b
Isopropanol	$(CH_3)_2CHOH$	0.8	82	12	399	b,c
Isopropyl ether	$((CH_3)_2CH)_2O$	0.7	68	−28	443	a,b
Kerosene	CH_4 series mixture	<1.0	149—316	38—71	229	a,b,c
Methanol	CH_3OH	0.8	65	11	464	a,b
Methylamine (aq)	CH_3NH_2	0.7	31	−18	384	a,b,c
Methylaniline	$CH_3NHC_6H_5$	0.8	151	49	533	a,b
Methylethyl ketone	$CH_3COCH_2CH_3$	0.8	79	−6	516	a,b,c
Methylethyl ketone peroxide	$C_4H_8O_2$	—	—	63	—	a,b
Naphtha (mixture)	—	0.8—0.9	149—216	38—46	277—496	a,b,c
Paraldehyde	(trioxane ring with three CH_3)	1.0	124	36	238	a,b,c
2-Pentanone	See ethylacetone	—	—	—	—	—
3-Pentene nitrile	$CH_3CH=CHCH_2C\equiv N$	0.83	14.5	40	—	b,c,d
Peracetic acid	CH_3COOOH	1.2	105	40	—	a,b,c
Petroleum ether	—	<0.7	38—79	<0	288	a,b,c
Propionaldehyde	CH_3CH_2CHO	0.8	49	8	207	a,b,c
Propylamine	$CH_3CH_2CH_2NH_2$	0.7	49	−37	318	a,b,c
Propylene glycol	$CH_3CHOHCH_2OH$	1.0	188	99	421	a,b
Sulfur chloride	S_2Cl_2	1.7	138	118	234	b,c
Sulfuryl chloride	$SOCl_2$	1.7	69	—	—	a,b,c
Tetrahydrofuran	(furan ring)	0.9	60	−17	321	a,b,c
Thionylchloride	$SOCl_2$	1.6	79	—	—	—
Toluene	$C_6H_5CH_3$	0.87	111	4	510	a,b,c
Triethanolamine	$(HOCH_2CH_2)_3N$	1.1	360	179	—	b,c,d
Triethylamine	$(C_2H_5)_3N$	0.7	89	7	—	a,b,c
Xylene (o-)	$C_6H_4(CH_3)_2$	0.88	144	32	463	a,b,c
Xylene (m-)	$C_6H_4(CH_3)_2$	0.86	139	29	527	a,b,c
Xylene (p-)	$C_6H_4(CH_3)_2$	0.86	138	39	529	a,b,c

[a] a, Alcohol foam; b, carbon dioxide; c, dry chemical extinguisher; d, water.

CHEMICAL CARCINOGENS

The following table contains data on chemicals often used in the analytical laboratory, which have come under scrutiny for their suspected or measured carcinogenicity.[1,2] The reader is advised to use these tables with care, as there is a great deal of volatility in the classifications as new data become available. It is suggested that the reader maintain a current file of data from the appropriate regulatory agencies.

REFERENCES

1. Chemicals/Processes Classified as Carcinogens, Public Health Service, Centers for Disease Control, National Institute for Occupational Safety and Health, U.S. Department of Health and Human Services, Cincinnati, April 1985.
2. **Ruth, J. H.**, Odor thresholds and irritation levels of several chemical substances: a review, *Am. Ind. Hyg. Assoc. J.*, 47, A-142, 1986.

CHEMICAL CARCINOGENS

Chemical name	CAS No.	IARC[1,a]	NTP[2,a]	NIOSH[3,a]	OSHA[4,a]	ACGIH[5,a]	Odor low (mg/m³)	Irritating conc (mg/m³)
2-Acetylaminofluorene	53-96-3	b	b		a			
Acrylonitrile	107-13-1	b	b	a	a	b	8.1	
Actinomycin D	50-76-0	c						
Adriamycin	23214-92-8	c						
Aflatoxins		b	b	a	d		0.2536	
Aldrin	309-00-2		b					
2-Aminoanthraquinone	117-79-3	a	a		a	a		
4-Aminobiphenyl	92-67-1		b		a	b		
1-Amino-2-methyl anthraquinone	82-28-0	c	b			b		
Amitrole	61-82-5		b					
o-Anisidine	90-04-0	c	b		d			
o-Anisidine hydrochloride	134-29-2		b					
Antimony trioxide		b				b		
Aramite	140-57-8	b	a					
Arsenic compounds		a	a	a	a			
Arsenic, inorganic		a		a	d	a		
Asbestos	1332-21-4	a	a	a	d	a		
Auramine, technical	2465-27-2	c						
Azathioprine	446-86-6	a						
Benz[a]anthracene	56-55-3	c	b		d			
Benzene	71-43-2	a	a	a	d	b	4.5	9,000
Benzidine	92-87-5	a	a	a	a	a		
Benzidine-based dyes			a	a				
Benzo[b]fluoranthene	205-99-2	c	b					
Benzo[j]fluoranthene	205-82-3	c						
Benzo[k]fluoranthene	207-08-9	c						
Benzo[a]pyrene	50-32-8	b	b			b		
Benzotrichloride	98-07-7	c	b					
Beryllium		b	b	a	a	b		
Beryllium compounds (certain)		b	b	a	a	b		
N,N-bis(2-chloroethyl)-2-naphthylamine (chlornaphazine)	494-03-1	a	a					

Chemical	CAS No.							
Bis-chloroethyl nitrosourea (BCNU)	154-93-9	c						
Bis-chloromethyl ether (BCME)	542-88-1	a			a	a		
1,3-Butadiene	106-99-0	a		a	d	b	0.352	
1,4-Butanediol dimethanesulfonate	55-98-1	c						
Cadmium		a	b					
Carbon black	1333-86-4	c		a	d			
Carbon Tetrachloride	56-23-5	c	b	a	d	b	60—300	
Chlorambucil	305-03-3	a	a					
Chloramphenicol	56-75-7	c	a					
1-(2-Chloroethyl)-3-cyclohexyl-1-nitrosourea (CCNU)	13010-47-4	c						
Chloroform	67-66-3	c	b	a	d	b	250	20,480
Chloromethyl methyl ether	107-30-02	a	a		a	b		
Chlorophenols		c					0.0189	6,801.18
Chloroprene	126-99-8				b	d		
Chromium		a				d		
Chromium compounds (certain)		a	a			d		
Chromium (VI) ions				a		d		
Chrysene	218-01-9			a		b		
Cisplatin	15663-27-1			a				
p-Cresidine	120-71-8		b					
Cupferron	135-20-6		b					
Cycasin	14901-08-7	a	b					
Cyclophosphamide	50-18-0	a	a					
Decarbazine	4342-03-4	c						
DDT	50-29-3	c		a	d		5.0725	
2,4-Diaminoanisole	615-05-4		b	a				
2,4-Diaminoanisole sulfate	39156-41-7		b					
2,4-Diaminotoluene	95-80-7		b					
Dibenz[a,h]acridine	226-36-8	c	b					
Dibenz[a,j]acridine	224-42-0		c	b				
Dibenz[a,h]anthracene	53-70-3	c	b					
7H-dibenzo[c,g]carbazole	194-59-2	c	b					
Dibenzo[a,e]pyrene	192-65-4	c	b					
Dibenzo[a,h]pyrene	189-64-0	c	b					
Dibenzo[a,i]pyrene	189-55-9	c	b					
Dibenzo[a,l]pyrene	191-30-0	c	b					

CHEMICAL CARCINOGENS (continued)

Chemical name	CAS No.	IARC[1,a]	NTP[2,a]	NIOSH[3,a]	OSHA[4,a]	ACGIH[5,a]	Odor low (mg/m³)	Irritating conc (mg/m³)
1,2-Dibromo-3-chloropropane (DBCP)	96-12-8		b		a		0.0965	1.93
3,3'-Dichlorobenzidine	91-94-1	c	b		a	b		
3,3'-Dichlorobenzidine salts					a			
Dieldrin	60-57-1			a	d			
Dienoestrol	84-17-3	c						
Diepoxybutane	1464-53-5		b	a				
Di(2,3-epoxypropyl) ether (DGE)	2238-07-5							
Di(2,3-ethylhexyl) phthalate	117-81-7		b	a	d			
Diethylstilbestrol (DES)	56-53-1	a	a					
Diethyl sulfate	64-67-5	b						
3,3'-Dimethoxybenzidine	119-90-4	c	b					
4-Dimethylaminoazobenzene	60-11-7		b		a	b		
Dimethylcarbamoyl chloride	79-44-7	c	b			b		
1,1-Dimethyl hydrazine	57-14-7			a	d	b		
Dimethyl sulfate	77-78-1	b	b	a	d			
1,4-Dioxane	123-91-1	c	b	a	d		0.0108	792
Direct black 38, technical	1937-37-7	c	b	a				
Direct blue 6, technical	2602-46-2	c	b	a				
Direct brown 95, technical	16071-86-6	c		a				
Epichlorohydrin	106-89-8	c		a	d		50	335
Estradiol-17B	50-28-2	c						
Estrone	53-16-7							
Ethinylestradiol	57-63-6	c						
Ethylene dibromide (EDB)	106-93-4	c	b	a	d	b	76.8	
Ethylene dichloride (EDC)	107-06-2		b	a	d		24	
Ethyleinimine	151-56-4				a		4	200
Ethylene oxide	72-21-8	c		a	a	b	520	
Ethylene thiourea	96-45-7	c	b	a				
Formaldehyde	50-00-0	c	b	a	d	b	1.47	1.50
2-(2-Furyl)-3-(5-nitro-2-furyl) acrylamide	3688-53-7	c						
Gyromitrin	16568-02-8	c						

Chemical	CAS No.							
Hexachlorobenzene	118-74-1		b			b		
Hexachlorobutadiene	87-68-3			a		b	12	
Hexachloroethane	67-72-1			a		b		
Hexamethyl phosphoramide	680-31-9					b		
Hydrazine	302-01-2	c	b	a	d		3	
Hydrazine sulfate	10034-93-2		b					
Hydrazobenzene	122-66-7		b					
Indeno[1,2,3,-cd]pyrene	193-39-5	c	b					
Lead(II) acetate	301-04-2		b					
Lead phosphate	7446-27-7	b	b					
Lindane (and other hexachlorocyclohexane isomers)	58-89-9		b		d	b		
Melhalan	148-82-3	a	a					
Mestranol	72-33-3	c						
5-Methylchrysene	3697-24-3	c						
4,4'-Methylenebis (2-chloroaniline) (MOCA)	101-14-4		b	a		b		
4,4'-Methylenebis (N,N-dimethyl) benzenamine	69522-43-6		b					
4,4'-Methylene dianiline	101-77-9				d	b		
Methyl bromide	74-83-9			a	d		80	
Methyl chloride	74-87-3			a			21	1,050
Methyl hydrazine	60-34-4			a		b	1.75	
Methyl iodide	74-88-4			a	d	b		21,500
Metronidazole	443-48-1	c						
Michler's ketone	90-94-8		b					
Mirex	2385-85-5		b					
Mustard gas	505-60-2	a	a				0.015	
α-Naphthylamine	134-32-7		a		a	a		
β-Naphthylamine	91-59-8		a	a	a			
Nickel carbonyl	13463-39-3			a	d	a	0.21	
Nickel		b		a	d			
Nickel compounds (certain)		b			d			
Nickel, inorganic				a				
Nitrilotriacetic acid	139-13-9	b						
5-Nitro-o-anisidine	99-59-2	b						
4-Nitrobiphenyl	92-93-3		b		a	a		
Nitrofen	1836-75-5	c	b			a		

CHEMICAL CARCINOGENS (continued)

Chemical name	CAS No.	IARC[1,a]	NTP[2,a]	NIOSH[3,a]	OSHA[4,a]	ACGIH[5,a]	Odor low (mg/m^3)	Irritating conc (mg/m^3)
Nitrogen mustard	55-86-7	b						
2-Nitropropane	79-46-9			a		b	17.5	
N-Nitrosodimethylamine	62-75-9	b			a	b		
N-–Nitrosodi-*n*-butylamine	924-16-3		b					
N-–Nitrosodiethanolamine	1116-54-7		b					
N-Nitrosodiethylamine	55-18-5		b					
p-Nitrosodiphenylamine	156-10-5		b					
N-Nitrosodi-*n*-propylamine	621-64-7		b					
N-Nitroso-*n*-ethyl urea	759-73-9		b					
N-Nitroso-*n*-methyl urea	684-93-5	b						
N-Nitrosomethylvinylamine	4549-40-0		b					
N-Nitrosomorpholine	59-89-2		b					
N-Nitrosonornicotine	16543-55-8		b					
N-Nitrosopiperidine	100-75-4		b					
N-Nitrosopyrrolidine	930-55-2		b					
N-Nitrososarcosine	13256-22-9		b					
Norethisterone	68-22-4	c						
Oxymetholone	434-07-1	b	b					
Phenacetin	62-44-2	b	b					
Phenazopyridine	94-78-0	c	b					
Phenazopyridine hydrochloride	136-40-3	c	b					
Phenoxyacetic acid derivatives		c			d			
N-Phenyl-β-naphthylamine	135-88-6			a		b		
Phenylhydrazine	100-63-0			a	d	b		
Phenytoin	57-41-0	c	b					
Phenytoin, sodium salt	630-93-3		b					
Polybrominated biphenyls (PBBs)	36355-01-8		b					
Polychlorinated biphenyls (PCBs)	1336-36-3	c	b	a	d			
Procarbazine	671-16-9	b	b					
Procarbazine hydrochloride	366-70-1		b					
Propane sultone	1120-71-4					b		

Chemical	CAS No.							
β-Propiolactone	57-57-8		b		a			
Propyleneimine	75-55-8				d	b		
Propylthiouracil	51-52-5	b	b			b		
Reserpine	50-55-5		b					
Saccharin	81-07-2		b					
Safrole	94-59-7		b				1.4586	
Selenium sulfide	7446-34-6		b		d			
Streptozotocin	18883-66-4		b					
Sulfallate	95-06-7	c	b					
2,3,7,8-Tetrachlorodibenzo-p-dioxin (TCDD)	1746-01-6	c	b	a				
1,1,2,2-Tetrachloroethane	79-34-5			a	d		21	1,302
Tetrachloroethylene	127-18-4			a	d		31.3561	710.2
Thioacetamide	62-55-5		b					
Thiourea	62-56-6		b					
Thorium dioxide	1314-20-1		a					
p-Tolidine	119-93-7		b	a				
o-Toluidine	95-53-4	b	b		d	b		
o-Toluidine hydrochloride	636-21-5		b					
p-Toluidine	106-49-0		b			b		
Toxaphene	8001-35-2		b		d		2.366	
Treosulphan	299-75-2	a						
1,1,2-Trichloroethane	79-00-5		a		d			
2,4,6-Trichlorophenol	88-06-2	c	b	a				
Trichloroethylene	79-01-6			a	d			
tris(Aziridinyl)-p-benzoquinone [triaziquinone]	68-76-8	c	b					
tris(2,3-Dibromopropyl) phosphate	126-72-7		b		d			
Tryptophan P1	62450-06-0	c						
Tryptophan P2	62450-07-1	c						
Uracil mustard	66-75-1	c						
Urethane	51-79-6		b					
Vinyl bromide	593-60-2		a	a		b		
Vinyl chloride	75-01-4	a	a	a	a	a		
Vinyl cyclohexene dioxide	106-87-6			a		b		
Vinyl fluoride	75-02-5			a				
Vinylidene chloride	75-35-4			a				
Vinylidene fluoride (monomer)	75-38-7			a			2,000	

CHEMICAL CARCINOGENS (continued)

a   Letters a through d refer to the following types of substances: a, known to be a carcinogenic; b, may reasonably be anticipated as carcinogenic to humans; c, carcinogenic to animals but not proven carcinogenic to humans; d, currently regulated by OSHA for reasons other than carcinogenesis.

1   Compiled from publications of the International Agency for Research on Cancer: (a) IARC Monographs on the Evaluation of the Carcinogenic Risk of Chemicals to Humans, Chemicals, Industrial Processes and Industries Associated with Cancer in Humans, IARC Monographs, Volumes 1 to 29, Supplement 4, 1982; (b) IARC Monographs on the Evaluation of the Carcinogenic Risk of Chemicals to Humans, Volumes 30 to 34, 1983 to 1984; and (c) IARC, Handling Chemical Carcinogens in the Laboratory, Problems of Safety, *IARC Int. Agency Res. Cancer Sci. Publ.* No. 33, f1979.

2   Compiled from the 3rd Annu. Rep. on Carcinogens, Summary, National Toxicology Program, U.S. Department of Health and Human Services, 1983.

3   Compiled from the NIOSH Recommendations for Occupational Safety and Health Standards.

4   Compiled from Title 29, Code of Federal Regulations, Part 1910.10000 *et seq.*, 1984.

5   Compiled from Threshold Limit Values (TLV) for Chemical Substances and Physical Agents in the Work Environment and Biological Exposure Indices with Intended changes for 1984—85, American Conference of Governmental Industrial Hygienists (ACGIH).

Index

X

Y

Z